ACS SYMPOSIUM SERIES **983**

Color Quality of Fresh and Processed Foods

Catherine A. Culver, Editor

Pepsi-Cola Company

Ronald E. Wrolstad, Editor

Oregon State University

Sponsored by the

ACS Division of Agricultural and Food Chemistry, Inc.

American Chemical Society, Washington, DC

Library of Congress Cataloging-in-Publication Data

Color quality of fresh and processed foods / Catherine A. Culver, editor, Ronald E. Wrolstad, editor ; sponsored by the ACS Division of Agricultural and Food Chemistry, Inc.

p. cm.—(ACS symposium series ; 983)

Proceedings of a symposium entitled "Color quality of fresh and processed foods" held in Atlanta, Georgia, Mar. 2006.

Includes bibliographical references and index.

ISBN 978–0–8412–7419–8 (alk. paper)

1. Color of food—Congresses. 2. Coloring matter in food—Congresses. 3. Food—Quality—Congresses.

I. Culver, Catherine A. II. Wrolstad, Ronald E., 1939- III. American Chemical Society. Division of Agricultural and Food Chemistry, Inc.

TP370.9C65 2007
064′.07—dc22

2007060564

The paper used in this publication meets the minimum requirements of American National Standard for Information Sciences—Permanence of Paper for Printed Library Materials, ANSI Z39.48–1984.

Distributed by Oxford University Press

PRINTED IN THE UNITED STATES OF AMERICA

Foreword

The ACS Symposium Series was first published in 1974 to provide a mechanism for publishing symposia quickly in book form. The purpose of the series is to publish timely, comprehensive books developed from ACS sponsored symposia based on current scientific research. Occasionally, books are developed from symposia sponsored by other organizations when the topic is of keen interest to the chemistry audience.

Before agreeing to publish a book, the proposed table of contents is reviewed for appropriate and comprehensive coverage and for interest to the audience. Some papers may be excluded to better focus the book; others may be added to provide comprehensiveness. When appropriate, overview or introductory chapters are added. Drafts of chapters are peer-reviewed prior to final acceptance or rejection, and manuscripts are prepared in camera-ready format.

As a rule, only original research papers and original review papers are included in the volumes. Verbatim reproductions of previously published papers are not accepted.

ACS Books Department

Contents

Beverages

Meats and Seafood

Oils, Emulsion, Cereals, and Dairy Products

Regulatory Aspects

Indexes

Preface

Consumers use color as an index for quality when purchasing foods and the color has to be "right". This is a given assumption in industry where considerable emphasis is given to color quality in purchase specification and quality assurance. In some instances, color assessment is done by qualitative visual inspection and sometimes with the aid of standardized color charts. Quantitative instrumental measurement of color is preferable in both research and industrial applications. Relatively simple spectrophotometric measurements are widely used, but are limited to homogenous transparent materials. More sensitive and accurate instrument measurements using CIEL*a*b indices are increasingly being used, in research and industrial practices. These test methods can be adapted for much broader range of test materials. Presently the natural pigments (such as carotenoids, anthoyanins, chlorophylls, and betalains) are being intensely investigated because of their potential health benefits. The focus of this book, however, is on their relationship to color quality and not on their biological activity, absorption, metabolism, or safety. While chemists take a disciplinary approach when investigating natural pigments and their relationship to food color quality, industry is typically organized along commodity lines. This book has taken a commodity approach in examining color quality, rather than examine pigments class by class.

This book is the end product of an American Chemical Society (ACS) sponsored symposium, *Color Quality of Fresh and Processed Foods,* held in Atlanta, Georgia in March 2006. Symposium participants, who were invited because of their expertise in the fields, wrote the chapters. This list of authors includes several international experts, and efforts were made to have representation from industry and government as well as academia. A number of color figures are in this book, which add to its cost. It was not difficult to justify the use of color figures, however, in a book on color quality. Efforts were made to select color figures that made for a clearer presentation of the author's concepts and not just for decorative purposes.

The book opens with a section on *Color Measurement* where the basic principles and practical issues of color measurement are presented by authors from manufacturers of color measurement instruments and supplies. The scope of the following two sections, *Fruit and Vegetables* (10 chapters) and *Beverages* (6 chapters) reflects the bias of the two co organizers of the symposium and coeditors of the book. The section on *Fruits and Vegetables* addresses both fresh and processed foods and includes the perspective of the plant breeder as well as that from industrial and academic scientists. The section on *Beverages* reaffirms the justification for taking a commodity approach in this book. The acceptable methods for measuring color of beer and wine, for example, are heavily influenced by recommendations from professional trade associations. The three chapters in the section *Meats and Seafood* emphasize the importance of color quality in the trade of those commodities, and also the difficulties in measuring color in those complex matrices. The section *Oils, Emulsions, Cereals, and Dairy Products* incorporates a broad range of commodities with the chapters ranging from the fundamentals of measuring emulsion appearance to understanding the relationships between pigment composition and color stability of different products. Food colorants are given intense regulatory scrutiny because of a centuries-long history of disguising foods and food ingredients of inferior quality. In the past, some the added food colorants included toxic substances and led to human sickness and death. Another reason may be the widely held belief that color is cosmetic and unnecessary. Several authors in the book present convincing evidence that such a premise is false. Food regulations vary considerable from country to country, both in basic principles for colorant regulation and in very specific rules for use and labeling of individual colorants. The six chapters in the final section *Regulatory Aspects* give a very comprehensive update on colorant regulations in the United States, Europe, Central and South America, and Asia. The viewpoint from regulatory agencies is given along with that from manufacturers and users of food colorants. Although regulatory agencies and professional associations are striving for harmonization of food colorant regulations to facilitate trade, it is unlikely that this goal will be achieved very soon. While colorant regulations will continue to undergo change, this section provides a very clear picture of food colorant regulations at the present time.

The symposium and the book would not have been possible without the generous financial contributions from organizations and companies.

The funds provided travel support for several of the speakers and contributions were also used to partially underwrite the cost of the color figures in the book. We thank the following contributors: The ACS Division of Agricultural and Food Chemistry, Inc., San-I Gen EFI, BASF, Color MAKER, Danisco USA, Food Ingredient Solutions, GNT Europa, the International Association of Color Manufacturers, and Pepsi-Cola Company. We also thank the authors and their affiliate organizations, as well as the many anonymous individuals who served as peer reviewers. Gordon Leggett and Arthur Lipman deserve special thanks, not only for their excellent chapters, but also for serving as invaluable sources during the discussion periods at the symposium. Also, a note of appreciation is expressed to Andreas Schieber and Florian Stintzing for being so timely with submission of their manuscripts.

Ronald E. Wrolstad
Department of Food Science and Technology
Wiegand Hall
Oregon State University
Corvallis, OR 07331

Catherine A. Culver
Pepsi-Cola Company
100 Stevens Avenue
Valhalla, NY 10595

Color Measurement

Chapter 1

Communicating Food Color Effectively with Physical Color Standards

Arthur C. Schmehling

GretagMacbeth LLC, 617 Little Britain Road, New Windsor, NY 12553

When working with color it is possible to define a color instrumentally for repeatable and consistent production of that color. But, there are times when an instrumental measurement is not possible and it becomes necessary to use a physical color standard for evaluation and comparison purposes. Physical Standards are a cornerstone of any good color program and they allow for the ability to specify and communicate a specific color accurately. This paper will go into the aspects that it takes to manage and create physical color standards to effectively communicate color in a global environment.

A few years ago, a leading ketchup producer developed several specialty colors to appeal to kids. While kids loved them, adults had mixed reactions. One might expect purple ketchup to taste differently than conventional red ketchup – however, it did not taste differently at all.

One is "taught" to expect a certain taste or flavor based on color. We associate red with sweet strawberries and yellow with sour lemons. We also associate a shift in color as potentially negative. For example, several bottles of red ketchup on a retail shelf with differing shades of red might indicate that some of the bottles have been on the shelf for a long time and perhaps are not as "fresh" as the other bottles. When the expected color changes, our perception also changes.

Color management programs ensure quality

In grading and packaging food products color often indicates quality. Therefore, a color management program is necessary not only for economic reasons, but also for brand quality and standardization.

An important part of a color management program includes the use of physical color standards to communicate color and establish acceptable visual deviations. This is especially important in the food industry where the colors of organic ingredients change quickly with time, temperature and light. Physical color standards are also essential for industries where the value of the product is determined by grades of color. Failure to accurately assess color could have a significant economic impact.

Color management involves the product cycle from grading and sorting to processing and production. Each of these phases can have a profound effect on the color of the final product. Establishing an effective color management program utilizing physical color standards requires an understanding of basic color principles.

Basic color principles – the foundation of color management

The three components required to see color are: (1) light source, (2) object, and (3) observer. An effective color management program considers how these three components can be controlled or monitored so that they do not adversely affect color evaluation.

Light source

All color is created from the interaction of light energy with an object that either reflects or absorbs the light, which is then viewed by an observer.

The light source is controlled by using a standardized lighting product, such as a SpectraLight® light booth or Examolite® overhead fixture that meets industry guidelines for color evaluation (ASTM D1729 (*1*), ISO 3664 (*2*), SAE J361 (*3*), DIN (*4*), ANSI (*5*) and BSI (*6*)). Standardized lighting usually involves color evaluation under daylight and at least one other light source, such as cool white fluorescent or incandescent. One's color perception can change dramatically based on the lighting condition. Standardized lighting simulates the environment under which various food products will be evaluated, such as noon sky daylight or a typical retail environment, such as fluorescent. Some retail environments require a light source that enhances food color. For example, grocery stores often illuminate the meat department with a lower color temperature, higher red energy light source to enhance the red tones in fresh meat.

The way in which each light source renders color is based on the energy present in the light source at each wavelength. This is characterized as the spectral power distribution (SPD). The SPD for daylight at D65, with a color temperature of 6500 Kelvin, indicates more evenly balanced amounts of spectral energy (red, orange, yellow, green, blue, indigo, and violet – commonly referred to as ROYGBIV). It is for this reason that daylight at D65 is often the primary light source under which products are evaluated.

Cool white fluorescent, on the other hand, has a SPD that spikes in the blue and green regions of the spectrum. This excess of blue and green energy tends to dull the appearance of red and is typically used as a secondary or tertiary light source to simulate color rendering in a retail environment.

Observer

Aside from perceptual differences in the way we interpret color, observers must be tested for physiological conditions, such as color blindness, which may prevent an observer from accurately evaluating color. Additionally, certain physiological differences may not completely undermine one's ability to evaluate color, but may hamper one's ability to discriminate between colors. For example, one observer may be less able to detect color shifts affecting red colors as opposed to green.

A simple 15-minute test, known as the Farnsworth Munsell 100 Hue Test, will indicate the degree to which one is able to discriminate color, in addition to ruling out color blindness. The Farnsworth Munsell 100 Hue Test meets the industry guides for use in evaluating the color vision of an observer as mentioned in ASTM D-1729 and D-1488 (*1*) for establishing visual color evaluation procedures. Screening of personnel who will be performing visual assessment of color is an important and often overlooked step when developing an effective color management program.

Object

As discussed previously, physical standards establish an ideal color and the acceptable variation in color. This is of crucial importance to the food industry, where the color of various ingredients can change based on environmental conditions. While instrumental measurements from a spectrophotometer monitor color digitally, often a digital standard is not practical for use in certain parts of the process. For example, grading or sorting freshly picked produce is more easily accomplished with a standard light source, physical standard and an evaluator with good color discrimination, than by measuring the sample on a spectrophotometer on site.

Many color management programs incorporate the use of spectrophotometric measurements, in addition to the physical samples, for detailed color analysis. This is more practical in a laboratory setting and still requires the use of a physical standard to establish benchmark colors.

Developing a physical color standard

When developing a physical color standard, it is important to identify the state in which the sample will be evaluated – for example, liquid or solid – and the properties such as fine or coarse powders, opaque or translucent solids or liquids, etc. This is the basis for developing a standard that is representative of the material being evaluated.

Today's physical standards can be produced from a wide range of materials that allow more flexibility in evaluating a sample. For example, a washable standard allows fluid samples to be poured directly on the standard for more accurate evaluation, instead of potential color distortion caused by a beaker. Physical standards also simulate physical characteristics of the material being evaluated such as translucency, gloss, and texture.

A physical standard may represent a single acceptable color or several color variations or tolerances within a given process. Some process variation is unavoidable, and color tolerance standards allow one to manage the deviation to an acceptable visual level, and reduce waste or downgraded product. An acceptable visual level is determined based on lightness-darkness, saturation (intensity) and hue shift.

In industries where grading is required, it is necessary to develop a color scale. Colors on the scale correspond to the grading system and ultimately help determine the product's value.

The standards development process itself must be tightly controlled and repeatable. Color formulations are derived from the least number of colorants that will produce an accurate match. This eliminates metamerism (a set of colors which match under one light source and not under another) and ensures a more repeatable process.

Additionally, a documented system of controls must be implemented. All color standards have a shelf life. Therefore, expiration dates, part numbers storage conditions, etc. must all be documented for tracking and certification purposes.

Use of physical color standards

Accurate physical color standards can provide the closest representation of the ideal sample. In some cases, this is the only means for quality control and production personnel to see the actual color target and provide adequate visual evaluation of samples.

Physical color standards not only help control the final product, but also the steps within a process. For example, a certain beverage is assessed following the pasteurization process. Unacceptable shifts in color may indicate out-of-control process variables such as temperature or cooling time.

When physical color standards are produced within a tightly controlled process as described previously, they enable effective color communication across a large manufacturing or supplier base.

The use of physical color standards in a color management program complements instrumental color control processes. The human eye remains the final arbiter of color, and therefore, visual evaluation in conjunction with instrumental evaluation provides a comprehensive and effective color management program.

Conclusion

Physical color standards are an important part of a color program. Developing accurate physical color standards requires an understanding of basic color principles applied to a tightly controlled process. While physical color standards enable the flexibility to work within a variety of production and manufacturing processes, their effectiveness is optimized when integrated with instrumental color control and a controlled viewing environment.

References

1. ASTM International Home Page. http://www.astm.org (accessed Mar 11, 2007)
2. International Organization for Standardization Home Page. http://www.iso.org (accessed Mar 11, 2007)
3. SAE International Home Page. http://www.sae.org (accessed Mar 11, 2007)
4. Deutsches Institut für Normung Home Page. http://www2.din.de (accessed Mar 11, 2007)
5. American National Standards Insititute Home Page. http://ansi.org (accessed Mar 11, 2007)
6. BSI Home Page. http://www.bsi-global.com (accessed Mar 11, 2007)

Chapter 2

Color Measurement Techniques for Food Products

Gordon J. Leggett

Hunter Associates Laboratory, Inc., 11491 Sunset Hills Road, Reston, VA 20190

The CIE (Commission Internationale de l'Eclairage) offers a well-defined system for the color measurement of opaque and transparent materials. What makes the color of food products especially challenging is that most food products exhibit light trapping or translucency effects which makes them highly dependent on choices of instrument geometry, as well as sample preparation and presentation techniques. This article summarizes the user choices involved in measuring the color of each of the optical categories of food products and provides the best options for color measurement of each.

CIE System of Color Measurement

The CIE system quantifies color, as a person perceives it. Human perception of color requires a source of white light, an object that modifies the light and a person who perceives color from the object stimulus. In the CIE system, each element of this triad is represented as numbers. The white light source is representated as a standardized set of numbers called an illuminant. Food products range from transparent to opaque with many falling in between as translucent or light trapping. In all cases, the product measurement will be quantified as a reflectance or transmission spectrum. A person is defined as a standard set of numbers called the CIE Standard Observer. CIE color values are calculated using a mathematical model based on a white light source, object and human observer that represent all colors in terms of L* lightness, a* redness-greenness, and b* blueness-yellowness.

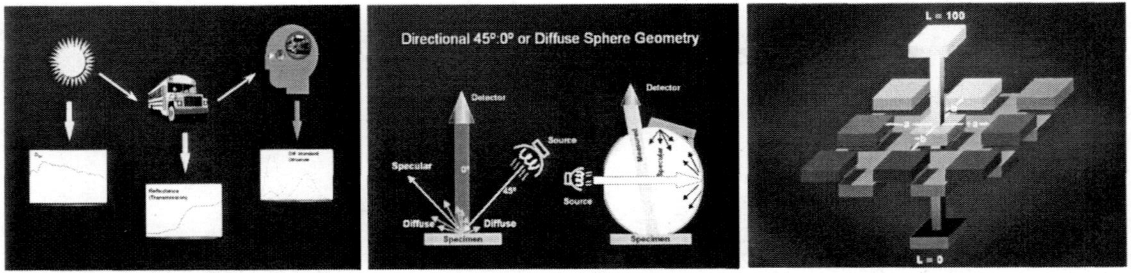

Figure 1. The CIE System of color measurement defines a mathematical model of measuring color as a person sees it. Instruments with CIE standardized geometries measure the product in transmittance or reflectance which is then used to calculate L, a*, b* values for all colors.* *(See page 1 of color inserts.)*

CIE instruments are further standardized in terms of instrument geometry – the arrangement of lamp source, sample place and detector used to measure the object. There are two CIE standard instrument geometries to measure all samples – a bi-directional 45/0 (or its inverse, 0/45), and a diffuse sphere. Both offer advantages with some types of samples. The diffuse sphere has versatility to measure in transmission for transparent samples and reflectance for opaque samples. The bi-directional 45/0 measures in reflectance only; best correlates to visual evaluation of samples; is more robust at measuring translucent samples and can measure a larger area of sample view appropriate for non-uniform food samples (*1*, *2*).

Categories of Food Products and their Color Measurement

While food products vary tremendously in their optical characteristics, there is a systematic method of separating all food applications into general categories and presenting them in a manner that is most uniform and consistent for color measurement.

Transparent Solids

While few food products are transparent solids, it is recommended that transparent solids be measured in Total Transmission (TTRAN) mode on a CIE sphere geometry instrument. As the transmission spectra and corresponding color measurements are dependent on the path length of the sample, the thickness must be kept constant as a condition of measurement.

Transparent Liquids

Transparent liquids have to be made effectively into a solid by measuring them in a glass or clear plastic transmission cell. A sphere instrument is standardized in TTRAN transmission using a clear cell filled with distilled water as a blank, negating the effects of the cell and solvent. The cell becomes a condition of the measurement.

The cell path length is selected based on the chroma of the sample – the more chromatic the sample is, the shorter the path length of the cell. Typical cell path lengths are 10-mm for highly absorbing, chromatic liquids; 20-mm as a general case for most colors and 50-mm for near colorless liquids.

The scattering of light due to non-soluble particles in the liquid sample is a separate optical phenomenon from the absorbance of the colorant that causes the perception of color, yet can affect the appearance of the food product. The

Figure 2. Transparent materials – solids, liquids, highly absorbing or haze, are measured in transmission on a sphere instrument. (See page 1 of color inserts.)

hemispherical collector of a sphere instrument collects both regular and slightly scattered signal, negating the effects of any minor scattering inherent even in clear samples.

A single measurement using a viewed area of the sample greater than a 15-mm diameter is generally sufficient for clear, transparent liquids.

Highly Absorbent Transparent Liquids

Samples such as soy sauce are typically so highly absorbing as not to allow any light through using standard path length cells. The compensating technique is to use a samples cell with a very thin path length cell, typically 2-mm or less, that will allow sufficient signal through the sample to separate color differences among lots. Standardize in TTRAN transmission using this "thin" 2-mm path length cell filled with distilled water to negate the cell and solvent effects prior to making sample measurements.

Hazy Transparent Liquids

Some beverages may have pulp present that may cause significant scattering. Use of a 10-mm path length cell in TTRAN transmission on a sphere instrument, a large area of sample view and the averaging of 2 – 4 readings with replacement of the liquid between readings ensures a repeatable color measurement of this highly scattering sample.

In addition to color, a colorimetric sphere instrument also has the ability to measure relative haze, quantifying the amount of pulp or clouding agents in the beverage.

Translucent Liquids

As the solids content increases, the sample moves from transparent to translucent where a color measurement choice must be made. Translucent samples can be made thinner by measuring in a thin 2-mm cell and measured in transmission, or made thicker and measured in reflectance.

If the solids content is high like tomato sauce, the sample becomes effectively opaque if you fill a sample cup to the top and the sample can then be measured in reflectance.

Translucent Liquids in Transflectance

Translucent liquid samples like mango juice have a low solids content such that even if the sample cup is filled to full height, this sample is not opaque. A ring-and-disk technique can be used to control the sample surround and make this translucent sample effectively opaque for reflectance measurement.

A black ring of fxed height, typically 10 – 15 mm, is inserted into the sample cup. Translucent liquid is poured into the cup to a level above the ring and a white disk is floated down on top of the ring.

In a "transflectance" measurement using the ring-and-disk set, most of the reflected signal comes from the sample. However, some measurement light passes through the sample, reflects off the white background of the disk and passes back through the sample again to the detector. This ring-and-disk technique makes a low-solids translucent liquid effectively into a solid for color measurement.

It is always recommended that transparent samples be measured in transmission. However, if a reflectance instrument is the only measurement option available, it is possible to obtain comparative measurements using the ring-and-disk technique described above with transparent liquids. The measurements are dependent on the ring-and-disk surround as a condition of measurement.

Translucent Semi-Solids

There are a large number of food sauces or purees that are effectively opaque but must be measured in a clear glass container to be made effectively opaque. Measurements can be made on a bi-directional 45/0 (most common) or sphere geometry instrument. If the sample is highly viscous like peanut butter, the sample may need to be pressed slightly to remove any air occlusions.

Translucent Solids

Usually translucent food samples tend to be in the form of sheets like pasta, and layering of multiple sheets is usually sufficient to make the sample effectively opaque for reflectance measurements on a bi-directional 45/0 (most common) or sphere geometry instrument.

Opaque Solids

While few in number, there are food applications like the color measurement of brick cheese where the sample is effectively opaque, smooth and solid. Measurements are possible on either a bi-directional 45/0 (most common) or sphere geometry instrument. The use of the largest area of sample view and

Figure 3. With translucent samples, it is necessary to control the thickness of the sample to make it effectively opaque for color measurement. (See page 1 of color inserts.)

Figure 4. For opaque food products, a key measurement issue making the sample more uniform for a repeatable color measurement. (See page 2 of color inserts.)

the averaging of multiple readings will negate the effects of any non-uniform color across the surface of the sample.

Loose Powders

Loose powders trap light between the particles, and can be thought of as sharing some of the optical characteristics of translucent samples. Pressing the powder into a flat, plaque makes the sample effectively into an opaque solid that can be measured on a directional 45/0 (most common) or diffuse sphere instrument.

Another option is to measure loose powder like flour through the bottom of a clear sample cup. Tap the cup lightly to tighten up the flour. A large area of sample view 25-mm in diameter or greater is preferred. A single reading per measurement is acceptable, but averaging two readings with replacement of the flour between readings is better.

Particulates

Medium-sized particulates or granules such peppercorns trap light in the interstitial cracks among the particles, and must be contained in a sample cell to be made effectively solid and opaque.

Crushing or grinding are potential techniques for achieving a more uniform sample and minimizing the light trapping variability. However, this also has the potential to mix exterior (what the customer sees) and interior color, and may distort the requirement of the measurement.

Bi-directional 45/0 instruments tend to be preferred over sphere instruments as correlating best to visual assessment of the product. Use of a large area-of-view instrument with a 25 – 50-mm diameter is helpful for area-averaging non-uniform color. The preferred measurement technique for particulates is to average multiple readings per measurement with dump-and-fill replacement between readings.

Flakes, Chunks and Large Particulates

Flakes, chunks and large particulates require an instrument with a large area of view, preferably 40-mm or larger, combined with averaging multiple readings (suggest 3 – 6) per measurement, with sample replacement between readings.

With some large particulates such as grapes, a second option is to average multiple readings of single grapes using a small area of view. (*2*, *3*)

Figure 5. As the sample gets larger and more non-uniform, the approach is to average out the variation in single measurement or as a group. *(See page 2 of color inserts.)*

Summary

By separating food products into optical categories and taking appropriate steps to prepare and present each sample in the most uniform form, a repeatable measurement in transmittance and reflectance can be made and used to quantify color using the CIE system.

References

1. *CIE Publication 15.2004 – Colorimetry 3rd Edition*; Commission Internationale de L'Eclairage, Kegelgasse 27A-1030, Wien, Austria, NET: http://www.cie.co.at; 2004.
2. MacDougall, D. B. ed.; *Colour in Food Improving Quality*; Woodhead Publishing Limited: Cambridge, England, 2002; pp 80-110.
3. Hunter, R. S.; Harold, R. W.; *The Measurement of Appearance, Second Edition*; John Wiley & Sons, Inc.: New York, NY, 1987; pp 316-333.

Fruits and Vegetables

Chapter 3

Color Quality of Fresh and Processed Strawberries

Ronald E. Wrolstad[1], Thao Ngo[1], Chad E. Finn[2], and Yanyun Zhao[1]

[1]Department of Food Science and Technology, Oregon State University, Corvallis, OR 97331–6602
[2]Horticultural Crops Research Laboratory, Agricultural Research Service, U.S. Department of Agriculture, 3420 NW Orchard Avenue, Corvallis, OR 97330

Color and appearance are very important quality attributes of fresh and processed strawberries (*Fragaria* × *ananassa* Duch. ex Rozier). In evaluating new germplasm, the plant breeder considers size, shape, uniformity, along with hue and whether the berries are too light or too dark. The attractive color of fresh strawberries is dramatically affected by processing. While some pigment degradation occurs with freezing, physical changes, which are manifested by drip loss, are largely responsible for deterioration of appearance. Canning and manufacture into jam, juice, and wine results in marked color degradation. Anthocyanin pigment degradation and accompanying browning reactions cause their color deterioration. Reaction of anthocyanins with enzymes, ascorbic acid, acetaldehyde and other reactive compounds account for much of the pigment degradation. L*, hue angle and chroma are very useful indices for monitoring color change, and they are complementary to measurement of total anthocyanins, polymeric color and anthocyanin pigment profiles.

No one will deny that the attractive color and appearance of fresh strawberries are major factors contributing to their widespread appeal. Figure 1 is a color plate showing three cultivars and three experimental selections grown at the North Willamette Research and Extension Center, Aurora, OR. Most of Oregon's strawberry crop is sold for processing rather than fresh market. The color and appearance of strawberries are changed markedly by processing, and OSU's Food Science and Technology Department has had a long-term cooperative project with horticulturists and plant breeders with the objective of selecting cultivars with improved color, flavor, and textural quality. This chapter will give a historical perspective of what we have learned from our efforts to produce processed strawberry products with improved color quality and stability.

Color Quality of Fresh Strawberries

Plant breeders when screening strawberry selections for potential release, need to give attention to a number of factors concerning color and appearance. Lightness, darkness, hue, brightness, color uniformity, pigment distribution within the fruit, shape, and size all need to be considered. Large-sized berries are preferable, not necessarily because consumers demand large berries, but rather because strawberries are hand-harvested, and picking small-sized berries is much more costly. In the field, the plant breeder in a medium sized program commonly evaluates 6-10,000 seedlings that represent 60-100 crosses (each cross is a single parental combination), and must make a decision whether to reject or save material for further study based on visual examination of a few fruit on a single plant. Typically a breeder keeps 0.5-1.0% of the seedlings evaluated. Promising selections are grown in subsequent seasons so that larger quantities of berries are available for objective evaluation of fresh and processed fruit. Instrumental measurement of color is very useful since humans have very poor color memory (*1*), making it very difficult to objectively compare visual color characteristics on fruit from different harvest days and different seasons.

We recently (*2*) investigated the color properties of strawberries and the changes they undergo when processed into frozen and canned fruit and made into jam. Table I shows the CIEL*a*b* indices for the same six strawberry genotypes shown in Figure 1. These six genotypes were chosen for this study, because in a 2005 cutting of frozen strawberries from the 2004 season, they exhibited a wide range in hue and lightness and darkness. Close examination of Figure 1 reveals considerable berry-to-berry color variation in some samples. The color measurements in Table I were taken using a circular cell 130 mm in diameter and 50 mm deep. From a pool of approximately 50 berries, the cell was filled, measured, emptied and refilled so that the values are means of 3 subsequent measurements. Berry-to-berry variation will be a major contributor to

Table I. CIE L*C*h Color Indices, Total Anthocyanin Pigment, and pH for Six Strawberry Genotypes.

Selection	*L**	*C**	*h°*	*Total ACN*[b]	*pH*
Ovation	29.1 (±1.4)	41.9 (±1.5)	32.6 (±2.0)	37.1 (± 4.4)	3.28 (±0.07)
Puget Reliance	22.5 (±0.9)	34.4 (±2.6)	26.3 (±1.2)	50.9 (± 2.8)	3.40 (±0.02)
Totem	21.9 (1.3±)	30.4 (±2.2)	25.2 (±2.0)	76.0 (± 4.0)	3.58 (±0.01)
1723-2	21.3 (±1.2)	24.3 (±2.2)	24.5 (±2.0)	122.3 (± 2.3)	3.46 (±0.02)
2273-1	23.4 (±1.6)	32.6 (±3.6)	30.0 (±1.9)	71.8 (± 2.3)	3.25 (±0.01)
2384-1	24.3 (±1.3)	24.8 (±2.6)	28.8 (±1.8)	62.1 (± 0.4)	3.80 (±0.09)

[a]Color measurements were taken on a HunterLab ColorQUEST instrument with 45/0 geometry, using D65 illuminant, a 10° viewing angle, and a view port 88.9mm in diameter. Numerical values are the mean and standard deviation of 3 readings taken from a pool of c.a. 50 berries. Berries were placed in an optical glass cell (130 mm dia x 50 mm ht), and emptied and refilled after each reading.

[b]Total monomeric anthocyanin pigment was determined by the pH differential method (*21*) and expressed as mg perlargonidin-3-glucoside per 100 g of fresh weight, using a molecular weight of 433.2 g/mol and a molar absorptivity of 22,400. Values are the mean and standard deviation for 2 replicate determinations.

SOURCE: The table is derived from data presented in reference 2 by permission. Copyright 2007.

Figure 1. Photographs of strawberries from Oregon State University's North Willamette Experiment Station, 2005 season. Varieties 'Totem',' Puget Reliance', 'Ovation', and selelctions ORUS 1723-2, ORUS 2273-1, and ORUS 2384-1. Reproduced with permission from reference 2. Copyright 2007. (See page 3 of color inserts.)

the standard deviations of the measurements. The total anthocyanin pigment content of the six samples is also listed in Table I. The relation between pigment content and lightness (L*) is evident if one compares 'Ovation', an eastern U.S. fresh market cultivar, with ORUS 1723-2, a selection for the Northwest processing market. ORUS 1723-2 contains over three times as much pigment (122 vs. 37.1 mg/100g) and has a much lower L* (21.3 vs. 29.1. A higher correlation between L* and pigment content for all samples would be more likely if it was not for differences in pigment distribution (white-centered vs. pigmentation throughout) and pH. The relationship between hue angle and color is evident from the orange-red color of Ovation (h° = 32.6) and the bluish-red color of 1723-2 (h° = 24.5). Measurements on these same fruits were taken on two other instruments, the Minolta CR-300 having d/0 geometry and a 8.0 mm diameter viewing port and the HunterLab LabScan II with 0/45 geometry and a 6.35 mm diameter viewing port. For the two instruments with small viewing ports, readings were taken on the surface of an individual berry, readings being taken on from 40-50 berries and averaged. Figure 2 compares the L*C*h° data for the six selections with different instruments. Differences between instruments are evident, particularly for L*. The data from the two instruments with small viewing parts had larger standard deviations. Strawberries do not have a flat surface, and the "pillowing effect" can result in distorted values. Readings on individual berries necessitated multiple readings, whereas the larger cell used with the ColorQUEST instrument allowed for average readings from a large sample of fruits.

Impact of Processing on Color Quality of Strawberry Products

Our laboratory has been involved with several projects concerned with the color quality of strawberries processed by freezing, canning, freeze-drying, and manufactured into preserves, juice, syrups and wines. The attractive appearance of fresh fruit is markedly changed by any of these processing methods. Furthermore, it is often difficult to know what fruit characteristics will be predictive of acceptable color in the processed product.

Color Quality of Frozen and Canned Strawberries

Frozen sliced strawberries were an important commercial product in Oregon in the 1960's. Many strawberry selections had unacceptable color quality after processing, the appearance being described as faded, dull, bluish or purplish. With an objective of determining what compositional information could be predictive of color quality, we processed and analyzed 40 lots of strawberries, consisting of 13 different selections and 5 cultivars (*3*). Compositional measurements of fresh fruit included total anthocyanin pigment, the amounts of pelargonidin-3-glucoside and cyanidin-3-glucoside (determined by densitometry of thin-layer chromatograms), ascorbic acid content, pH, titratable acidity, and °Brix. Gardner L, a, and b values were determined on fresh and thawed frozen berries. Detailed sensory evaluation of the processed berries was conducted. For acceptable color quality it was concluded that the total anthocyanin content should range between 45-70 mg/100g. A surprising finding was that pH had the highest correlation ($r = 0.78$) with overall color quality. The pH ranged from 3.21 -3.81 which is essentially the same as the range for the six samples in the 2005 study (Table I). Figure 3 is a visible spectral scan of strawberry juice adjusted to pH 3.2 and 3.8. The differences in absorbance intensity and slight shift in wavelength of maximum absorption will have an impact on color. This can be explained by the well-known equilibrium between the colored flavylium form that dominates at pH 1 and the colorless hemiketal form that dominates at pH 4.5 (Figure 4). The pK_A for this transformation with anthocyanin-3-glycosides is approximately 3.0; this translates to having 37% of the pigment in the colored flavylium form at pH 3.2 but only 13% at pH 3.8. All strawberries in these two studies were grown in the same experimental plots under very similar environmental conditions. One would expect commercial strawberries from different geographic locations to show an even wider pH range. This project was re-visited 15 years later (*4*) where 45 lots of ripe strawberries representing 14 different selections and 14 cultivars were investigated. Measurements of total anthocyanin, which ranged from 13.3-57.3 mg/100g and pH (3.20-3.80) were helpful in predicting color quality, however, physical measurements were even more useful. Drip loss of thawed berries that ranged from 6.6-33.3 mL/100g of fruit and penetrometer readings (range = 185-490 g) correlated well with color acceptability. A general trend was the lower the drip loss and the firmer the fruit the more acceptable the color.

Strawberry anthocyanins degrade during long-term frozen storage. A common industrial practice has been to "cap" barrels of strawberries with sugar to preserve color and flavor. We (*5*) analyzed samples of strawberries that had been packed with 0, 10, 20 and 40% by weight added sucrose and stored at -15°C for 3 years. Sucrose had a small but statistically significant protective effect on anthocyanin pigment content and also retarded browning and polymeric color formation. Possible mechanisms are competitive inhibition of anthocyanin degrading enzymes, steric interference with condensation reactions, and provision of a partial oxygen barrier.

A commercial lot of 'Totem' strawberries were processed into frozen and canned berries, and also processed into jam (*2*). Table II compares total anthocyanin pigment, % polymeric color, and L*, C* and h° values for the samples. Total anthocyanin pigment content is actually higher in the frozen sample than in the fresh fruit. This apparent increase may partially be due to

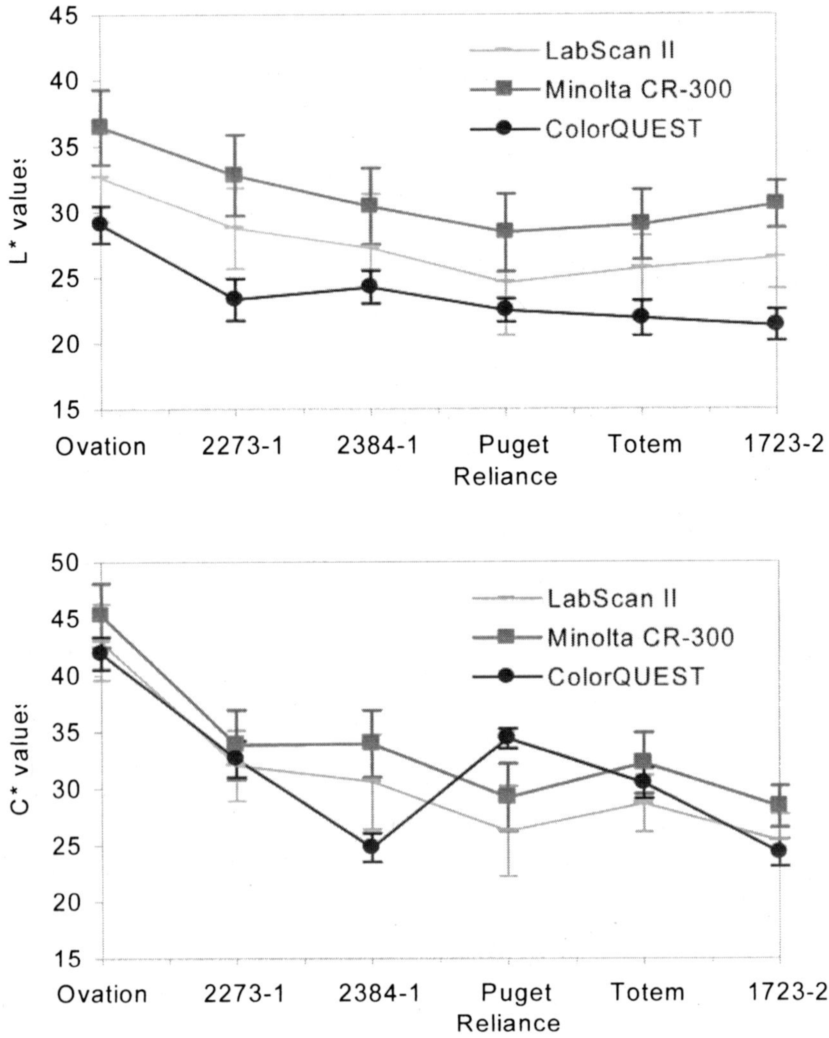

*Figure 2. CIE L*C*h° color indices for six strawberry genotypes measured by three different instruments. The HunterLab ColorQUEST has 45/0 geometry and a viewing port with 88.9 mm diameter. The HunterLab LabScanII has 0/45 geometry and a 6.35 mm diameter viewing port. The Minolta CR-300 has 0/0 geometry and an 8.00 mm diameter viewing port. Reproduced by permission from reference 2. Copyright 2007.*

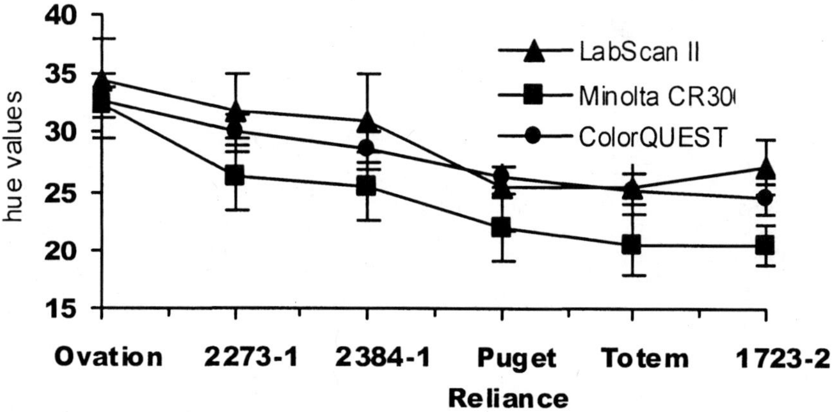

Figure 2. Continued.

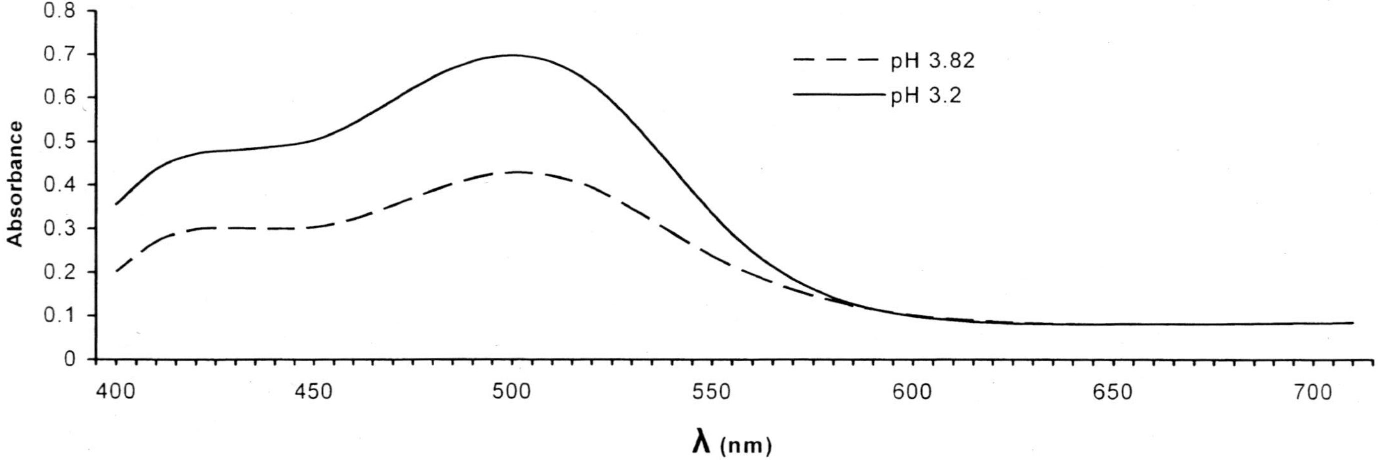

Figure 3. Visible spectrum of strawberry juice adjusted to pH 3.20 and 3.82.

Table II. Total Anthocyanin Pigment Content, % Polymeric Color, and CIE L*a*b* Color Indices[a] of Fresh, Frozen and Canned 'Totem' Strawberries, and Strawberry Jam.

Sample		*Total ACN*[b]	*% Polymeric*[c]	*Browning Index*[d]	*L**	*C**	*h°*
Fresh		65.1			23.3 (±2.2)	31.1 (±0.7)	27.4 (±0.2)
Frozen							
	Puree	69.7	7.2	1.1	21.6 (±0.4)	44.7 (±0.2)	33.7 (±0.4)
	Berries	20.7			30.6 (±0.6)	30.6 (±0.1)	24.8 (±0.1)
	Drip	49.0			66.5 (±0.2)	89.0 (±0.1)	49.4 (±0.1)
Canned							
	Total	21.7					
	Berries	8.7	33.2	2.1	24.0 (±0.3)	28.6 (±0.5)	31.7 (±0.1)
	Liquid	13.0	27.4	1.9	80.9 (±2.8)	38.4 (±4.7)	35.8 (±0.6)
Jam		20.6	27.7	5.0	3.47 (±0.3)	12.5 (±0.6)	12.5 (±0.9)

[a]Lightness (L*), Chroma (C*) and hue angle (h°) values are mean and standard deviations of two replicate measurements.
[b]Total monomeric anthocyanin pigment was determined the pH differential method, and measured as mg pelargonidin-3-glucoside/100g fruit. Note: For canned berries and jam, results are based on fruit content rather than weight of final product.
[c]% Polymeric color is a spectral determination of the proportion of the pigment in the extract that is resistant to bleaching by sodium metabisulfite.
[d]Browning Index= absorbance units/100g of initial fruit in product.
SOURCE: The table is derived from data presented in reference 2 by permission. Copyright 2007.

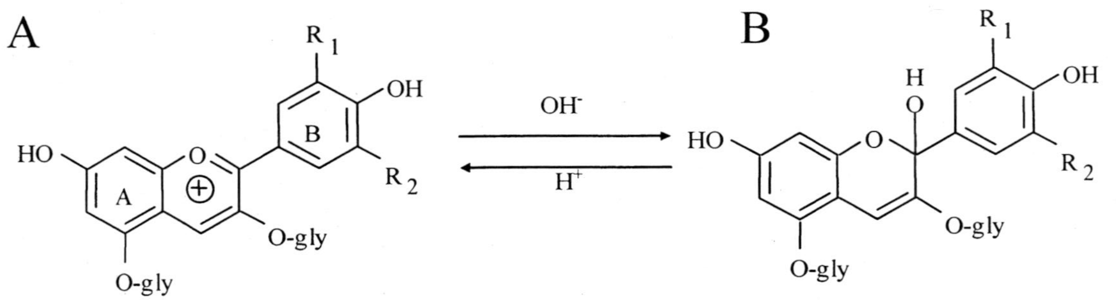

Flavylium cation: orange to purple

pH =

Hemiketal form: colorless

pH =

Figure 4. Reversible transformation of anthocyanin flavylium and hemiketal forms.

sampling differences due to berry-to-berry variation, but may also be because of increased extraction efficiency due to cellular rupture during freezing and thawing. The drip loss was measured to be 19.7 g/100 g fruit with 70% of the anthocyanins being in the drip. Much of the color change is due to pigment transfer rather than pigment degradation. L*, C* and h° measurements do not adequately describe the change in appearance, since cellular breakdown with pigment transfer, loss of gloss, and decreased brightness are better explanations for the observed appearance change. Canned strawberries show marked change in color and appearance, which partially explains their declining popularity as food item. Breakdown of cell structure occurs, with pigment transferring to the draining liquid. In contrast to frozen berries, there was a 69% loss in total anthocyanins, with 19% being in the liquid and 13% remaining in the berries. The canned fruit was analyzed after 70 days of storage at 25°C, so pigment degradation can be attributed to both processing and storage. Adams and Ongley (*6*) investigated anthocyanin destruction in canned strawberries and found that most of the pigment degradation occurred during storage of the canned product.

Color Degradation in Strawberry Jams and Preserves

The J.M. Smucker Co. supported a project in our laboratory that compared the color quality of strawberry jams made from 'Hood' and 'Tioga' (*7*). From the Company's experience, 'Hood' was known to give a product with better appearance and color stability than 'Tioga'. A major objective of the project was to determine what compositional factors were responsible for the differences in color quality. Table III compares the berry composition and Table IV compares the total monomeric anthocyanin pigment content and polymeric color of the jams during storage. Jam made from 'Tioga' berries had lower anthocyanin pigment content and showed a much greater increase in % polymeric color during storage. One conclusion was that color degradation was not just due to anthocyanin pigment degradation, but also caused by the formation of polymeric browning pigments. 'Hood' was higher in total anthocyanins, and lower in leucoanthocyanins, flavanols, and total phenolics. Presumably the reducing sugar levels in the two jams were very similar, so that amino acid and ascorbic acid levels should relate to nonenzymatic browning rates. Amino acids and ascorbic acid levels, however, were higher in 'Hood' than 'Tioga'. This led us to speculate that degradation of reactive phenolics accounted for the greater browning in 'Tioga' Jams. The reduced water activity of jams actually favors anthocyanin pigment stability, since anthocyanin degradation rate decreases as water activity lowers. Refer to Table V (*8*), which shows the 1st order rate constants and half-lifes for degradation of pelargonidin-3-glucoside in glycerol-water model systems. In contrast, Maillard browning reactions are greatly accelerated in intermediate moisture foods (*9*).

Table III. Compositional Differences between 'Hood' and 'Tioga' Strawberries.

	'Hood'	*'Tioga'*
Total anthocyanins, mg/100g	37.7	27.6
Leucoanthocyanins, absorbance units / g	4.69	5.65
Flavanols absorbance units / g	0.404	0.876
Total phenolics, mg/100g	205	312
Ascorbic acid, mg/100g	48.2	18.8
Amino acids, µmoles/100g	540	442

SOURCE: Adapted from reference 7 by permission. Copyright 2007.

Table IV. Varietal Comparison ('Hood' vs. 'Tioga') of Total Anthocyanins and % Polymeric Color in Strawberry Jam Stored at 21°C.

Jam Sample	*Total Anthocyanins, mg/100g*	*% Polymeric Color*
'Hood'		
2 Weeks Storage	18.0	10
19 Weeks Storage	6.5	20
26 Weeks Storage	6.0	24
'Tioga'		
2 Weeks Storage	12.0	10
19 Weeks Storage	5.5	38
26 Weeks Storage	4.0	40

SOURCE: Adapted with permission from reference 7 by permission.

In our recent study (*2*) on the effects of processing on the color quality of strawberry products, 'Totem' fruit were processed into jam using an open kettle process where the ingredients were concentrated to 67 °Brix. The temperature reached 104°C and the process took nearly 20 minutes. The anthocyanin content of the jam (Table II) represents a 70% loss. Garcia-Viguera and others (*9*) used a vacuum pan in making strawberry jam, their milder heating conditions resulting in 36% pigment destruction. The low L* and C* values (Table II) in the open-kettle jam are indicative of a very dark-colored product. The high % polymeric color and browning index are believed to be the result of nonenzymatic browning reactions as well as anthocyanin degradation. Figure 5 shows the changes in total anthocyanin pigment, % polymeric color, and browning index for the 'Totem' jam stored at 10, 21 and 38°C for 9 weeks. Temperature accelerates anthocyanin destruction and the formation of polymeric brown pigments.

Color Quality of Strawberry Juice, Juice Concentrate, Syrups and Wines

While physical factors such as cell rupture, pigment transfer, and light scattering within a solid-liquid matrix complicates color problem solving in the previously discussed strawberry products, in the case of juices, syrups and wines one is dealing with the chemistry of solutions and dispersed systems. Commercial processors of these strawberry products are plagued by problems of poor color stability. With the hypothesis that native enzymes played a major role in color degradation in strawberry products, we investigated what effect microwave blanching would have on the color and composition of strawberry juice and concentrate (*11*). Juice and concentrate were made from microwave-blanched and unheated strawberries (mixture of 'Hood' and 'Benton'). Composition and color parameters were monitored during 8 weeks of storage at 20°C. Blanching resulted in improved color stability and it had a protective effect on anthocyanin pigments, leucoanthocyanins, flavanols, and ascorbic acid. Blanched strawberry concentrate browned at a lower rate than the control. While blanching had a statistically significant impact on color, all treatments showed pronounced anthocyanin degradation with over 90% pigment destruction after 55 days storage. Anecdotal information from juice processors characterizes their experiences with strawberry juice color as being erratic and unpredictable. For example, one production run from the same lot of fruit might give acceptable color, while a subsequent run using fruit from the same lot and the same processing conditions would not. In such cases, microbial contamination may be the source of the problem. Strawberries are particularly susceptible to mold contamination, and molds posses a broad spectrum of oxidative and glycosidase enzymes that can catalyze pigment destruction. To demonstrate the impact of

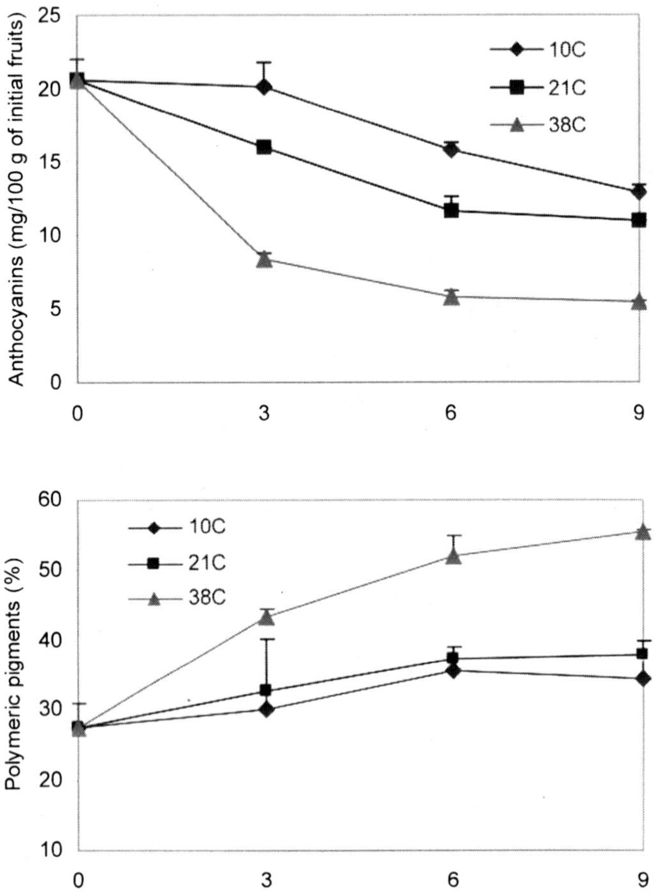

Figure 5. Changes in total anthocyanin pigment, % polymeric color, and browning index during storage of 'Totem' strawberry jam at 10, 21 and 38°C for 9 weeks. Reproduced with permission from reference 2. Copyright 2007.

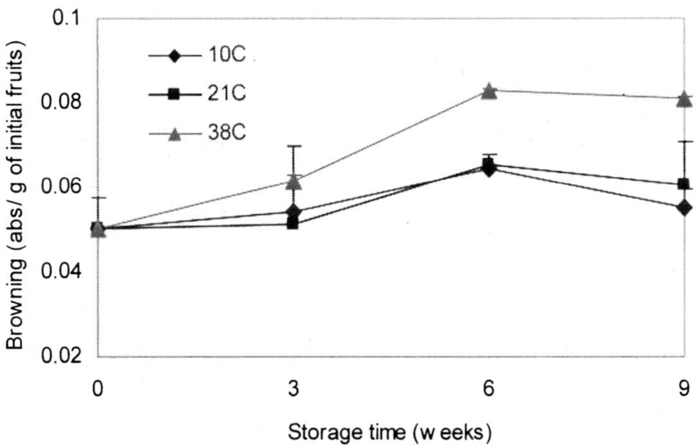

Figure 5. Continued.

Table V. Degradation of Pelargonidin-3-glucoside in Glycerol-Water Model Systems at Different Water Activity (A_w).

A_w	1st Order Rate Constant	Half-life (days)
1.00	3.74×10^{-3}	186
0.90	2.08×10^{-3}	332
0.66	7.48×10^{-4}	934

SOURCE: Adapted from reference 8 by permission. Copyright 2001.

mold contamination, we made strawberry juice and concentrate from three lots of strawberries: good quality fruit, and fruit spiked with 7.5% and 15% moldy berries (*12*). Mold contamination was deleterious to color quality with increased polymeric color, browning, and haze formation all being statistically significant. Strawberries are low in total anthocyanin pigment content when compared to other berry fruits such as blueberries, blackberries, cranberries, and raspberries (*13*), and this is one factor accounting for the poor color stability of strawberry products. In a recent investigation, we (*14*) processed strawberries ('Totem') into juice and juice concentrate, and fortified samples with purified anthocyanin pigments so that the pigment concentration approximately doubled. Degradation rates of anthocyanins were monitored at 25°C for 26 days. The 1st order rate constants and half-lifes are shown in Table VI. Fortifying the juices with additional pigment did increase the half-lifes, however, they are still extremely short. The stability of purified pigments in model systems (Table V) is far greater, suggesting that pigment degradation in strawberry products is not due to inherent anthocyanin instability, rather other reactants in the matrix must play a major role.

Blackcurrant and strawberry cordials are popular beverages in northern Europe. They are made by fortifying a clarified fruit juice with sugar (c.a.50° Brix) and ascorbic acid, and then heat pasteurizing the product. They are consumed throughout the year after diluting the cordial with hot or cold water. A quality issue with strawberry cordial is color degradation during storage. This problem was investigated (*15*) by making blackcurrant and strawberry syrups where portions of the strawberry syrup were fortified with pure strawberry anthocyanins so that the pigment concentration was equivalent to that of blackcurrant, fortifying strawberry syrup with ascorbic acid, and fortifying strawberry syrup with both strawberry anthocyanins and ascorbic acid. Pigment and color measurements were monitored over 6 months storage at 20°C. Figure 6 shows the % retention of anthocyanins. The rate of anthocyanin degradation in strawberry syrup fortified with anthocyanins was the same as that for blackcurrant syrup. This demonstrated that total anthocyanin pigment concentration was more important than differences in qualitative composition. Blackcurrant anthocyanins are the 3-glucosides and 3-rutinosides of cyanidin and delphinidin while strawberry consists primarily of pelargonidin-3-glucoside. Ascorbic acid fortification promoted anthocyanin destruction in both of those syrups. A number of workers have shown that ascorbic acid will accelerate anthocyanin pigment degradation (*16*). The change in hue angle of the syrups during storage (Figure 7) clearly shows the change in visual appearance. Blackcurrant's bluish-red color was reasonably stable during storage, as was the red color of strawberry syrup fortified with strawberry anthocyanins. Strawberry syrup changed to an orange color, and strawberry syrup fortified with ascorbic acid became an unacceptable yellow-orange.

Table VI. Degradation Rates of Anthocyanins[a] in Strawberry Juice and Juice Concentrate Stored at 25°C.

System[b]	*1st Order Rate Constant*	*Half-life (days)*
Juice (8° Brix)	7.3×10^{-2}	8
+ Pgd-3-Glu	4.7×10^{-2}	12
+ Pgd-3-Soph	3.8×10^{-2}	12
+ Acyl-Pgd-3-5-Gly	5.0×10^{-2}	12
Concentrate (65° Brix)	8.0×10^{-2}	3.5
+ Pgd-3-Glu	5.4×10^{-2}	4
+ Pgd-3-Soph	4.7 x 10-[2]	5
+ Acyl-Pgd-3-5-Gly	3.6×10^{-2}	5

[a] Total anthocyanins was determined the pH differential method.
[b] Abreviations: Pgd = pelargonidin; Glu = glucoside; Soph = sophoroside; Acyl = Acylated with *p*-coumaric, ferulic and malonic acids; Gly = 3-sophoroside-5-glucoside.
SOURCE: Adapted from reference 14 by permission. Copyright 2002.

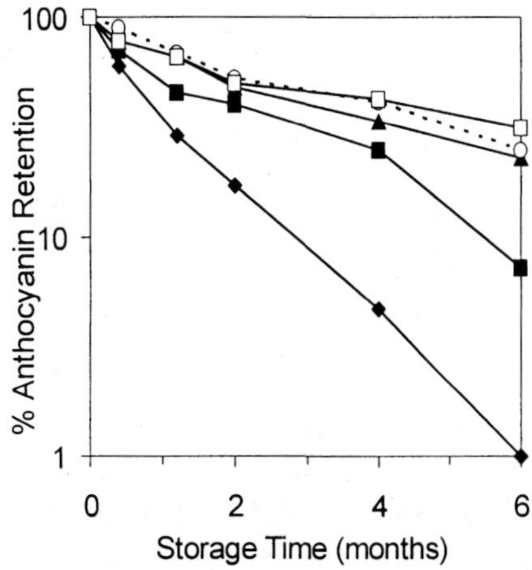

Figure 6. Retention (%) of anthocyanins in strawberry and blackcurrant syrups during storage. -■- Strawberry juice, -◆- Strawberry juice + ascorbic, -□- Strawberry juice + anthocyanin, -▲- Strawberry juice + anthocyanin + ascorbic, -O- Black Currant. Adapted with permission from reference 15. Copyright 1992.

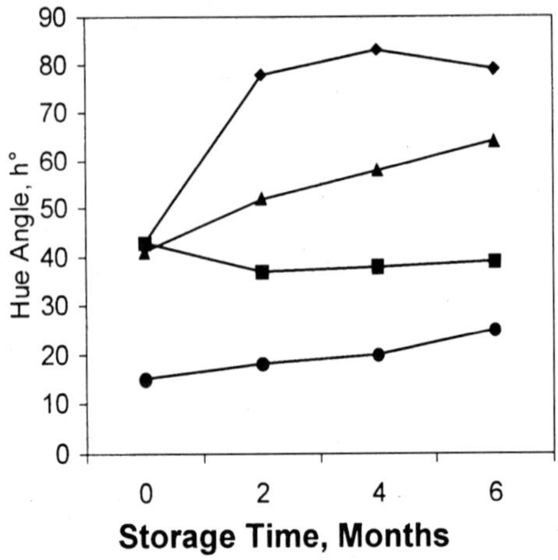

Figure 7. Change in hue of strawberry and blackcurrant syrups during storage. -▲- Strawberry, -■- Strawberry + anthocyanin, -◆- Strawberry + ascorbic, -●-Black Currant. Adapted with permission from reference 15. Copyright 1992.

Oregon fruit wine-makers advised us that a major problem in their industry was color degradation of strawberry wine. They told us that cultivars with high anthocyanin content gave wines with better color, and another observation of theirs was that fruit harvested early in the season tended to give better color stability than late-season fruit. With that background information we investigated the influence of cultivar ('Totem' vs. 'Benton'), maturity (full-ripe vs. overripe), and mold contamination on color of strawberry wine (*17*). Care was taken during picking and sorting to ensure that overripe berries (darker colored and softer texture) were not contaminated with mold. Compositional analyses showed that 'Totem' was higher in anthocyanins, flavanols, leucoanthocyanins and ascorbic acid whereas 'Benton' fruit was higher in polyphenoloxidase (PPO), total phenolics and titratable acidity. Overripe fruit was higher in anthocyanins, flavanols, leucoanthocyanins, total phenolics, pH and °Brix, but lower in PPO, ascorbic acid and titratable acidity than fully ripe fruit. The influence of mold was tested by adding 15% moldy berries by weight to full-ripe berries. Fruit was first processed into juice with a recovery of about 50% of the anthocyanins in the juice. Tremendous destruction of anthocyanins took place during fermentation, only 3-9% of the anthocyanins being retained. Overripe fruit gave wines with better color than fully ripe fruit, and wines from 'Totem' fruit had better color than those from 'Benton'. Mold contamination accelerated color degradation, caused "stuck" fermentation, and increased juice viscosity. Changes in total anthocyanins, % polymeric color, and Hunter L values in the wines were monitored over 6 weeks storage at 25°C (Figures 8, 9 and 10, respectively).

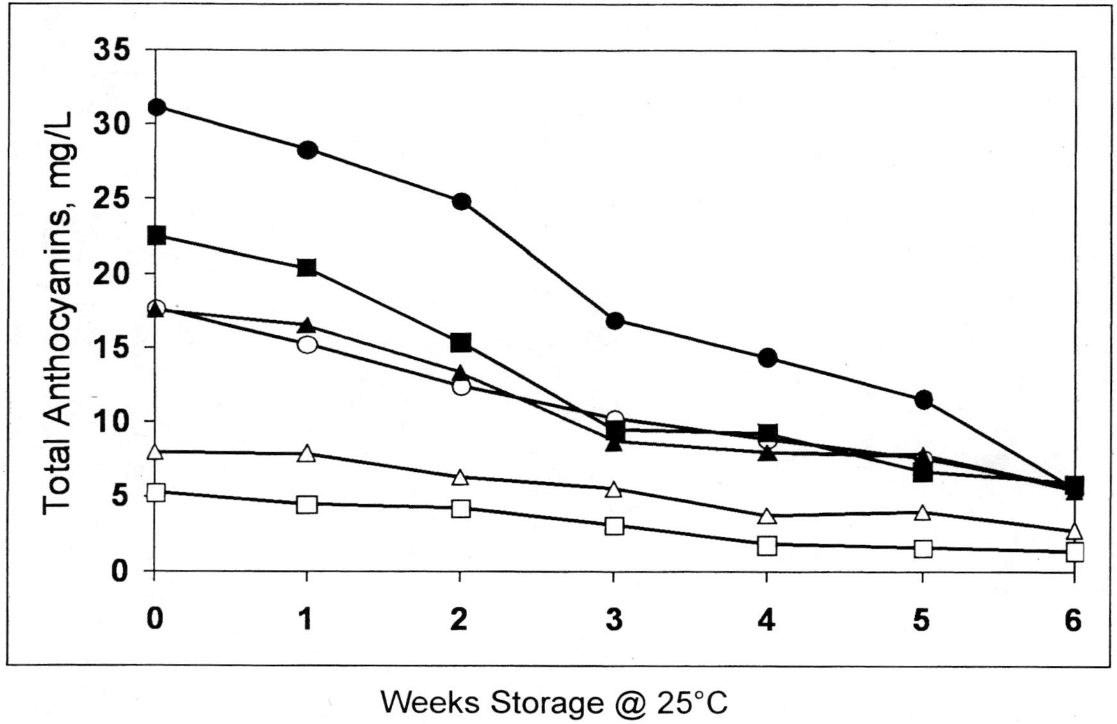

Figure 8. Changes in total anthocyanins (mg/L) in strawberry wines during six weeks storage at 25°C. -□- 'Benton' full-ripe, -△- 'Benton' full-ripe + mold, -○- 'Benton' over-ripe, -■- 'Totem' full-ripe, ·▲· 'Totem' full-ripe + mold, -●- 'Totem' over-ripe. Adapted with permission from reference 17. Copyright 1985.

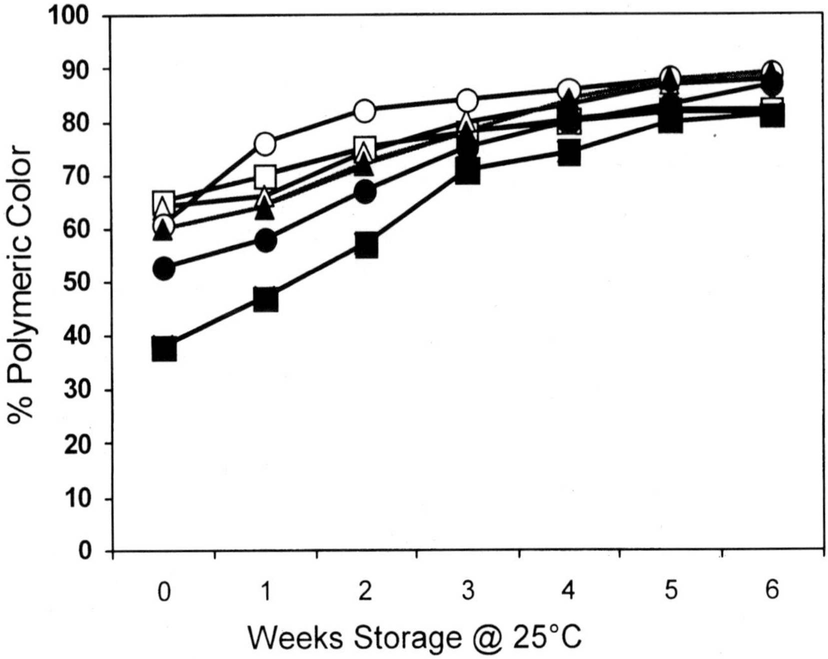

Figure 9. Changes in % polymeric color in strawberry wines during six weeks storage at 25°C. -□- 'Benton' full-ripe, -△- 'Benton' full-ripe + mold, -○- 'Benton' over-ripe, -■- 'Totem' full-ripe, ·▲· 'Totem' full-ripe + mold, -●- 'Totem' over-ripe. Adapted with permission from reference 17. Copyright 1985.

Sensory evaluation of the wines after six weeks storage (data not shown) showed total anthocyanins and Hunter L values to be highly correlated with wine color quality.

What Have We Learned?

In attempting to determine the causes of poor color stability in strawberry products, a chemist's bias tends to focus on the structure and stability of the anthocyanins. The anthocyanin pigment profile of all strawberry cultivars is dominated by pelargonidin-3-glucoside. Figure 11 shows the anthocyanin pigment profile for 'Totem', minor pigments being cyanidin-3-glucoside,

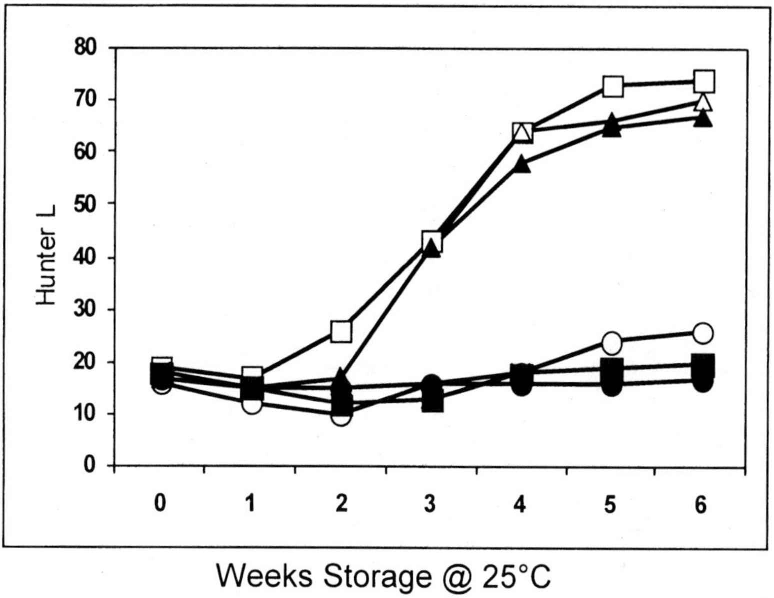

Figure 10. Changes in Hunter L values in strawberry wines during six weeks storage at 25°C. -□- 'Benton' full-ripe, -△- 'Benton' full-ripe + mold, -○- 'Benton' over-ripe, -■- 'Totem' full-ripe, ·▲· 'Totem' full-ripe + mold, -●- 'Totem' over-ripe. Adapted with permission from reference 17. Copyright 1985.

pelargonidin-3-rutinoside, and pelargonidin-3-glucoside acylated with malonic and succinic acids (*18*). All of these pigments are labile to acid hydrolysis, which results in rapid pigment destruction. All cultivars we have examined show a similar profile, suggesting that classical plant breeding is unlikely to result in marked changes in pigment composition. Substitution of anthocyanins with di- and tri-glycosides, and more importantly, acylation with cinnamic acids would very likely increase pigment stability. This could be possible through genetic engineering technology. Selecting cultivars with high anthocyanin content is a valid recommendation, particularly for fruit to be used for juice, syrups and wines. Strawberries are a rich source of ascorbic acid, which unfortunately, accelerates pigment destruction in many products. Procyanidins and other reactive phenolics also react with anthocyanins in deleterious reactions. Fermentation products such acetaldehyde and pyruvic acid react with anthocyanins in the formation of polymeric and vitisin-type pigments. Native peroxidase and polyphenoloxidase enzymes may play important roles in color

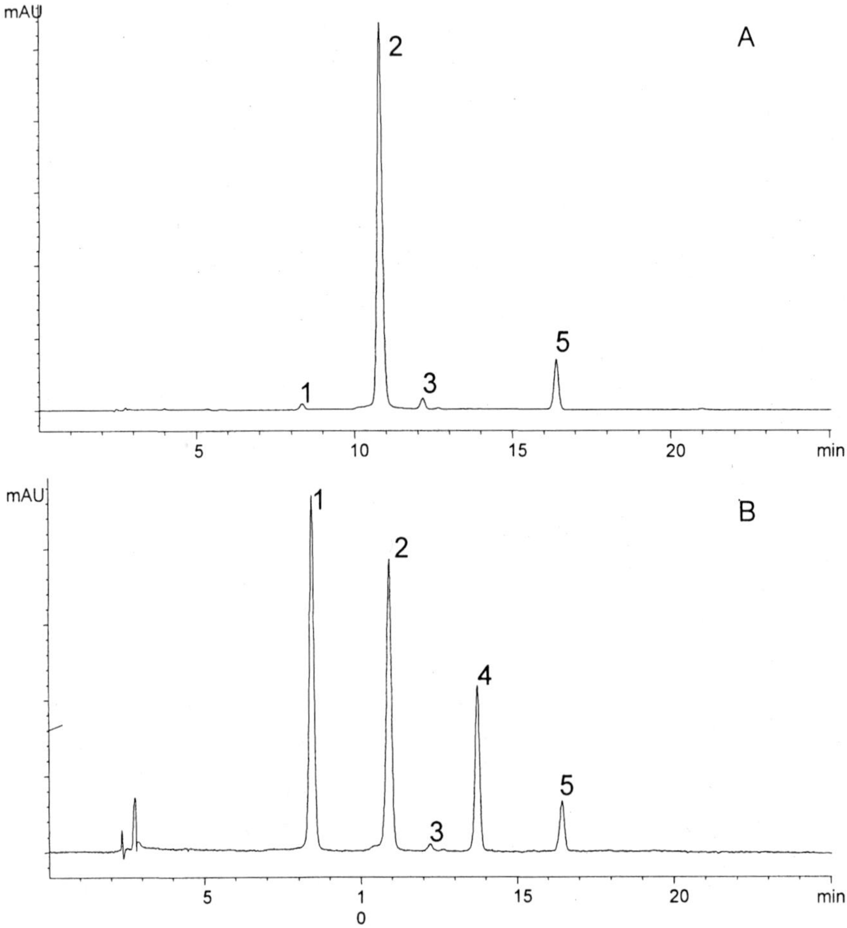

Figure 11. HPLC chromatogram of strawberry anthocyanins recorded at 520 nm. Peak 1 = cyanidin-3-glucoside; 2 = pelargonidin-3-glucoside; 3 = pelargonidin-3-rutinoside; 4 = pelargonidin-3-malonyl-glucoside; 5 = pelargonidin-3-succinyl-glucoside. Reproduced by permission from reference 18. Copyright 2005 American Chemical Society.

degradation in many products. The textural fragility of strawberries is responsible for undesirable changes in color and appearance for frozen strawberries and also is also a major factor in canned strawberries. Strawberries are very susceptible to mold contamination, and molds are great sources for oxidase and glycosidase enzymes that will rapidly degrade anthocyanins. Juice processing enzymes are another potential source of glycosidase enzymes that can lead to pigment destruction (*19*). In these investigations, we have found that measurement of CIEL*C*h values combined with total anthocyanin pigment determination (pH differential method), anthocyanin pigment profiles by HPLC, and spectral determination of polymeric color and browning to be effective for tracking color and pigment changes (*20*).

What advice can be offered? From a chemist's perspective it is known that high anthocyanin concentration, low concentration of reactive phenolics (procyanidins), low ascorbic acid, low pH and reduced water activity will favor anthocyanin stability. Accompanying Maillard browning reactions may be just as important as anthocyanin degradation. But for many applications, this understanding does not provide a solution to the problem. From a technologist's perspective, giving attention to raw fruit quality (eliminating mold contamination), blanching to inactivate enzymes, and storing products at low temperatures in the dark can help. The chemist and the technologist will often refer the solution to the plant breeder to develop cultivars with the composition and characteristics that will give superior color and appearance. The objective often is to have a "universal" cultivar that will be superior for all product applications. This is unrealistic since different chemical and physical properties are desired for fresh market and the many different processed products.

References

1. Hutchings, J. B. *Food Colour and Appearance*; Blackie Academic and Professional: Glasgow, UK, 1994; pp 96-98.
2. Ngo, T.; Wrolstad, R. E.; Zhao, Y. *J. Food Sci.* **2007**, *72*, C25-32.
3. Wrolstad, R. E.; Putnam, T. P.; Varseveld, G. W. *J. Food. Sci.* **1970**, *35*, 448-452.
4. Varseveld, G. V.; Martin, L. W.; Wrolstad, R. E.; Richardson, D. G. OSU Agricultural Experiment Station. Unpublished data. 1985.
5. Wrolstad, R. E.; Skrede, G.; Lea, P.; Enersen, G. *J. Food Sci.* **1990**, *55*, 1064-1065, 1072.
6. Adams, J. B.; Ongley, M. H. *J. Food Technol.* **1973**, *8*, 139-145.
7. Abers, J. E.; Wrolstad, R. E. *J. Food Sci.* **1979**, *44*, 75-78.
8. Garzon, G. A.; Wrolstad, R. E. *Food Chem.* **2001**, *75*, 185-196.
9. Villamiel, M.; del Castillo, M.D.; Corzo, N. In *Food Biochemistry & Food*

Processing; Hui, Y.H., Ed.; Blackwell Publishing: Ames, IA, 2006; pp 80-82.

10. García-Viguera, C.; Zafrilla, P.; Romero, F.; Abellán, P.; Artés, F.; Tomás-Barberán, F. A. *J. Food Sci.* **1999**, *64*, 243-247.
11. Wrolstad, R. E.; Lee, D. D.; Poei, M. S. *J. Food Sci.* **1980**, *45*, 1573-1577.
12. Rwabahizi, S.; Wrolstad, R.E. *J. Food Sci.* **1988**, *53*, 857-861,872.
13. Skrede, G.; Wrolstad, R. E. In *Functional Foods. Vol 2.Biochemical and Processing Aspects*; Shi, J.; Mazza, G.; Le Maguer, M. Ed.; CRC Press: Boca Raton, FL, 2002; pp 71-133.
14. Garzon, A.; Wrolstad, R. E. *J. Food Sci.* **2002**, *67*, 1288-1299.
15. Skrede, G.; Wrolstad, R. E.; Lea, P.; Enersen, G. *J. Food Sci.* **1992**, *57*, 172-177.
16. Poei-Langston, M. S.; Wrolstad, R. E. *J. Food Sci.* **1981**, *46*, 1218-1222,1236.
17. Pilando, L. S.; Wrolstad, R. E.; Heatherbell, D. A. *J. Food Sci.* **1985**, *50*, 1121-1125.
18. Aaby, K. G.; Skrede, G.; Wrolstad, R. E. *J. Agric. Food Chem.* **2005**, *53*, 4032-4040.
19. Wrolstad, R. E.; Wightman, J. D.; Durst, R. W. *Food Technol.* **1994**, *48*, 90-97.
20. Wrolstad, R. E.; Durst, R. W.; Lee, J. *Trends Food Sci. Technol.* **2005**, *16*, 423-428.
21. Lee, J.; Durst, R. W.; Wrolstad, R. E. *J. Assoc. Off. Anal. Chem.* **2005**, *88*, 1269-1278.

Chapter 4

Color Quality of Maraschino Cherries

M. Monica Giusti[1] and Ronald E. Wrolstad[2]

[1]Department of Food Science and Technology, The Ohio State University, 2015 Fyffe Road, Columbus, OH 43210
[2]Department of Food Science and Technology, Oregon State University, 100 Wiegand Hall, Corvallis, OR 97331–6602

Maraschino cherry processing involves immersion of cherries in a brine solution to extend their shelf life, resulting in color loss. Color is typically replaced with Allura Red (FD&C Red No.40) or erythrosine (FD&C Red No.3), depending on the market. However, manufacturers of maraschino cherries have sought natural colorants which could serve as acceptable alternatives to the use of artificial dyes. This task was challenging due to the residual SO_2 from manufacturing and pH (3.5) which favor pigment degradation. We evaluated red radish anthocyanin extracts (RAE) to color primary and secondary bleached cherries. Color analysis (CIELch), showed that RAE imparted color extremely close to that of FD&C Red No. 40, for ~6 mo storage. The high pigment stability (half-lives of 29-33 wk) was attributed to the acylated anthocyanins in RAE. Color quality depended on the bleaching process, anthocyanin concentration and exposure to light. Anthocyanin-rich maraschino cherries may have added value as a functional food because of their more natural character and their high phytonutrient content.

Maraschino Cherries History and Production

Maraschino cherries are a type of candied fruit made from fresh cherries. The term "Maraschino Cherries" originally referred to Marasca cherries that were preserved in liqueur, a practice begun in Dalmatia centuries ago. Today, this term is used to describe cherries which have been dyed with a food colorant, impregnated with sugar and packed in sugar syrup flavored with oil of bitter almonds or a similar flavor. The cherries are also preserved with small amounts of sodium benzoate and/or potassium sorbate.

Maraschino cherries, as we know them today, differ considerably from the original product prepared many years ago in Dalmatia. The delicately colored and flavored liqueur was made from fruit, ground pits, bark, and leaves of the Marasque cherry tree. This extract was then used to flavor and color cherries similar to the Royal Anne variety cultivated in the USA (*1*). The Marasca cherry was consumed as a delicacy by royalty and the wealthy. These cherries made their way into the United States in the late 1800s and were served in fine hotels and upscale restaurants. By the turn of the century, American producers were experimenting with different ways to produce similar cherries in the USA using a variety of liqueurs and flavorings. Most of the production was done in the Eastern USA using cherries imported from Italy.

Around that time, cherries were already growing in the western Coast of the USA as well as in Michigan. However, these new cherries did not match the texture characteristics of the Italian brined cherries that had dominated the market. Therefore in the late 1920's Ernest Wiegand, a researcher at Oregon State University, developed a brining process (*2*), that made possible the commercialization of a fruit otherwise highly perishable, in a very attractive and stable form. Although the maraschino process was inspired by the European product, the brining process provided a higher quality, more uniform cherry, which USA manufacturers colored and flavored to produce the distinctive commercial product.

The cherry-producing districts of Oregon soon experienced an increasing interest in the preparation of cherries for maraschino use. During the 1930 season, approximately 10,000 barrels (1,136 Tons) of cherries were bleached in Oregon. Some of these were shipped to the East coast, furnishing an outlet for a great portion of the Northwest's increasing Royal Ann cherry crop (*3*). In 1992, the State of Oregon was the second major producer of sweet cherries (52,000 Tons) in the USA, after Washington (97,000 Tons), from a nation-wide production of 205,400 Tons (*4*), and more than 50% of the Oregon production (28,000 Tons) was destined for brining. For 2004, the major producers were Washington, California, Oregon and Michigan and the total cherry production in the USA added to 282,060 Tons (*5*). About 30% of those cherries were brined.

Maraschino Cherry Processing

Among the hundreds of cherry varieties around the world, the ones used for brining are usually the light sweet cherries, such as the Royal Ann, Rainier or Gold varieties. These varieties typically exhibit a yellow/blush color on their skin, as compared to other deep red varieties. However, most cherries are suitable for maraschino cherry processing. Sweet cherries destined for maraschino or similar processing are harvested either by hand or machine before full ripeness and placed in a brine solution containing between 1 and 1.5% sulfur dioxide and 3000 to 5000 ppm of calcium salts. The brine bleaches the fruit to a pale yellow color, and it also acts as a preservative during subsequent storage. Calcium acts as a hardening agent, increasing fruit firmness (*6,7*). After a period of several weeks to two years, the fruit is removed and "finished" into the final product (*8*).

Cherries are a very delicate fruit and they bruise easily and non-uniform discoloration from incompletely bleached cherries were frequently obtained. To improve the color quality of maraschino cherries, Oregon State University food scientists Beavers and Payne developed a secondary bleaching process that completely eliminated bruise marks and other dark skin discolorations from brined sweet cherries. After the first brining process, cherries are leached in water and placed in acidified sodium chlorite. This process made it possible to produce high quality, brightly colored maraschino and fruit cocktail cherries from fruit otherwise considered undesirable for these products (*7-9*).

After the bleaching process the cherries are ready for finishing into maraschino cherries. They are removed from the brine, rinsed with water and graded. The fruit is leached in running water to remove most of the sulfur dioxide. At this point of the process cherries are firm, and lack cherry color and flavor. Therefore flavoring (cherry and/or almond food flavors), coloring and sweetening agents are added. Potassium sorbate and sodium benzoate are used as preservatives, and citric acid is used to adjust the pH to acidic conditions (pH usually between 3.4 and 3.8). The cherries are finally bottled and pasteurized (*1, 6*).

Coloring Maraschino Cherries

Maraschino cherries are typically red, althought many other colors can also be produced. For a long time, FD&C Red No. 4 (Ponceau SX) was the colorant of choice for maraschino cherry producers. This synthetic colorant had a brilliant red color with unusual resistance to the destructive influences of food ingredients and heat (*10*). However, the use of this colorant was banned in the USA in1976 because of unresolved safety questions (*11*), and FD&C Red No. 3 (erythrosine) and FD&C No. 40 (Allura Red) have been used since, FD&C Red No. 40 being the colorant of choice for maraschino cherries because of its solubility properties.

There is considerable demand for food colorants from natural sources that can serve as alternatives to the use of synthetic dyes due to both legislative action and consumer concerns over the use of synthetic additives. Interest in the use of natural extracts as coloring agents has intensified because of their possible health benefits. However, finding a natural red colorant that can effectively replace FD&C Red No. 40 has proved to be a difficult task, because few natural materials have its bright red color unmixed with other tones (*12*). The use of natural colorants such as cochineal, concord grape extract, other anthocyanins, and some carotenoids has been studied (*12-14*). Relatively good stability has been obtained (between 3 and 6 months of storage); however, limitations have been found trying to reproduce the desired hue.

Major Challenges Faced by Natural Colorants

Manufacturers of maraschino cherries have sought a natural colorant which could serve as acceptable alternative to the use of FD&C Red No. 40, which gives the product its attractive color. This is a challenging task since processors wanted to match the hue of the Allura Red colored cherries and have a reasonable shelf life for a product that can be packed in glass and stored at ambient temperatures. Also, the matrix offers it own challenges since the pH is 3.5 (higher than the typical pH ranges for anthocyanin applications) and the residual SO_2 after processing, which can lead to pigment degradation.

Acylated anthocyanins, particularly those with cinnamic acid acylations, have been found to possess increased stability *(15-17)*. These pigments may impart desirable color and stability for commercial food products. Examples of suitable acylated anthocyanin sources are red radishes, red potatoes, red cabbage, black carrots, and purple sweet potatoes. Among these, radishes and red potatoes stand out as potential alternatives for the use of FD&C Red No. 40 (Allura Red) because of the hue that can be produced with these extracts in solution. Radish and potato extracts imparted color characteristics to model juices at pH 3.5 extremely close to those of Allura Red (*15*).

Advantages of Radish Anthocyanins as Natural Alternatives to Color Maraschino Cherries

Our worked focused on the evaluation of radish anthocyanins as potential alternatives to the use of FD&C Red No. 40 to color maraschino cherries. This source was selected based on the information available in the literature regarding their pigment composition. The pigments in radish were reported as acylated pelargonidin derivatives. Among all anthocyanin aglycones, pelargonidin is the one that posses the shortest wavelength of maximum absorption in the visible range, with a hue closer to orange / orange-red. Addition of acylation has been shown to cause a bathochromic shift (an increase in the wavelength of maximum absorption). For pelargonidin, this would result on a change toward a more red color, while we would expect a more purple red color on most other acylated anthocyanins. In addition, the acylation would be expected to increase the stability, protecting the oxonium ring from the attack of water and sulfites, and ameliorating the effects of the increase in pH to 3.5.

The chemical structure of radish anthocyanins was elucidated by use of chromatography, spectral characteristics, molecular mass and nuclear magnetic resonance (*18, 19*). All this combined information allowed us to determine the identity of the pigments as pelargonidin-3-sophoroside-5-glucoside mono acylated with p-coumaric or ferulic acid or di-acylated with p-coumaric or ferulic acid plus a malonic acid (Figure 1). In addition, we determined the close proximity between hydrogens of the pyrilium ring of the anthocyanidin and hydrogens in the cinnamic acid acylating the pigment. This supports the hypothesis that acylation may enhance anthocyanin stability by folding of the molecule so that the acid can protect the pyrilium ring, or by stacking of the molecules with a similar result.

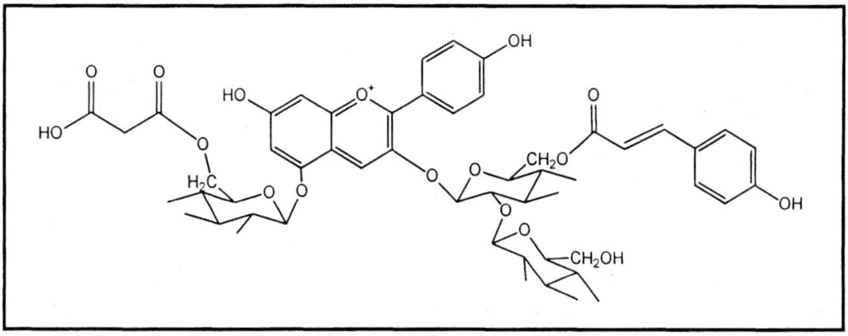

Figure 1. Chemical structure of pelargonidin-3-sophoroside-5-glucoside acylated with p-coumaric acid and malonic acid found in radish.

The color characteristics of the pelargonidin aglycones as well as the glycosylated and acylated derivatives were evaluated, and we were able to confirm that acylated pelargonidin derivatives, or more specifically, the pigments in radish closely resemble the color characteristics of FD&C Red No. 40 (*17*).

Color Quality of Maraschinos Colored With a Radish Extract

Red radish anthocyanin extract (RAE) was used to color brined cherries as an alternative to FD&C Red No. 40 *(17, 20)*. Primary and secondary bleached cherries were colored with two different concentrations of radish anthocyanins (600 and 1200 mg anthocyanin/L syrup) to compare the color quality to that obtained with FD&C Red No. 40 (200 ppm). Color and pigment stability of secondary bleached cherries and syrup colored with RAE were evaluated during storage (25°C) in the dark and exposed to light.

Important Considerations:

The processing of maraschino cherries was done in collaboration with the maraschino cherry industry to assure the use of practices that would be applicable to the industry. The pH of maraschino cherries can range from 3.5 to 4.0. Since anthocyanin tinctorial power and stability are favored by low pH values, we chose to manufacture cherries at pH 3.5, the lowest pH within the typical ranges used by the industry. Prior to coloring, brined cherries were washed several times to drastically decrease the free sulfur dioxide level (from 3,500 and 2000 ppm for primary and secondary bleached cherries, respectively, to less than 480 ppm after washing) to minimize the detrimental effects on color and also to reduce the risks for allergic reactions. This level was further reduced during the following steps of the manufacture of maraschino cherries.

After washing, the sweetening process begins. It is very important that this process be done gradually to avoid damage to the cell integrity and texture characteristics of the cherry. This process is done under controlled temperature conditions (40°C) to favor infusion of the sugar at a rate of 3 degree brix every 12 hours. At the end of the sweetening process, flavors and colors are added, and allowed to equilibrate for 4 days. Finally, cherries and syrup are bottled, pasteurized and stored at room temperature (Figure 2).

Color characteristics (CIELAB, chroma and hue angle) of cherries were measured on a ColorQuest Hunter spectrophotometer set up for reflectance specular included measurements using illuminant C and 10 degree observer angle. Cherries were drained and placed in a 5 cm pathlength optical glass cell and 4 repeated measurements were taken for each sample. Color characteristics and haze of syrup were measured on the same instrument using a 1 cm pathlength optical glass cell under transmittance mode using illuminant C and 10 degree observer angle.

Radish anthocyanin extracts imparted cherries and syrup color characteristics extremely close to those obtained with FD&C Red No. 40 (Figures 3 and 4). The color characteristics of primary and secondary bleached cherries were markedly different, with primary bleached cherries showing a high

Figure 2. Major changes on cherry color during manufacturing of maraschino cherries.
**Photograph by Lynn Ketchum, Oregon Agricultural Experiment Station (See page 4 of color inserts.)*

chroma (26.8) and a yellow hue, while secondary bleached cherries were close to white in color with a chroma of only 2.4. The lower concentration (600 mg anthocyanin/L syrup) of radish anthocyanins gave cherries and syrup the closest color characteristics to FD&C Red No. 40.

Monomeric anthocyanin degradation followed first-order kinetics, with half-lives of 29 and 33 wk for syrups colored with RAE C1 and RAE C2, respectively

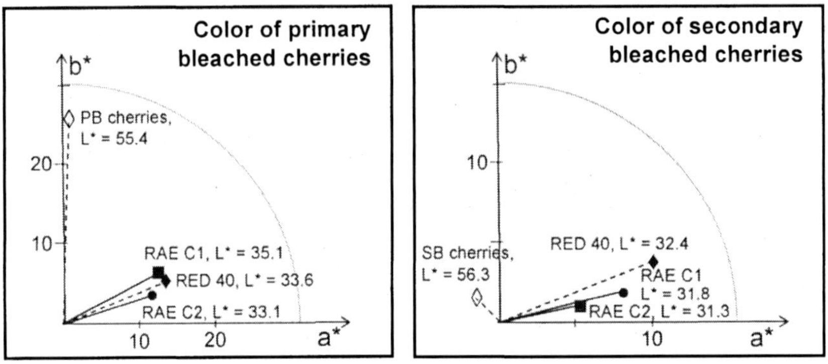

Figure 3. Color characteristics of primary and secondary bleached cherries before and after coloring with radish anthocyanins or FD&C Red No. 40. Reflectance specular included mode, Illuminant C and 10 ° observer angle.

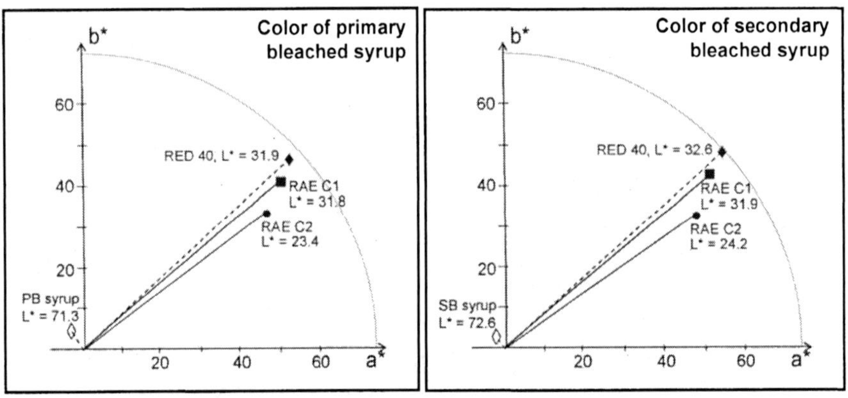

Figure 4. Color of syrup of primary and secondary bleached cherries before and after coloring with radish anthocyanins or FD&C Red No. 40. Transmittance measurements, Illuminant C, 10 ° observer angle, 1 cm pathlength.

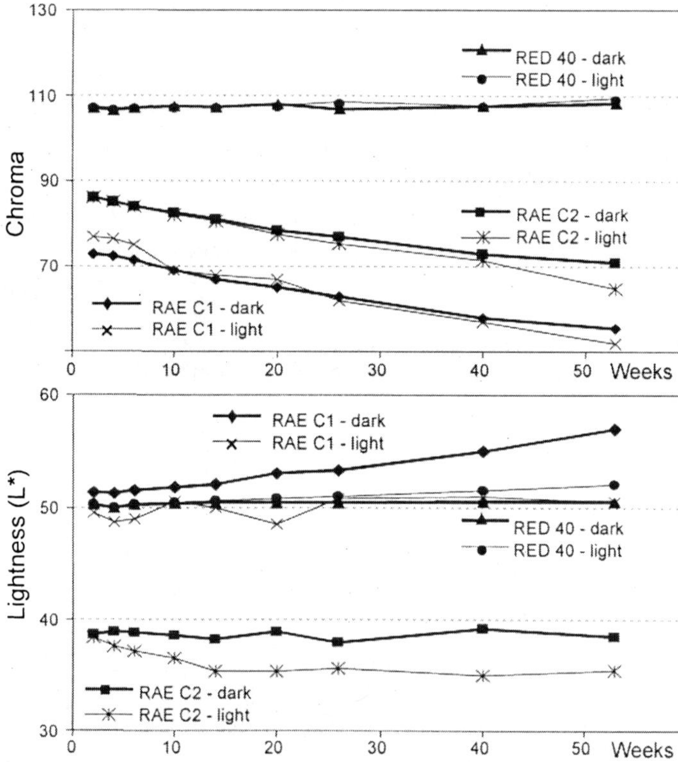

Figure 5. Changes in chroma and lightness of syrup samples colored with radish anthocyanin extracts (600 mg anthocyanin/L = RAE C1; 1200 mg anthocyanin/L = RAE C2) or FD&C Red No. 40 (200 ppm, RED 40) over a year of storage under the dark or exposed to light.

(*17, 20*). Higher anthocyanin concentration exerted a protective effect on color stability. Exposure of syrup samples to light slightly increased monomeric anthocyanin degradation. However, the half lives of syrups colored with RAE C1 were not significantly affected by light exposure, while the half life of syrups colored with RAE C2 were 1.5 times shorter than when stored in the dark. In samples exposed to light a precipitated material was observed after 50 weeks of storage, suggesting that light exposure favored formation of larger polymers that came out of solution.

Color stability of secondary bleached cherries and syrup were monitored during a year of storage under dark condition or exposed to light. Color characteristics remained similar to FD&C Red No. 40 for about 6 months of storage at room temperature, and light exposure had little effect on color stability (Figures 5 and 6).

Maraschino Cherries With Added Value

Interest in anthocyanin-rich foods and extracts has intensified because of their possible health benefits. Anthocyanins are potent antioxidants and may be chemoprotective. Optimizing health and performance through the diet is believed to be one of the largest and most lucrative markets in the US, and throughout the world. Findings of acylated anthocyanins with increased stability have shown that these pigments may impart desirable color and stability for commercial food products. Maraschino cherries with bright attractive and stable red color were obtained with radish extract. Radish imparted color characteristics to model juices extremely close to those of Allura Red. The increased stability of these pigments together with their added value due to potential beneficial effects opens a new window of opportunity for use of these extracts in a variety of food applications.

References

1. Filz, W. F.; Henney, E. N.. *Home Preparation of Maraschino Cherries*; Station Bulletin 497; Agric. Exp. Stn. Oregon St. College: Corvallis, OR **1951**; pp 3-11.
2. Rose, S. *Capitol Press*, Salem, OR, August 29, 1975.
3. Bullis, D. E.; Wiegand, E. H. *Bleaching and Dyeing Royal Ann Cherries for Maraschino or Fruit Salad Use*; Station Bulletin 275; Agric. Exp. Stn, Oregon St. Agric. College: Corvallis, OR **1931**, pp 4-29.
4. OSU Extension Service. Commodity data sheet. Sweet Cherries. Extension Economic Information Office, Oregon St. Univ.: Corvallis, OR **1994**, 5110-5194.
5. National Agricultural Statistics Service (NASS), USDA, 2005
6. Wiegand, E. H.; Bullis, D. E. *Maraschino Cherries Methods for Bleaching and Dyeing*; Station Bulletin 32; Agric. Exp. Stn., Oregon St. Agric. College: Corvallis, OR **1930**, pp 1-6.
7. Beavers, D. V.; Payne, C. H. *Food Technol.*, **1969**, *23*, 175-177.
8. Anonymous. *Oregon's Agric. Prog.* **1968**, *15*, 2.
9. Beavers, D. V.; Payne, C. H; Milleville, H. P. *Procedure for Secondary Bleaching Brined Cherries with Sodium Chlorite*; Circular of Information 632; Agric. Exper. Stn., Oregon State University: Corvallis, OR, **1970**, 1-7.

10. Yang, H. Y.; Ross, E.; Brekke, J. E. *Cherry Brining and Finishing*; Circular of Information 624; Agric. Exper. Stn., Oregon State University, Corvallis, OR **1966**, 1-7.
11. Rumore, M. M. *Pharm. Technol.*, **1992**, 1*6*, 68-82.
12. LaBell, F. *Food Process.*, **1993**, *June*, 88-89.
13. McLellan, M. R.; Cash, J. N.. *J. Food Sci.*, **1979**, *44*, 483-487.
14. Sapers, G. M. *J. Food Sci.*, **1994**, *59*, 135-138.
15. Rodriguez-Saona, L. E.; Giusti, M. M.; Wrolstad, R. E. *J. Food Sci.*, **1999**, *64*, 451-456.
16. Giusti, M. M.; Wrolstad, R. E. *Biochem. Eng. J.,* **2003**, *14*, 217-225.
17. Giusti, M. M.; Wrolstad, R. E. *J. Food Sci.*, **1996**, *61*, 688-694.
18. Giusti M. M.; Wrolstad R. E. *J. Food Sci.*, **1996**, *61*, 322-326.
19. Giusti, M. M.; Ghanadan, H.; Wrolstad, R. E. *J. Agric. Food Chem.*, **1998**, *46*, 4858-4863.
20. Giusti Hundskopf, M. M. MS Thesis, Oregon State University, Corvallis, OR, 1995.

Chapter 5

Color Quality of Cranberry Products

David G. Cunningham, Antelmo F. Santos, and Rodney A. Serres

Ocean Spray Cranberries, Inc., 1 Ocean Spray Drive, Lakeville, MA 02349

The American cranberry, *Vaccinium macrocarpon* Aiton, is a fruit with a number of unique and desirable attributes that has found its way into a wide variety of food products. In addition to its growing reputation as a healthy fruit, cranberries are known for their distinctive taste and color. Whereas the taste of cranberries is highly acidic and consumer preference testing has necessitated the development of products with adjusted Brix to acid ratios, the crimson red color of cranberries is naturally appealing, from their aesthetics in the field to when they are finally consumed. Next to crop yield, fruit color is the attribute paid most attention to by growers. Growers strive to grow and deliver fruit with high color, while handlers and processors strive to manage fruit inventory and processing to optimize color usage in their products. The recent development and popularity with consumers of white cranberry juice beverages has introduced additional and entirely new challenges to cranberry growers, handlers and processors. This chapter will discuss grower, handler and processor practices that enable the delivery of cranberry products with high and consistent color quality to consumers.

Introduction

Cranberries are the small, red fruit of the perennial *Vaccinium macrocarpon*. The fruits are typically 1-2 centimeters in diameter, weigh 1-2 grams, and are known for their bright crimson red color. *Vaccinium* is classified under the Ericaceae family. Other *Vaccinium* species include the European cranberry (*V. oxycoccus*), the highbush cultivated blueberry (*V. corymbosum*), the lowbush blueberry (*V. angustifolium*) and the lingonberry (*V. vitis idaea*). Cranberries are one of the few cultivated berries of significant commercial value that are indigenous to North America, along with blueberry and fox grape (*Vitis labrusca*), known for the Concord grape cultivar. The name cranberry evolved from craneberry, which originated with colonists who thought the shape of the plant's flower resembled the neck, head and bill of a crane (*1*). Figure 1 shows the cranberry flower and ripened fruit. Cranberry plants are propagated by sticking cuttings into the soil. The cuttings root and grow by sending out lateral shoots from which new vertical shoots (uprights) develop. Flowers form on the uprights.

Figure 1. Cranberry Flower (left) and Fruit (right) (See page 4 of color inserts.)

While there are many native selections and cultivars of cranberries, commercially nine cultivars predominate and just two, 'Early Black' and 'Stevens', account for approximately 60% of acreage in production. 'Early Black' is a native selection dating back to 1857 and 'Stevens' is a hybrid introduced by the United States Department of Agriculture (USDA) in 1950 (*1*). While considered a minor crop, the worldwide production of cranberries has averaged around 800 million pounds per year over the past three years (2003-2005) (*2,3*). The United States accounted for about 78% of the crop, while 20%

was produced in Canada, and 2% was grown in Chile. The Ocean Spray Cranberry cooperative was formed over 75 years ago to market the fruit grown by its members, currently 648 growers in Massachusetts, New Jersey, Wisconsin, Oregon, Washington and the Canadian provinces of British Columbia and Quebec, who together currently deliver about 65% of the worldwide crop.

The cranberry industry represents a $2.5 billion worldwide market (2006 estimate) whose products include fresh fruit, juices, sauces and dried fruit. There is also a growing business for cranberry ingredient products such as frozen fruit, juice concentrate, juice concentrate powder and nutraceutical extract powders. Color is a desirable and important quality attribute for many of these products.

Proximate Composition

The composition of single strength (7.5° Brix) cranberry juice is shown in Table I (*4*). Carbohydrates, primarily sugars (3.7 g) and organic acids (3.1 g), account for 97% of the total solids. Pigments represent about 0.6 % of the total solids. As with most agricultural products, fruit composition varies due to factors such as variety, maturity, growing location and environmental stresses, while processing effects can further impact the composition of the materials derived from the fruit. Of these factors, growing location appears to have the greatest impact on pigment content.

Table I. Proximate Composition of 7.5 °Brix Juice (100 g)

Water	92.9 g
Solids	7.1 g
- Carbohydrates	6.9 g
- Minerals	96 mg
- Pigments	40 mg
- Vitamin C	2 mg
- Protein	< 0.1 g
- Fat	< 0.1 g

Anthocyanins

Anthocyanins are the red pigments found in cranberries. The basic anthocyanin chemical structure is shown in Figure 2. The anthocyanins found in cranberries are the arabinose, galactose and glucose 3-*O*-glycosides of cyanidin and peonidin (*5*). At the naturally occurring low pH of cranberries and cranberry products the anthocyanins are found in the stable red AH^+ form. Table II lists the amounts of the individual anthocyanins quantified in a random sampling (n=12) of cranberry juice cocktail (*6*). The total anthocyanin (TAcy) content of cranberries varies by growing area as seen in a graph (Figure 3) of the TAcy values for harvested fruit averaged over 14 years. The effect growing area has on anthocyanin level is primarily due to differences in the length of their respective growing seasons. While the total anthocyanin amount may vary, there is less variation in the ratios of the individual anthocyanins among the common commercial varieties. Table II also lists average ratios of cranberry anthocyanins normalized to cyanidin-3-*O*-galactoside. This distinctive anthocyanin profile is one factor that can be used to authenticate cranberry juice. Figure 4 is a chromatogram of a reverse phase HPLC anthocyanin separation depicting the typical cranberry anthocyanin profile. The elution order shown is the same as listed in Table II. However, recent work found that these anthocyanins profiles do vary significantly among 252 clonal accessions of *V. macrocarpon* (*7*).

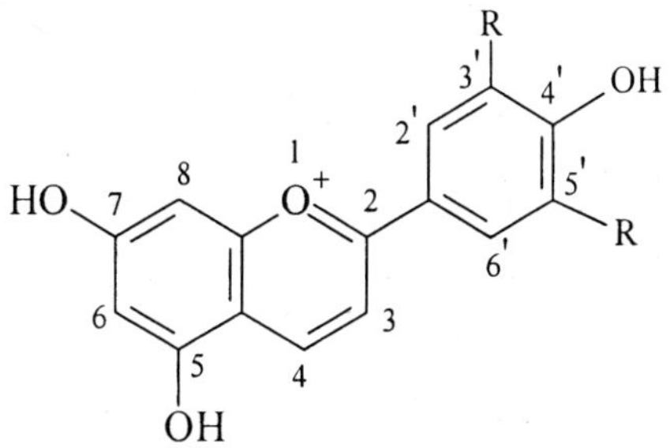

Figure 2. Anthocyanidin aglycone

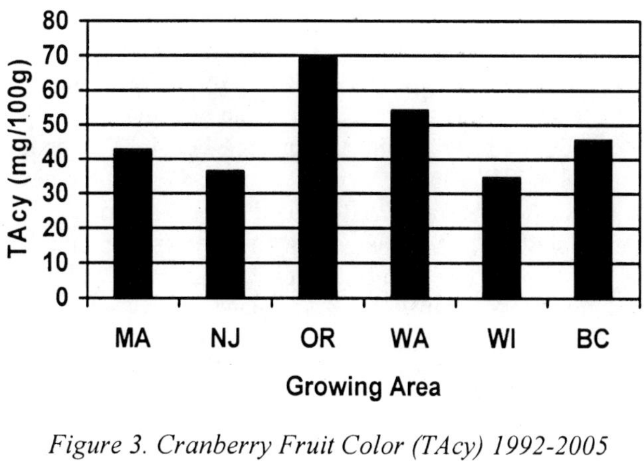

Figure 3. Cranberry Fruit Color (TAcy) 1992-2005

Table II. Anthocyanins in Cranberry Juice Cocktail

Anthocyanin	Amount (mg/L)	Relative Ratio
Cyanidin-3-*O*-galactoside	2.0 ppm	1.0
Cyanidin-3-*O*-glucoside	0.1 ppm	0.05
Cyanidin-3-*O*-arabinoside	1.4 ppm	0.9
Peonidin-3-*O*-galactoside	2.8 ppm	1.4
Peonidin-3-*O*-glucoside	0.3 ppm	0.2
Peonidin-3-*O*-arabinoside	1.1 ppm	0.6

Flavonols

The flavonols are the yellow pigments found in cranberries. The basic flavonol chemical structure is shown in Figure 5. The flavonols found in cranberries are quercetin and myricetin and their 3-*O*-glycosides. Francis and Clydesdale reported the presence of six flavonols in cranberry (*5*). Table III lists the amounts of five flavonols, quercetin and the quercetin 3-*O*-glycosides of rhamnose (quercitrin), galactose (hyperin) and arabinose (avicularin) and myricetin, measured in a random sampling (n=12) of cranberry juice cocktail (*6*). The flavonols are not as thoroughly characterized as the anthocyanins and thus not as useful for authenticating cranberry juice. In addition to their function as yellow pigments, flavonols also function as co-pigments that help to stabilize and enhance the color of the anthocyanins (*8,9*).

Cranberry Color Development

Cranberry bloom and fruit set occurs from mid-May through the middle of July depending on the growing region. Cranberry color development begins in earnest in August. Figure 6 shows the progression of fruit color development for the Stevens cultivar collected from a cranberry bog in Massachusetts at 1-2 week intervals, from mid-August to mid-October. Fruit color development is triggered by the onset of a cycle of cool nights and warm days. Once color development has started, good color quality is the result of the length of the remainder of the growing season. Cranberry processors will pay growers a premium for highly colored fruit. Growers manage the overall color of their crop by balancing the desire for color development against the logistical need to harvest their crop before winter and below freezing temperatures set in.

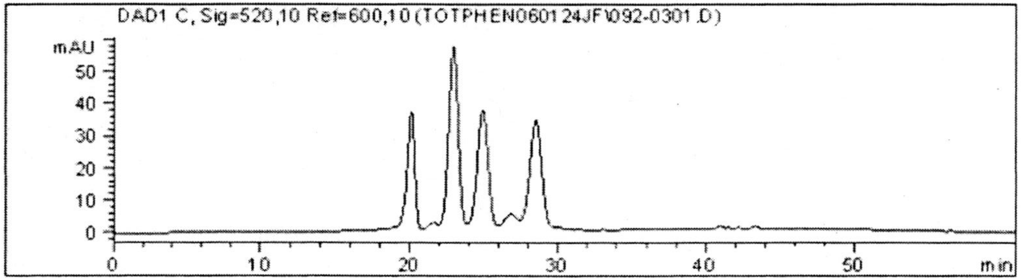

Figure 4. Cranberry Anthocyanin HPLC Chromatogram

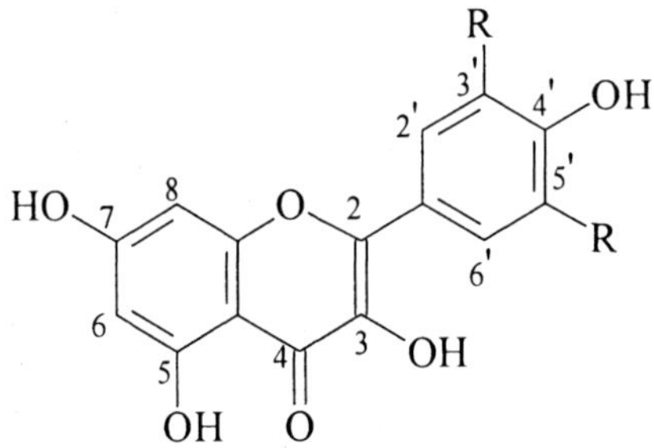

Figure 5. Flavonol aglycone

Table III. Flavonols in Cranberry Juice Cocktail

Flavonol	Amount (mg/L)
Quercetin	13.0 ppm
Quercitrin	5.2 ppm
Hyperin	23.2 ppm
Avicularin	1.8 ppm
Myricetin	5.3 ppm

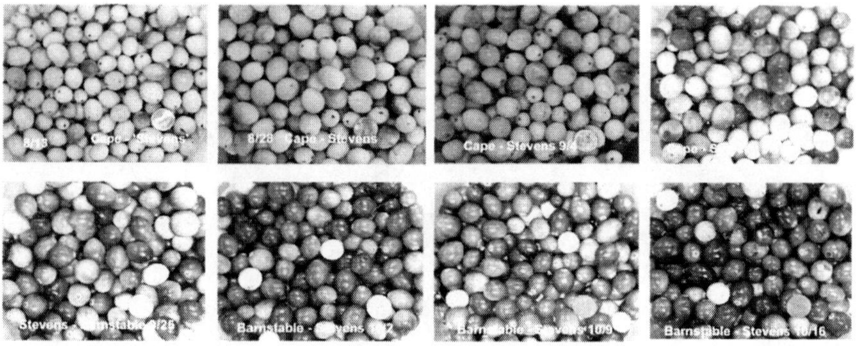

Figure 6. Cranberry Color Development (See page 4 of color inserts.)

Pigmentation develops in the outer skin layer of the fruit as can be seen in the cranberry cross section shown in Figure 7. The flesh of the fruit is white, but pigment can diffuse into the flesh at the later stages of maturation.

Figure 7. Cranberry Fruit Cross-Section (See page 5 of color inserts.)

The presence of white fruit can also be noted in Figure 6. White fruit appears to be a stage in cranberry color development associated with the loss of production and increased degradation of chlorophyll that occurs prior to the onset of anthocyanin synthesis and the reddening of the fruit. This point where the fruit turns white provides a narrow window in which cranberries can be harvested to deliver fruit that can then be used for some of the recently introduced white cranberry juice products.

Cranberry Uses

While there is a seasonal fresh market for cranberries, most cranberries are processed and find their way into an ever increasing number of products. In fact less than 6% of the crop is utilized as fresh (*2*), the remainder is used by industry to produce juice, juice concentrate, sauce, dried fruit and a range of cranberry based ingredients. In actuality, even fruit sold on the fresh market to consumers is primarily processed into such products as homemade cranberry sauce and various home baked goods, because fresh cranberries are rarely consumed as such due to their low sugar and high acid content relative to other fresh fruits.

Fresh Fruit

Fresh fruit is usually dry harvested by plucking the berries off the vine. Traditionally this has been done by hand using a cranberry scoop, but is generally now accomplished using mechanical harvesters. The basic design of both incorporates a set of wooden or metal tines that are combed through the vine to pluck off the fruit. For the cranberry scoop, the cranberries collect in the box part of the scoop and are periodically dumped into a box or bin. For the mechanical harvester, the cranberries are conveyed up a belt and into a bag hanging from the back, which is subsequently dumped into a box or bin when full. Fresh fruit may also be wet harvested by some growers, a technique that is more commonly used for harvesting fruit for the process market. When fruit destined for the fresh market is wet harvested, it is dried using air blowers located in a facility at the farm to ensure good keeping quality.

Fresh fruit is graded according to a U.S. No. 1 Grade standard (36 CFR 16893) for fresh cranberries. Cranberries meeting this grade must be free from damage caused by bruising, frost, mold (rotten, soft or decayed fruit), scald and insects, be clean and have no adhering foreign matter, be at least 13/32" in diameter and have no more than 115 berries in a cup count, and finally, from a color standpoint, cannot be green or speckled. Fresh fruit that doesn't meet these standards is downgraded and may be utilized as process fruit.

Fresh fruit keeping quality is managed by storing the fruit in coolers after it has been received and awaiting packaging for the fresh market. Under these conditions the fruit color may actually continue to develop to some degree for certain varieties. The Early Black cultivar is especially notable in this regard.

In packaging fresh fruit for sale, the fruit is sorted using automated and hand sorting techniques as depicted in Figure 8, to eliminate defective fruit and foreign matter, such as dirt, leaves and vines. The color quality of fresh fruit is ensured through the elimination of any green, white or speckled fruit.

Figure 8. Hand and Automated Color Sorting (See page 5 of color inserts.)

Process Fruit

Fruit destined for use by the process industry is usually wet harvested. Wet harvesting takes advantage of the fact that cranberries are buoyant due to four air-filled locules, which are a significant part of the internal structure of the fruit. In the wet harvesting technique, cranberry bogs are flooded with water and the buoyant berries rise up over the plant canopy, but remain attached to it by long, thin pedicels. A mechanical water reel is used to agitate the water and break the pedicels (Figure 9), allowing the fruit to float to the surface of the water. The cranberries are corralled on the surface using booms and pumped or conveyed off the surface of the water into bins or large trailers by the side of the bog. The fruit is then trucked to a centralized receiving station.

Upon delivery to the receiving station, the truckload of fruit is representatively sampled for analysis. In addition to determining the excess

Figure 9. Mechanical water reel and corraled cranberries (See page 5 of color inserts.)

water, trash and percent rot of the load, the TAcy content of the fruit is measured using the method published by Francis (*10*). The method has been modified to eliminate the organic solvent due to hazardous waste disposal concerns. In the method, color extracts are prepared by blending the fruit with acidified water (0.01 N HCl) solution. The color extracts are filtered, diluted and the absorbance of an aliquot read at 515 nm. The resulting TAcy content of the fruit is reported as milligrams of anthocyanin per 100 grams of fruit. The TAcy result of the sample is used to determine a grower payment premium for delivering fruit with good color. Conversely, the grower may be penalized for delivering fruit with poor color if the TAcy result is too low.

After the delivery sample has been taken, the fruit is dumped into hoppers or pools at the receiving station to be cleaned, binned and sent to a freezer for storage. Before the fruit is shipped to the freezer, representative samples are taken for another TAcy determination. This second TAcy result is used to categorize the load of fruit going to the freezer into color grades that help to manage and optimize process fruit utilization. The assigned color grades are used to selectively pull the process fruit loads in inventory for specific product color needs. Increasingly, fruit may go directly from the receiving station to a juice concentrate production facility. In this case there is no opportunity to selectively blend different fruit color grades. However, some blending is achieved due to the large scale of the juice concentrate operation.

Processing and Products of Cranberries

Cranberries can be processed into juice and juice concentrate for use in manufacturing cranberry juice cocktail (CJC) and a wide variety of other beverage products. Juice concentrate provides a stable and economical way to store and ship cranberry juice to bottling plants. The majority of process fruit, 55-60%, is currently utilized for this purpose. In addition, cranberries can also be processed into a number of food products including cranberry sauce and sweetened dried cranberries (SDCs), utilizing another 35-40% of the process fruit crop. Finally, cranberries can be processed by a number of technologies to produce extracts and colorants, which along with concentrate, SDCs, and frozen cranberries, can be used as ingredients in other food products, accounting for the remaining 5-10% of the process fruit crop.

Juice and Juice Concentrate Production

Frozen fruit is processed or converted into juice by either traditional fruit juice pressing methods or by water extraction. Pressing is done on a batch basis compared to water extraction, which can be a batch or continuous process. The juice, if not used at the pressing location, may be semi-concentrated for shipping to beverage bottling locations. Each production lot of juice or semi-concentrate is tested for brix, titratable acidity, haze and color. The color measurement procedure is designed to assess the color of the juice at its usage level in CJC, which is typically a 27% single strength (7.5° Brix) cranberry juice beverage. The procedure simply requires the dilution of the juice with distilled water to 0.2° Brix and reading the absorbance of an aliquot of the diluted juice at 515 nm. The absorbance value is used "as is" to report juice color. The color of each lot of juice produced can be controlled by the selection and blending together of fruit of different color grades.

Cranberry juice concentrate is produced by the hot mash depectinization of fresh or frozen cranberries. The mash is either pressed or filtered to recover the juice, which is then concentrated to 50° Brix as an industry standard. As with juice, each production lot of concentrate is tested for Brix, titratable acidity, haze and color. Color is again tested for by diluting a sample to 0.2° Brix, reading its absorbance at 515 nm and reporting this reading as juice color.

The shipping of juice and juice concentrate from cranberry receiving and conversion locations to beverage manufacturing and bottling locations provides a further opportunity to manage color in the production of CJC. This enables the production of a product that is of uniform color no matter when or where it is purchased.

CJC and White CJC Production

Cranberry juice cocktail is a cranberry juice beverage typically formulated to contain 27% single (7.5° Brix) strength cranberry juice. In addition to cranberry juice, water and sweetener are added to dilute the high acidity and sweeten the beverage to achieve a Brix to acid ratio (BAR) that is preferred by consumers and more typical of other fruit juices such as apple and grape, as can be seen in the Figure 10 (*11*).

Depending on the circumstances, cranberry juice, juice concentrate, or a blend of the two may be used to formulate the beverage. The lots of juice or juice concentrate used are selected based on the color values that were determined and assigned during their production to achieve a finished product color target. This target can be adjusted when the overall color of the cranberry crop changes due to year to year variances. The use of a beverage color target enables the production of product with consistent color at various bottling facilities throughout the year.

In addition to controlling color quality in production, the use of tristimulus color measurement has been used as a research tool in product development to quantify the change in color as a function of shelf life in accelerated storage studies. Hunter "a" values have been found to be a good indicator of shelf life as a result of product browning and loss or red color. CJC usually has a 6-12 month

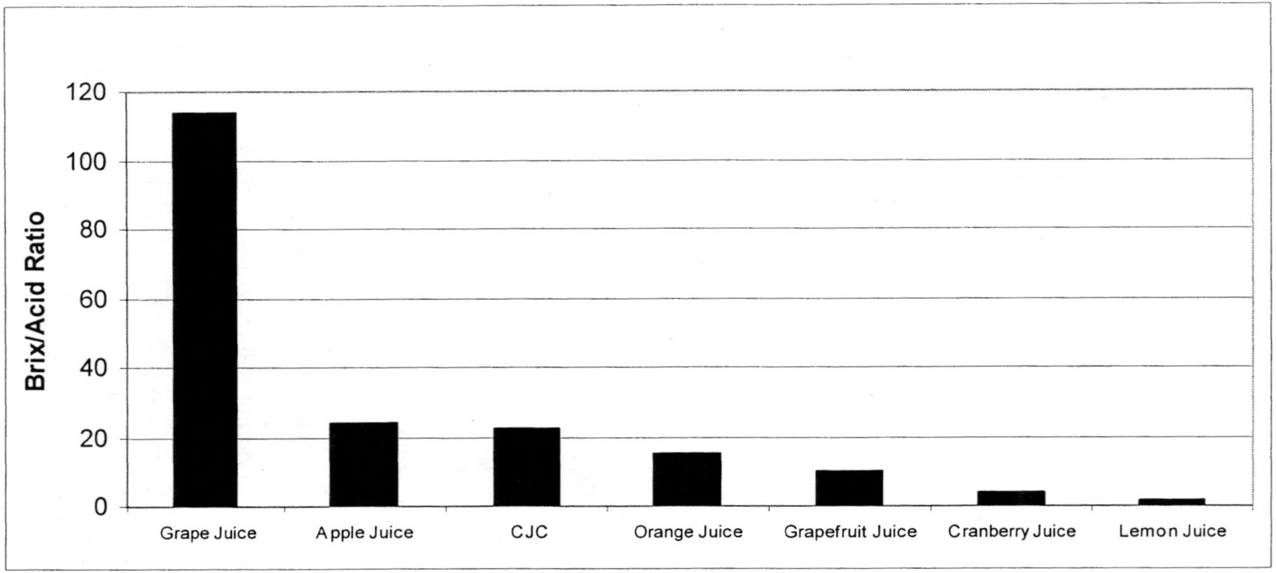

Figure 10. Brix/Acid ratio of various fruit juices

shelf life, depending on the packaging material, package size and bottling line conditions.

While CJC has been a staple product of the cranberry industry since 1930, White CJC was introduced to the market only a few years ago. White CJC utilizes juice and juice concentrate produced from cranberries harvested during the stage when the berries are white, as was discussed in the section on fruit color development. In order to ensure the color of the finished product, fruit that is harvested, delivered and accepted for white CJC production purposes can not exceed a designated (proprietary information) TAcy result. Growers receive an incentive payment to deliver white fruit due to a loss of yield, since white cranberries are slightly smaller in size, and the loss of a red color incentive payment.

Cranberry Sauce Production

Cranberry sauce is produced by the cooking of whole cranberries in a kettle with the addition of cranberry presscake, a juice production byproduct, as a source of added pectin for jelling purposes. As a result of this process, almost all the available pigment in the fruit is recovered. This actually necessitates that fruit of lower color be used in order to avoid producing a cranberry sauce that is too dark in color, which is not preferred by the consumer.

Sweetened Dried Cranberries Production

Sweetened dried cranberries are produced from cranberries that are infused with sugar, cranberry juice concentrate or possibly another fruit juice concentrate and then dried to produce the finished product. As with cranberry sauce, almost all the color is recovered and lower colored fruit is more desirable in order to avoid producing SDCs that look too dark.

Ingredients

There are a growing number of cranberry based ingredients being developed and sold to customers who are utilizing them for various consumer products. In addition to the health benefit that may be imparted to their product by incorporating a cranberry ingredient, customers are also buying these ingredients for their color quality. For example, cranberry juice concentrate powder is being sold for incorporation into products such as teas and nutraceutical products. The natural pinkish red color of the cranberry powder imparts a desirable and stable color to these products and is usually specified as such in cranberry powder technical specifications and certificates of analysis.

Summary

As suggested by the National Cancer Institute's 5-A-Day Program, a key nutrition guidance message is to incorporate foods of various colors into your diet, since the various natural pigments inherent in fruit and vegetables are believed to contribute to health in different ways. Cranberries and cranberry products are one way to accomplish this goal and thus their color quality is a key attribute to many consumers. With an increased demand for cranberry products, the cranberry industry must continue to efficiently mange the crop from field production through processing in an effort to recover the valuable color components of cranberry. Newly developed cranberry cultivars, such as Crimson Queen and HyRed, promise to deliver more color per acre of fruit and new conversion methods, such as water extraction, allow processors to recover and utilize that color.

References

1. *The American Cranberry*; Eck, P.; Rutgers University Press; New Brunswick, NJ, 1990; pp 2-3.
2. Noncitrus Fruits and Nuts 2005 Summary, USDA NASS, July 2006, 42.
3. William Frantz, Ocean Spray Cranberries, Inc., Private Communication
4. Kuzminski, L. N. *Nutr. Rev.* **1996**, *54*, S87-S90.
5. *Food Colorimetry: Theory and Applications*; Francis, F.J.; Clydesdale, F.M.; The AVI Publishing Company Inc.; Westport, CT, 1975; pp 229-230.
6. Cunningham, D.G.; Vannozzi, S.; Turk, R.; Roderick, R.; O'Shea, E.; Brilliant, K. In *Nutraceutical Beverages: Chemistry, Nutrition and Health Effects*; Shahidi, F.; Weerasinghe, D.K. Eds.; American Chemical Society, Washington, DC, 2004; pp 35-51.
7. Vorsa, N.; Polashock, J.; Cunningham, D.; Roderick, R. *J. Amer. Soc. Hort. Sci.* **2003**, *128*, 691-697.
8. Brouillard, R. In *Anthocyanins as Food Colors*; Markakis, P. Ed.; Academic Press; New York, NY, 1982; pp 1-38.
9. Osawa, Y. In *Anthocyanins as Food Colors*; Markakis, P. Ed.; Academic Press; New York, NY, 1982; pp 41-65.
10. Francis, F.J. *Proc. Am. Soc. Hort. Sci.* **1957**, *69*, 296-301.
11. Leahy, M.; Roderick, R.; Brilliant, K. *Nutr. Today* **2001**, *36*, 254-265.

Chapter 6

Anthocyanin Colorants from Fruits and Vegetables

S. Hake and J. Quinn

GNT USA, Inc., 660 White Plains Road, Tarrytown, NY 10591

Understanding the relationship between the various internal and external factors that influence the color governance of new natural colors is important when choosing a color. This chapter will focus on the industries' criteria for natural colors and explain the driving factors influencing the demand for all natural ingredients, specifically focusing on natural red colors from fruits and vegetables. Two important external factors that influence the color governance are the regulatory or FDA policies and the demand from the customer or marketplace. In addition to research and development, an important internal factor is the relationship between nature, growers, and location.

Today's consumers are more educated than ever about food ingredients and additives. With obesity on the rise, especially in children, focus on portion size and nutritional value in food and beverages is getting a lot of attention from lawmakers in Washington. Retailers are also starting to educate their consumers. Stores such as Whole Foods and Trader Joe's have a strong consumer following because of their commitment to provide healthy, organic, and natural food products. For example, Whole Foods Market's internal quality standards team has developed a list of unacceptable food ingredients. Ingredients on this list include saccharin, FD&C artificial colors, bleached flour, monosodium glutamate, sulfites, and artificial flavors.*(1)* The retail products in these cases give the consumer the confidence that the products on the shelves will be nutritious and healthy.

The demand for natural and organic ingredients has been on the rise over the last couple of years making the organic and natural products sector one of the fastest growing in the industry. In the past, the food and beverage industry has seen many trends including fat free, sugar free, soy, and low carb. More natural and healthy products are no longer a trend but rather a revolution to clean up existing products from artificial additives. New products are now being formulated with emphasis on all-natural ingredients. Consumers are also going through lifestyle changes to improve their eating habits.

Food colors play an important role in formulating foods. Traditionally, mainstream products have been colored with artificial FD&C colors, however recently, the demand for an all-natural color has been on the rise. Color manufacturers and users now need to consider numerous factors. In this paper we will focus on red colors derived from fruits and vegetables. We will discuss the multiple steps that one should take into consideration.

As shown in Figure 1, the foundation for developing a new color is Research & Development. There are also two outside factors that influence the Business Governance of new natural colors. The first factor is Regulatory / FDA and the second factor is the demand from the Customer / Marketplace. Additionally, there are some driving factors (People, Funding, Marketing) influencing the demand for all natural ingredients. One important factor is the influence that location and growers have on the naturally occurring red colors. While this factor will be discussed in more detail, this chapter will not discuss in depth the importance that Marketing and Funding play with Business Governance.

Research and Development

Anthocyanin molecules are responsible for the natural red, blue, violet, and magenta colors found in various fruits, vegetables, and plants. The anthocyanin structure is a glycoside of flavylium or 2-phenylbenzopyrylium salts and are

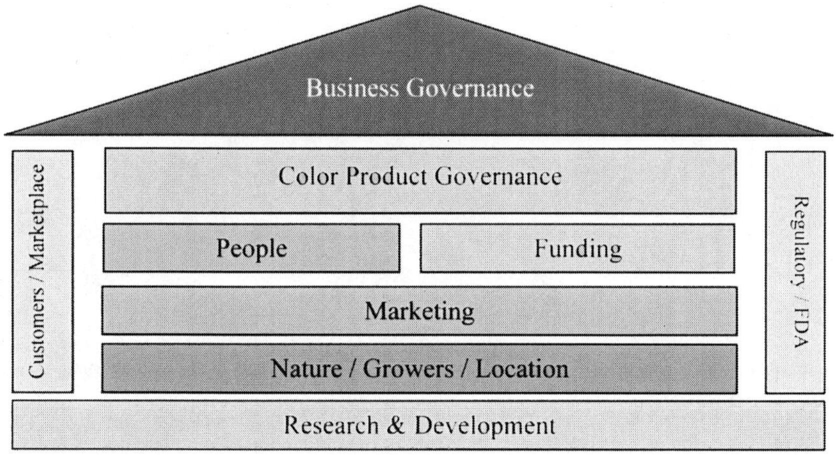

Figure 1. Business model for color governance. (See page 6 of color inserts.)

most commonly based on six anthocyanidins: pelargonidin, cyanidin, peonidin, delphinidin, petunidin, and malvidin*(2)* The sugar molecule on this structure, may also be acylated with either a phenolic or aliphatic acid *(2)*. There are currently over 275 identified anthocyanin structures which are closely related to nearly 5,000 flavonoid compounds *(3)*. Figure 2 shows some common anthocyanin structures found in nature.

Extraction of anthocyanins for colorants can be done naturally using just water and physical means. This can yield a highly concentrated product that can be used for coloring food substances around the world. When this method is utilized, other naturally occurring substances (vitamins, minerals, phytonutrients) in the fruit or vegetable are also retained. If the extraction process is carried out further by the use of solvents, a pure anthocyanin extract can be achieved. While a pure anthocyanin extract is highly concentrated, it is not currently permitted for use as a colorant in the United States.

Anthocyanins act as pH indicators changing in both color shade and color intensity as the pH varies. At a pH of 1, anthocyanins will exhibit a very red color shade. As the pH increases up to 4.0, structural transformations will alter the color to a more red-blue hue. At a pH above 8.0, anthocyanins can shift to a blue, green, yellow, or colorless compound, with the most intensity seen at the lower pH values. While all anthocyanins exhibit a color shade shift with pH variation, the exact color hue will be dependent on the raw material source. With most food applications having a pH value between 3.0 – 5.0, fruit and vegetable juices for color can provide a wide range of natural red, pink and purple shades.

Natural red colors from fruits and vegetables can be used in a variety of applications, including but not limited to, fruit preparations, beverages, dairy

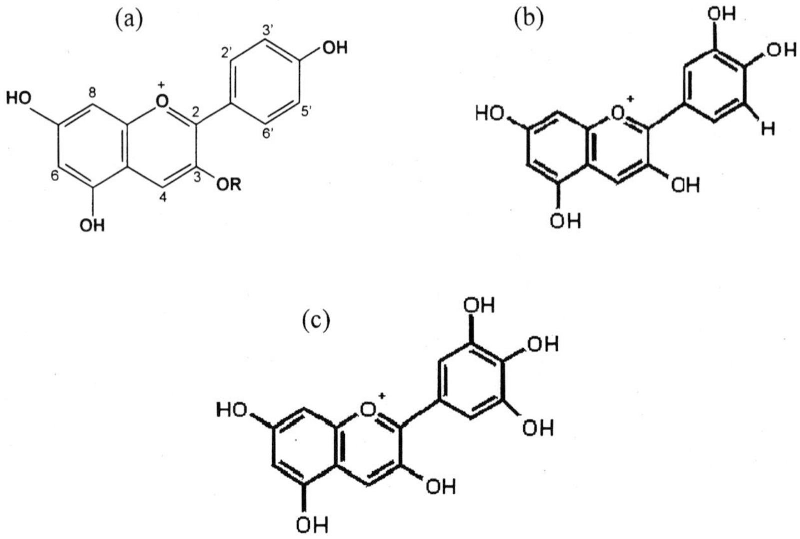

Figure 2. Commonly found anthocyanin molecules. (a) Pelargonidin-3-glycosid: main anthocyanin in strawberries, (b) Cyanidin: main anthocyanin in Roses, (c) Delphinidin: main anthocyanin in blue flowers

products, dry mixes, confections, vinegars, and frozen products. Because high concentrations can be achieved, when used as food colorant, they typically do not impart any flavor to the finished product.

The stability of the natural colorant will depend on a variety of factors including the raw material source, processing parameters, pH of product, presence of sulfur dioxide, and available oxygen. While anthocyanin colorants are stable in most food applications, they can slowly oxidize over time or react with sulfur dioxide to form a colorless product. An experienced color supplier will be able to guide one in choosing the correct color source for ones application.

Naturally occurring anthocyanin compounds can be found in numerous food products such as: red onion, strawberry, purple sweet potato, mango, cherry, plum, blueberry, red radish, tamarillo, and lychee *(4)*. For commercial purposes the color yield, in combination with manufacturing procedures that are not too complex, time-consuming, or costly, must be examined. Traditionally grapes were used as a primary source for extracting anthocyanin based colors because of the wide availability *(2)*, however, because this product is dependent on the wine industry, the price and quantity will vary seasonally. As a result, natural color companies have examined the use of other raw material sources. The most popular raw materials utilized today include elderberry, aronia, red cabbage, black carrot, and hibiscus. Photos of these raw materials can be seen on Plate 1.

Regulatory/FDA

Research shows us that anthocyanin molecules impart red, blue, violet, and magenta colors to food and beverage applications. Before these products can be manufactured commercially, it is important to understand the FDA regulation of colors. FDA does not recognize the term "natural color". Any ingredient used to impart color is considered an additive and must be labeled as a colorant. For example: if a cherry juice imparts a red color to a strawberry yogurt, the cherry juice is considered a color additive. The labeling of the cherry juice must be in compliance with 21 CFR 101.22 (k). Hence, anything that is used for the purpose of adding color to a food product becomes an additive and is therefore regulated as a color by the FDA.

Under the color regulation there are certified colors and colors exempt from certification. Certified colors are FD&C colors such as FD&C Red# 40. Each batch of these colors must be certified by the FDA. Colors exempt from certification are not required to have every batch certified by the FDA, however, good-manufacturing practices apply. FDA legal classification for colors derived from fruits and vegetables with a history of edible use is as follows: Color Exempt from Certification: Fruit Juice color 21CFR 73.259 and Vegetable juice color 21CFR 73.260.

The color regulations mentioned above were amended by GNT in 1994 to include the use of dry materials. Previous to this amendment, the regulations allowed for the fruit or vegetable juice color to be prepared by expressing the juice, using water only (no solvents) from mature varieties of fresh, edible fruits or vegetables only. The juice can only be extracted from the portion of the fruit or vegetable that is typically consumed. With the new amendement, color manufacturers now have the option of using dried fruits and vegetables, with the help of water, to make fruit and vegetable juice colors under 21CFR 73.259 and 21CFR 73.260. Unlike other areas of federal regulation, the statutory provisions with respect to colors contain no clear mechanisms or avenues for the FDA to pursue a "flexible" approach to the regulation of colors.

The current color regulations do not allow for selective extraction of anthocyanins and therefore only fruits and vegetables with a history of edible use would be permitted. This now eliminates anthocyanin sources from plants such as roses. Since selective extraction is not permitted, one can focus on extracting color from the whole fruit and vegetable. When doing this, it is important to have a good understanding of other naturally occurring substances (in addition to anthocyanins) found in fruits and vegetables that will have influence on color hue and stability. Before we focus on the growing challenges of fruits and vegetables, one needs to understand the Customers and Marketplace.

Customers/Marketplace

The food and beverage industry uses color to increase shelf life and to improve the appearance of products. The color appearance of a product catches the consumer's eye and can help move a product off the shelf more quickly. In the past, natural colors were used only in speciality foods aimed at a niche market but because of consumer awareness and increasing health concerns, natural colors are now widely used in mainstream products developed for a wider range of the population.

Because today's natural colors from fruits and vegetables can be found in mainstream products, the color industry must be able to produce supply in large volumes. Customers not only demand consistency in color strength and color hue, but also quality of the product. Since nature does not ensure a consistent supply, it is important to work with a supplier that has the capabilities of growing a product in multiple regions and understands how to hedge against these inconsistencies. The marketplace for mainstream products also looks for price stability. A natural color being used in a common food item must provide the price stability that the manufacturer provides to the end consumer.

Since color is added to increase the shelf life of a product, stability of the natural color in the end product is essential. The stability of each fruit and vegetable color depends on the way the anthocyanin interacts with other ingredients. An example of this is products containing high doses of vitamin C (greater than 60mg per 8oz serving) could have a negative effect on the stability of the anthocyanin. Stability also depends on a variety of factors such as the raw material source. Various anthocyanin structures found in fruits and vegetables have a different stability in different applications. For optimum stability the right combination of fruits and vegetables should be used and an experienced color supplier can ensure that you are using the correct combination.

Nature/Growers/ Location

Through research and development it is shown that various anthocyanins found in fruits and vegetables offer different stability, color shades, and color hues to food and beverages. Because FDA regulation only permits the use of fruits and vegetables with a history of edible use, we can say that the key in delivering consistent, price stable natural colors lies in the raw material source.

Under the current regulation, a fruit or vegetable juice color can be concentrated approximately 20 to 30 times. It should be noted that a common mainstream product could require one million pounds of concentrated product. For example, if a company launches a new product with red grape juice color, 20 to 30 million pounds of grapes would be needed. This would have a significant effect not only on the supply and demand on grapes, but also on the price stability. To increase the supply of grapes to meet this demand would take five to seven years. A more efficient way to supply natural colors to mainstream products is to grow fruits and vegetables where the supply can easily be increased from one year to another. Additionally, a serious supplier of natural color from fruits and vegetables has to have the capability of growing the product in multiple regions.

Since anthocyanin molecules are responsible for the natural red colors found in various fruits and vegetables, the relative anthocyanin content of these red fruits and vegetables has a significant effect on price. Our research found that the anthocyanin content in different fruits fluctuates depending on variety, location, and harvest time. In Figure 3, we measured the anthocyanin content in seven different fruits *(5)*. We found the highest percentage of anthocyanins were in Elderberry. This percentage was then used as a benchmark. It was also seen that while elderberries have the highest percentage of anthocyanins, they also have the highest level of fluctuation. The relative anthocyanin content was also high in chokeberries and black currants.

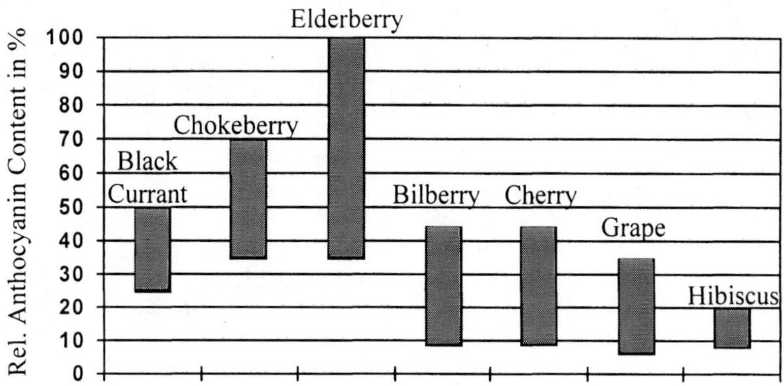

Figure 3. Relative anthocyanin content in various fruits and vegetables.

In Figure 4 we examined the effect of harvesting date on anthocyanin content in different varieties of red fruits *(5)*. While some varieties showed small fluctuations, others showed a fluctuation of over 30% indicating that understanding the variety, location, and harvest time is all essential to producing a high quality natural color.

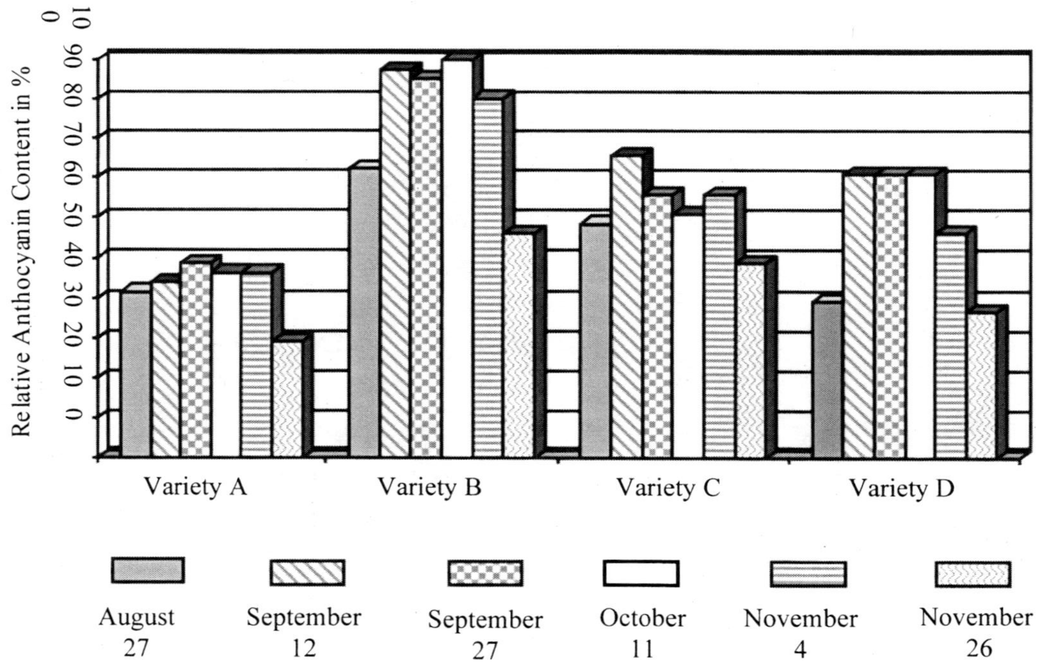

Figure 4. Influence of harvesting date on relative anthocyanin content in different varieties of anthocyanin containing fruits and vegetables.

Having control over the harvesting of fruits and vegetables can also play an important role in price stability. Typically, a fruit or vegetable bought in a grocery store has 30 to 40 % less anthocyanins than one that is grown commercially for the purpose of color. This is mainly due to the fact that fruits found in most stores are picked before they are ripe.

Since nature varies but the food and beverage industry wants a natural color that is consistent from year to year, natural color manufacturers must take the inconsistency from nature and standardize the end product. This is where Art meets Science. While the color variation between crops and growing seasons is unpredictable, it can be monitored. Cultivation conditions from soil, weather, and harvesting time, can all influence the color outcome as shown with the various shades of red in Figure 5 *(5)*.

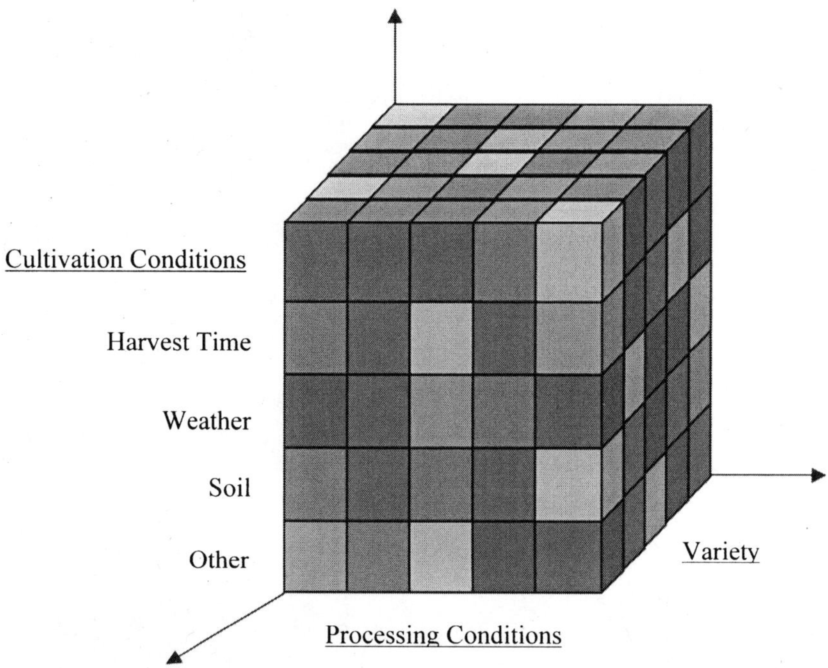

Figure 5. Affect of cultivation conditions, processing conditions, and variety on the color shade and color hue. (See page 6 of color inserts.)

(a) Red Cabbage (b) Black Carrot

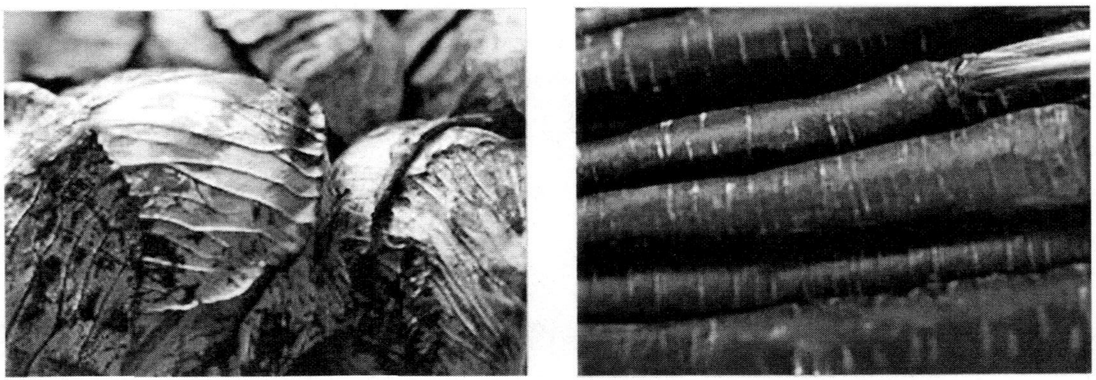

(c)Elderberry

(d) Hibiscus

Plate 1. Fruits and vegetables commonly used for the production of natural food colors. (a) Red Cabbage, (b) Black Carrot, (c) Elderberry, and (d) Hibiscus (See page 7 of color inserts.)

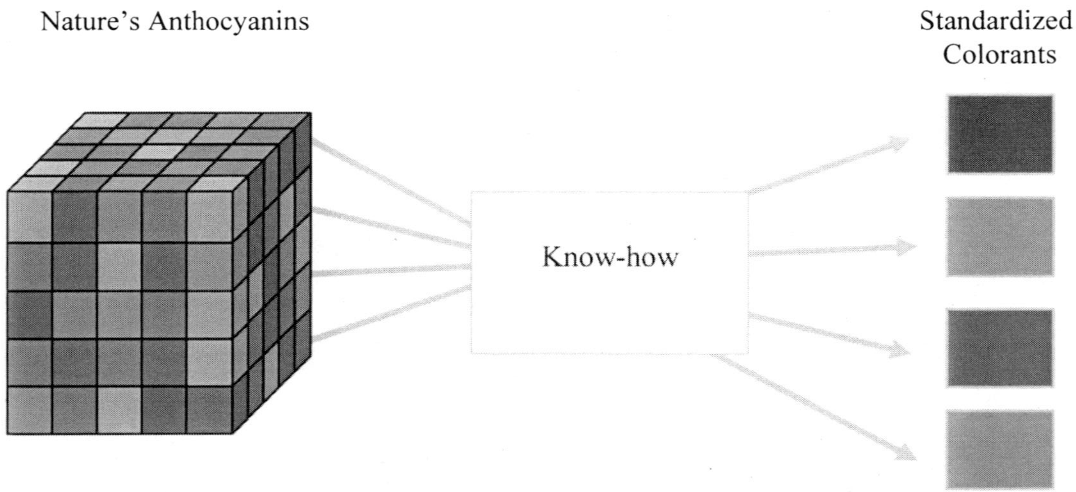

Figure 6. Converting nature's color variation into a standardized natural colorant.
(See page 7 of color inserts.)

Art is taking all these variables and mixing them into one standardized color as shown in Figure 6 *(5)*. Making a standardized color from the various starting materials involves influencing the color by optimizing processing conditions. It is a color manufacturer's responsibility to convert the variation into a standardized color that can be used by the food industry.

Conclusion

With good Color Product Governance one can achieve a consistent, price stable, natural color from fruits and vegetables. Natural red colors can be found in many mainstream food and beverage applications throughout the United States. As the demand for natural ingredients increases, so will the demand for these natural colors. As homework, the natural color industry must now look into developing a supply of natural blue and green colorants which are currently not available.

References

1. Whole Foods Market Home Page. http:www.wholefoodsmarket.com (accessed Mar 13, 2007)
2. Henry, B.S. Natural Food Colors in *Natural Food Colorants;* Hendry, G. A. F., Houghton, J. D., Eds; Chapman & Hall: New York, 1996; pp 40–79.
3. Francis, F.J. Food Colorings in *Colour in Food;* MacDougall, D. B., Ed.; Woodhead Publishing Ltd.: Cambridge, UK, 2002; pp 297-331.
4. Jackman, R.L.; Smith, J. L. 1996. Anthocyanins and Betalins In *Natural Food Colorants;* Hendry, G. A. F, Houghton, J. D., Eds.; Chapman & Hall: New York; pp 244-310.
5. Stich, E. 2004. *Unpublished.* GNT Europa. Kackertstrasse 22, Aachen, Germany.

Chapter 7

Betalain Pigments and Color Quality

Florian C. Stintzing, Kirsten M. Herbach, Markus R. Mosshammer, Florian Kugler, and Reinhold Carle

Institute of Food Technology, Plant Foodstuff Technology, Hohenheim University, 70599 Stuttgart, Germany

Betalain pigments are responsible for imparting yellow, orange, red and purple colors to flowers, grains, vegetables and fruits of a comparatively limited number of species all belonging to the suborder *Chenopodiniae* within the *Caryophyllales* (*1*). Due to their limited distribution and the even more restricted occurrence in edible plants, betalains have received little attention. Currently, however, betalains are witnessing a renewed interest, especially in the food sector (*2*). Among the presently available food sources, red beet represents the most widespread betalain crop: It exhibits high pigment contents and is easy to grow so that it is well introduced on the market. Right now, alternative betalain sources are receiving consideration, such as yellow beet, red-purple pitayas or cactus pears complementing the narrow color range of red beets. In order to fully benefit from the potential of betalains, the present review will describe the inter-relationships between color and structure of both betacyanins and betaxanthins and will also discuss how these principles govern the appearance of the respective food.

Introduction

Betalains are known from ornamentals such as *Bougainvillea* bracts (*Bougainvillea* sp.), globe amaranth (*Gomphrena globosa.*), feathered amaranth (*Celosia argentea* var. *plumosa*), common cockscombs (*Celosia argentea* var. *cristata*), purslane (*Portulaca grandiflora*), midday flowers (*Lampranthus* sp.), and also occur in pokeberries (*Phytolacca americana*). The latter had previously been applied to reinforce faint red wine color, but are no longer allowed for food use (*2*). Food sources are rather limited and comprise yellow and red beet, as well as Swiss chard (*Beta vulgaris* sp.), cactus pears (*Opuntia* sp.) and pitayas (*Hylocereus* sp.). The colors encountered range from bright yellow as in yellow beet and yellow Swiss Chard to yellow-orange in cactus pears and orange Swiss chard petioles to red such as in cactus fruits and some red beet cultivars and even purplish tones as found in pitaya fruits and red beet (*3*, *4*).

Betalains are easily differentiated from anthocyanins on a structural basis with the former exhibiting a nitrogen-based skeleton. On an evolutionary scale, the anthocyanins appeared earlier (*5*) and it is intriguing that the two pigment classes are mutually exclusive, never occurring in the same plant tissue (*6*). Hence, from a plant physiological standpoint both pigment groups should take over the same tasks such as attracting insects for pollen transfer and seed dispersers for securing plant propagation (*2*). Colorwise, however, betalains are quite different from the anthocyanins because they are stable from pH 3 to 7 while the anthocyanins are not. It is exactly this feature which makes the betalains interesting from a technological point of view because anthocyanins not only fade upon increasing pH values through hemiketal formation, but also change their color considerably by chalcone and quinonoidal base transformation reactions (*2*, *7*, *8*). The principles governing the broad palette of betalainic color shades will be detailed below.

A Structural Approach

The principle chromophore of all betalains is the 1,7-diazaheptamethin cation exhibiting a bright yellow appearance (Figure 1). This electron resonance system is easily recognized in betalamic acid, the common building block of all betalain structures (Figure 2). Extension of betalamic acid after conjugation with amino acids or amines results in yellow-orange betaxanthins while condensation with *cyclo*-Dopa yields betanidin, the general precursor of betacyanins. Betanidin may be further extended by glycosylation at C5 or C6 and aliphatic or aromatic acylation.

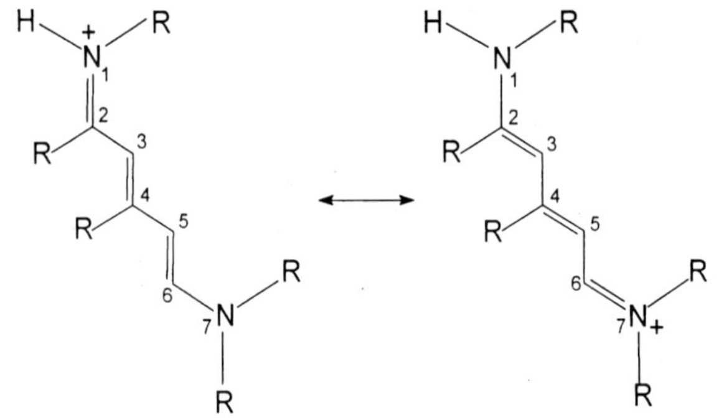

Figure 1. The pentasubstituted 1,7-diazaheptamethin cation – the common betalain chromophore

Betacyanins

Betacyanins are the red, sometimes violet betalains with betanidin as their common backbone (Figure 2). They may vary in the type of sugars attached, but also with respect to their glycosylation site, yielding C5- or C6-glycosidic derivatives. The typical sugars are glucose (C5: betanin; C6: gomphrenin I), sophorose (C5: bougainvillein r-I; C6: bougainvillein v) or glucuronosyl-glucose (C5: amaranthin). Interestingly, the betacyanins exhibit two stereogenic centers, namely C2 and C15. While C2-isomers appear to be indistinguishable by current separation techniques, C15-isomers usually co-exist with the 15*S* commonly predominating compared to the corresponding 15*R*-stereomers. However, visual appearance will not change upon isomerization. Moreover, additional glycosylation will not change the absorption properties, neither will aliphatic acid substitution (*2*). In contrast, modulation may be achieved by 6- instead of 5-glycosylation, *i.e.* betanin vs. gomphrenin I and aromatic acylation induces a bathochromic shift by 4-6 nm. Acylglycosides may be generated following esterification with aliphatic or aromatic acids. In the former case, sulfuric acid (prebetanin, rivinianin), malonic acid (phyllocactin), 3-hydroxy-3-methylglutaric acid (hylocerenin, iresinin I) are common, while ferulic (celosianin II, gomphrenin III, lampranthin II), *p*-coumaric (celosianin I, gomphrenin II, lampranthin I) are the most frequent acyl moieties, followed by caffeic acid. Interestingly, sinapic or hydroxybenzoic acids have not yet been reported to constitute betacyanins (*4*, *9*, *10*). As is well known for the anthocyanins, multiple acylations may occur, having beenreported in *Bougainvillea* bracts (*11*). If further acylation occurs, an additive bathochromic shift takes effect (*11-13*). Since the betanidin aglycone will generally not contribute greatly to *in vivo* color, only aromatic acylation will alter the appearance of betacyanic plant tissues. A notable color change may also be induced if betanin is dehydrogenated yielding neobetanin (14,15-dehydro-betanin; Figure 2). However, neobetanin is a minor plant pigment only randomly found in red beet or cactus pears (*14-17*). Therefore, its overall impact on appearance remains marginal.

Betaxanthins

The structure of betaxanthins is also quite variable because in theory any amino acid or amine may be condensed with the betalamic acid chromophore (*3, 18-20*; Figure 2). The absorption maxima of the yellow-orange betaxanthins (bx) is bathochromically shifted by 50-70 nm compared to their common precursor. Increasing the chain length by one CH_2-unit as is true for the asparagine-bx and glutamine–bx or valine-bx and leucine-bx pair does not alter the absorption maximum and is only slightly increased from glycine-bx (468 nm) to alanine-bx (472 nm). However, a two CH_2-unit difference such as in glycine-bx and γ-aminobutyric acid-bx brings about a hypsochromic shift by 7 nm (*3*). While a hydroxy- instead of an amino-moiety as for glutamine-bx and glutamic acid-bx or additional hydroxylation (phenylalanine-bx and tyrosine-bx) do not alter the absorption maxima, methoxylation of a phenolic hydroxy moiety affords an increase by 3 nm. The most notable changes, however, are observed after decarboxylation (i.e., dopa-bx and dopamine-bx, tryptophan-bx and tryptamine-bx, tyrosine-bx and tyramine-bx) by a decrease of up to 10 nm being less important for histidine-bx and histamine-bx (4 nm; *3, 19*), presumably due to the missing phenolic π-electron system extension in the latter betaxanthin pair. While the lowest absorption maximum has been reported for γ-aminobutyric acid-bx (459 nm) and then tyramine-bx (461 nm), the highest are usually achieved by proline-bx (477 nm) and hydroxyproline-bx (477 nm) with others such as glutamine-bx (469 nm) being intermediate (*3, 18-20*). Although absolute values differ depending on the pH and the solvent system, the reported tendencies remain valid.

A Quantitative Approach

Genuine Color Blends in Betalainic Plants – the *in vivo* approach

Free betalamic acid and betanidin (Figure 2) have been reported to occur in betalainic plants, although in low amounts (*9, 16, 21-23*). Since the molar extinction coefficient of betalamic acid ($\varepsilon_{betalamic\ acid}$ = 25,000 $L{*}mol^{-1}cm^{-1}$; *20*) is

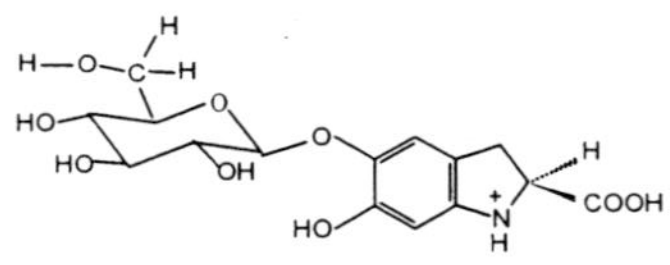

cyclo-Dopa-glucoside
colorless

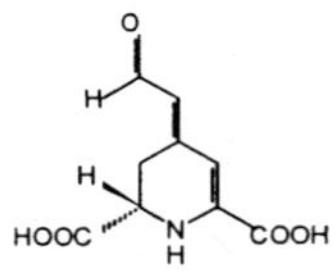

betalamic acid
bright yellow

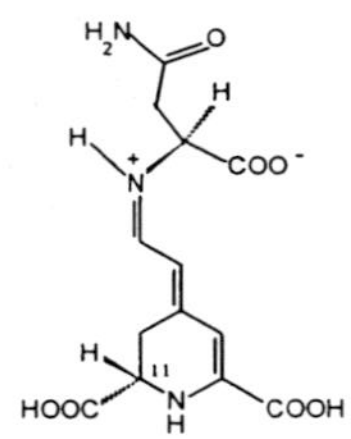

vulgaxanthin I
[glutamine-betaxanthin]
yellow

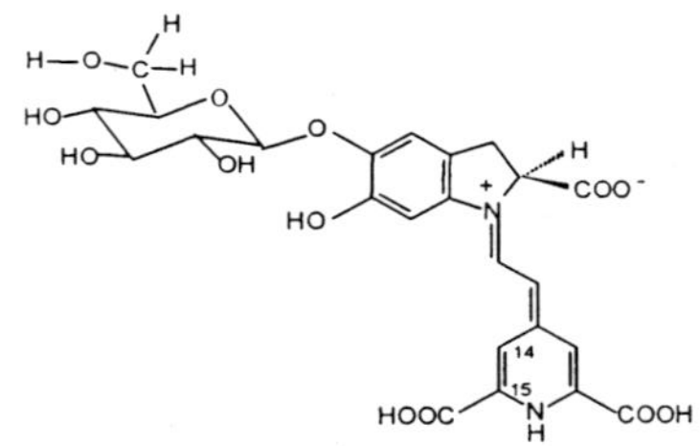

neobetanin
[14,15-dehydrobetanin]
yellow-orange

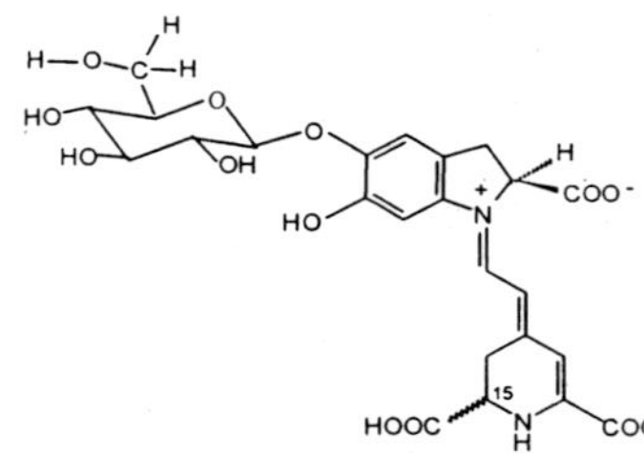

betanin, isobetanin
red

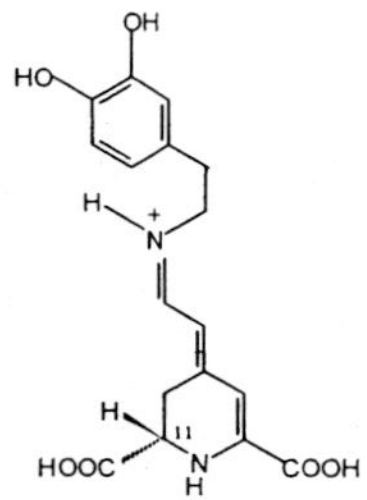

miraxanthin V
[dopamine-bx]
yellow

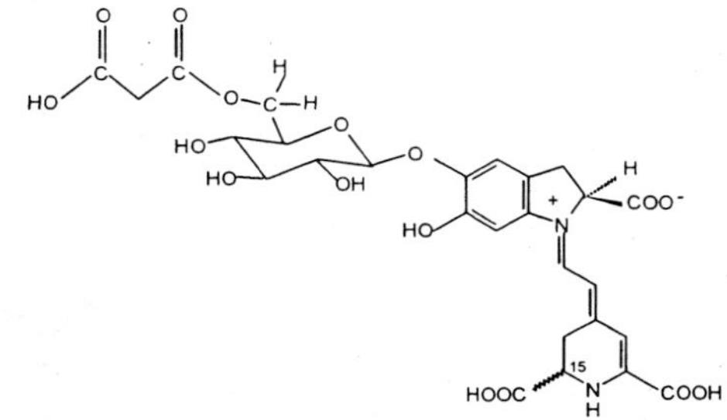

phyllocactin, isophyllocactin
[malonyl-(iso)-betanin]
red

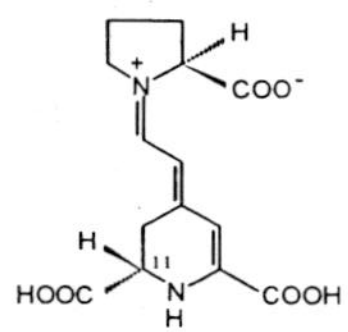

indicaxanthin
[proline-betaxanthin]
yellow-orange

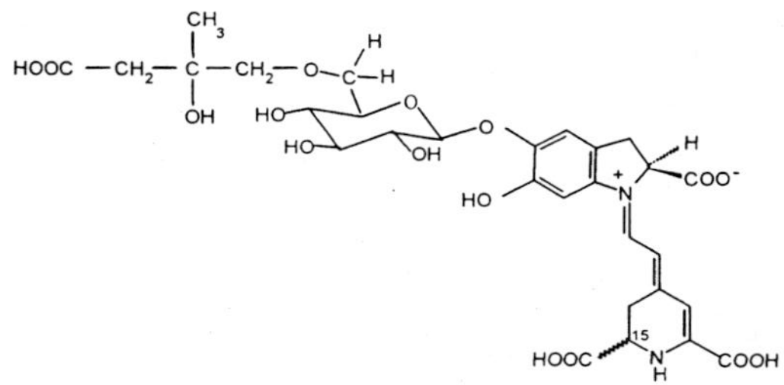

hylocerenin, isohylocerenin
[3-hydroxy-3-methyl-glutaryl-(iso)-betanin]
red

Figure 2. Chemical structures of cyclo-Dopa-glucoside, betalamic acid, neobetanin, betanin, phyllocactin, hylocerenin, vulgaxanthin I, miraxanthin V, and indicaxanthin

comparatively small, its potential color impact is negligible when betaxanthins (ε_{mean}=48,000 L*mol^{-1}cm^{-1}; *24*) and betacyanins (ε_{mean}=60,000 L*mol^{-1}cm^{-1}; *25*) are present at the same time. Although betanidin ($\varepsilon_{betanidin}$=49,000 L*mol^{-1}cm^{-1}; *26*) exhibits a similar extinction coefficient as betanin, it appears to be a minor compound in plants (*9*).

In addition to individual structures, the overall appearance of betalainic tissues will be governed by the total color content with a darkening effect through higher pigment contents, i.e. a yellow shade will appear orange if more concentrated, the same being true for red turning to a purplish tone.

Red Beet, Yellow Beet, and Swiss Chard

Typically, the main pigments in red beet are betanin and its C_{15}-stereoisomer isobetanin for the betacyanins and vulgaxanthin I (glutamine-bx) for the betaxanthins (Figure 2). Both their total contents and their specific ratios will determine the resulting hue, also reported as the so-called color shade, *i.e.* the quantitative betaxanthin-betacyanin ratio. This pigment proportion will mainly be affected by cultivars (Table I; *27-32*) and the harvest date (Table II; *33*, *34*), while the time of planting (*35*, *36*), and fertilizing treatments (*37*, *38*) may also alter the final shade. Noteworthy, early harvested beets tend to contain less betaxanthins which results in purplish hues (*33*, *34*), while nitrogen fertilization enforces betaxanthin accumulation (*37*). It is interesting to note that yellow beet is devoid of betacyanins with vulgaxanthin I predominating (*19*, *39*).

The rainbow colors of Swiss chard petioles have been shown to be due to specific quantitative proportions of betaxanthins and betacyanins to produce yellow, orange, red and red-purple shades. Again, the predominant betacyanins were betanin and isobetanin, while the major betaxanthins were characterized as miraxanthin V (dopamine-betaxanthin) and vulgaxanthin I (*3;* Figure 2). With increasing betaxanthin contents, higher lightness values resulted, whilst chroma were maximum either for the yellow and the purple petioles with lower scores for orange and red-orange samples, respectively.

Cactus Fruits

The pigment patterns of cactus pear fruits (*Opuntia* sp.) and pitayas (*Hylocereus* sp.) are well characterized (*16*, *19*, *40-43*). The former may range from yellow-orange to purple, with betanin and indicaxanthin (proline-bx) predominating the betacyanin and betaxanthin spectrum, respectively. Specific ratios between red and yellow pigments define the color shade which may vary depending on the cultivar (Table III; *44, 45*) and the date of harvest (Table IV; *16*, *47-49*).

Table 1. Betalain contents and color shades of red and yellow beet cultivars (according to *27-30*; fw: fresh weight; * dry weight)

Cultivar	*Betaxanthin content [g/kg fw]*	*Betacyanin content [g/kg fw]*	*Betaxanthin/ Betacyanin content (Color shade)*
Forono, peel *	1.8	10.7	0.17
Long Season	0.33	1.32	0.25
Egyptische Platronde, peel *	1.4	5.0	0.28
Rubia, peel *	2.2	7.6	0.29
Detroit Dark Red MT	0.11	0.32	0.34
Burpee's Red Ball	0.61	1.68	0.36
Little Ball	0.11	0.30	0.37
King Red	0.68	1.84	0.37
Crosby Green Top	0.52	1.36	0.38
Mono King Burgundy	0.84	2.15	0.39
Asmer Beethoven	0.11	0.28	0.39
Mono King Explorer	0.91	2.23	0.41
Detroit Dark Red ST	0.13	0.32	0.41
Okragly Ciemnoczerwony	0.11	0.26	0.42
Slowbolt R-2289	0.15	0.36	0.42
Detroit Dark Red	0.68	1.61	0.42
Rubidus	0.10	0.23	0.43
Gladiator	0.12	0.28	0.43
Early Wonder	0.13	0.30	0.43
Formanova	0.47	1.07	0.44
Ruby Queen	0.68	1.54	0.44
Little Ball, peel *	4.3	9.6	0.45
Podzimniaja 0474	0.15	0.31	0.48
Detroit Nero RS	0.16	0.33	0.48
Egyptische Platronde, flesh *	1.5	2.93	0.51
Rubia, flesh *	2.3	4.4	0.52
Little Ball, flesh *	1.9	3.62	0.52
Uniball	0.21	0.39	0.53
Holmes Fireball	0.81	1.54	0.53
Iowa	0.15	0.28	0.54
Red E403	0.19	0.35	0.54
Rubin	0.39	0.71	0.54
Crveno	0.15	0.27	0.55
Detroit Sluis	0.17	0.31	0.59
Detroit	0.65	0.41	1.58

Continued on next page.

Table 1. *Continued.*

Cultivar	*Betaxanthin content [g/kg fw]*	*Betacyanin content [g/kg fw]*	*Betaxanthin/ Betacyanin content (Color shade)*
Choghundur	0.18	0.30	0.60
Favorit	0.44	0.73	0.60
Bonel	0.48	0.77	0.62
Polso	0.17	0.27	0.63
Bordo 237	0.22	0.35	0.63
Nero	0.35	0.54	0.64
Spangsbjerg	0.17	0.25	0.68
Forono, flesh *	4.0	5.6	0.71
Burpee's Golden	0.61	0.03	20.3
Golden	0.77	0.03	25.7

Interestingly, purple pitayas such as *Hylocereus polyrhizus*, fleshy many seeded fruits from vine cacti originating from Latin and South America, are devoid of or only contain traces of betaxanthins (*50*). Their red betalains are characteristic and differ greatly from other betalain containing crops, because in addition to betanin and isobetanin, phyllocactin (malonyl-betanin) and hylocerenin (3-hydroxy-3-methy-glutaryl-betanin) together with their respective C_{15}-isomers occur in appreciable amounts (*40-43*).

Color Blends from Betalanic Plants- the *in vitro* Approach

Driven by the enormous chromatic range offered by plants, it is tempting for food technologists to attempt to imitate this natural paintbox. Therefore, both unpurified and purified cactus juices from red-violet pitaya and yellow-orange cactus pear were blended at definite ratios and the resulting chromatic data were registered. As observed for Swiss chard petioles, chroma (synonymous to color brillance or color purity) reached highest values with either increasing red or yellow proportions and a minimum at a 50:50 ratio (*51*). Juice blends covered a similar range compared to purified samples, making juices the better choice for commercial applications. In continuation of these investigations, blends from red and yellow beet juices were produced covering an even broader range in the yellow region (Stintzing et al., unpublished data; Figures 3 and 4). Tailor-made hues can be obtained from yellow and red beet preparations with predictable C*- and L*-values as shown in Figure 5.

Table II. Betalain contents of red beet depending on the date of harvest

Cultivar	*Betaxanthin content [g/kg fw]*	*Betacyanin content [g/kg fw]*	*Betaxanthin content / Betacyanin content (Color shade)*
Detroit Dark Red Short Top, early harvest	0.482	1.201	0.40
Garnet, early harvest	0.537	1.355	0.40
Detroit Dark Red Medium Top, early harvest	0.487	1.126	0.43
Detroit Dark Red Morse' Strain, early harvest	0.464	0.899	0.52
Detroit Dark Red Ferry's strain, early harvest	0.547	1.018	0.54
Firechief, early harvest	0.731	1.284	0.57
Ruby Queen NK, early harvest	0.601	1.037	0.58
Nero, early harvest	0.524	0.873	0.60
Ruby Queen FM, early harvest	0.545	0.899	0.61
Ruby Queen PW, early harvest	0.605	0.938	0.64
Gladiator, early harvest	0.658	1.016	0.65
Redpack, early harvest	0.607	0.868	0.70
Garnet, late harvest	0.899	1.269	0.71
Detroit Dark Red Short Top, late harvest	0.818	1.132	0.72
Detroit Dark Red Medium Top, late harvest	1.031	1.379	0.75
Firechief, late harvest	0.964	1.174	0.82
Detroit Dark Red Morse' Strain, late harvest	0.785	0.937	0.84
Redpack, late harvest	0.929	1.102	0.84
Detroit Dark Red Ferry's strain, late harvest	0.930	1.091	0.85
Ruby Queen Fm, late harvest	0.938	1.109	0.85
Ruby Queen PW, late harvest	1.016	1.156	0.88
Gladiator, late harvest	0.924	1.025	0.90
Nero, late harvest	1.152	1.233	0.93
Ruby Queen NK, late harvest	1.221	1.198	1.01

NOTE: Values calculated based on fresh weight (fw)
SOURCE: Data from Reference 33

Table III. Betalain contents and color shades of various cactus pear fruits

Opuntia sp.	*Betaxanthin content [mg/kg]*	*Betacyanin content [mg/kg]*	*Betaxanthin content / Betacyanin content (Color shade)*
O. sp_3	0	568	0
O. decumbrens	0	211	0
O. sp_1	7	1008	0
O. sp_2	3	417	0
O. acculata	0	16	0
O. sherri	0	79	0
O. microdasys	0	5	0
O. robusta-robusta	54	459	0.12
O. curvispina	116	838	0.14
O. robusta	72	231	0.31
O. ficus indica cv. 'Rossa' (pulp)	26.1	51.2	0.5
O. ficus-$indica_3$	17	9	1.9
O. ficus indica cv. 'Gialla' (pulp)	84.2	10.4	8.1
O. ficus $indica_1$	70	2	35
O. ficus $indica_2$	140	3	47

SOURCE: Data from references 44 and 45

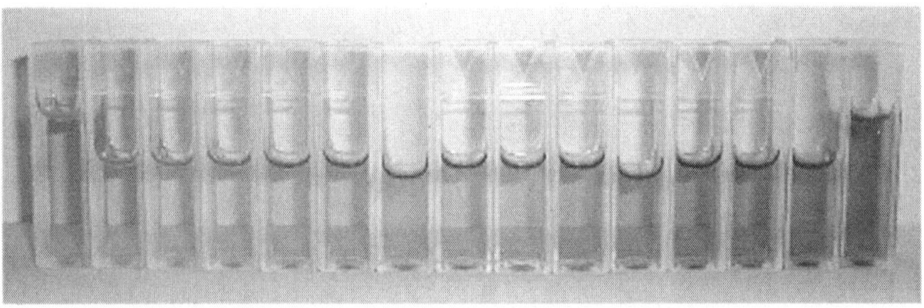

Figure 3. Tailor-made color blends from yellow and red beet juice (far left: 100% yellow beet juice; far right: 100% red beet juice) (See page 8 of color inserts.)

Table IV. Betalain contents and color shades of cactus pear fruit juice depending on the harvest date

Opuntia ficus-indica	*Betaxanthin content [mg/L]*	*Betacyanin content [mg/L]*	*Betaxanthin content / Betacyanin content (Color shade)*
cv. 'Rossa'	36.4	73.9	0.50
cv. Purple (Jan)	195.8	431.0	0.45
cv. Purple (Nov)	151.4	303.3	0.50
cv. Red (Jan)	67.9	120.0	0.57
cv. Red (Nov)	52. 2	68.3	0.76
#1281	98.8	106.9	0.92
#1240	581.0	615.6	0.94
#1379	34.3	34.9	0.98
cv. Orange (Nov)	84.4	11.1	7.60
cv. Orange (Jan)	76.3	6.6	11.56
#1320	179.6	15.4	11.66
cv. 'Gialla'	48.3	1.3	37.10

NOTE: Harvest dates are January (Jan) and November (Nov)
SOURCE: Data from references 16 and 47

Heat-induced Color Changes- the Technological Approach

Manufacturers tend to prefer crops that exhibit defined color qualities irrespective of harvest time and year. However, since genotypic and external factors will govern the final shade considerably, technologists will have to chose the adequate tools to modify color tints. Early studies by von Elbe and co-workers (*52*, *53*) demonstrated that betanin was cleaved into its biosynthetic precursors betalamic acid and *cyclo*-Dopa glucoside accompanied by fading of the solution. In addition, decarboxylation at C_{15} was reported for the first time (*54*). It was also recognized at that time that cool storage after processing was instrumental in restoring color (*52*, *53*).

Betacyanins

Recent studies on the processing stability of betalainic foodstuffs revealed that after extensive heating of red beet (*54*) or red-purple pitaya (*55*), a color shift towards red-orange was registered, *i.e.* the background color (*55*) was

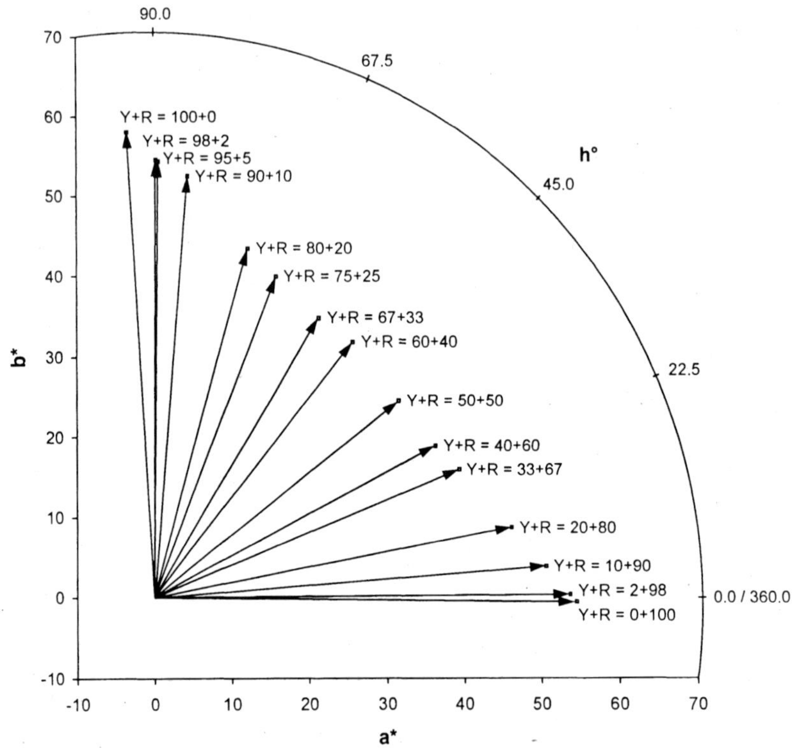

Figure 4. Chroma (C, arrow length), hue angle (h°), a* and b* indices of different blends from yellow(Y) and red beet(R) at pH 4*

changed from blue to yellow (Figure 6). Since this observation had not been reported before, it was intriguing to track the underlying structural pigment alterations. As a consequence, purified pigment isolates of betanin, phyllocactin and hylocerenin were heated in aqueous solutions and exhibited similar color shifts as the juices. Extensive mass spectrometric studies in connection with objective color measurements revealed that multiple reactions were triggered by heating to different extents (Figure 7, Figure 8): i.) isomerization at C_{15}, ii.) hydrolytic cleavage, iii.) dehydrogenation at C_{14}-C_{15}, iv.) decarboxylation at C_2, v.) decarboxylation at C_{15}, vi.) decarboxylation at C_{17}, and vii.) combinations thereof (*54*, *55*, *58-60*). By comparison of the pigment patterns in purified solutions and food samples, an impact of the matrix on specific degradation trails was discovered (*54, 56, 58*). Heating in ethanolic rather than aqueous solutions favoured 17-decarboxylation as compared to C_2-decarboxylation (*60*). Although differences among pigments were found, the food matrix has been shown to generally counteract hydrolytic cleavage, i.e. color fading. Consequently, in

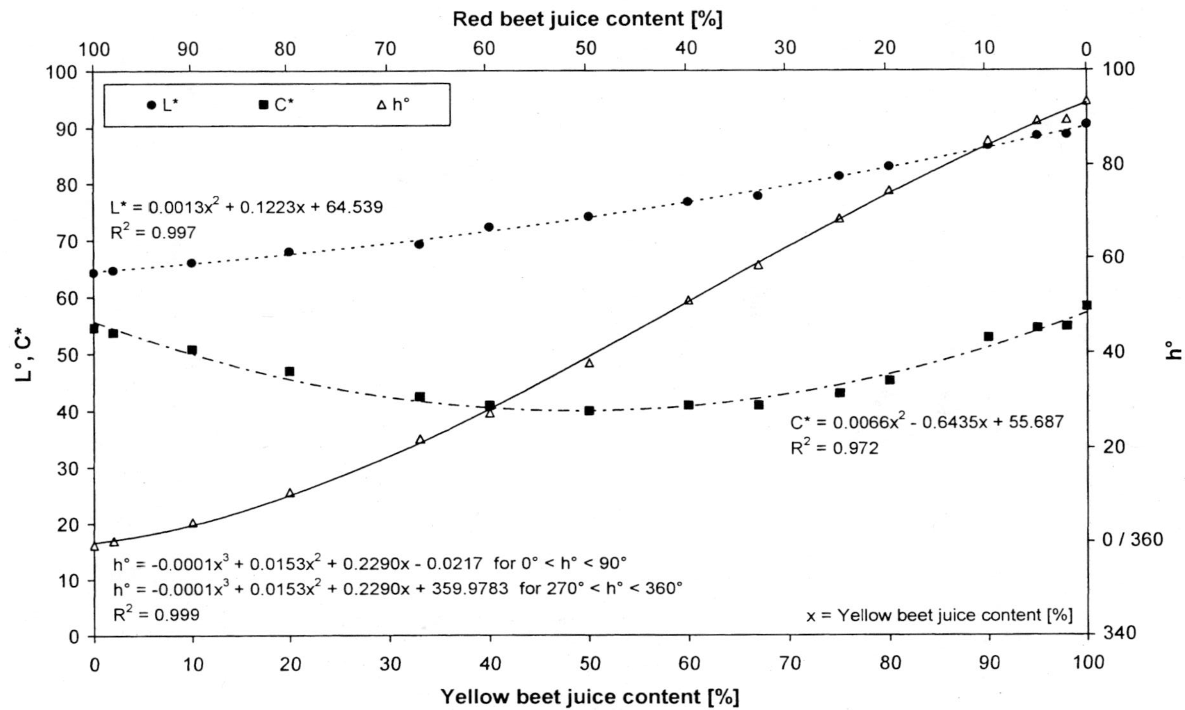

Figure 5. Course of h°, C, and L* with varying ratios of red and yellow beet juices at pH 4*

Figure 6. Color alterations of purple pitaya juice (most left sample) upon heating for 1, 2 , 3, 4, and 5 hours, respectively (See page 8 of color inserts.)

heated foods, dehydrogenation and decarboxylation reactions were stronger than in purified solutions. This clearly demonstrated that the matrix will modulate degradation as compared to pigment isolates (*54, 55, 57, 58*).

As mentioned before, hydrolytic cleavage ii.) will bring about a pigment loss which is partly reversible under optimum conditions. Since C_{15}-isomerization (i.) and decarboxylation at C_2 (iv.) and C_{15} (v.) will not alter the π-electron extension, no color changes were induced (Figure 7, Figure 8). In contrast, CO_2 loss at C_{17} (vi.) shortened the resonance system bringing about a hypsochromic shift of 33 nm (*54, 55*) leading to red-orange structures. The most dramatic color alteration, however, was due to dehydrogenation at C_{14}-C_{15} resulting in neo-derivatives exhibiting similar absorption properties to betaxanthins (*54, 55, 57-60*). While neo-formation (14,15-dehydrogenation) together with C_2 and C_{17} decarboxylation were the most important reactions for purified phyllocactin and hylocerenin, betanin was predominantly hydrolyzed into betalamic acid and *cyclo*-Dopa-glucoside (*57, 58*). Interestingly dehydrogenated and decarboxylated compounds were more heat stable than their parent compounds explaining why pitaya juice was more stable than red beet preparations (*54, 55*). Another possible structure alteration that needs to be addressed is betacyanin deglycosylation which may be intentionally induced by fermentation in the presence of β-glucosidase. This approach has been considered viable to change red to purple tones in red beet preparations. While a notable bathochromic shift of 4-6 nm may thus be induced, the aglycones are very prone to further degradation and a concomitant color deterioration by oxidative events leading to a brownish tint reducing the value of this technique for general application. The relevant positions for modulating betacyanin color are depicted in Figure 7.

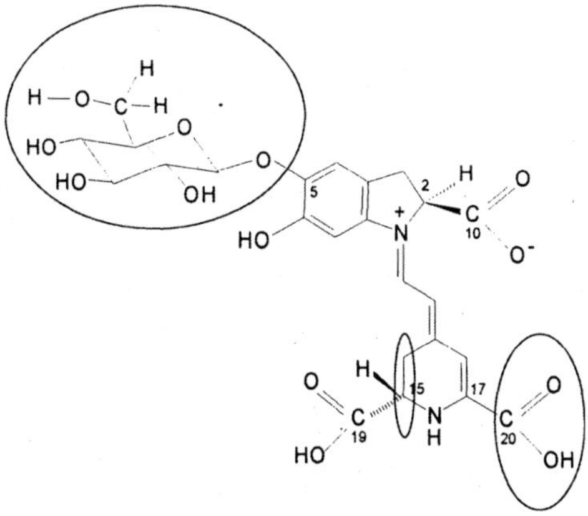

Figure 7. Relevant positions for structural alterations in betacyanins inducing color changes (here: betanin)

Betaxanthins

Knowledge about betaxanthin degradation is scarce. According to a previous report (*61*) betaxanthins will be more easily degraded by hydrolytic cleavage to yield their biosynthetic building blocks as compared to betacyanins. A recent technological study revealed that indicaxanthin was cleaved following heat exposure (*62*), accompanied by isomerization at C_{11}, the reaction being reversible as discussed for the betacyanins before (*52*, *63*). To get a clearer picture of betaxanthin degradation and structural modification upon processing, continuing investigations are currently underway.

A complexity diagram summarizing betalain color alterations upon heating is shown in Figure 8.

Conclusions

The apperance of betalainic food crops such as red and yellow beets, Swiss chard petioles, and cactus fruits depend on cultivar, the edaphic factors at the site of cultivation, the time of planting and harvest, but also fertilization measures. These factors will govern the total betalain contents but also modulate the respective color shade. Following processing and betalain extraction for food coloring purposes, blends may be produced mimicking the natural color palette. Finally, dehydrogenation and 17-decarboxylation of betacyanins induced upon heat exposure will alter the tint considerably thus presenting a viable tool to

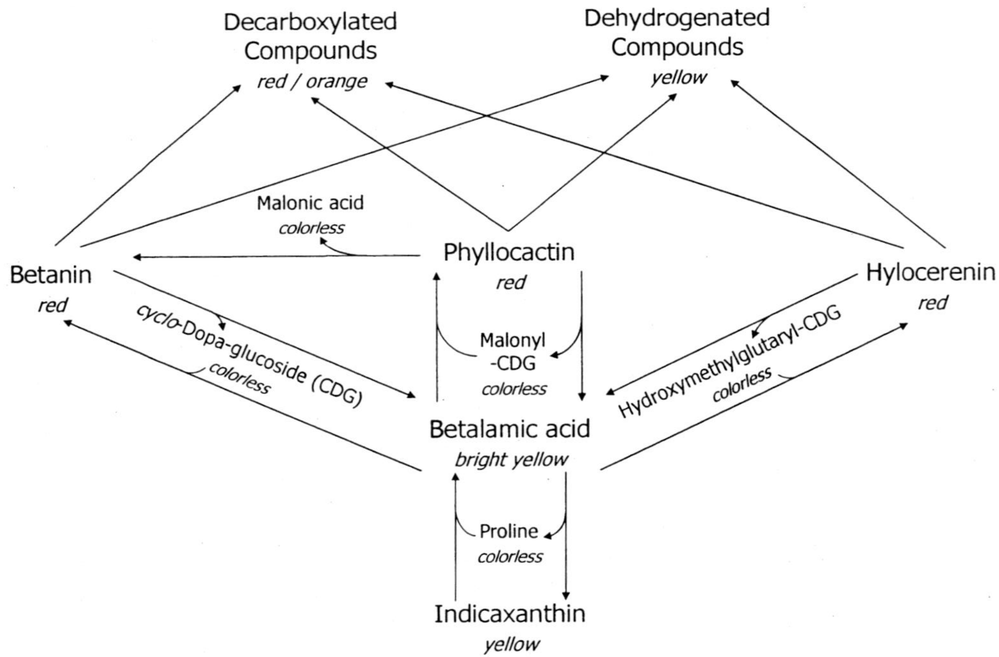

Figure 8. Complexity diagram of betalain degradation upon heating

intentionally modify the final color shade. Since foods contain colorless phenolics that accompany the betalains, the brilliance might be affected by oxidative browning of phenolics masking betalain color thus resulting in a less vivid appearance. Specific countermeasures are thus recommended such as inhibition of polyphenoloxidase activity by heat or acidification. Exclusion of oxygen is also quite effective, but may not be realized on a technological scale.

It is therefore concluded that the potential of betalains can only be fully exploited when the entire production chain starting from crop cultivation and subsequent technological processing including storage are well understood.

Acknowledgements

The authors would like to thank Mrs Judith Bretag for providing data for Figure 4 and Figure 5 which are part of her diploma thesis.

References

1. Clement, J. S.; Mabry, T. J. *Bot. Acta* **1996**, *109*, 360.
2. Stintzing, F. C.; Carle, R. *Trends Food Sci. Technol.* **2004**, *15*, 19.
3. Kugler, F.; Stintzing, F. C.; Carle, R. *J. Agric. Food Chem.* **2004**, *52*, 2975.
4. Strack, D.; Vogt, T.; Schliemann, W. *Phytochemistry* **2003**, *62*, 247.
5. Vogt, T. *Planta* **2002**, *214*, 492.
6. Stafford, H. A. *Plant Sci.* **1994**, *101*, 91.
7. Cevallos-Casals, B. A.; Cisneros-Zevallos, L. *Food Chem.* **2004**, *86*, 69.
8. Torskangerpoll, K., Andersen, O. M. *Food Chem.* **2005,** *89*, 427.
9. *The Alkaloids - Chemistry and Pharmacology*; Brossi, A., Ed.; Academic Press Inc.: San Diego, CA, 1990; p 1.
10. *Plant Pigments and Their Manipulation.* Davies, K. M., Ed.; CRC Press: Boca Raton, FL, 2004; p 185.
11. Heuer, S.; Richter, S.; Metzger, J. W.; Wray, V.; Nimtz, M.; Strack, D. *Phytochemistry* **1994**, *37*, 761.
12. Heuer, S.; Wray, V.; Metzger, J. W.; Strack, D. *Phytochemistry* **1992**, *31*, 1801.
13. Schliemann, W.; Strack, D. *Phytochemistry* **1998**, *49*, 585.
14. Alard, D.; Wray, V.; Grotjahn, L.; Reznik, H.; Strack, D. *Phytochemistry* **1985**, *24*, 2383.
15. Kujala, T.; Loponen, J.; Pihlaja, K. *Ztschr Naturforsch C/J. Biosci.* **2001**, *56*, 343.
16. Stintzing, F. C.; Herbach, K. M.; Moßhammer, M. R.; Carle, R.; Yi, W.; Sellapan, S.; Akoh, C. C.; Bunch, R.; Felker, P. *J. Agric. Food Chem.* **2005**, *53*, 442.

17. Strack, D.; Engel, U.; Wray, V. *Phytochemistry* **1987**, *26*, 2399.
18. Schliemann, W.; Kobayashi, N.; Strack, D. *Plant Physiol.* **1999**, *119*, 1217.
19. Stintzing, F. C.; Schieber, A.; Carle, R. *J. Agric. Food Chem.* **2002**, *50*, 2302.
20. Trezzini, G. F.; Zryd, J.-P. *Phytochemistry* **1991**, *30*, 1901.
21. Mabry, T.; Kimler, L.; Larson, R. A. *Hoppe Seyler Ztschr. Physiol. Chem.* **1972**, *353*, 127.
22. Reznik, H. *Ztschr. Pflanzenphysiol.* **1978**, *87*, 95.
23. Schliemann, W.; Cai, Y.; Degenkolb, T.; Schmidt, J.; Corke, H. *Phytochemistry* **2001**, *58*, 159.
24. Girod, P.-A.; Zryd, J.-P. *Plant Cell Tiss. Org. Cult.* **1991**, *25*, 1.
25. Wyler, H.; Meuer, U. *Helv. Chim. Acta* **1979**, *62*, 1330.
26. Wilcox, M. E.; Wyler, H.; Dreiding, A. S. *Helv. Chim. Acta* **1965**, *48*, 1134.
27. Gasztonyi, M. N.; Daood, H.; Hájos, M. T.; Biacs, P. *J. Sci. Food Agric.* **2001**, *81*, 932.
28. Ng, T. J.; Lee, Y.-N. *HortSci.* **1978**, *13*, 581.
29. Kujala, T. S.; Vienola, M. S.; Klika, K. D.; Loponen, J. M.; Pihlaja, K. *Eur. Food Res. Technol.* **2002**, *214*, 505.
30. Sapers, G. M.; Hornstein, J. S. *J. Food Sci.* **1979**, *44*, 1245.
31. Nilsson, T. *Swed. J. Agric. Res.* **1973**, *3*, 187.
32. Goldman, I. L., Austin, D. *Theor. Appl. Genet.* **2000**, *100*, 337.
33. Watson, J. F.; Gabelman, W. H. *J. Am. Soc. Hort. Sci.* **1982**, *107*, 713.
34. Wolyn, D. J.; Gabelman, W. H. *HortSci.* **1986**, *21*, 1339.
35. Gaertner, V.; Goldman, I. L. *J. Am. Soc. Hort. Sci.* **2005**, *130*, 424.
36. Takács-Hájos, M. *Acta Agronom. Hungar.* **1994/1995**, *43*, 203.
37. Michalik, B.; Grzebelus, D. *Acta Hortic.* **1995**, *379*, 205.
38. Ugrinovic, K. *Acta Hortic.* **1999**, *506*, 99.
39. Strack, D.; Reznik, H. *Ztschr. Pflanzenphysiol.* **1979**, *94*, 163.
40. Stintzing, F. C.; Schieber, A.; Carle, R. *Food Chem.* **2002**, *77*, 101, 517.
41. Wybraniec, S.; Platzner, I.; Geresh, S.; Gottlieb, H. E.; Haimberg, M.; Mogilnitzki, M.; Mizrahi, Y. *Phytochemistry* **2001**, *58*, 1209.
42. Wybraniec, S.; Mizrahi, Y. *J. Agric. Food Chem.* **2002**, 50, 6086.
43. Stintzing, F. C.; Conrad, J.; Klaiber, I.; Beifuss, U.; Carle, R. *Phytochemistry* **2004**, *65*, 415.
44. Butera, D.; Tesoriere, L.; Di Gaudio, F.; Bongiorno, A.; Allegra, M.; Pintaudi, A. M.; Kohen, R.; Livrea, M. A. *J. Agric. Food Chem.* **2002**, *50*, 6895.
45. Odoux, E.; Domínguez-López, A. *Fruits* **1996**, *51*, 61.
46. Castellar, R.; Obón, J. M.; Alacid, M.; Fernández-López, J. A. *J. Agric. Food Chem.* **2003**, *51*, 2772.
47. Stintzing, F. C.; Schieber, A.; Carle, R. *Eur. Food Res. Technol.* **2003**, *216*, 303.

48. Felker, P.; Soulier, C.; Leguizamon, G.; Ochoa, J. *J. Arid. Environ.* **2002**, *52*, 361.
49. Felker, P.; del C. Rodriguez, S.; Casoliba, R. M.; Filippini, R.; Medina, D.; Zapata, R. *J. Arid. Environ.* **2005**, *60*, 405.
50. Stintzing, F.C.; Carle, R. *Fruit Process* **2006**, *16*, 166.
51. Moßhammer, M. R.; Stintzing, F. C.; Carle, R. *Food Res. Int.* **2005**, *38*, 975.
52. Huang, A. S.; von Elbe, J. H. *J. Food Sci.* **1985**, *50*, 1115, 1129.
53. Schwartz, S. J.; von Elbe, J. H. *Ztschr. Lebensm. Unters. Forsch.* **1983**, *176*, 448.
54. Herbach, K. M.; Stintzing, F. C.; Carle, R. *J. Food Sci.* **2004**, *69*, C491.
55. Herbach, K. M.; Stintzing, F. C.; Carle, R. *Eur. Food. Res. Technol.* **2004**, *219*, 377.
56. Vincent, K. R.; Scholz, R. G. *J. Agric. Food Chem.* **1978**, *26*, 812.
57. Herbach, K. M.; Stintzing, F. C.; Carle, R. *J. Agric. Food Chem.* **2006**, *54*, 390-398.
58. Herbach, K. M.; Stintzing, F. C.; Carle, R. *Rapid Comm. Mass Spectrom.* **2005**, *19*, 2603.
59. Wybraniec, S. *J. Agric. Food Chem.* **2005**, *53*, 3483.
60. Wybraniec, S.; Mizrahi, Y. *J. Agric. Food Chem.* **2005**, *53*, 6704.
61. Savolainen, K.; Kuusi, T. *Ztschr. Lebensm. Unters. Forsch.* **1978**, *166*, 19.
62. Moßhammer, M. R.; Stintzing, F. C.; Carle, R. *Innov. Food. Sci. Technol.* **2005**, *6*, 221.
63. Czapski, J. *Ztschr. Lebensm. Unters. Forsch.* **1990**, *191*, 275.

Chapter 8

Achieving Phytonutrient Enhancement in a Potato by Breeding for Increased Pigment

Charles R. Brown[1], David Culley[2], Ronald E. Wrolstad[3], and Robert W. Durst[3]

[1]Agricultural Research Service, U.S. Department of Agriculture, Prosser, WA 99350
[2]Batelle Pacific Northwest Laboratory, Richland, WA 99352
[3]Department of Food Science and Technology, Oregon State University, Corvallis, OR 97331

Potatoes have great genetic diversity in content and type of anthocyanins and carotenoids. Both pigment types are antioxidants. The carotenoids in potato are the same as those in the human retina and have been implicated as nutritional therapies for macular degeneration and cataracts. Anthocyanins, which belong to the very large group of phenolic compounds, may function as nutritional antagonists to processes leading to heart disease and certain cancers. The content of carotenoids and anthocyanins shows a wide range, a level of variation that augurs well for development of varieties where the salient feature will be nutraceutical in nature.

The potato is a modified underground stem that should be classified as a vegetable. The content of nutrients is as varied as in the case of other vegetables and it should be judged along side other vegetables. Having originated in South America, it was not known outside of this continent until European contact and colonization. Today there is still a large and highly genetically variable germplasm under cultivation in South America and residing in various collections around the world. One of the features of cultivars grown in the center of origin is the high frequency of pigmented skin and flesh compared to the modern varieties developed outside of the point of origin. It is not commonly known, for instance, that potatoes contain carotenoids, lutein and zeaxanthin, which are constituents of the human retina. Carotenoids have been found to stimulate processes involved in the immune system of animal models (*1*), and lutein has been found to inhibit breast cancer development in mice (*2*). Anthocyanins are present in the skin and flesh of potatoes and like other phenolic compounds the amount present in the diet is correlated with lower incidences of heart disease and cancer (*3, 4*). Both carotenoids and anthocyanins are antioxidants.

Genetics of Anthocyanins and Carotenoids in Potato

The natural variation of cultivated potato germplasm includes types that are red and purple pigmented due to the presence of flavonoids in the skin and/or flesh. Anthocyanins are among the many flavonoids that may be found in potato tubers. A series of single genes control presence and absence of red and blue pigments. Different genetic systems controlling pigment expression have been identified for diploid cultivated versus tetraploid cultivated potatoes, (*5, 6, 7*). De Jong (*8*) and Van Eck and co-workers (*9*) have argued that the genes appear to be synthetic and should be regarded as belonging to one genome. The symbol D denotes a single gene controlling synthesis of red pigment, located on chromosome 2; the symbol P stands for a single gene on chromosome 11 controlling blue pigment synthesis, while I, of undetermined location, epistatically controls presence and absence of tuber skin and flesh pigmentation even when P and D are present. Gebhardt et al. (*10*) reported a locus controlling purple skin color, Psc, on chromosome 4. The single gene Pf, linked to I, determines whether pigment is present beyond the periderm in the interior tissues of the tuber (*8, 9, 11*). The pigments have been determined to be various types of acylated anthocyanidin glycosides (*12, 13*). The gene Ac is imputed to control acylation of anthocyanins. Diploid cultivated potatoes display both acylated and non-acylated forms while only acylated anthocyanins are present in the tetraploid cultivars (*14*). Potatoes have acylated glucosides of several aglycons: pelargonidin, petunidin, malvidin, and peonidin. Although genetic control of presence and absence of anthocyanins is monogenic, the completeness of anthocyanin distribution in pigmented flesh may be under complex genetic control (*8, 15*).

Outside of the center of origin of cultivated potato in the Andes of South America, it is rare to find varieties with anthocyanin pigments conferring red or purple flesh. However much of the world's production is occupied by yellow flesh potatoes which have higher total carotenoid than the white flesh varieties of North America and Great Britain. Potatoes synthesize and store in the tuber flesh xanthophyll type carotenoids, including predominantly lutein, violaxanthin and zeaxanthin (*13, 15, 16, 17, 18, 19*). White versus yellow flesh is thought to be under single gene control, while gene maps agree on the placement of this yellow flesh factor (*Y/y*) on homolog 3 (*10, 20*). White and yellow flesh potatoes have similar composition of carotenoids, however the yellow color of the latter group is due to higher concentrations of certain xanthophylls (*21, 22*). Evidence points to a gene encoding beta-carotene hydroxylase (BH) as the putative candidate for the *Y/y* gene. In *Papa Amarilla* germplasm, discussed later, an allele of BH shows significant regression with total carotenoid content with an $R^2 = 0.46$ (*23*).

Carotenoid Content in Potato

White and yellow flesh potato have xanthophyllous carotenoids. Yellow color intensity is a determinant of xanthophyll content up to a point. In Figure 1 we see the relationship of a Yellow Index (*24*) to the concentration of total carotenoids. Up to about 1000 μg per 100 g FW yellowness is correlated with content, but above these levels no further increase in yellowness is measured.

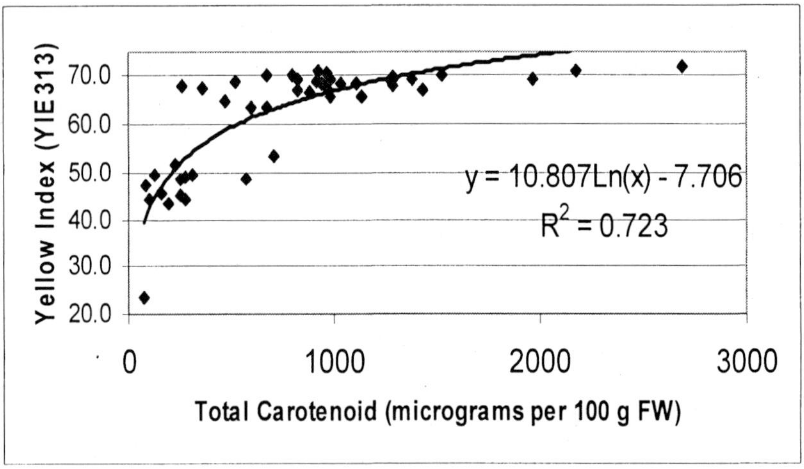

Figure 1. Relationship of yellow index to total carotenoid content

The total carotenoid content of white cultivars and breeding lines ranges from 50 to 100 μg per 100 g FW. Yellow flesh cultivars may have contents ranging up to 270 μg, while more intensely yellow breeding clones are will range up to 800 (Table I).

Although it is sometimes not directly observable, solidly red or purple flesh due to high anthocyanin concentration may be accompanied by higher total carotenoids as was noted in the red-yellow flesh clones in Table I. Levels of total carotenoid exceeding 2000 μg per 100 g FW have been reported in a number of studies (*22, 23, 25*).

There is a class of potato cultivars in South America called Papa Amarilla (PA)(= yellow potato) which have exceedingly high carotenoid values. We have found that certain crosses made between PA parents produce progeny that exceed either of the parents by more than two population standard deviations (Figure 2). Three progeny exceeded 2, 400 μg per 100 g FW despite the fact that the mid parent value is 900 μg. This is an indication of transgressive segregation. It also indicates that it may be possible to breed intensely yellow cultivars with exceptionally high levels of total carotenoids.

Anthocyanin Content in Potato

Anthocyanins are red to purple pigments ubiquitous in the plant kingdom. Anthocyanins are water soluble and are potent antioxidants. The general public

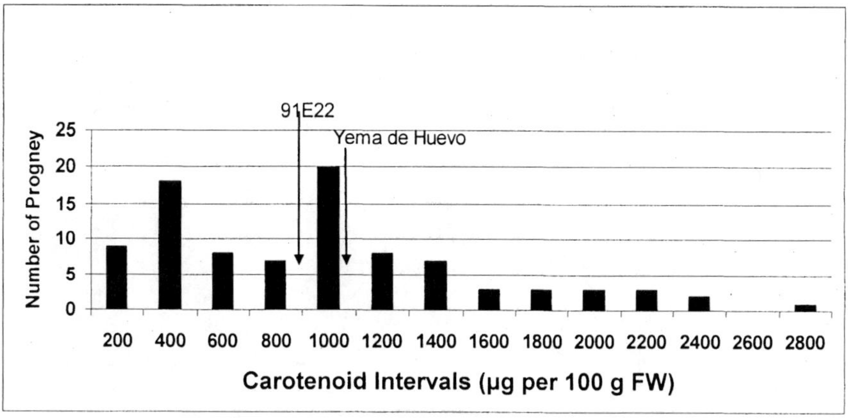

Figure 2. Distribution of total carotenoids contents in progeny of a cross between two Papa Amarilla types (91E22 and Yema de Huevo [= Egg Yolk]

Table I. Total carotenoid content in yellow and white flesh named varieties and experimental lines

Cultivar or Breeding line	*Skin/Flesh Type*[\1]	*Total Carotenoid µg / 100 g FW*	*Significance*
Light Yellow flesh cultivars and breeding lines			
Adora	W/Y	227	cdefg
Divina	W/Y	271	cdef
Fabula	W/Y	179	cdefg
Ilona	W/Y	176	cdefg
Morning Gold	W/Y	101	defg
Provento	W/Y	191	cdefg
Satina	W/Y	248	cdefg
Yukon Gold	W/Y	194	cdefg
POR00PG4-2	W/Y	250	cdefg
Dark Yellow flesh breeding lines			
91E22	W/DY	795	a
PA99P11-2[\w]	PR/DY	509	b
PA99P1-2[\w]	PR/DY	525	b
PA99P2-1[\w]	PR/DY	738	a
POR00PG4-1	W/DY	634	ab
Red and Yellow breeding lines			
POR00PG9-1[\w]	PR/R&Y	299	cd
POR00PG9-2[\w]	PR/R&Y	307	cd
POR00PG9-3[\w]	PR/R&Y	109	defg
POR00PG9-5[\w]	PR/R&Y	273	cde
POR00PG9-6[\w]	PR/R&Y	327	c
White flesh cultivars and breeding lines			
Norkotah	RT/W	40	g
Ranger	RT/W	71	efg
Burbank	RT/W	58	fg
A8893-1	RT/W	56	g
A9014-2	RT/W	55	g
A90586-11	RT/W	99	defg
A9045-7	RT/W	64	efg
A90490-1	RT/W	101	defg
A91790-13	RT/W	75	efg
A92030-5	RT/W	54	g
A93157-6LS	RT/W	66	efg

NOTE: RT/W = russet skin/white flesh, W/Y = white skin/light yellow flesh, W/DY = white skin/dark yellow flesh, PR/R&Y = Partially red skin/red and yellow flesh, PR/DY = partially red skin/dark yellow flesh. Means not sharing the same letter are significantly different at the $\underline{P} < 0.05$ level, using Duncan's Multiple Range Test.

is familiar with red skin potatoes. The skin is red due to a high concentration of red anthocyanins in the epidermal layer. However, much higher levels of anthocyanins are present in the clones with pigmented flesh. The degree of pigmentation can vary from streaks or blotches of pigment generally associated with to solid dark degrees of pigment (Figure 3)

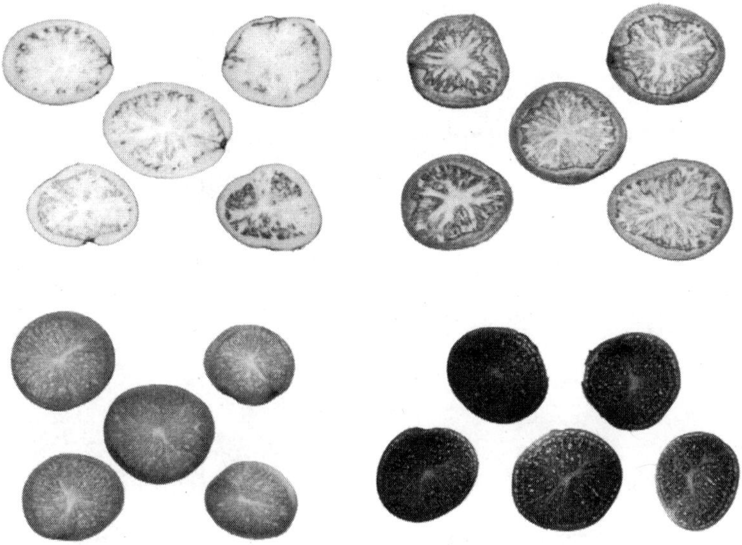

Figure 3. Different patterns and degrees of anthocyanin pigmentation in potato. The degree of pigmentation is under polygenic control, while presence and absence of pigment in the flesh is under single gene control. *(See page 9 of color inserts.)*

The concentration of anthocyanins can have a large range. The concentration of anthocyanin in skin tissue is quite high. However the skin is such a small volume of the whole tuber that generally a red skinned white fleshed potato has no more than 1.5 mg per 100 g FW when skin and flesh are extracted together. However, potatoes with anthocyanin in the flesh range from 15 to nearly 40 mg per 100g FW (Table II). We have found that red flesh potatoes contain predominantly acylated glucosides of pelargonidin. Purple flesh potatoes have a more complex content of acylated glucosides of pelargonidin, petunidin, cyanidin, and malvidin (*15*).

Red-skinned and purple-skinned potatoes are familiar in the marketplace. Pigmented flesh is new to people outside of the Andean countries of South America. The degree of pigmentation varies from flecks, lines and circles

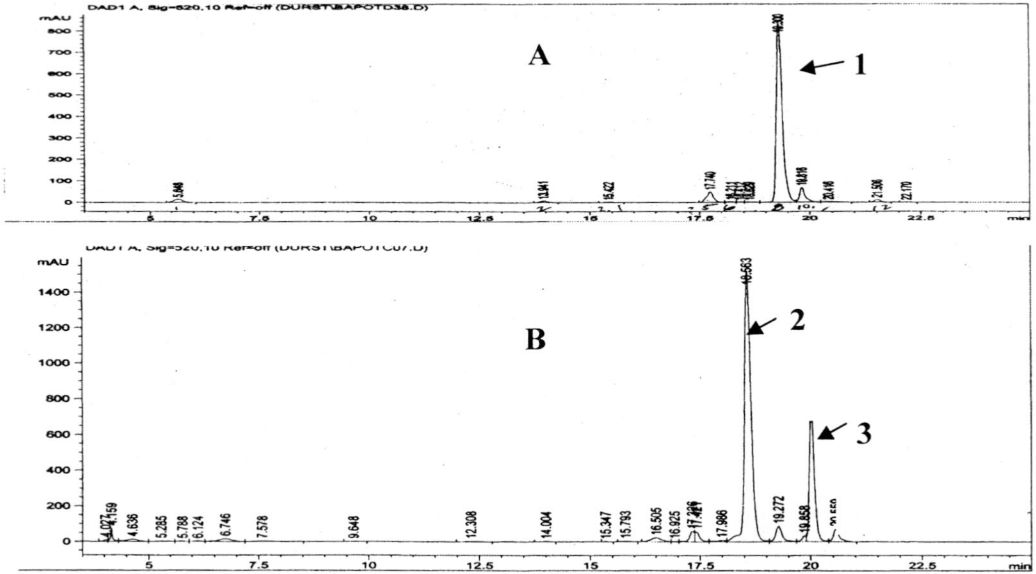

Figure 4. HPLC separation of anthocyanin extracts from red flesh (A) and purple flesh potatoes (B). Peak 1 labeled in panel A, was identified by saponification, hydrolysis and HPLC analysis to be an acylated pelargonidin glucoside. Peaks 2 and 3 in panel B were identified by the same means as acylated glucosides of petunidin and peonidin, respectively.

associated with particular tuber tissues to solid pigmentation. The antioxidant potential of potatoes is correlated to the anthocyanin concentration. Other compounds are antioxidants and contribute to the total Oxygen Radical Absorbance Capacity (ORAC) (*26*) value. This is apparent in Figure 4 where the clones lacking any anthocyanin show a broad range of ORAC values.

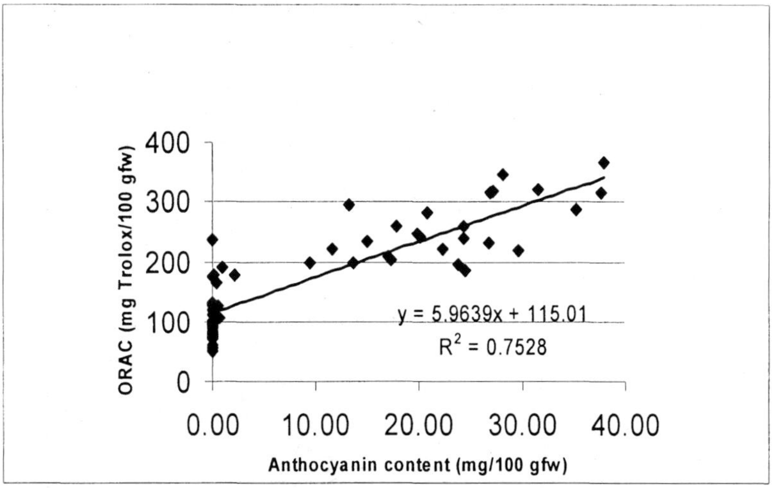

Figure 5. Regression between anthocyanin content and antioxidant value (ORAC) is significant.

Breeding Objectives

Development of new potato cultivars with specific non-traditional traits designed to appeal to a diet conscious populace is a relatively young endeavor. These potato have yet to find a steady market. However, the genetic diversity and nutritional bonuses embodied by the genetic diversity in pigment types and concentration have captured a faithful audience. Home gardeners, and small and large scale producers are watching this phenomenon. Ultimately the consumer will determine which direction the industry goes.

However, selection programs are looking for attractive skin that retains a bright color even after extended storage. It appears that there is a market for below four ounce size tubers. At the same time, total yield is important, and it is likely that a high yield somewhat evenly divided between small and medium size tubers might be the most advantageous combination. It is likely that the crop destined for specialty markets will need to be closely managed for size. In this

Table II. Total anthocyanin and associated antioxidant level hydrophilic Oxygen Radical Absorbance Capacity (ORAC) for purple flesh and red flesh breeding lines.

Breeding lines	*Skin/Tuber flesh Type*[z]	*Total Anthocyanins mg / 100 g FW*	*Signi-ficance*	*Hydrophilic ORAC μmole / 100 g FW*[y]	*Signi-ficance*
Purple skin / Purple flesh					
PA97B29-2	P/P	17.0	de[w]	800	c
PA97B29-4	P/P	20.1	cde	930	bc
PA97B29-6	P/P	17.3	de	840	bc
Red skin / red flesh					
NDOP5847-1	R/R	37.8	a	1420	a
PA97B35-1	R/R	24.3	bcde	1100	abc
PA97B36-3	R/R	15.0	e	850	bc
PA97B37-7	R/R	31.5	abc	1410	abc
PA99P9-2	R/R	26.9	abcde	1210	abc

PA99P9-4	R/R	24.5	bcde	790	c
PA99P10-2	R/R	27.2	abcde	1150	abc
PA99P20-1	R/R	20.8	bcde	1040	abc
PA99P20-2	R/R	26.8	abcde	980	bc
PA99P32-5	R/R	23.8	bcde	850	bc
POR00PG1-4	R/R	35.1	ab	1020	bc
POR00PG2-1	R/R	22.2	bcde	950	bc
POR00PG2-7	R/R	28.1	abcde	1160	abc
POR00PG2-11	R/R	29.6	abcd	1100	abc
POR00PG2-16	R/R	24.3	bcde	1160	abc
POR00PG3-1	R/R	19.8	cde	1020	bc

\z Key to skin and tuber flesh types: R = red, P = purple.

\y Hydrophilic ORAC = trolox equivalents

\w Means not sharing the same letter are significantly different using the Duncan's Multiple Range Test at $P \leq 0.05$

regard the overall yield is least likely to suffer if a heavy set of small tubers is the innate yield characteristic of the variety. Otherwise, the only way to limit size in more traditional plant types is to stop the growth by killing the foliage quite early in the growing season, reducing yield by a considerable amount.

Consumers may prefer flesh pigmentation that maximizes nutritional benefits by having the highest possible concentration of anthocyanins or carotenoids. Alternatively, intriguing and attractive patterns of partial anthocyanin pigmentation may be highly appealing in fresh and processed products.

References

1. Lee, C.M.; Boileau, A. C.; Boileau, T. W. M.; Williams, A. W.; Swanson, K. S.; Heintz, K A.; Erdman, J. W. *J. Nutr.* **1999**, *129*, 2271-2277.
2. Park, J. S.; Chew, B. P.; Wong, T. S. *J. Nutr.* **1998**, *128*, 1650-1656.
3. Hertog, M. G. L.; Feskens, E.; Hollman, P.; Katan M.; Kromhout, D. *Lancet* **1993**, *342*, 1007-1011.
4. Wang, H.; Nair, M. G.; Strasburg, G. M.; Chang, Y. C.; Booren, A. M.; Gray, J. I.; DeWitt, D. L. *J. Nat. Prod.* **1999**, *62*, 294-296.
5. Dodds, K. S.; Long, D. H. *J. Genetics* **1955**, *53*, 136-149.
6. Dodds, K. S.; Long, D. H. *J. Genetics* **1956**, *54*, 27-41.
7. Lunden, A. P. *Euphytica* **1960**, *9*, 225-234.
8. De Jong, H. *Am. Potato J.* **1991**, *68*, 585-593.
9. Van Eck, H.J.; Jacobs, J. M. E.; Van den Berg, P. M. M. M.; Stiekema, M. J.; Jacobsen, E. *Heredity* **1994**, *73*, 410-421.
10. Gebhardt, C.; Ritter, E.; Debener, T.; Schnachtschabel, U.; Walkemeier, B.; Uhrig, U.; Salamini, F. *Theor. Appl. Genet.* **1989**, *78*, 65-75.
11. De Jong, H. *Am. Potato J.* **1987**, *64*, 337-343.
12. Harbourne, J. B. *Biochem. J.* **1960**, *74*, 262-269.
13. Rodriguez-Saona, L. E.; Giusti, M. M.; Wrolstad, R. E. *J. Food Sci.* **1998**, *63*, 458-465.
14. Swaminathan, M. S.; Howard, H. W. *Bibliographia Genet.* **1953**, *16*, 1.
15. Brown, C. R.; Wrolstad, R.; Durst, R.; Yang, C.-P.; Clevidence, B. A. *Am. J Potato Res.* **2003**, *80*, 241-250.
16. Fossen, T.; Andersen; Ø. M.. *J. Hort. Sci. Biotechnol*. **2000**, *75*, 360-363.
17. Fossen, T; Øvstedal, D O.; Slimestad, R.; Andersen, Ø. M. *Food Chem.* **2003**, *81*, 433-437.
18. Iwanzik W.; Tevini, M.;Stute, R.; Hilbert, R. *Potato Res.* **1983**, *26*, 149-162.
19. Mazza, G.; Miniati, E. *Anthocyanins in Fruits, Vegetables and Grains.* CRC Press: Boca Raton, FL, USA. 1993; pp 269-272.

20. Bonierbale, M. W.; Plaisted, R. L; Tanksley, S. D. *Genetics* **1988**, *120*, 1095-1103.
21. Gross, J. *Pigments in Vegetables: Chlorophylls and Carotenoids.* Van Nostrand Reinhold: New York, NY, USA.1991; pp208-216.
22. Brown C. R.;Edwards, C. G.; Yang, C.-P.; Dean, B. B. *J. Amer. Soc. Hort. Sci.* **1993**, *118*, 145-150.
23. Brown, C. R; Kim, T.S.; Ganga, Z.; Haynes, K.; De Jong, D.; Jahn, M.; Paran, I.; De Jong, W. *Am J. Potato Res.* **2006**, *83*, 365-372.
24. Haynes, K. G.; Potts, W. E.; Chittams, J. L.; Fleck, D. L. *J. Amer. Soc. Hort. Sci.* **1994**, *119*, 1057-1059.
25. Lu, W.H.; Haynes, K.;Wiley, E.; Clevidence, B. *J. Am. Soc. Hort. Sci.* **2001**, *126*, 722-726.
26. Prior, R. L.; Cao, G. *Free Radical Biol. and Med.* **1999**, *27*, 1173-1181.

Chapter 9

Expanding the Potato Industry: Exotic–Colored Fleshed Tubers

Luis E. Rodriguez-Saona[1], M. Monica Giusti[1], and Ronald E. Wrolstad[2]

[1]Department of Food Science and Technology, The Ohio State University, 2015 Fyffe Road, Columbus, OH 43210
[2]Department of Food Science and Technology, Oregon State University, 100 Wiegand Hall, Corvallis, OR 97331

The potato industry is expanding their lines of products to respond to the opportunities for value added products and the changing needs of society. Colored potato varieties, such as red- and purple-fleshed potatoes, can capture new market demands for product extensions that will provide new flavors, exotic colors and premium natural products for consumers. The potential application of colored-fleshed potato cultivars as source of natural colorants and for manufacture of colored snacks will be discussed. Qualitative anthocyanin composition, pigment content and phenolic composition were screened on 33 cultivars. Red-fleshed potato extracts imparted desirable orange-red color and adequate stability and may serve as alternative sources of natural food colorants. Glycoalkaloids (SGAs) in color extracts were detected by ESMS. Alkaline treatment precipitated ~90% SGA with minimal anthocyanin degradation. Growth conditions and location impact anthocyanin and SGA accumulation in tubers. Frying of colored potato slices yield chips with bright attractive color ranging from light pink to dark red and pleasant flavor. Factors that will affect the quality of chips include reducing sugar and amino acid levels and processing conditions, influencing color development and the formation of the potentially toxic acrylamide. Current activities are dealing with the antioxidant capacity and health benefits of anthocyanin-containing products. There is a great opportunity for expanding the market of snacks through novelty chips with added value.

Introduction

Potato tubers and roots have formed part of the diet of the Andean people for several thousands of years. The area of origin and diversity of the cultivated potato lies in the central Andes of Peru and Bolivia, and for centuries have served as the primary food source of the Andean people. Native potato varieties have been selected and grown by Andean farmers for over 8000 years to bring diversity to their tuber dependent diet. Native potatoes range in color from yellow and white to pink and dark purple, and come in a broad range of shapes and sizes. Over the centuries these diverse varieties have been important to the Andean peoples not only for food, but also as an integral part of their culture, including ritual and medical practice (*1*).

The potato was introduced into Europe by the Spanish in the 16th century. As a strange plant, it was subjected to nearly 100 years of botanical curiosity and slow acceptance as a food crop (*2*). The spread of potatoes in Europe was due to the interplay of several factors, hunger during war being the most important. Potato yielded better than cereals, grew on marginal soils, and its subterranean location prevented pilferage and destruction by armies. Potato became the "bread of the poor" of thousands of people suffering from starvation (*3*, *4*). The lack of genetic diversity of European potato varieties led to failure of the potato crop in 1845 and 1846 due to the attack by the fungus *Phytophthora infestans* (blight or late blight) and caused one of the worst famines in European history. The potato crop re-emerged in Europe after 1859 thanks to breeding and selection of new potato varieties resistant to diseases (*4*, *5*). From Europe, the potato spread all over the world primarily through the colonial powers. Thus, in a remarkably brief period of 3 centuries, the potato attained its current role as one of the four major food crops in the world, along with rice, wheat and maize.

Today, potatoes (*Solanum tuberosum* L.) are the mainstay in the diet of populations in many parts of the world because of their high yield, low cost of production, adaptability to a wide variety of soils and climates, secure production under stress or short crop seasons, wide variety of cultivars, acceptance as daily food, and excellent nutritive value. Because of culinary preferences and processing traits, breeding of high-yield, easy to peel white flesh varieties with low water and sugar contents have been selected. Nevertheless, potatoes offer valuable diversity of native varieties, including over 4,000 varieties grown by Andean farmers that provide a wide range of flavors, shapes and colors. Nowadays there is an increasing interest for novel products and specialty potatoes with colored skin and flesh are of special appeal for growers and food processors alike including the colorant and snack industries as well as gourmet restaurants. Colored potato tubers have been shown to have high antioxidant capacity, comparable to sweet potatoes, spinach and Brussels sprouts (*6*) which may contribute to lowering the risk of developing certain chronic diseases associated with the consumption of commodities rich in phytochemicals. Phenolic compounds and anthocyanins in particular have been associated with the high antioxidant capacity in potato tubers and their contribution to health-promoting effects is under current investigations in many studies (*7*).

Compounds Responsible for the Color of Potatoes

Anthocyanins are responsible for the attractive blue, purple, violet, pink, and red color of potato tubers while the deep yellow and orange flesh colors are due to the presence of carotenoid pigments. We will focus our discussion on the chemistry and abundance of anthocyanin pigments in colored potatoes that show pigmentation in the skin and flesh of the tubers.

Anthocyanins are part of the flavonoid group of compounds which are characterized by the $C_6C_3C_6$ carbon skeleton and same biosynthetic origin (*8*). Although flavonoids are generally colorless, anthocyanins occur in the cell sap in chemical states that strongly absorb visible light (*9*). Anthocyanins are glycosides of anthocyanidin (aglycon) chromophores, these being polyhydroxy and polymethoxy derivatives of 2-phenylbenzopyrylium (flavylium) salts (*8*, *10*). There are only 6 major anthocyanidins found in nature in spite of the great variety of plant colors (*11*). The sugar residues can be acylated with aromatic acids including *p*-coumaric, caffeic, ferulic, sinapic, gallic or *p*-hydroxybenzoic acids, and/or aliphatic acids such as malonic, acetic, malic, succinic or oxalic acids (*8*) which may impart an important stabilizing effect *via* intramolecular copigmentation. Under acidic conditions, the color of non- and mono-acylated anthocyanins (Figure 1) is determined by the number and position of hydroxyl and methoxyl groups on the B-ring of the aglycon, the nature, number and position of sugars attached to the aglycone, the nature and number of the acylating groups attached to the sugar, and physiochemical conditions (*8*, *12*, *13*). Mazza and Miniati (*14*) reported that increased hydroxyl substitution on the B-ring results in a shift of the visible absorption maximum to longer wavelengths to yield a bluer hue.

Investigation of *Solanum tuberosum* L. anthocyanin composition was first reported by Chmielewska (*15*) who studied the pigments of a purple black cultivar called “Negresse” (also called “Congo”). These authors identified negretein [Mv-3-(*p*-coumaroyl-rutinoside)-5-glucoside] as the major anthocyanin

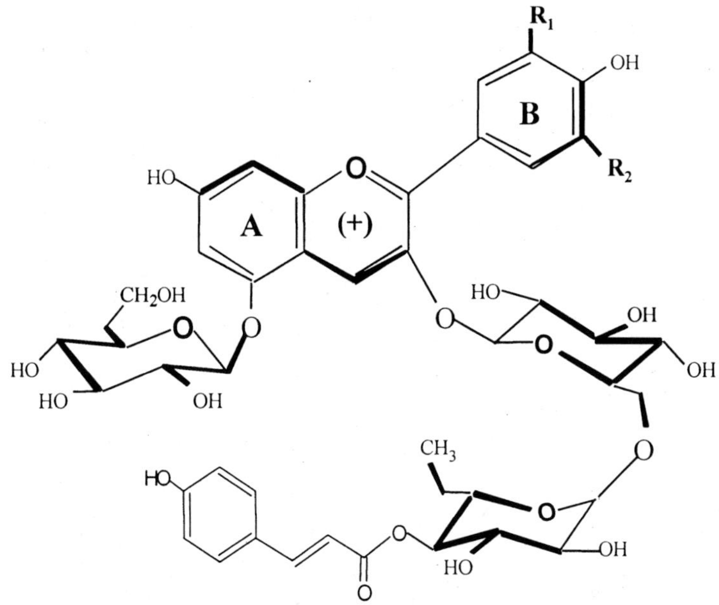

Figure 1. Structure of p-coumaroyl-3-rutinoside-5-glucoside anthocyanin derivatives in potatoes. Pelargonidin: R_1 and R_2 = H; Malvidin: R_1 and R_2 = OCH_3; Petunidin: R_1 = OCH_3 and R_2 = OH; Peonidin: R_1 = OCH_3 and R_2 = H.

present in the cultivated potato tubers. Later, Harborne (*16*) described the isolation and characterization of ten anthocyanins from diploid and tetraploid potato species and depending on the color, the *p*-coumaroyl-3-rutinoside-5-glucoside derivatives of all six common anthocyanins were identified. Additional studies have confirmed the predominance of acylated anthocyanins (Fig. 1) in pigmented potato varieties (*17-24*). The purple-skinned, purple-fleshed potato cultivars contain varying ratios of Pt- and Mv -3-(*p*-coumaroyl-rutinoside)-5-glucoside (*20*, *24*) while the major anthocyanin pigment present in red-skinned, red-fleshed tubers is Pg -3-(*p*-coumaroyl-rutinoside)-5-glucoside (*19-21*).

The reported anthocyanin content of different colored-fleshed potato cultivars is presented in Table I. Overall, the potato selections with dark purple flesh have been associated with the highest pigment content (Table I) and the cultivar "Urenika" has shown levels of up to 230 mg anthocyanin/100 g FW in the flesh (*20*). It is difficult to compare literature values because of the different methods used for extraction and quantification (i.e. pH differential *vs* LC external standard calibration curve) of anthocyanins which will result in marked differences in the estimation of the levels of these pigments. Nevertheless, the pigment content of highly pigmented potato selections are comparable to those reported for such commodities as red radish (30 - 60 mg/100 g root), eggplant (86 mg/100g), red onions (48 mg/100g), Concord grapes (120 mg/100 g), strawberries (15-30 mg/100 g), cranberries (140 mg/100 g) and red raspberry (90 mg/100 g) (*25-27*). On the other hand, higher levels have been reported for red cabbage (320 mg/100 g) from which an approved "natural" colorant is obtained, blackberries (300 mg/100 g), blueberries (365 mg/100 g) and chokeberries (1480 mg/100 g) (*25*). However, since the major anthocyanins in berries are non-acylated (*8*) the stability of these pigments after extraction has been a limiting factor.

Potential Application as a Food Colorant

As a result of toxicological studies, safety issues and legislative action, the number of artificial dyes suitable for foods has decreased (*29*, *30*). There is a growing world market for natural alternatives to synthetic dyes because of a "healthy eating" marketing approach by the food industry and as a consequence of consumer preferences (*29*, *31*, *32*) towards food products incorporating natural ingredients, leading to the research for new potential natural sources of colorants.

According to US legislation, color additives exempt from certification include colors obtained from natural sources (vegetables, animals or minerals) and nature identical compounds which are synthetic counterparts of natural derivatives (*33*). Uncertified colorants are commonly referred as "natural" colorants; however, the FDA has rejected the description of "natural color" on label declarations and advocated the use of terms such as "artificially colored", "artificial color added" or similar statements, since the color has been added to the food product for a specific purpose. The use of the term "artificial" can be avoided, provided that the declaration identifies the color by its common name or by function (*29*, *33*). Use of "natural" colorants has increased as a consequence of perceived consumer preference as well as legislative action which have continued the de-listing of approved artificial dyes (*31*).

Anthocyanins are considered as potential replacements for banned dyes because of their bright attractive colors and water solubility that allows their incorporation into aqueous food systems as well as possible health benefits (*14*, *34*). Preparations of anthocyanin-based extracts from wild blueberries and European bilberries are marketed as potent phytochemicals (*27*). Acylation of the anthocyanin molecule improves pigment stability during processing and storage (*31*, *35-37*). This increased stability is chemically attributed to the stacking of the acyl groups with the pyrylium ring of the flavylium cation, thereby reducing the susceptibility of nucleophile attack of water and subsequent formation of a pseudobase or a chalcone (*9*, *10*). These findings have encouraged research on acylated anthocyanin-based food colorants (*38*, *39*) such as red cabbage extract (*35*, *40*, *41*), red radish (*26*, *42*), black carrots (*43*), and sweet potatoes (*34*, *44*).

Challenges and Opportunities of a Potato-Based Red Colorant

There is potential for the application of red-fleshed potato anthocyanin extracts as an alternative to artificial dyes (*45*). The use of potato anthocyanin extracts (45 mg/100mL) to color model juices (pH 3.5) imparted the desirable orange-red color (L*=53.84, C*=63.07, h*= 29.51, haze= 1.14) comparable to that obtained with 1.5 mg/100 mL artificial dye FD&C Red # 40 (L*=67.29, C*=65.45, h*=37.73, haze= 1.2). A comparison between red-fleshed potato anthocyanins and a chemically related pelargonidin-based anthocyanin from radish showed that color and pigment degradation depended on storage temperature, anthocyanin structure and method of pigment extraction (aqueous *vs* solid phase extraction). At room temperature, higher stability was obtained on juices colored with C_{18} column-purified radish anthocyanins (22 wks half-life) and lowest with potato juice concentrate (10 wks half-life). Refrigeration (2°C) greatly increased the half-life of the pigment to over a year.

Since red-fleshed potato anthocyanin extracts can impart the desirable orange-red color and adequate stability for different food applications, an aqueous extraction using physical separation operations was devised. Anthocyanin concentrates obtained by physical means and using water as extracting solvent are permitted as food colorants under the category of fruit (21 CFR 73.250) or vegetable (21 CFR 73.260) juice color (*46*). It may be possible to prepare a relatively stable potato anthocyanin extract by adsorption of the anthocyanin pigments onto a C_{18} silica resin, however, such extractions will require toxicological and safety tests before such colorant extract could be approved by the FDA. The first challenge encountered in processing the potato tubers was their high polyphenoloxidase (PPO) activity which reacts with the anthocyanin pigments accelerating degradation and the development of undesirable browning of the extracts. Steam blanching of tuber slices at 100°C for 5 min before extraction prevented pigment degradation by inactivating PPO but it promoted starch gelatinization leading to the formation of a thick paste that markedly limited pigment extraction. An alternative method that involved blending of the tubers using acidified water (pH 3.0) and showed excellent pigment quality with minimal degradation due to PPO activity and allowed the rapid filtration of the vegetable juice extract by using a multiple pad filtration

Table I. Anthocyanin Levels in Flesh-Colored Potato (*Solanum tuberosum*) cultivars

Culltivar	Anthocyanin[a]	Content (mg/100 g, FW)	Source
Purple skin/purple flesh			
I53	Pt-crg[b]	50	Lewis et al. (20)
Stage II Blue	Pt-crg[b] Mv-crg[b]	22	Lewis et al. (20)
Urenika	Pt-crg[b] Mv-crg[b]	183	Lewis et al., (20)
All Blue	Pt-crg[b] Mv-crg[b]	60	Reyes and Cisneros-Zevallos (28)
Hermanns Blaue	Pt-rg[c] Pt-crg[b]	d	Eichhorn and Winterhalter (24)
Shetland Black	Pn-rg[c] Pt-crg[b]	d	Eichhorn and Winterhalter (24)
Vitelotte	Mv-crg[b] Mv-rg[c]	d	Eichhorn and Winterhalter (24)

Red skin/red flesh			
NDC4069-4	Pg-crg[b]	35	Rodriguez-Saona et al. (19)
NDOP5847-1	Pg-crg[b]	40	Rodriguez-Saona et al. (19)
Red Flesh	Pg-crg[b]	11	Lewis et al. (20)
Highland burgundy Red	Pg-crg[b] Pg-rg[c]	d	Eichhorn and Winterhalter (24)

[a] Pt: Petunidin; Mv: Malvidin; Pn: Peonidin; Pg: Pelargonidin

[b] crg: -3-(p-coumaroyl-rutinoside)-5-glucoside

[c] rg: -3-rutinoside-5-glucoside

[d] yield of anthocyanin-rich extracts ranged from 65 to 115 g/100g

unit. The filtered juice was blanched (90°C for 2 min) and concentrated by using a Centritherm CT-1B centrifugal film evaporator with operating conditions set at vacuum -94 kg/cm2, jacket temperature 65-70°C, head temperature 35-40°C, and feed rate 0.5 L/min.

Table II shows the recoveries obtained in the production of the red-potato colorant. The pigment level in the final concentrate was limited by the relatively high °Brix of our initial vegetable juice (~5°Brix). Use of special concentration techniques (direct osmosis, membrane concentration or ultrafiltration) to remove compounds that might accelerate anthocyanin degradation could improve the final quality of the pigment extracts, and its stability in food systems.

Table II. Pigment yields in colorant production from ND4069 red-fleshed variety

	ND4069
Anthocyanin Content (mg ACN/100 g root)	35
% Recovery in Juice	85
	Evap
Final °Brix of concentrate	~40
Monomeric anthocyanin (mg/100mL)	~300
% Polymeric color	~10

Evap: Centritherm centrifugal film evaporator

Analysis of the potato anthocyanin extracts by electrospray mass spectroscopy (ESMS) revealed the presence of alkaloids (α-solanine and α-chaconine) (Fig. 2). The levels of steroidal glycoalkaloid (SGAs) in the red-fleshed potato breeding clones (ND4069-4 and NDOP5847-1) used for pigment extraction ranged from 4.5 to 15.5 mg/100 g FW tuber. All Blue, a purple fleshed variety, showed alkaloid levels of 27 mg/100 g FW tuber (*47*). The variability in SGAs levels could be attributed to inherited differences of potato varieties or to environmental factors such as growing conditions, planting location, pesticide application, climate and storage conditions (*47*).

Potato breeding typically introduces desired traits (frost, disease and pest resistance) but may result in increased concentrations of SGAs or incorporation of new alkaloids (*48*). Alkaloids are natural toxins that accumulate in the peel of potato tubers that probably evolved as protective compounds in response to tissue invasion. Alkaloids have shown antimicrobial, insecticidal and fungicidal

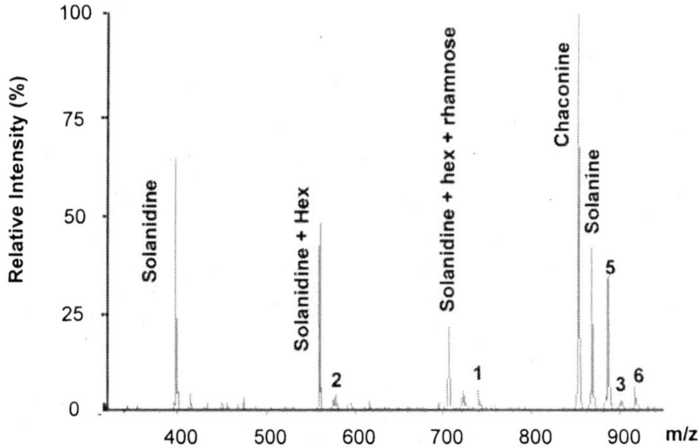

Figure 2. Electrospray Mass Spectroscopy of potato pigment extract. The molecular ions correspond to the alkaloids and anthocyanins (identified by numbers) detected in the extract (adapted from Rodriguez-Saona et al. (47)). (See page 9 of color inserts.)

properties that provide resistance against several insect pests and hervibores (*49, 50*). Nevertheless, alkaloids also posses pharmacological and toxicological effects against humans (*51, 52*). Since potatoes constitute an important nutritional source in human diet, SGAs along with cyanogenic glycosides are responsible for more of human illnesses and death than any other plant toxicants (*53*). Toxicity of SGAs is due to their significant anticholinesterase activity and disruption of cell membranes (*54*). A safety level of SGAs in potatoes was established at 200 mg/kg for acute toxicity, but these levels do not account for possible sub-acute or chronic effects. An upper limit of 60 to 70 mg/kg has been proposed for cultivars to be selected for human consumption (*51, 55*).

The major SGAs in potato tubers, α-solanine and α-chaconine, are readily soluble in water and will be concentrated along with the anthocyanins during preparation of a color extract from potatoes. This emphasizes the importance for mandatory surveillance by FDA of all additives to ensure they meet current government specifications (21 CFR Part 73) and are used in accordance with all the provisions of its regulation (*56*). These include category of use, specified uses and restrictions, identity, levels of usage, specifications, and labeling requirements (*46*).

SGAs are fairly heat-stable compounds, slightly affected by steaming, boiling, baking, frying, cooking and microwaving of potatoes (*51, 57*). However, SGAs will precipitate under basic conditions and this property is used in purification of alkaloids (*58*). However, the stability of anthocyanins is affected with basic pH which increases the rate of destruction (*59*, *60*). We have devised an effective method to precipitate SGAs by treating the anthocyanin-containing juices with KOH to a final pH of 8.0 precipitating up to 90% of the alkaloids without severe anthocyanin degradation (ca 30%) or changes in pigment profile (*47*). It was also noted that application of the alkaline treatment in the concentrated colorant further minimized anthocyanin degradation possibly due to improved pigment stability intermolecular co-pigmentation. Our results have shown that red-fleshed potatoes could be used as a potential source of natural red colorant providing the right hue, acceptable stability, no undesirable odors and very low alkaloid levels, which make potato anthocyanin extract a suitable alternative to FD&C Red # 40 for food applications.

Naturally Colored Potato Chips

Potato chips have become America's favorite snack with annual US retail sales in excess of $6 billion, leading the total retail market for salted snacks followed by tortilla and corn-based chips (*61*). Moderate annual growth of the potato chip category is expected due to the introduction of new flavors, premium products (gourmet chips) with higher price tags and expansion of niche markets (i.e. Organic consumers). Naturally colored potato chips can provide the opportunity of being marketed as healthful snacks with attractive and bright colors. Anthocyanins have been associated with disease prevention and there is evidence that these pigments may contribute to the chemo-protective effect of fruits and vegetables (*62-67*). The International Potato Center is promoting the cooperation between the Aymara farmer community and the industry to market "Jalca" chips produced from native Andean colored potatoes (*67*). Another chip product made from colored potatoes includes Terra Blues™ Potato Chips that are marketed as premium quality chips with vibrant bluish-purple color and a slightly nutty flavor.

In order to produce high quality chips from colored tubers, potato varieties with good frying properties need to be selected from the available genetic pool. Frying in oil is a dehydration process being potato frying the most intense of all snack food processing since potatoes used for chipping are ideally about 75% water (25% total solids) and need to be reduced to about 1.5% water. The final product will adsorb oil at approximately 30% of its weight. To produce uniform, high quality chips frying time and temperature need to be controlled and monitored. Frying time and temperature depend on the chip's composition and thickness, and can influence oil content. Oil temperatures typically range from 340 to 360°F which will turn potato slices into chips in less than four minutes. Too high frying temperature can burn chips and break down the oil while too low temperature will produce moist chips. Thicker slices will require longer cooking times (*61*). Preliminary data (Rodriguez-Saona, unpublished) has suggested that a rapid frying process minimizes anthocyanin degradation probably due to the rapid removal of water which prevents hydrolytic reactions leading to improved stability of the pigments (Fig. 3).

Nevertheless, excessive browning during frying produces chips results in undesirable color and unacceptable bitter taste (*68*). The formation of brown colored products and bitter compounds upon frying, due to Maillard reactions, are critical for the potato processing industry because it severely decreases the quality of the end products. The levels of sugars, in particular reducing sugars, and amino acids in potato tubers have an important effect on color development of fried potatoes. Reducing sugars are normally the limiting factor in color development (*68*, *69*). A maximum tolerable level of sucrose and glucose of 100 and 35 mg/100 g respectively, has been suggested for acceptable potato chip color (*70*). Rodriguez-Saona and collaborators (*71*) developed a model system to evaluate the contribution of potato constituents to color development in fried chips and confirmed the major role of reducing sugar concentration to the development of chip color (Fig. 4). In addition, at low concentration of reducing sugars (<40 mg/100 g), ascorbic acid and sucrose levels affected color quality (chroma and hue angle) of fried chips. There is limited available information on the composition of colored-fleshed potato tubers with regards to sugars and free amino acid content but an informal screening of commercial colored chips showed considerable variability in color with several chips showing marked browning (Fig. 3).

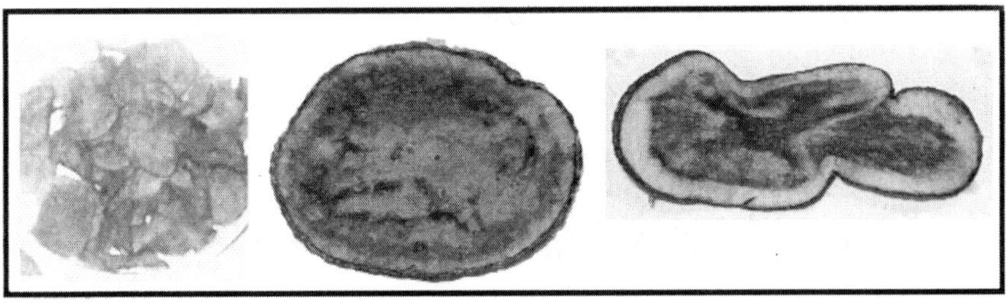

Figure 3. Colored Potato Chips showing browning defects
(See page 10 of color inserts.)

The International Potato Center (*1*) reported that chips made from colored native potatoes have lower water contents than conventional potato varieties and less energy is expended when fried. These native potato chips have been shown to absorb up to 25 percent less oil than regular potato chips, and because these potatoes are not peeled, it provides the consumers all the nutritional value (phenolics, minerals, vitamins and fiber) found in skins.

The safety of colored chips with regards to the accumulation of glycoalkaloids and acrylamide needs to be determined. Table III shows the

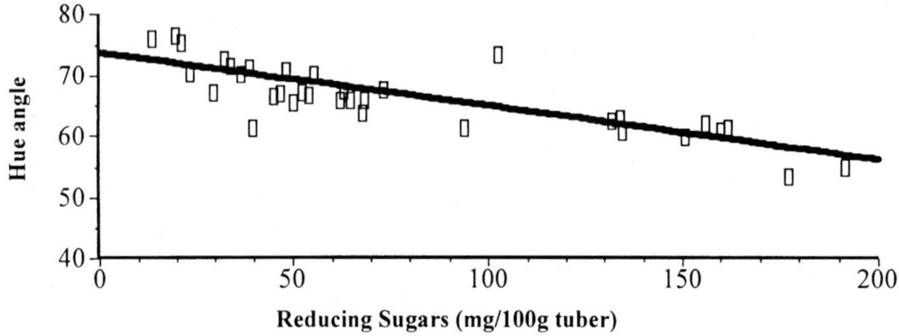

Figure 4. Effects of reducing sugar concentration on potato chip color (adapted from Rodriguez-Saona et al. (71)).

Table III. Acrylamide Levels in Snack Foods (modified from Friedman (*73*); Zyzak (*75*))

Food product	Acrylamide (ppb)
Corn Chips	34 - 416
Tortilla Chips	97
Banana Chips	125
Pretzels	196
Baked Potato Chips	317
Potato puffs, deep fried	1270
Potato chips, crisps	170 - 3700
Kettle Potato Chips	3400
Blue Potato Chips	736
Snacks other than potato	30 - 1915

reported levels of acrylamide for some snack foods. Acrylamide in food is largely derived from heat-induced reactions between the amino group of the free amino acid asparagine and the carbonyl group of reducing sugars such as glucose during baking and frying. Foods rich in both of these precursors are largely derived from plant sources such as potatoes and cereals (barley, rice, wheat) but apparently not animal foods such as poultry, meat, and fish (*72*). According to Friedman (*72*), the observed wide variations in levels of acrylamide in different food categories as well as in different brands of the same food category (e.g., French fries; potato chips) appear to result not only from the amounts of the precursors present but also from variations in processing conditions (e.g., temperature; time; nature of frying oil; nature of food matrix). The content of reducing sugars in the tuber determines the level of acrylamide formed after frying (*73*) and selection of "cold-chippers" varieties with low glucose and fructose levels throughout storage could produce fried products with appetizing colors and low acrylamide levels (*74*).

Conclusions

There is a great opportunity for expanding the market of "exotic" potatoes through the development of novelty products with added value such as colored potato chips or "natural" colorants that could develop new consumer demands and premium prices. Nevertheless, more research is needed on evaluating varieties with suitable traits, developing adequate processing parameters that addresss potential safety concerns and determining potential health benefits.

References

1. Natural Colored Potato Chips, Centro Internacional de la Papa. http://www.cipotato.org/Papandina/publicaciones/Chips%20flyer%20 Eng.pdf (accessed Mar 11, 2007)
2. Niederhauser, J. S. *Am. Potato J.* **1993**, *70*, 385.
3. Brown, C. R.. *Am. Potato J.* **1993**, *70*, 363.
4. Leszczynski, W. Potato Tubers as a Raw Meterial for Processing and Nutrition. In *Potato Science and Technology*. Lisinska, I. G., Leszczynski, W., Eds., Elsevier: Sussex, UK, 1989; pp 34-76.
5. Burton, W. G. *The Potato*, 3rd Ed., Longman: Essex, UK, 1989.
6. Brown, C. R.; Culley, D.; Yang, C. P.; Durst, R.; Wrolstad, R. *J. Amer. Soc. Hort. Sci.* **2005**, *130*, 174-180.
7. Brown, C. R. *Am. J. Potato Res.* **2005**, *82*, 163-172.

8. Jackman R. L.; Smith J. L. Anthocyanins and Betalains. In *Natural Food Colorants*, 2nd Ed., G. A. F.Hendry, J. D. Eds., Houghton Blackie: London, 1996; pp 244-280.
9. Brouillard, R. *Phytochem.* **1983**, *22*, 1311-1323.
10. Brouillard, R. Chemical structure of anthocyanins. In *Anthocyanins as Food Colors*. P. Markakis, Ed., Academic Press: New York, 1982; pp 1-40.
11. Goto, T. *Prog. Chem. Org. Nat. Prod.* **1987**, *52*, 113-158.
12. Giusti, M. M.; Ghanadan, H.; Wrolstad, R. E. *J. Agric. Food Chem.* **1998**, *46*, 4858-4863.
13. Mazza, G.; Broillard, R. *Phytochem.* **1990**, *29*, 1097-1102.
14. Mazza, G.; Miniati, E. *Anthocyanins in Fruits, Vegetables, and Grains*; CRC Press: Boca Raton, F, 1993.
15. Chmielewska, Z. *Bull. Soc. Chim.* **1936**, *5*, 1575-1588.
16. Harborne, J. B. *Biochem. J.* **1960**, *74*, 262–269.
17. Sachse, J. *Z. Lebensm. Unters. Forsch.* **1973**, *153*, 294-300.
18. Andersen, Ø. M.; Opheim, S.; Aksnes, D. W.; Frøystein, N. A. *Phytochem. Anal.* **1991**, *2*, 230-236.
19. Rodriguez-Saona, L. E.; Giusti, M. M.; Wrolstad, R. E. *J. Food Sci.* **1998**, *63*, 458-465.
20. Lewis, C. E.; Walker, J. R. L.; Lancaster, J. E.; Sutton, K. H. *J. Sci. Food Agric.* **1998**, *77*, 45-57.
21. Naito, K.; Umemura, Y.; Mori, M.; Sumida, T.; Okada, T.; Takamatsu, N.; Okawa, Y.; Hayashi, K.; Saito, N.; Honda, T. *Phytochem.* **1998**, *47*, 109-112.
22. Fossen, T.; Andersen, O. M. *J. Hortic. Sci. Biotech.* **2000**, *75*, 360-363.
23. Fossen, T.; Øvstedal, D. O.; Slimestad, R.; Andersen, Ø. M. *Food Chem.* **2003**, *81*, 433-437.
24. Eichhorn, S.; Winterhalter, P. *Food Res. Int.* **2005**, *38*, 943-948.
25. Wu, X.; Beecher, G.; Holden, J. M.; Haytowitz, D.; Gebhardt, S. E.; Prior, R. L. *J. Agric. Food Chem.* **2006**, *54*, 4069-4075.
26. Giusti, M. M. and Wrolstad, R. E. *J. Food Sci.* **1996**, *61*, 322-326.
27. Timberlake, C.F. *NATCOL Quarterly Bulletin*. **1988**.
28. Reyes, L. F.; Cisneros-Zevallos, L. *J. Agric. Food Chem.* **2003**, *51*, 5296-5300.
29. Spears, K. *Trends in Biochem. Technol.* **1988**, *6*, 283-288.
30. Noonan, J. E. *Cereal Foods World* **1985**, *30*, 265-267.
31. Giusti, M. M.; Wrolstad, R. E. *Biochem. Eng. J.* **2003**, *14*, 217-225.
32. Lauro, G. J. *Am. Assoc. Cereal Chemists* **1991**, 36, 949-953.
33. Dziezak, J. D. *Food Technol.* **1987**, *41*, 78-88.
34. Bridle, P.; Timberlake, C. F. *Food Chem.* **1997**, *58*, 103-109.
35. Murai, K.; Wilkins, D. *Food Technol.* **1990**, *44*, 131.
36. Teh, L. S.; Francis, F. J. *J. Food Sci.* **1988**, *53*, 1580-1581.

37. Bassa, I. A.; Francis, F. J. *J. Food Sci.* **1987**, *52*, 1753-1754.
38. Malien-Aubert, C.; Dangles, O.; Amiot, M. J. *J. Agric. Food Chem.* **2001**, *49*, 170-176.
39. Francis, F. J. *Cereal Foods World.* **2000**, *45*, 208-213.
40. Dyrby, M.; Westergaard, N.; Stapelfeldt, H. *Food Chem.* **2001**, *72*, 431-437.
41. Idaka, E.; Suzuki, K.; Yamakita, H.; Ogawa, T.; Kondo, T.; Goto, T. *Chem. Lett.* **1987**, *1*, 145-148.
42. Giusti, M. M.; Wrolstad, R. E. *J. Food Sci.* **1996**, *61*, 688-694.
43. Kammerer, D.; Carle, R.; Schieber, A. *Eur Food Res Technol.* **2004**, *219*, 479-486.
44. Shi, Z.; Bassa, I. A.; Gabriel, S. L.; Francis, F. J. *J. Food Sci.* **1992**, *57*, 755-757, 770.
45. Rodriguez-Saona, L. E.; Giusti, M. M.; Wrolstad, R. E. *J. Food Sci.* **1999**, *64*, 451-456.
46. Lipman, A. L. Current regulations for certification exempt color additives in the USA. In *The Second International Symposium on Natural Colorants* INF/COL II. The Hereld Organization. Acapulco, Mexico. 1996.
47. Rodriguez-Saona, L. E.; Wrolstad, R. E.; Pereira, C. *J. Food Sci.* **1999**, *64*, 445-450.
48. van Gelder, W. M. J.; Tuinstra, L. G. M.; van der Greef, J.; Scheffer, J. J. C. *J. Chrom.* **1989**, *482*, 13-22.
49. Roddick, J. G.; Rijnenberg, A. L.; Weissenberg, M. *Phytochem.* **1990**, *29*, 1513-1518.
50. Tingey, W. M. *Am. Potato J.* **1984**, *61*, 157-167.
51. van Gelder, W. M. J. Chemistry, Toxicology, and Occurrence of Steroidal Glycoalkaloids: Potential Contaminants of the Potato (*Solanum tuberosum* L.). In *Poisonous Plant Contamination of Edible Plants*. Abbel-Fattah, M., Rizk, M. Eds.. CRC Press: Boca Raton, FL, 1991; pp 117-156.
52. Jadhav, S. J.; Sharma, R. D.; Salunkhe, D. K. *Crit. Rev. Toxicol.* **1981**, *9*, 21-104.
53. Hall, R. L. *Food Technol.* **1992**, *46*, 109-112.
54. Roddick, J. G.; Rijnenberg, A. L. *Phytochem.* **1987**, *26*, 1325-1328.
55. Valkonen, J.; Keskitalo, M.; Vasara, T.; Pietila, L. *Crit. Rev. Plant Sci.* **1996**, *15*, 1-20.
56. Marmion, D.M. *Handbook of U.S. Colorants, 3rd Edition;* Wiley: New York, 1991; pp 3-122.
57. Friedman, M.; McDonald, G. M. *Crit. Rev. Plant Sci.* **1997**, *16*, 55-132.
58. Friedman, M.; Dao, L. *J. Agric Food Chem.* **1992**, *40*, 419-423.
59. Wrolstad, R. E. Colorants. In *Food Chemistry: Principles and Applications;* Christen, G. L., Ed.,. AFNC Press: Cutten, CA., 1997; pp 215-239.
60. Francis, F. J. Food colorants: anthocyanins. *Crit. Rev. Food Sci. and Nutr.* **1989**, *28*, 273-314.

61. Berry, D. Chip Celebration. http://foodproductsdesign.com/archive/2003/0803CS.htm (accessed Mar 12, 2007)
62. Stintzing, F.C.; Carle, R. *Trends Food Sci. Technol.* **2004**, *15*, 19-38.
63. Kong, J.M.; Chia, L.S.; Goh, N.K.; Chia, T.F.; Brouillard, R. *Phytochem.* **2003**, *64*, 923-933.
64. Singletary, K. W.; Stansbury, M. J.; Giusti, M.; van Breemen, R. B.; Wallig, M.; Rimando, A. *J. Agric. Food Chem.* **2003**, *51*, 7280-7286.
65. Liu, M.; Li, X. Q.; Weber, C.; Lee, C. Y.; Brown, J.; Liu, R. H. *J. Agric. Food Chem.* **2002**, *50*, 2926-2930.
66. He, J.; Magnuson, B. A.; Giusti, M. M. *J. Agric. Food Chem.* **2005**, *53*, 2859-2866.
67. Bernet, T.; Amoros, W. *LEISA Magazine* **2004**, *20*, 18-19.
68. Roe, M. A.; Faults, R. M.; Belsten, J. L. *J. Sci Food Agric.* **1990**, *52*, 207-214.
69. Marquez, G.; Añon, M. C. *J. Food Sci.* **1986**, *51*, 157-160.
70. Sowokinos, J. R. and Preston, D. A. In *Station Bulletin 586*. Minnesota Agricultural Experiment Station. University of Minnesota. 1988.
71. Rodriguez-Saona, L. E.; Wrolstad, R. E.; Pereira, C. *J. Food Sci.* **1997**, *62*, 1001-1005, 1010.
72. Friedman, M. *J. Agric. Food Chem.* **2003**, *51*, 4504-4526.
73. Wicklund, T.; Ostlie, H.; Lothe, O.; Knutsen, S. H.; Brathen, E.; Kita, A. *LWT--Food Sci. Technol.* **2006**, *39*, 571-575.
74. Olsson, K.; Svensson, R.; Roslund, C.A. *ISHS Acta Horticulturae 684*: Meeting of the Physiology Section of the European Association for Potato Research. pp. 159-164. 2005.
75. Zyzak, D. *Presentation at the Acrylamide Food Advisory Committee Meeting*. U. S. Food and Drug Administration, Center for Food Safety and Applied Nutrition. Feb. 24-25.
http://www.cfsan.fda.gov/~dms/acryzyza.html (accessed Mar 11, 2007)

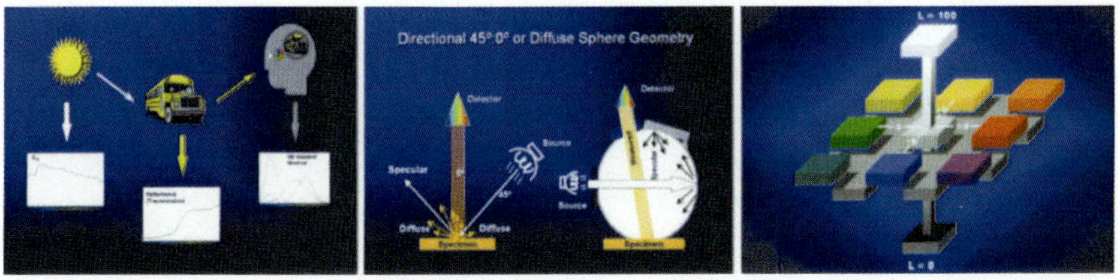

Figure 2.1. The CIE System of color measurement defines a mathematical model of measuring color as a person sees it. Instruments with CIE standardized geometries measure the product in transmittance or reflectance which is then used to calculate L, a*, b* values for all colors.*

Figure 2.2. Transparent materials – solids, liquids, highly absorbing or haze, are measured in transmission on a sphere instrument.

Figure 2.3. With translucent samples, it is necessary to control the thickness of the sample to make it effectively opaque for color measurement.

Figure 2.4. For opaque food products, a key measurement issue making the sample more uniform for a repeatable color measurement.

Figure 2.5. As the sample gets larger and more non-uniform, the approach is to average out the variation in single measurement or as a group.

Figure 3.1. Photographs of strawberries from Oregon State University's North Willamette Experiment Station, 2005 season. Varieties 'Totem', 'Puget Reliance', 'Ovation', and selections ORUS 1723-2, ORUS 2273-1, and ORUS 2384-1. Reproduced with permission from reference 2. Copyright 2007.

*Figure 4.2. Major changes on cherry color during manufacturing of maraschino cherries. *Photograph by Lynn Ketchum, Oregon Agricultural Experiment Station*

Figure 5.1. Cranberry Flower (left) and Fruit (right)

Figure 5.6. Cranberry Color Development

Figure 5.7. Cranberry Fruit Cross-Section

Figure 5.8. Hand and Automated Color Sorting

Figure 5.9. Mechanical water reel and corraled cranberries

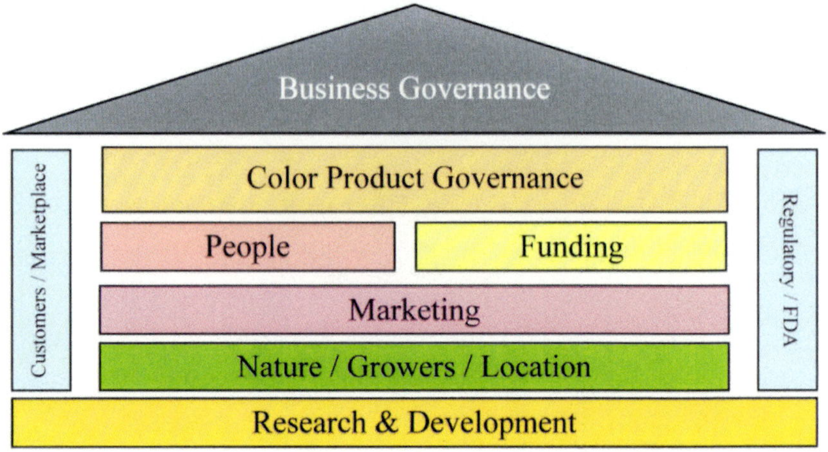

Figure 6.1. Business model for color governance.

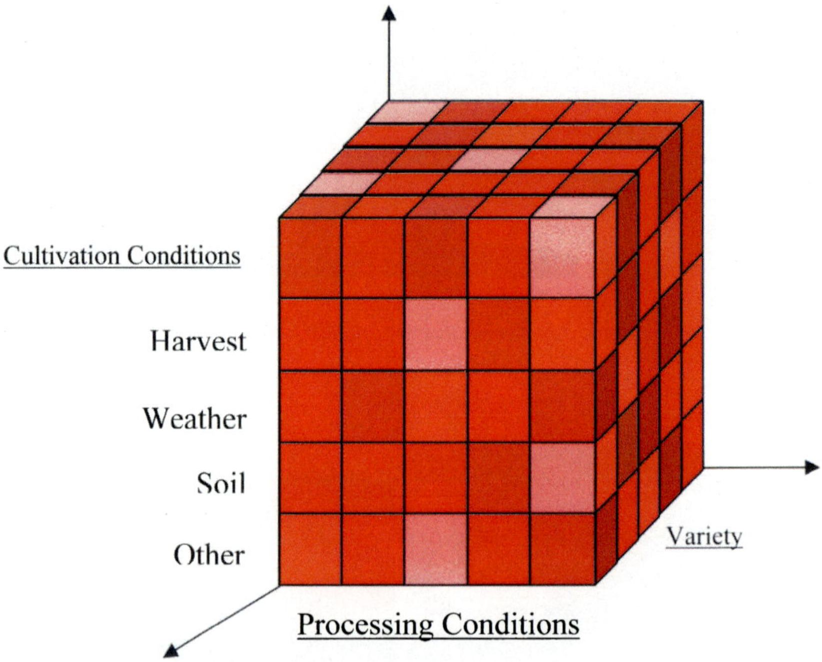

Figure 6.5. Affect of cultivation conditions, processing conditions, and variety on the color shade and color hue.

Plate 6.1. Fruits and vegetables commonly used for the production of natural food colors. (a) Red Cabbage, (b) Black Carrot, (c) Elderberry, and (d) Hibiscus

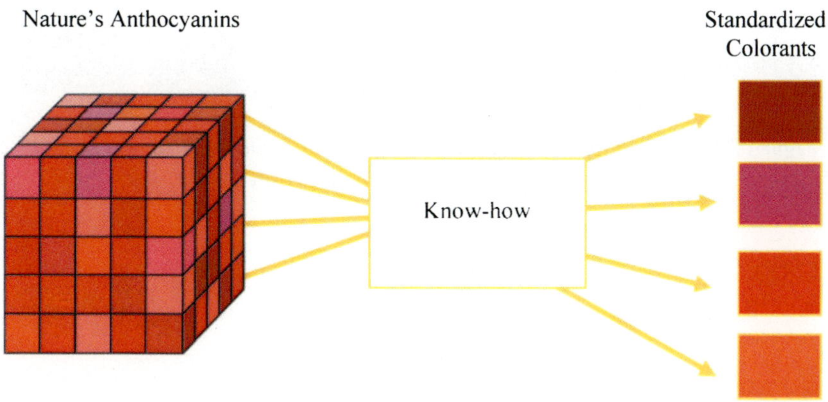

Figure 6.6. Converting nature's color variation into a standardized natural colorant.

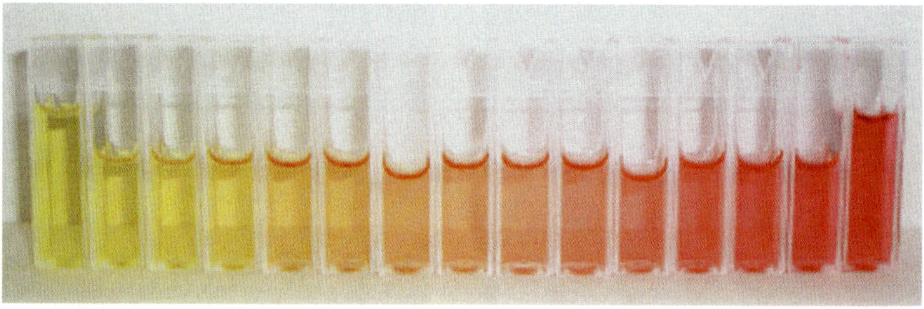

Figure 7.3. Tailor-made color blends from yellow and red beet juice (far left: 100% yellow beet juice; far right: 100% red beet juice)

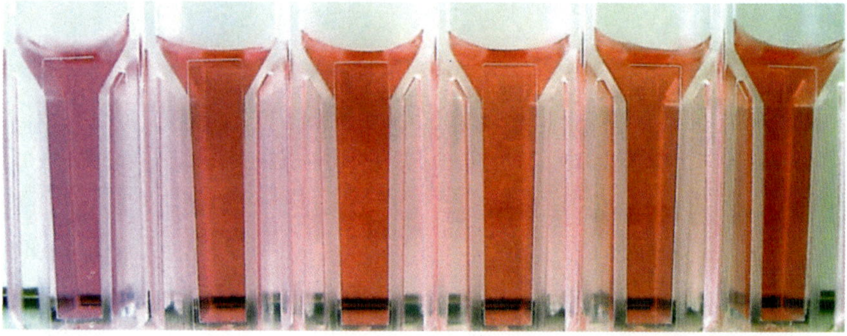

Figure 7.6. Color alterations of purple pitaya juice (most left sample) upon heating for 1, 2 , 3, 4, and 5 hours, respectively

Figure 8.3. Different patterns and degrees of anthocyanin pigmentation in potato. The degree of pigmentation is under polygenic control, while presence and absence of pigment in the flesh is under single gene control.

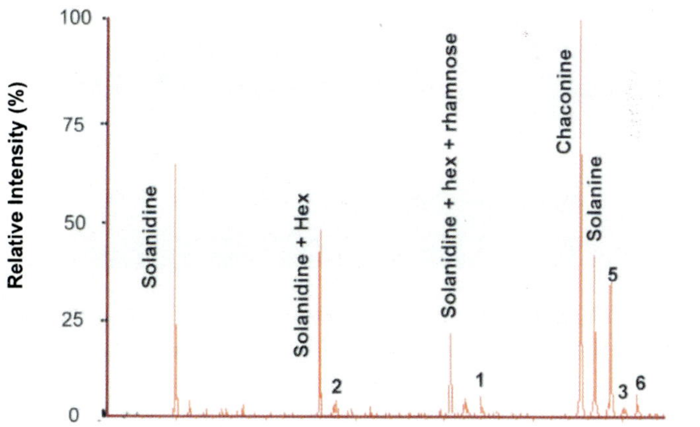

Figure 9.2. Electrospray Mass Spectroscopy of potato pigment extract. The molecular ions correspond to the alkaloids and anthocyanins (identified by numbers) detected in the extract (adapted from Rodriguez-Saona et al. (47)).

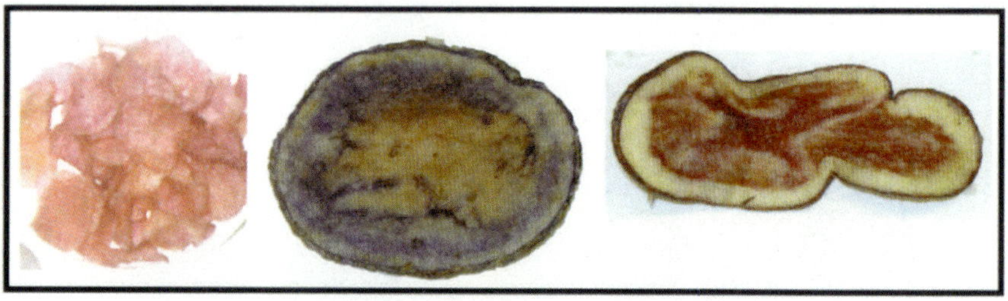

Figure 9.3. Colored Potato Chips showing browning defects

Figure 10.1. Tomato puree in a glass sample cup, placed on the port of a HunterLab LabScan X™ instrument. (Photo courtesy of Gordon Leggett, Hunter Lab.)

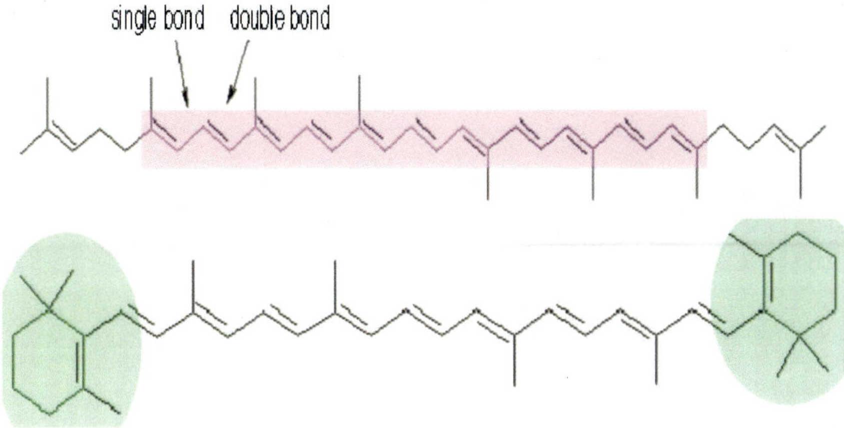

Figure 10.2. Chemical components responsible for tomato color. Lycopene (top) and β-carotene (bottom).

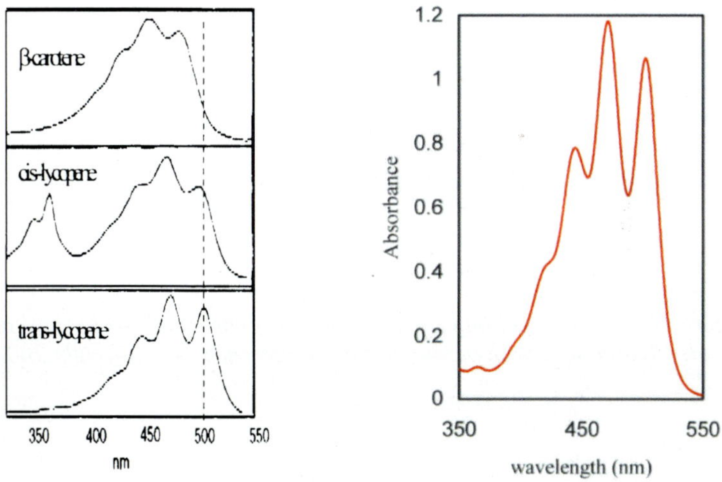

Figure 10.3. The absorbance spectra of β-carotene and cis- and trans- lycopene (left). The dashed line indicates 503 nm. The absorbance spectrum of a hexane extract of tomato (right).

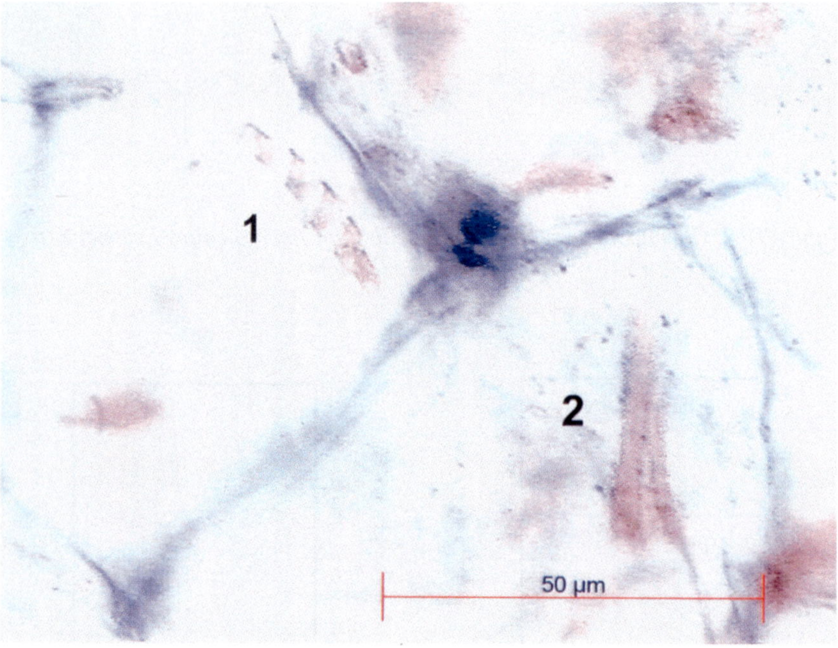

Figure 11.3. Helical [1] and ribbon-shaped [2] chromoplasts in raw carrot roots (Reproduced with permission from reference 17. Copyright 2004.)

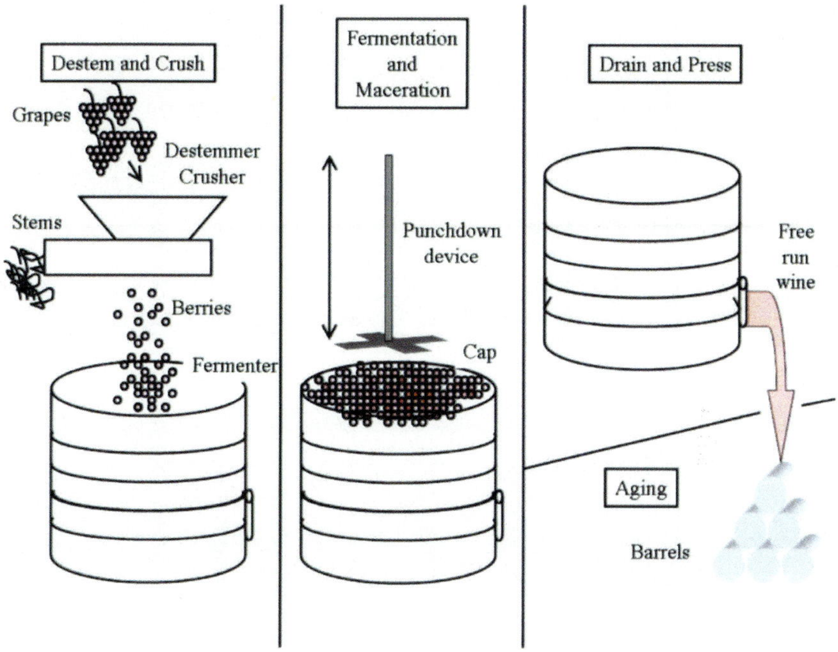

Figure 13.1. General overview of the major steps in red wine production.

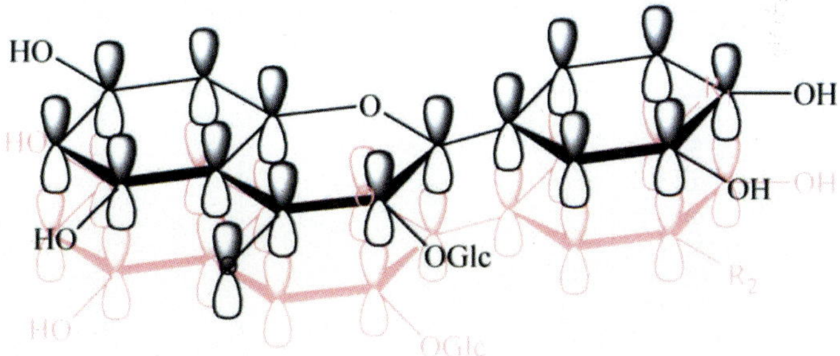

Figure 13.5. Copigmentation between an anthocyanin (gray) and a cofactor (black, flavonol) demonstrating noncovalent interaction between π orbitals.

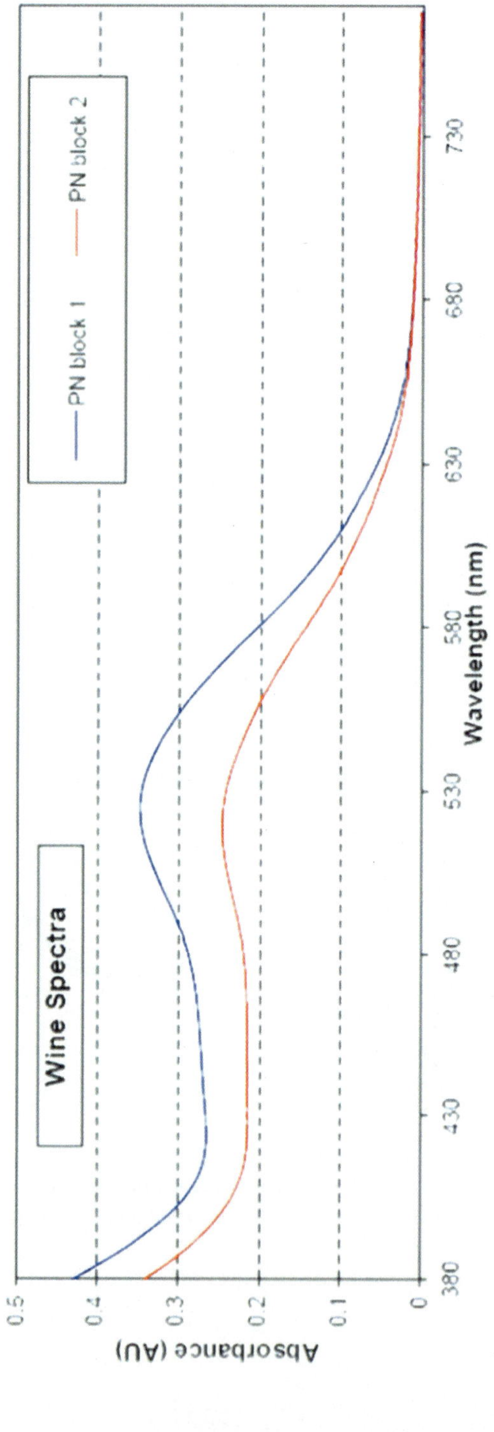

Figure 14.1. Spectral plot of two Pinot Noir wines. (See page 14 of color inserts.)

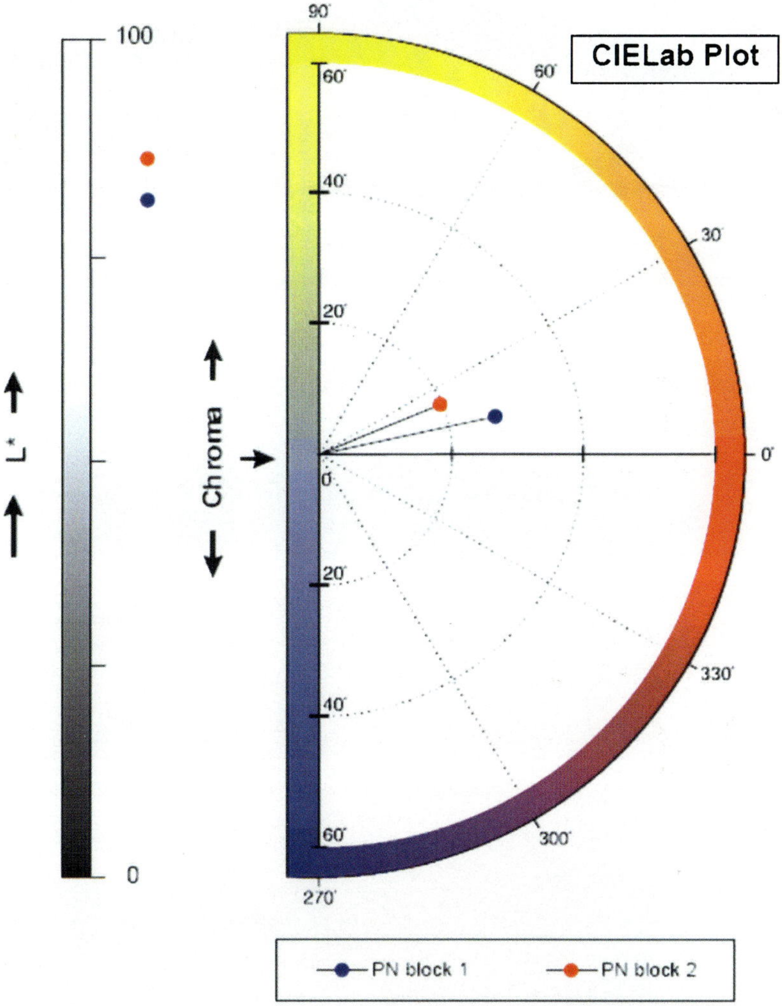

Figure 14.2. Plots of CIELab data L and a* vs. b*.*

Figure 19.1. Farmed Atlantic salmon (Salmo salar L.).

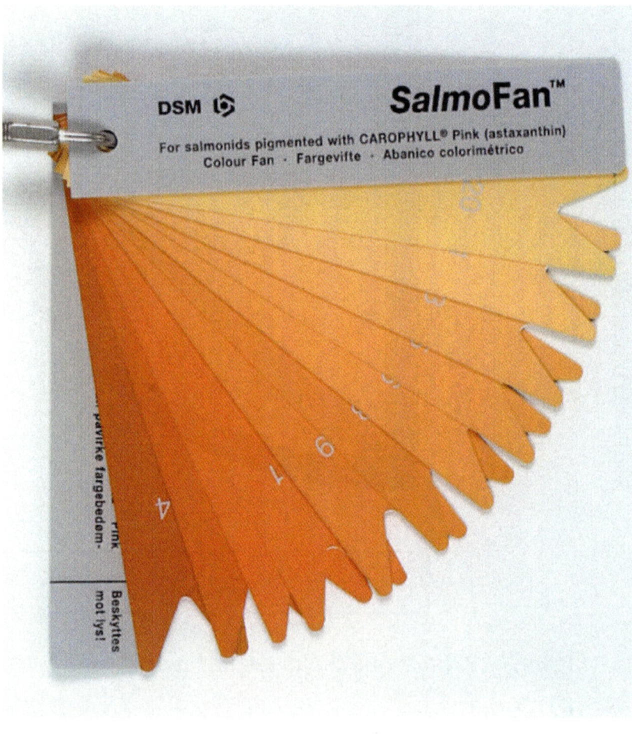

*Figure 19.2. Salmo*Fan[TM] *used for evaluating color of salmon world wide.*

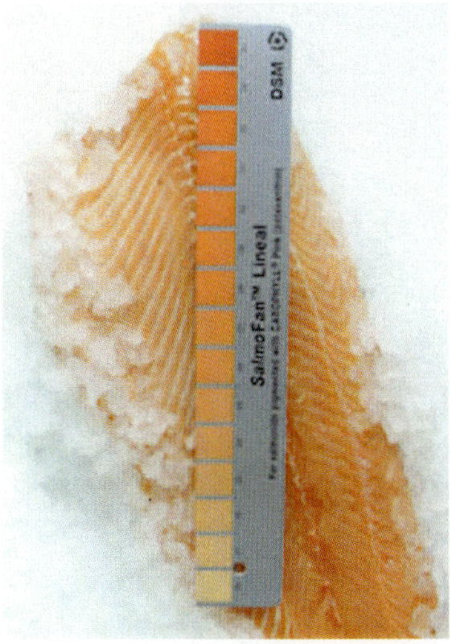

Photo: Annica Thomsson

*Figure 19.3. Salmo*Fan™ *Lineal applied in evaluating salmon fillet color.*

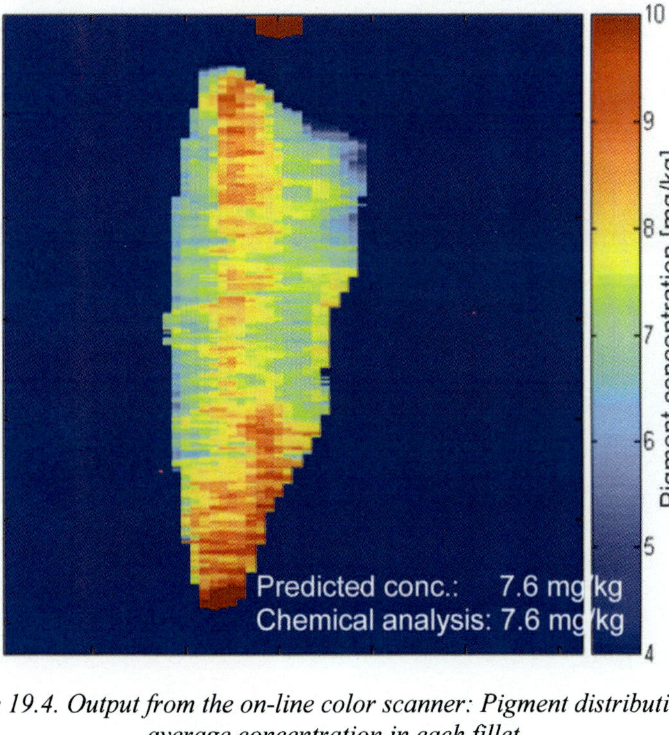

Figure 19.4. Output from the on-line color scanner: Pigment distribution and average concentration in each fillet.

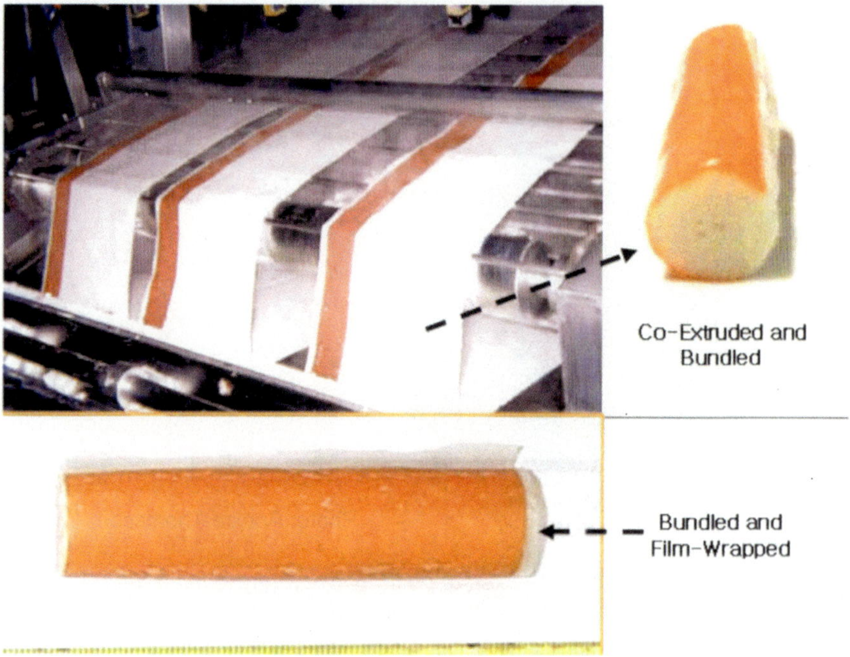

Figure 20.2. Two different methods of color applications.

Figure 22.6. Typical Spanish-style green table olives

Figure 22.1. Changes in the color of olive fruits during ripening. A: green (unripe) stage, B: turning color stage, C: black (ripe) stage.

Figure 22.7. Spanish-style green table olives (Gordal variety) affected by the green-staining alteration.

Figure 22.8. Bottle extra virgin olive oil with its characteristic greenish color.

Figure 23.1. Spanish red pepper (C. annuum) cv. Agridulce cultivated for paprika production at La Vera region (Cáceres, Spain).

Figure 23.5. Spanish paprika from La Vera region (Cáceres, Spain).

Figure 23.6. Paprika oleoresin prepared by solvent extraction from Spanish paprika.

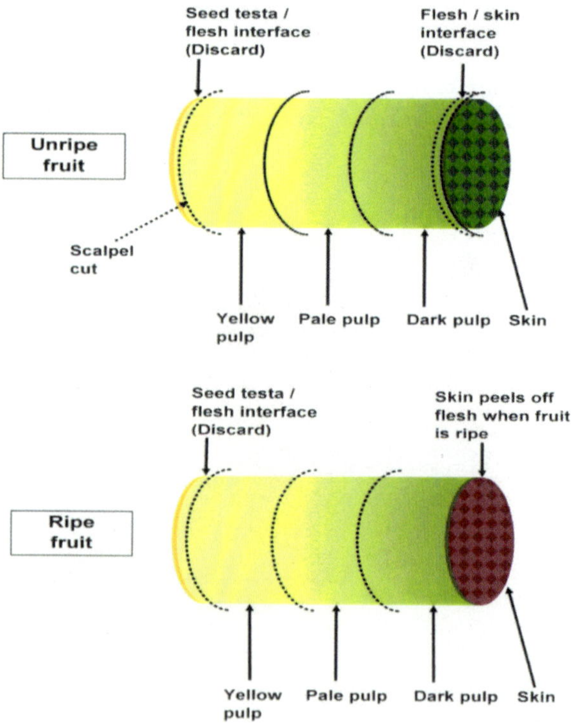

Figure 24.2. Tissue sections from ripe and unripe 'Hass' avocados (Reproduced from reference 17. Copyright 2006 American Chemical Society.)

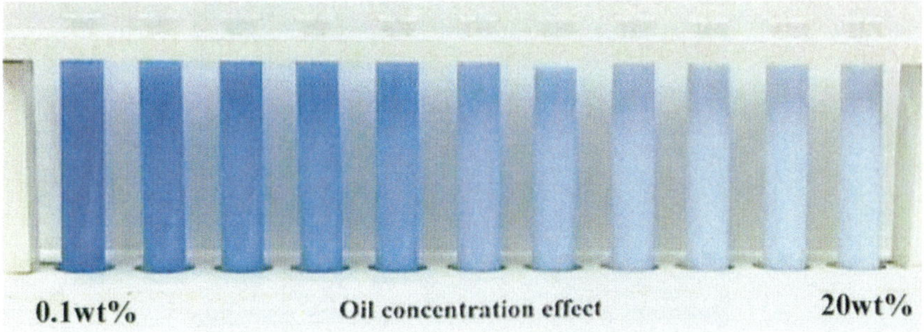

Figure 26.2. A series of n-hexadecane oil-in-water emulsions with different droplet concentrations (0.1 to 20 wt%), but the same mean droplet diameter (d_{32} = 0.3 μm) and dye content (0.005 wt%).

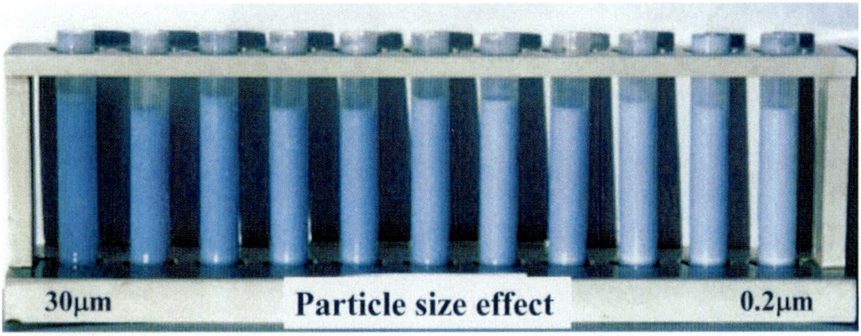

Figure 26.3. A series of n-hexadecane oil-in-water emulsions with the same droplet concentration (9.5 wt%) and dye content (0.005wt%), but different droplet diameters (d_{32} = 0.2 to 30 μm).

Figure 26.6. Two series of 4 wt% n-hexadecane oil-in-water emulsions with the same median droplet diameter (1 μm) but a range of different n_d/n_{aq} ratios (0.97-1.07), in the absence and presence of red dye (0.002 wt%).

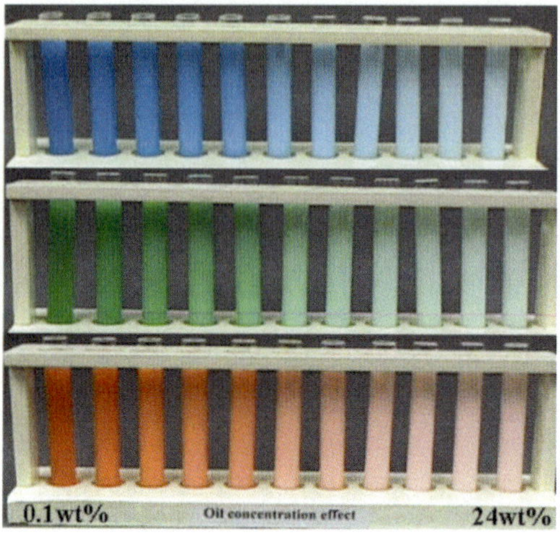

Figure 26.11. Three series of n-hexadecane oil-in-water emulsions with the same mean droplet diameter (0.3 μm), dye concentration (0.1 wt% for red, 0.079 wt% for green, and 0.071 wt% for blue), but different dye type (red, green, or blue) and droplet concentration (0 to 24 wt%).

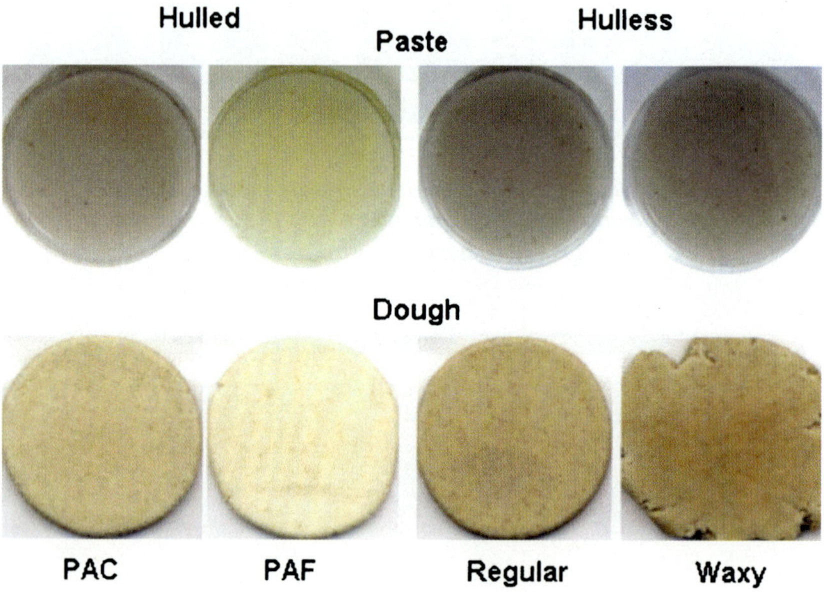

Figure 27.1. Color of pastes and doughs prepared from flours of various types of barley, including hulled, hulless, proanthocyanidin-containing (PAC), proanthocyanidin-free (PAF), regular starch endosperm and waxy endosperm.

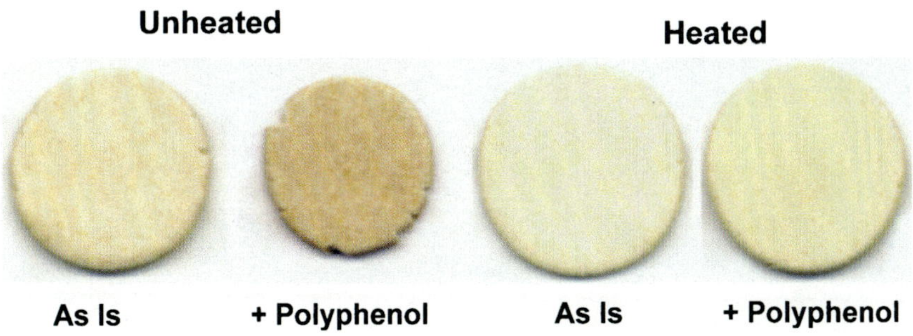

Figure 27.4. Color of doughs prepared from flours of cv. Radiant before or after heat treatment to inactivate polyphenol oxidase and with or without addition of polyphenol extract.

Figure 32.1. From left: 100% orange juice; juice with 5% Mandarin juice; with beta-carotene; and 100% Mandarin juice. (Photographs by Cathy Culver)

Figure 32.2. Beverage base formulated with different commercially-available beta-carotene preparations. (Photographs by Cathy Culver)

Figure 32.3. Beverages prepared with artificial and natural red colors. (Photographs by Cathy Culver).

Figure 32.4. Beverages prepared with 300, 100, and 50 ppm tartrazine (FD&C Yellow #5; Photographs by Cathy Culver)

Figure 32.5. Beverages containing Brilliant Blue and: Allura Red, Ponceau 4R, carmoisine, amaranth, and cochineal. (Photographs by Cathy Culver)

Chapter 10

Color Quality of Tomato Products

Diane M. Barrett and Gordon E. Anthon

Department of Food Science and Technology, University of California, Davis, CA 95616

The vibrant red color of tomatoes is due to the presence of the carotenoid, lycopene. Beta-carotene may also contribute to the color profile, particularly in the case of immature or orange pigmented tomatoes. Tomato color may be determined instrumentally using simple instruments employing filters or light-emitting devices or more complex tri-stimulus colorimeters and spectrophotometers. In California, processing tomato cultivars are evaluated using all of these instruments, and values may prove difficult to correlate. Raw color measurements may be used to predict the color of finished tomato products. The University of California – Davis has for years generated a "soft tomato standard" that is used to calibrate instruments and allows for incorporation of the parameter of translucency. We have attempted to use color measurements to estimate lycopene content and will discuss this and other color-related research.

Why Measure Tomato Color?

The color and appearance of products are the first quality attributes to stimulate us to purchase, consume and enjoy them. Tomatoes are known for their vibrant red color, which indicates not only maturity and therefore level of desired flavor, but also relative content of the beneficial antioxidant lycopene. Tomatoes that are deep red in color, as compared to those that are lighter red or pink, are usually more mature fruit with desirably sweet flavor and a high content of lycopene.

The USDA Processed Products Standards and Quality Certification program has developed color standards showing minimum color for grades "A" and "C" in tomato juice, puree, paste and catsup. Canned tomato color may be judged in accordance with grade "C" or better. Color is such an important quality attribute in tomato products that 30 points of the total 100 in the grade are specifically allocated to the color quality of the fruit.

In this chapter we discuss changes in the measurement of tomato color over time, instrumental measurements of tomato color, industry practices for determining tomato color in California, the chemical components responsible for tomato color and efforts to correlate color measurements to lycopene content in tomatoes.

Changes in Tomato Color Measurement over Time

Prior to 1972, the color acceptability of raw tomatoes for processing was determined visually, and state inspectors at receiving stations compared fruit of questionable minimum color to standardized U.S. Department of Agriculture color discs (*1*). The United States Department of Agriculture recently designed bi-color vinyl tiles to be used in place of the original color discs for visual determination of the minimum color allowed for different grades of products.

While sensory evaluation is the optimal means of determining color, individuals differ in their assessment of color, and for that reason sensory evaluation may be both time-consuming and challenging to quantify. In many cases instrumental measurements of color may be used to approximate sensory color determination. In 1972, the California tomato industry decided that a system of color determination was needed in which human judgment was at an absolute minimum. A research project carried out by the University of California and the California Department of Food and Agriculture developed a method for evaluating tomato color using an Agtron E-5M instrument.

Instrumental Measurements of Tomato Color

Instrumental tomato color measurements were originally made using a spinning disk in a light booth. Maxwell's spinning disk was one of the first devices used for a semi-quantitative determination of the color of various foods.

Evaluation of Color Measuring Instrumentation for USDA Processed Tomato Scores

In the late 1970's, the USDA and the University of California – Davis worked with the California tomato industry and instrument suppliers to correlate visual scoring of tomato product quality to instrumental color measurements. A report published in 1980 (*2*) established equations for correlation of color measured on Gardner, Agtron and Hunter (HunterLab, Reston, VA) instruments, as compared to the D25 A™ from HunterLab, which was housed at UC Davis. The D25 A™ was considered by the US Department of Agriculture to be the 'gold standard' or standard instrument for measuring tomato color. All other instruments were referenced to this one. A "C" illuminant was used with a 2° observer and directional 45/0 conditions. All samples were measured through the bottom of a glass sample cup in reflectance mode (Figure 3).

Figure 1. Tomato puree in a glass sample cup, placed on the port of a HunterLab LabScan X™ instrument. (Photo courtesy of Gordon Leggett, Hunter Lab.) (See page 10 of color inserts.)

The following equations were generated for tomato paste and puree score (TPS), sauce (TSS), catsup (TCS) and juice (TJS) products using the HunterLab D25 A™:

Tomato Paste and Puree = TPS = $-46.383 + 1.0211(a) + 10.607(b) - 0.42198(b^2)$
Tomato Sauce = TSS = $-154.39 + 1.1142(a) + 22.596(b) - 0.86736(b^2)$
Tomato Catsup = TCS = $-74.937 + 7.5172(a) - 0.1278(a^2) - 0.8051(b)$
Tomato Juice = TJS = $29.6000 + 0.88354(a) - 1.8553(b)$

Since this study was published in 1980, most of the instruments used for measuring color have been replaced by models that are more sensitive, accurate and robust. In addition, the HunterLab D-25 A™ instrument originally used as the 'gold standard' for establishing tomato scores was replaced at the University of California – Davis by a Hunter Lab Scan 5100™. Therefore, the original study was repeated for Hunter Lab and BYK Gardner instruments and approved by the USDA in 2003 (*3*). Equations to correlate the Minolta CR-410 to the Hunter LabScan 5100 are currently being developed.

A two-part study was conducted to compare color measuring capabilities of five different color measuring devices (ColorFlex™, LabScan XE ™ and D25 A™ from HunterLab; and Color Guide and Color View from BYK Gardner) to the UC Davis Reference LabScan 5100™. The goals of Part I were (1) to establish sample cup variability, (2) to establish measurement variability for a single instrument and (3) to generate data to determine USDA Processed Tomato Scores using the different color measuring instruments and 30 different samples each of tomato juice, sauce/puree, catsup and paste diluted to 8.5° Brix.

Part II of the study was a more comprehensive evaluation based on Part I, using more samples and replicate instruments. The goals of Part II were (1) to establish measurement variability, (2) to generate data to validate tomato scores using ten different color measuring instruments and fifty different samples each of tomato juice, sauce/puree, catsup and paste diluted to 8.5° Brix and (3) to establish instrument variability. For this part of the study, two ColorFlex™, two LabScan XE™, two Color Guide and two Color View color measuring devices were used in addition to the D25™ and LabScan 5100™.

All color measuring instruments except the Color Guides and D25 A™, which were used independently of computer, were interfaced to IBM-compatible computers. These consisted of an optical sensor and used a directional orientation of 45°/0°, except for the D25 A™, which was not interfaced to a computer and uses a directional orientation of 0°/45. The illuminant used for the color study was CIE illuminant C. The standard observer was 1931 2° Standard Observer and the color scale was Hunter L, a, b (Hunter) or L*a*b* (BYK Gardner).

Equations for TPS, TSS, TCS and TJS were determined statistically to correlate the LabScan XE™, ColorFlex 4500L™, re-qualified D25A-9000™, Color View and Color Guide back to the USDA/UC – Davis Reference LabScan 5100™ (serial number 12379) in the measurement of Processed Tomato Scores. As an example, the new equations to be used for tomato paste scores by the various instruments are as follows:

UC Davis Reference LabScan 5100 TPS $= -46.383 + 1.0211(a) + 10.607(b) - 0.42198(b^2)$

HunterLab LabScan XE TPS $= -40.926 + 1.061(a) + 9.473(b) - 0.376(b^2)$

HunterLab ColorFlex TPS $= -81.582 + 1.069(a) + 15.390(b) - 0.591(b^2)$

Hunter D25A-9000 $= -58.296 + 1.093(a) + 12.120(b) - 0.480(b^2)$

BYK Color Guide $= -304.741 + 1.134(a) + 46.595(b) - 1.687(b^2)$

BYK Color View $= -2.63270 + 13.822(\log b) - 7.442(\log L) + 0.0234(a) - 1.002(b) + 0.295(L)$

Similar equations were generated and validated for tomato sauce, catsup and juice. Following completion of the study, data and statistical analysis was submitted to the USDA for review. The HunterLab LabScan XE™ and Color Flex™ were approved, and the D25A-9000 was accepted by USDA for use in the measurement of processed tomato paste, sauce, catsup and juice color. The BYK Color View was also approved for all of the tomato products, while the Color Guide was only approved for measurement of tomato juice and sauce.

California Processing Industry Practices

Processing tomatoes comprise the largest volume of vegetable harvested in the state, typically 10 to 11 million tons annually. Color is an important quality attribute and is therefore determined both in the raw fruit and in the processed product.

Raw tomato inspection

Raw tomato color is determined by the Processing Tomato Advisory Board (PTAB), a neutral third party inspection agency, on every load of tomatoes harvested. Growers receive a financial incentive for producing tomatoes that are of good character, color and flavor, and defect free. California is the only state in the U.S. that does not rely on U.S.D.A. inspectors to carry out grade inspection. Two fifty pound samples of tomatoes are randomly taken from each 25,000 pound truckload of tomatoes.

In the late 1970's, the Agtron E-5M instrument, which used green and red filters, was used to evaluate color on a homogenized sample. In 1996, PTAB started using an instrument that utilized light emitting devices or LEDs, developed by Dr. David Slaughter, Dept. of Biological and Agricultural Engineering at UC Davis. The LED instrument employs an array of green and red LEDs to illuminate tomato juice and the ratio of green to red reflected light is measured. Although this instrument is rugged and has functioned well in raw tomato color measurement for the industry for 10 years, processed product color is most often measured by Quality Control departments using L, a and b values and PTAB is considering moving to this scale. This would allow growers, processors and inspection agencies to use the same color scale for measurement of both the raw material and finished tomato product.

Finished tomato product color

The California tomato processing industry uses USDA 'approved' instruments (see discussion above) for measurement of processed tomato paste, sauce, catsup and juice color. Instruments are standardized against white and black tiles, and a green tile supplied by the factory is used as a further check, using illuminant D65, 10° observer and either the XYZ (Hunter) or CIELAB (BYK Gardner) scale. The illuminant used for the color study was CIE illuminant C. Color measurement is made using the 1931 2° Standard Observer and the color scale used is Hunter L, a, b (Hunter) or L*a*b* (BYK Gardner).

During the tomato processing season, sample bags or containers are pulled from the processing line at least hourly and read in duplicate. Processors have their own particular specifications which define whether color is out of the acceptable range or not. Color quality may be inadequate due to either poor raw material color or excessive thermal exposure. The raw fruit color may suffer if immature tomatoes are utilized, or if tomatoes have been sunburned. During the tomato paste process, there is opportunity for non-enzymatic browning to occur during the hot break or concentration steps. Likewise, tomato juice may have undergone browning reactions if it is exposed to high heat.

Use of the University of California - Davis Soft Tomato Standard

Reflectance color measurements, while useful, are not able to measure all of the properties of translucent samples such as tomato sauce, juice and paste. Light emitted by an instrument not only reflects back, but also penetrates into the product to a certain degree, bounces around and then is reflected back into the instrument. In order to standardize instruments for use with translucent samples, the University of California – Davis developed a 'soft tomato standard' to address the issue of translucency. These standards are produced on an annual basis under standard processing conditions. A sub-sample is evaluated using the 'gold standard' instrument. It is necessary to produce the standard each year because the color degrades with storage time in the can. Each year, the industry need for soft standard is determined, that amount of tomato sauce is canned and a statistically valid sub-set of the canned soft is measured using the 'gold standard' instrument. Cans of soft standard are labeled with the L, a and b values obtained with that instrument and distributed to the California tomato processing industry.

Tomato Industry Interest in Measuring Lycopene

Processing tomato industry interest in the measurement of lycopene, in addition to color, began in the early to mid 1990's. At that time scientific research indicated the benefits of lycopene consumption in preventing both cardiovascular disease and cancer. Tomatoes are the primary source of lycopene in the American diet, so the fresh and processing tomato industries took note.

During the period from 1999 to 2005, the California tomato processing industry funded research to begin determining lycopene content of new tomato varieties, to evaluate the range of lycopene content in these varieties grown in different counties throughout California *(4)*, and to develop a standard rapid method for determining lycopene.

Chemical Components Responsible for Tomato Color

In mature red tomatoes, lycopene comprises ninety percent of the pigment responsible for the red color. Small amounts (<5%) of ß-carotene exist as well. Lycopene is a 40 carbon molecule with alternating single and double bonds (Figure 1). Structurally similar to lycopene, ß-carotene also contains 40 carbons, but has a 6 carbon ring structure at each end of the molecule, with a 28 carbon straight chain between them.

The concentration of lycopene in a solvent extract from tomatoes can be determined either spectrophotometrically or by high performance liquid chromatography (HPLC). Figure 2 illustrates the absorbance spectrum for *cis-* and *trans-*lycopene, as well as β-carotene, as adapted from Ishida et al., 2001 (*5*). Lycopene has three peaks of absorbance at 444, 471 and 503 nm. This third peak is at a wavelength where the absorbance of β-carotene (and other minor carotenoids) have relatively low absorbance and thus cause very little interference. Spectrophotometric measurements of absorbance at 503 nm are thus very good estimates of the lycopene content.

We have developed a standard method for lycopene determination in mature red processing tomatoes for the California processing tomato industry (*6,7*). This method, based on the solvent extraction procedure developed by Sadler et al. (*8*), involves adding a mixture of hexane:ethanol:acetone at 2:1:1 to an aliquot of homogenized tomato juice, mixing and waiting for 10 min to several hours, then adding water, mixing and letting the sample stand to separate the polar and non-polar phases. The carotenoids remain in the upper hexane layer, while the cellular debris is separated into the ethanol:water:acetone layer below.

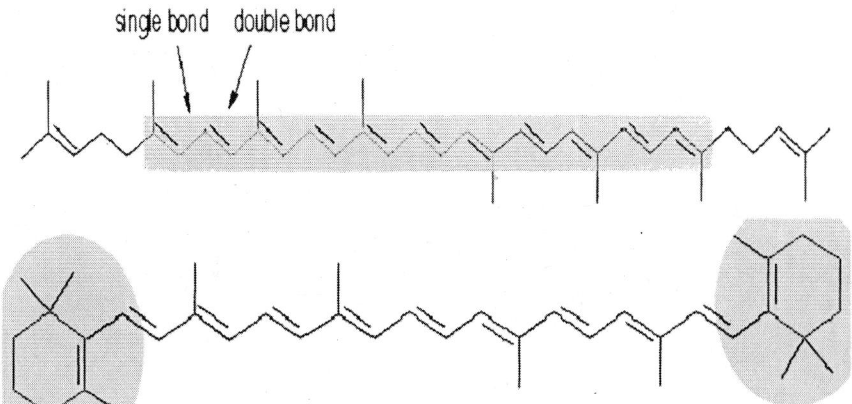

Figure 2. Chemical components responsible for tomato color. Lycopene (top) and β-carotene (bottom). (See page 11 of color inserts.)

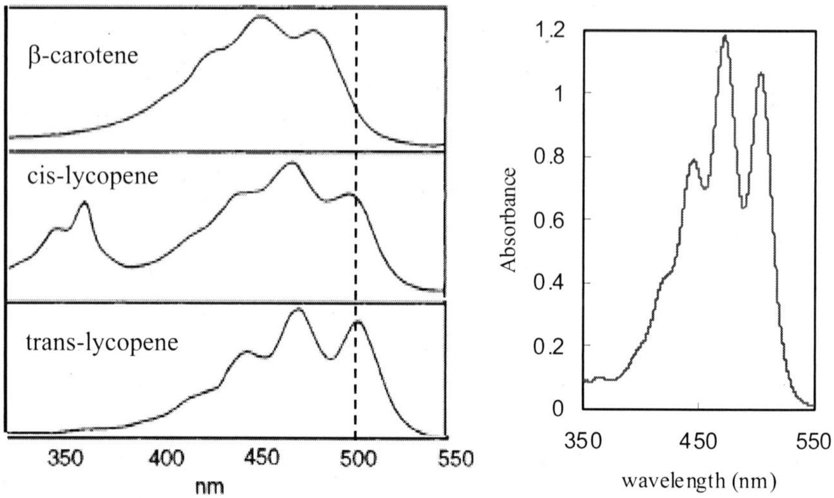

Figure 3. The absorbance spectra of β-carotene and cis- and trans- lycopene (left). The dashed line indicates 503 nm. The absorbance spectrum of a hexane extract of tomato (right). (See page 11 of color inserts.)

The hexane layer is removed and absorbance at 503 nm is read as a measurement of lycopene content.

Correlation of Color Measurements to Lycopene Content in Tomatoes

Determination of lycopene using a wet chemistry approach, e.g. by extraction and measurement using a spectrophotometer, is often not possible for tomato breeders, growers, processors and other parties. The spectrophotometric method standardized in our laboratory, while time-saving in comparison to HPLC procedures, still requires capital investment in an instrument, and a skilled analyst to carry out the measurement. For this reason, we have attempted for a number of years to develop a correlation between L, a and b values measured by a colorimeter and chemical measurements of lycopene.

Our early efforts to correlate lycopene content to L, a, b, hue, chroma, value, ΔE and other common colorimetric measurements were unsuccessful. More recently, we have undertaken evaluations of the Hunter LabScan™ and the Hunter UltraScan XE™ for their ability to predict lycopene levels in tomatoes (*6, 7*). Using the Hunter LabScan™ with a tomato juice sample diluted 1:10 with water, we obtained a roughly linear correlation between the a/b value and the lycopene content of the juice samples, determined by spectrophotometric measurements of a hexane extract (Figure 4).

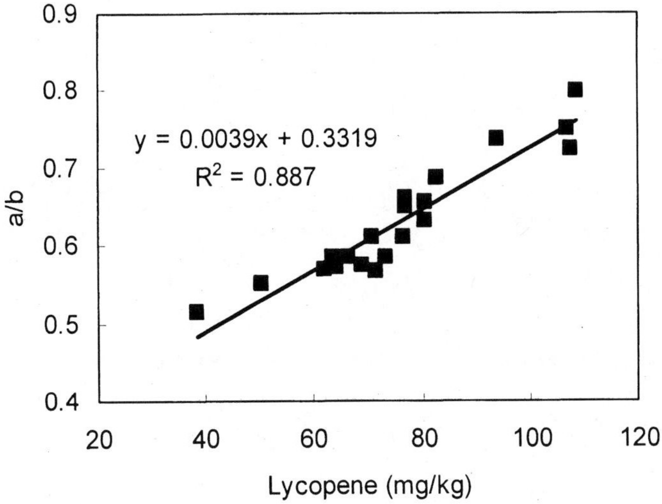

Figure 4. Relationship between the a/b ratio of a 10-fold diluted juice sample and the lycopene content of the tomato juice. The lycopene values given are for the undiluted juice.

References

1. O'Brien, M; Leonard, S.J; G.L. Marsh. Color determination for loads of processing tomatoes. *Trans. ASAE* **1975**, *18*, 745-747.
2. Marsh, G.; Buhlert, J.; Leonard, S.; Wolcott, T.; Heil, J. 1980. Color scoring tomato color measurement. Report to the United States Department of Agriculture. 30 pp. Unpublished.
3. Barrett, D.M. 2003. Protocol for evaluation of color measuring instrumentation for USDA processed tomato scores. Report to the United States Department of Agriculture. 19 pp. Unpublished.
4. Barrett, D.M. and G.Anthon. Lycopene content of California-grown tomato varieties. *Acta Horticulturae* **2001**, *542*, 165-174.
5. Ishida, B.K.; Ma, J.; Chan, B. A simple, rapid method for HPLC analysis of lycopene isomers. *Phytochem. Anal.* **2001**, *12*, 194-198.
6. Anthon, G.E.; D.M. Barrett. 2004. Lycopene method standardization. Report to the California League of Food Processors. 17 pp. *Unpublished.*
7. Anthon, G.E.; D.M. Barrett. 2005. Lycopene method standardization. Report to the California League of Food Processors. 18 pp. *Unpublished.*
8. Sadler, G.; Davis, J.; Dezman, D. Rapid extraction of lycopene and β-carotene from reconstituted tomato paste and pink grapefruit homogenates. *J. Food Sci.* **1990**, *55*, 1460-1461.

Chapter 11

Stability of Carotenoids in Vegetables, Fruits, Functional Foods, and Dietary Supplements with Particular Reference to *trans-cis*-Isomerization

Andreas Schieber and Reinhold Carle

Institute of Food Technology, Chair Plant Foodstuff Technology, Hohenheim University, August-von-Hartmann-Strasse 3, D–70599 Stuttgart, Germany

Carotenoids are widespread plant pigments occurring predominantly in their all-*trans* configuration. Processing and storage, especially heat treatment, light exposure, and the presence of photosensitizers may lead to the formation of carotenoid *cis*-isomers which exhibit different physical, chemical and biological properties. Despite a large number of studies carried out during the past two decades, the physiological relevance of *cis*-isomers has not completely been understood. This review summarizes our recent investigations on the effects of processing on carotenoid stability in vegetables, fruits, functional foods, and dietary supplements. Particular attention was given to methods for the determination of carotenoid stereoisomers and to the role of the physical state of carotenoids and the plant matrix for their stability.

Carotenoids are widespread pigments which are biosynthesized by higher plants (*1*), algae, fungi, and bacteria (*2, 3, 4*). Their presence in animal tissues, e.g. egg yolk, bird feathers, and the exoskeleton of invertebrates is attributed to ingestion via the food chain followed by accumulation in these tissues. In plants, carotenoids are accumulated and sequestered in chloroplasts and chromoplasts (*5, 6*). Carotenoids are broadly classified into carotenes, which are strict hydrocarbons, and the more polar xanthophylls (or oxycarotenoids), which contain one or more oxygen atoms. Prominent members of the carotenes are α-carotene, β-carotene and lycopene, whereas the most important xanthophylls are zeaxanthin, lutein, β-cryptoxanthin, astaxanthin, and canthaxanthin. Carotenoids containing a hydroxyl group may be esterified with fatty acids. The structures of selected carotenoids are shown in Figure 1. Among the 600 carotenoids known so far, only approximately 50 are active as provitamin A precursors, the most important and most efficient being β-carotene with its two β-ionone rings.

Carotenoids predominantly occur in their all-*trans* configuration. However, mainly due to the development of novel C_{30} stationary phases and advances in hyphenation techniques, in particular LC-MS and LC-NMR spectroscopy, there is abundant proof for the presence of *cis*-isomers in fruits (*7*), vegetables and physiological samples (*8, 9, 10*). Apart from naturally occurring in foods, heat treatment, light exposure, and the presence of triplet sensitizers and electrophilic compounds are the main reasons for the formation of *cis*-isomers. Exposure to light predominantly leads to the formation of 9-*cis*-β-carotene, whereas 13-*cis*-β-carotene is mainly formed by thermal treatment. Their structures are shown in Figure 2. The *cis*-isomers possess different physicochemical properties, such as lower melting points and increased solubility in non-polar solvents compared to their all-*trans* counterparts. Furthermore, a new maximum often referred to as '*cis*-peak' is observed in the UV spectrum. Since *trans-cis*-isomerization is accompanied by a hypsochromic shift in the λ_{max} and smaller extinction coefficients, this transformation may also lead to a decreased color intensity. It is also well known that isomerization results in a decreased provitamin A activity of carotenoids containing a β-ionone ring. However, the physiological relevance of the presence of *cis*-isomers in the diet is not completely understood. For a more comprehensive treatise also including basic principles and consequences of *trans-cis*-isomerization, we refer to a recent review by Schieber and Carle (*11*). The objective of this contribution is to summarize our studies on carotenoid stereoisomer analysis and the effects of processing on the stability of carotenoids.

Carotene Stereoisomers in Fortified Drinks

Fortified drinks containing ascorbic acid (vitamin C), tocopherols (vitamin E), and β-carotene (provitamin A), hereinafter referred to as ATBC drinks, have

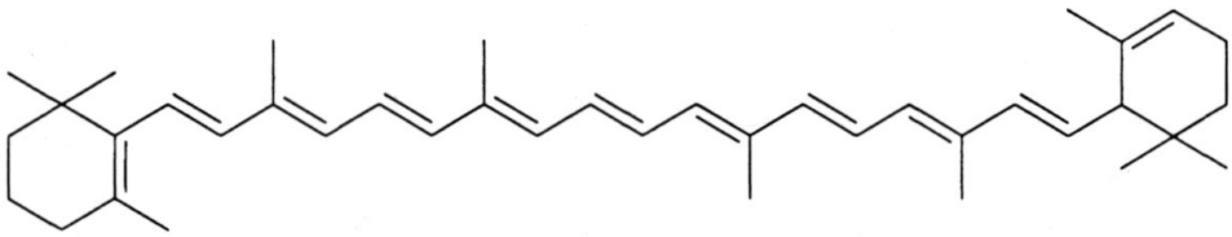

all-*trans*-α-carotene

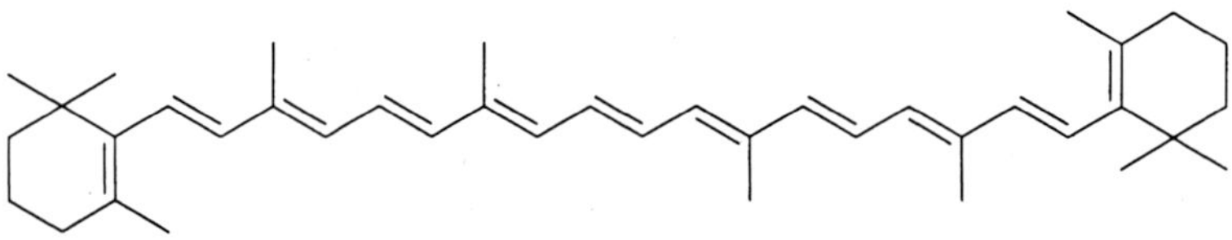

all-*trans*-β-carotene

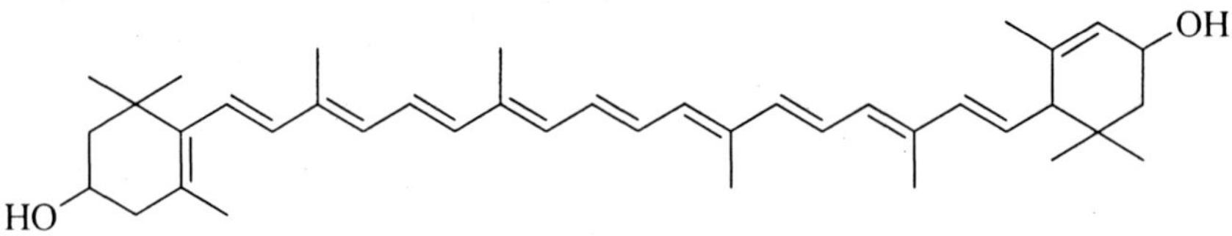

all-*trans*-lutein

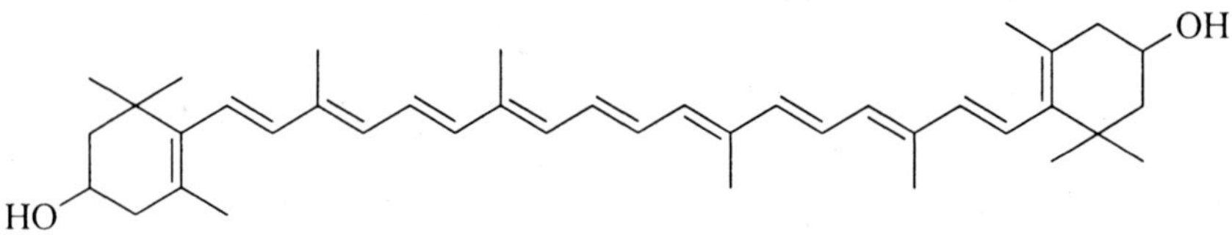

all-*trans*-zeaxanthin

Figure 1. Structures of selected carotenes and xanthophylls

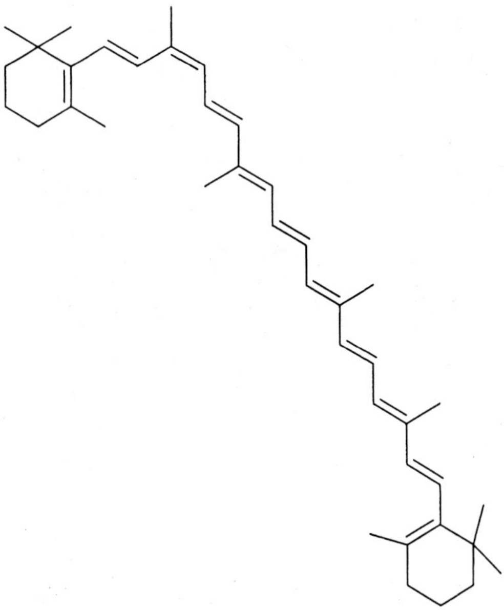

9-*cis*-β-carotene

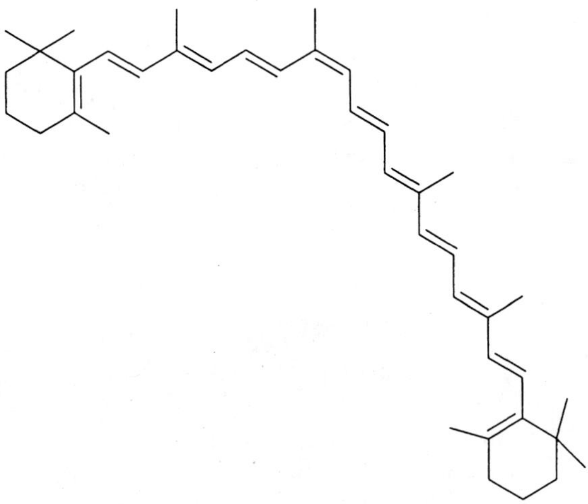

13-*cis*-β-carotene

Figure 2. Structure of 9-cis- and 13-cis-β-carotene

experienced growing popularity. The provitamin A moiety is usually provided as synthetic β-carotene but may also originate from carrot juice as a natural source of carotenes. Prior to the investigation of commercial carrot juices and ATBC drinks, we have established an efficient method for the separation of the most important carotene stereoisomers including all-*trans*-α-, all-*trans*-β-, 9-*cis*-β-, and 13-*cis*-β-carotene. Since lipids co-extracted from carrot juices did not interfere with the compounds of interest, a saponification step could be omitted (*12*). Later, the analytical system was extended to the simultaneous extraction and determination of carotene stereoisomers and tocopherols. Remarkably, the separation of δ-, γ-, β-, and α-tocopherol as well as α-tocopheryl acetate could be accomplished without compromising the resolution of the carotene stereoisomers (*13*).

α-Carotene could be detected only in those ATBC drinks containing carrot juice as the carotene source. Their relative amounts of *cis*-isomers, calculated as percentage of all-*trans*-β-carotene, ranged from 6.7 % to 13.6 %, with 13-*cis*-β-carotene being the predominant *cis*-isomer (1.1-2.5 mg/l). In contrast, extraordinarily high relative amounts (31.8-44.5 %) were found in those samples with added synthetic β-carotene (*12*). These findings may be explained by the particular technology used for the production of the the ATBC basic material. Usually, commercial β-carotene preparations containing micro-crystalline all-*trans*-β-carotene, together with vitamin E and weighting agents (e.g. sucrose acetate-isobutyrate), are dissolved in a hot mixture of lipophilic solvents. The lipophilic phase is finely dispersed by homogenization in the aqueous phase, which contains a hydrocolloid solution, a syrup or fruit juice concentrate, and antioxidants (*14*). Since isomerization does not appear with crystalline carotene, hot dissolution of all-*trans*-β-carotene is considered the crucial step leading to the formation of *cis*-isomers. Excessive heating may also cause de-esterification of α-tocopheryl acetate, which is comparatively stable towards oxidative degradation and therefore frequently used in supplements (*13*).

Effects of Processing on *trans-cis*-Isomerization of β-Carotene in Carrot Juices

Whereas ATBC drinks represent a comparatively simple matrix, a variety of parameters affect the chemical stability of carotenes in carrot juices. Our preliminary investigations on the presence of carotene stereoisomers in commercial carrot juices had revealed that the contents of all-*trans*-α-carotene ranged from approximately 20 mg/l to 50 mg/l and those of all-*trans*-β-carotene from 33 mg/l to 85 mg/l. With respect to *cis*-isomers, up to 4 mg/l and 10 mg/l of 9-*cis*- and 13-*cis*-β-carotene were found, with relative amounts ranging from

4.0 % to 16.2 %, respectively (*12*). The *cis*-isomers of α-carotene were not included in this study. Since *cis*-isomers could not be detected in raw carrot roots, the effects of processing on *trans-cis*-isomerization of β-carotene during carrot juice production were investigated in detail. For this purpose, carrot juices were produced on pilot-plant scale according to a standard process (*15*) which included blanching, comminution, and dejuicing using a decanter. The juices were either sterilized or acidified and pasteurized, and finally packaged aseptically. Time-temperature regimes for blanching and sterilization were varied. Another modification of the basic process consisted in the addition of grape seed oil to the coarse mash to investigate the effects of solubilization on carotene stability (*16*).

Unheated juices produced from carrots blanched at 80°C for 10 min were devoid of *cis*-isomers, and heat preservation during the standard process resulted in only weak isomerization (2-5 %). Only extensive blanching at time-temperature regimes not applied during production of carrot juice (80-100 °C for 30-60 min) led to the formation of 13-*cis*-β-carotene at levels between 1.8 and 10 %. Microscopic investigations revealed that raw and moderately blanched carrot roots contained helical and ribbon-like chromoplasts (Figure 3) (*17*), while excessive blanching caused the formation of yellow-colored lipid droplets containing solubilized carotenes. In pasteurized juices slightly higher levels of 13-*cis*-β-carotene were found compared to unheated juices. Sterilization had an additive effect on isomerization and resulted in an average increase in 13-*cis*-β-carotene of approximately 5 %. Both extended sterilization at 130 °C and the addition of grape seed oil to the coarse mash followed by heat preservation led to enhanced isomerization and caused formation also of 9-*cis*-β-carotene. UV/Vis spectroscopic investigations of carotene-containing particles obtained by density-gradient centrifugation of carrot juice demonstrated the presence of crystalline carotene. It is assumed that in carrot juice crystalline carotene is suspended and covered by polar lipids. Hydrocolloids of protein and/or polysaccharide nature may protect carotenes from being dissolved in neutral lipids during thermal preservation. However, it appears that at temperatures exceeding 120 °C a significant destruction of the protective matrix occurred which caused dissolution and isomerization of carotenes (*16*).

Isomerization of β-Carotene in Mango Fruits

Whereas carrots are important sources of provitamin A in countries of the western hemisphere, mango fruits substantially contribute to the β-carotene supply in tropical countries. Although they are mainly consumed as fresh fruits, the economical importance of processed mango products such as canned mango

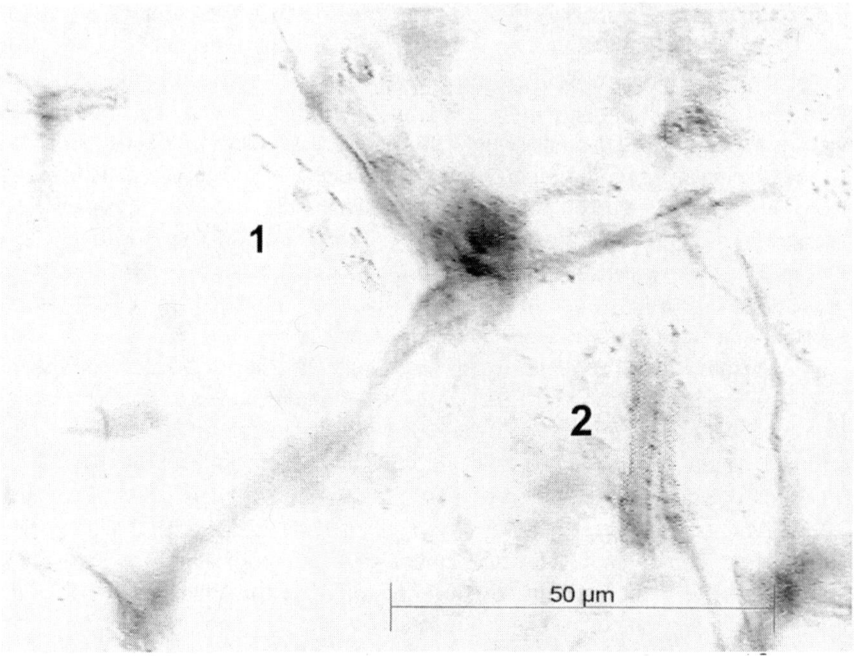

Figure 3. Helical [1] and ribbon-shaped [2] chromoplasts in raw carrot roots (Reproduced with permission from reference 17. Copyright 2004.) (See page 12 of color inserts.)

slices, purée, and nectar has considerably increased. Furthermore, dehydrated products like mango fruit bars and slices represent an efficient means to improve the pro-vitamin A status of the population especially during off season. Two processes for the dehydration of mango fruits were assessed for their effects on *trans-cis*-isomerization of β-carotene. The first process was performed using an overflow dryer which was operated for 3-3.5 h at 75 °C and an air velocity of 1 m/s. In the second process, the mangos were dried in a solar tunnel dryer at 60-62 °C for 7-8 h. Surprisingly, *cis*-isomers of β-carotene could be detected even in fresh mango fruits at relative contents of 19-27 %, which increased to 37 % upon thermal drying. In those samples dried in the solar tunnel dryer relative contents ranging from 51 % to 64 % were found. The profile of *cis*-isomers was clearly dependent on the process technology used for mango drying. Whereas 13-*cis*-β-carotene was the predominant *cis*-isomer formed after thermal treatment, solar tunnel drying mainly caused the formation of 9-*cis*-β-carotene (*18*).

Determination of Xanthophyll Stereoisomers in Thermally Processed Vegetables and Dietary Supplements

The dihydroxy carotenoids lutein and zeaxanthin have been identified as the major constituents of the macular pigment of the human retina. There is a growing body of evidence that they act as antioxidants in the macular region and display a protective role in the prevention of age-related macular degeneration (*19*). Therefore, a higher intake especially of lutein through either xanthophyll-rich vegetables and fruits or dietary supplements may have beneficial effects on the visual performance of people suffering from age-related eye disorders. While green leafy vegetables such as spinach, lettuce, and kale are the most important dietary sources of lutein, zeaxanthin is mainly ingested with sweet corn and orange paprika. Since only limited information on the effects of processing on the stability of lutein and zeaxanthin was available, spinach and sweet corn were used as model vegetables to assess the influence of blanching and canning on the isomeric profile of these xanthophylls. For this purpose, an HPLC method for the simultaneous determination of lutein and zeaxanthin stereoisomers was developed. The canning process of sweet corn (T_{max} = 121 °C; $F = 5$) resulted in a decrease of total lutein from 2.3 to 1.7 mg/100 g and of total zeaxanthin from 0.9 to 0.7 mg/100 g. In contrast, the moderate blanching conditions (100 °C; 2 min) led to a relatively large degradation of lutein in spinach (44 *vs.* 37 mg/100 g). Zeaxanthin could be detected only in trace amounts. Pronounced differences were also observed for the profile of *cis*-isomers in both vegetables. Whereas heat treatment of sweet corn resulted in a significant increase in the 13-*cis*-isomers of lutein and zeaxanthin, blanching of spinach mainly decreased lutein *cis*-isomers (*20*). Subequent investigations on the effects of heating and illumination on *trans-cis*-isomerization and degradation of β-carotene and lutein in isolated spinach chloroplasts indeed revealed that carotenoid stability needs to be evaluated for every single pigment in its genuine environment and that stability data obtained from model experiments may not necessarily be portable to complex food matrices (*21*).

In contrast to functional foods, dietary supplements are very similar to pharmaceutical products and frequently marketed as oral dosage formulations like gelatin capsules, dragées, tablets or powders. As a consequence, the applicability of procedures for the extraction of carotenoids from fruits, vegetables and functional foods to dietary supplements is limited. We have therefore developed methods for the determination of carotenoid stereoisomers from the above formulations comprising enzymatic digestion of the gelatin capsule using papain, extraction with acetone-hexane, and separation by HPLC. Whereas all-*trans*-β-carotene could be detected in all samples investigated, all-*trans*-α-carotene was found only in eight out of 11 preparations, in part at very low levels (*22*). Both 9-*cis*- and 13-*cis*-β-carotene were observed in all samples. Carotenes applied in dietary supplements often originate from microalgae such as *Dunaliella salina*, which contain all-*trans*-β-carotene and 9-*cis*-β-carotene at a ratio of approximately 3:2 (*23*). Consequently, two samples containing extracts of *Dunaliella salina*, as specified on the label, showed higher levels of 9-*cis*-β-carotene. In contrast, no evidence for the presence of *cis*-isomers of lutein and zeaxanthin was obtained. Although only small quantities of carotenoids could be detected in the capsule shell, enzymatic digestion of the gelatin capsules is recommended to ensure complete extraction and accurate quantification (*22*).

Isolation of Carotenoids by High-Speed Counter-Current Chromatography (HSCCC)

The limited availability of secondary plant metabolites as reference compounds is a major problem in analytical chemistry, also hampering *in vitro* and *in vivo* studies on possible health effects. High-Speed Counter-Current Chromatography (HSCCC) has emerged as a powerful technique for the isolation of various classes of compounds (*24*) including also carotenoids such as lycopene from tomato paste (*25*), crocin and its glycosides from the fruits of *Gardenia jasminoides* (*26*), and canthaxanthin from the microalga *Chlorella zofingiensis* (*27*). Compared to crude plant extracts, dietary supplements represent a simple matrix and contain carotenoids in a relatively pure form. Very recently, we have developed HSCCC methods for the isolation of lutein and zeaxanthin from soft gelatin capsules in high purities (> 90 %). *cis*-Isomers could be detected only in very low quantities (*28*). It is expected that a similar approach may also be suitable for the isolation of biologically active compounds other than carotenoids, e.g. polyphenols and phytosterols.

Conclusions

Mainly due to advances in analytical chemistry, considerable progress has been made to understand the phenomenon of carotenoid isomerization in foods. The physical state of carotenoids and the nature of the matrix have been identified as crucial factors affecting their stability and the extent of isomerization. However, in order to obtain a more comprehensive knowledge, further investigations, e.g. into the role electrophilic compounds such as quinones play in carotenoid isomerization (*29*), would be desirable. The development of methods for the isolation of carotenoids should also facilitate studies on their bioavailability, metabolism, distribution, and possible health effects.

References

1. van den Berg, H.; Faulks, R.; Granado, H. F.; Hirschberg, J.; Olmedilla, B.; Sandmann, G.; Southon, S.; Stahl, W. *J. Sci. Food Agric.* **2000**, *80*, 880.
2. Echavarri-Erasun, C.; Johnson, E. A. *Appl. Mycol. Biotechnol.* **2002**, *2*, 45.
3. Bhosale, P. *Appl. Microbiol. Biotechnol.* **2004**, *63*, 351.
4. Dufossé, L.; Galaup, P.; Yaron, A.; Malis Arad, S. ; Blanc, P. ; Chidambara Murphy, K. N. ; Ravishankar, G. A. *Trends Food Sci. Technol.* **2005**, *16*, 389.
5. Sitte, P. *Biol. Unserer Zeit.* **1977**, *7*, 65.
6. Vishnevetsky, M.; Ovadis, M.; Vainstein, A. *Trends Plant Sci.* **1999**, *4*, 232.
7. Godoy, H. T.; Rodriguez-Amaya, D. B. *J. Agric. Food Chem.* **1994**, *42*, 1306.
8. Dachtler, M.; Kohler, K.; Albert, K. *J. Chromatogr. B* **1998**, *720*, 211.
9. Dachtler, M.; Glaser, T.; Kohler, K.; Albert, K. *Anal. Chem.* **2001**, *73*, 667.
10. Updike, A. A.; Schwartz, S. J. *J. Agric. Food Chem.* **2003**, *51*, 6184.
11. Schieber, A.; Carle, R. *Trends Food Sci. Technol.* **2005**, *16*, 416.
12. Marx, M.; Schieber, A.; Carle, R. *Food Chem.* **2000**, *70*, 403.
13. Schieber, A.; Marx, M.; Carle, R. *Food Chem.* **2002**, *76*, 357.
14. Carle, R. *Fruit Process.* **1999**, *9*, 342.
15. Reiter, M.; Stuparic, M.; Neidhart, S.; Carle, R. *Lebensm. Wiss. Technol.* **2003**, *36*, 165.
16. Marx, M.; Stuparić, M.; Schieber, A.; Carle, R. *Food Chem.* **2003**, *83*, 609.
17. Schieber, A.; Marx, M.; Carle, R. In *Proceedings of the 3rd International Congress on Pigments in Food: Pigments in Food – more than colours...* (Dufossé, L., Ed.); Pigments Publishing, Quimper, France, 2004; pp 142-144.
18. Pott, I.; Marx, M.; Neidhart, S.; Mühlbauer, W.; Carle, R. *J. Agric. Food Chem.* **2003**, *51*, 4527.
19. Snodderly, D.M. *Am. J. Clin. Nutr.* **1995**, *62*, 1448.
20. Aman, R.; Biehl, J.; Carle, R.; Conrad, J.; Beifuss, U.; Schieber, A. *Food Chem.* **2005**, *92*, 753.
21. Aman, R.; Schieber, A.; Carle, R. *J. Agric. Food Chem.* **2005**, *53*, 9512.
22. Aman, R.; Bayha, S.; Carle, R.; Schieber, A. *J. Agric. Food Chem.* **2004**, *52*, 6086.
23. Orset, S.; Leach, G.C.; Morais, R.; Young, A.J. *J. Agric. Food Chem.* **1999**, *47*, 4782.

24. Ito, Y. *J. Chromatogr. A* **2005**, *1065*, 145.
25. Wei, Y.; Zhang, T.; Xu, G.; Ito, Y. *J. Chromatogr. A* **2001**, *929*, 169.
26. Degenhardt, A.; Winterhalter, P. *J. Liq. Chromatogr. Relat. Technol.* **2001**, *24*, 1745.
27. Li, H.-B.; Fan, K.-W.; Chen, F. *J. Sep. Sci.* **2006**, *29*, 699.
28. Aman, R.; Carle, R.; Conrad, J.; Beifuss, U.; Schieber, A. *J. Chromatogr. A* **2005**, *1074*, 99.
29. De Rigal, D.; Gauillard, F. ; Richard-Forget, F. *J. Sci. Food Agric.* **2000**, *80*, 763.

Chapter 12

Role of Color and Pigments in Breeding, Genetics, and Nutritional Improvement of Carrots

P. W. Simon[1], S. A. Tanumihardjo[2], B. A. Clevidence[3], and J. A. Novotny[3]

[1]Agricultural Research Service, U.S. Department of Agriculture, Vegetable Crops Research Unit, Department of Horticulture, University of Wisconsin at Madison, WI 53706
[2]Department of Nutritional Sciences, University of Wisconsin at Madison, Madison, WI 53706
[3]Agricultural Research Service, U.S. Department of Agriculture, Diet and Human Performance Laboratory, Beltsville Human Nutrition Research Center, Beltsville, MD 20705

The color of carrots was an important attribute during its domestication as a root crop. Modern carrot researchers continue to study color, and carrot genetic stocks have been developed with not only orange, but also distinctive dark orange, red, yellow and purple color. Genes for 22 carotenoid biosynthetic enzymes have been mapped and cloned, and the α- and β-carotene in typical orange and dark orange carrots, lycopene in red carrots, lutein in yellow carrots, and anthocyanins in purple carrots have been demonstrated to be bioavailable. The function of carrot color genes largely remains unknown and the sources of wide variation in pigment absorption are unexplained, but carrot has been demonstrated to be a sustainable source of dietary provitamin A and other phytonutrients of interest for researchers and consumers.

History of Carrot Color

Carrot is known around the world today as an orange vegetable rich in provitamin A carotenoids. While this correctly characterizes modern carrot, the first domesticated carrots of about 1000 years ago were only found in Central Asia, its center of diversity, and they were either yellow or purple in color (*1, 2, 3*). As traders and farmers introduced carrots west and east of carrot's center of diversity, both purple and yellow carrots were known across the Middle East and North Africa and in both Europe and China by the 15th century. Then, in the 16th century orange carrots were first reported in Europe and eventually in Asia. Thus, orange carrots are a relatively recent development.

Two other carrot colors became popular shortly thereafter – white carrots in Europe and red carrots in Asia. It is interesting to note that a broad array of carrot color was known 400 years ago, but orange came to be the preferred color in most of the world soon after their first appearance. The yellow, orange and red colors of carrot roots are attributable to carotenoids – pigments important in photosynthesis, while purple carrot color is attributable to anthocyanins. Neither carotenoids nor anthocyanins have a known function in plant roots so their abundance in carrots is likely a consequence of human selection over the course of domestication. While we do not know why yellow, orange, red, and purple carrots became popular, their nutritional implications are now known to be significant.

Biochemistry and Genetics of Carrot Pigments

Carotene and Anthocyanin Biosynthesis

The carotenoid biosynthetic pathway is lightly conserved in plants, fungi, and photosynthetic bacteria with very similar enzymes and genes for those enzymes across a very diverse array of organisms. (*4, 5, 6*). The structural genes for carotenoid biosynthetic enzymes are well characterized in terms of gene sequence, action, and location in the genome in several plants. The anthocyanin biosynthetic pathway is also well-conserved in higher plants and structural genes in this pathway have been characterized. In some cases variation in plant color is associated with variation in their structural genes. For example, the *y1* gene of maize that accounts for yellow versus white kernel color is associated with a carotenoid biosynthetic gene, phytoene synthase (*7, 8*); genes accounting for several bell pepper colors are associated with carotenoid structural genes, (*9*); and several of the *A1, A2, Bz1, Bz2, C1, C2, Pr1*, and *R1* genes of maize, that account for red and purple kernel, leaf, and root color are associated with anthocyanin structural genes (*10*). However, some color differences in maize, pepper, tomato, and cauliflower due to accumulation of carotenoid and anthocyanins are not associated with structural genes involved in the biosynthesis of these pigments, but rather with regulatory genes elsewhere in the genome (*11, 12, 13*).

Carrot Color Genetics

At least ten major genes that influence carotenoid and anthocyanin accumulation have been identified in carrot (*3*) but only two of these genes have been placed in the genetic map; along with approximately 15 more minor QTL (quantitative trait loci, genes associated with a continuously variable trait) (*14*). These major carrot color genes, Y and Y_2, account for the distinctive color differences: white (no carotenoids), yellow (primarily lutein), and orange (primarily α- and β-carotene). The QTL primarily accounts for genetic variation in α- and β-carotene concentration ranging from 10 ppm to over 600 ppm. Typical carrots in the U.S. today contain 170 ppm carotenoids (over 95% α- and β-carotene) and account for about half of the provitamin A carotenoids in the food supply (*15*). We recently placed 22 carotenoid biosynthetic enzymes on the carrot map and associated them with Y and Y_2 (*16*). Genetic analysis of carrot pigment accumulation is confounded by environmental and developmental variation. The development of DNA markers to track genetic variation without confounding non-genetic influences will set the stage for more efficient progress in breeding for altered carrot pigment content.

Color and Nutritional Value of Carrot Carotenoids

Carotenoids and Health Effects

Carotenoids are a group of yellow, orange, and red phytochemicals found in all plants and some microbes that are assumed to be important for overall human health (*17, 18*). Although hundreds of carotenoids are present in nature, only a few carotenoids circulate in appreciable amounts in human blood and specific carotenoids are localized in certain tissues. For example, lutein and zeaxanthin, which are xanthophylls (oxygen containing carotenoids), are concentrated in the macular region of the eye. Thus, lutein might be an important compound in the prevention of macular degeneration, the leading cause of blindness in the elderly. Over the past decade, several epidemiologic and clinical studies have suggested that carotenoid consumption is associated with lower risk of cardiovascular diseases and cancers; especially lung, oral cavity, pharyngeal, prostate and cervical; as well as eye diseases. However, intervention trials with pharmacological doses of isolated β-carotene have shown either no effect or harmful effects on lung cancer risk among smokers. This suggests that other carotenoids or other components in fruits and vegetables may be responsible for protective influences observed in epidemiological studies. Thus, the overall evidence suggests that diets high in fruit and vegetables are important for optimal health and reduced risk of disease. Carotenoids are one of several components that may confer health benefits yet are not considered essential nutrients (*19*). Research is very active in this area and identification of non-invasive ways to assess carotenoid status will be important in moving research forward.

In addition to making the world a more colorful place, carotenoids have known nutritional value and antioxidant properties. The essential nutrient that is derived from carotenoids is retinol or vitamin A. The predominant provitamin A carotenoids found in humans are β-carotene, α-carotene and β-cryptoxanthin. The vitamin A value of provitamin A carotenoids is under debate. The most recent conversion factors that have been assigned by the Institute of Medicine to β-carotene, α-carotene and β-cryptoxanthin are 12, 24, and 24 μg to 1 μg of retinol, respectively (*20*). Recent research in humans and Mongolian gerbils are in accord with these estimates, yet many factors affect these ratios (*17*). Most notably, these factors include the vitamin A status of the host, the amount or concentration of provitamin A carotenoid fed, and the food matrix.

Carotenoids are thought to act as potent antioxidants to neutralize free radicals formed from the natural metabolic processes of cells. Free radicals damage tissues and cells through oxidative processes. Environmental factors such as smoking and pollution can increase free radical concentrations. Carotenoids may counter these influences by functioning as antioxidants and quenching oxygen-containing free radicals. As components of lipoproteins, carotenoids may regenerate the antioxidant form of vitamin E as well as protect it from oxidation.

Carrots of Various Colors as Sources of Carotenoids

Carrots are a common vegetable and have gained popularity in the past decade in the United States due to the introduction of prepackaged "cut & peel" carrots. Typical orange carrots contain predominantly α- and β-carotene (Table I). High α-carotene serum concentrations uniquely indicate carrot consumption. The red carrot color can be attributed to high lycopene content, but α- and β-carotene are also present in appreciable quantities depending on the type of red carrot. The yellow carrot is generally low in carotenoids and the yellow color can be attributed to lutein, a carotenoid that does not have provitamin A activity, and small amounts of β-carotene.

Table I. Concentrations of carotenoids and anthocyanins in raw carrots of various colors [a,b]

	High β-carotene Orange	*Orange*	*Purple*	*Red*	*Yellow*
			mg/100 g ± SD		
Total carotenoids	28 ± 0.8	15 ± 4.1	18 ± 7	9.8 ± 1.4	0.71 ± 0.38
α-carotene	3.1 ± 2.4	2.2 ± 0.8	4.1 ± 1.2	0.11[c]	0.05[c]
β-carotene	19 ± 2.8	13 ± 3.3	12 ± 5	3.4 ± 0.9	0.18 ± 0.17
lycopene	1.7 ± 0.8	nd[d]	nd	6.1 ± 0.6	nd
lutein	0.4 ± 0.1	0.3 ± 0.1	1.1 ± 0.7	0.3 ± 0.3	0.51 ± 0.27
Total anthocyanins[e]	---[f]	---	133 ± 20	---	---

[a] Carotenoid and anthocyanin data are expressed as means ± SD of three determinations on a fresh weight basis.

[b] Carotenoids data are from reference 26.

[c] Carotenoid values were found in only one of the three carrots.

[d] nd, not detected

[e] Anthocyanins are cyanidin derivatives, reference 43.

[f] ---, not determined

Bioavailability of Carrot Carotenoids in Humans

Vegetables in general are often dismissed as a source of vitamin A because of factors that affect carotenoid bioavailability (*17*). The carrot matrix has a negative effect on β-carotene bioavailability compared with β-carotene beadlets in the ferret (*20*) and in humans (*22*). Thus, a series of studies was performed to compare the relative bioavailability of carotenoids from pigmented carrots to carotenoids from other vegetables or from supplements containing equivalent amounts of carotenoids.

The first study in young adults determined whether lycopene in red carrots is bioavailable by feeding carrot muffins at 5 mg lycopene/day for 11 days (*23*). The second study determined the effect of carrot fiber on lycopene absorption by mixing tomato paste with white carrot and feeding 5 mg lycopene/day for 11 days. The results showed that both lycopene and β-carotene are bioavailable from red carrots, but lycopene absorption was negatively affected by carrot fiber. Making inferences from both studies, the lycopene in the red carrot is about 44% as bioavailable as that from tomato paste and the serum concentration of lycopene begins to plateau at about 20 mg dietary lycopene/day. Red carrots provide an alternative to tomato paste as a good dietary source of lycopene and provide more β-carotene than typical red tomatoes.

Interestingly, yellow carrots predate orange carrots. Although the lutein concentration in carrots is not high, the bioavailability of 1.7 mg lutein fed as yellow carrot for seven days was tested against an equalized lutein-in-oil supplement in young adults (*24*). The lutein derived from the yellow carrot was 65% as bioavailable as the supplement. Lutein from this novel food source resulted in a significant increase in serum concentrations and did not result in the decline in β-carotene that accompanied administration of the lutein supplements. A subsequent analysis of a crystalline lutein supplementation trial (*25*) revealed that in this form, bioavailability varies greatly both within and between subjects.

A sustainable intervention to improve vitamin A status may be the promotion of common vegetables that have enhanced β-carotene concentrations. In agriculture, orange carrots may have up to a 5-fold variation in β-carotene concentrations. Carrots are also the most abundant dietary source of α-carotene (*26*). Orange and high-β-carotene dark orange carrots in muffins providing 2.6 and 7.0 mg β-carotene/day, respectively, were chronically fed for 11 days to healthy young adults (*27*). Although serum α- and β-carotene concentrations both responded to carrot treatment, the concentrations did not differ over time making the results difficult to interpret (*28*). Future studies need to be conducted in groups of people who have marginal vitamin A status in order to assess the impact of different carotenoid concentrations of carrots in at-risk populations.

Bioavailability of Carrot Carotenoids in Gerbils

Not all research questions concerning the specialty carrots can be answered with human studies. Therefore, a series of studies was performed in the Mongolian gerbil so that liver storage of the carotenoids could be determined. Freeze-dried carrot powder was fed in some studies and carotenoid supplements in others. In two studies (*29, 30*), the use of the gerbil as a model for the carotenoid lutein was dismissed as lutein-in-oil supplements were not efficiently absorbed and stored. However, the utility of the model for α- and β-carotene was confirmed.

The bioavailability of α- and β-carotene from carrots was determined in gerbils. Liver stores of β-carotene and vitamin A in the gerbils did not differ between the orange and purple carrot treatments and carrots resulted in higher liver vitamin A than β-carotene supplements equalized to the β-carotene in the carrots (*31*). Feeding dark orange carrots, as compared with typical orange carrots, resulted in more than two times higher β-carotene content in liver but only 10% greater vitamin A liver stores. Using the vitamin A utilization rate (i.e., 2.5 μg retinol/day) from another study in gerbils (*32*), conversion factors are estimated to be 9 to 11 μg β-carotene to 1 μg retinol for the typical orange carrots and ~23 μg β-carotene to 1 μg retinol for the dark orange carrots. It is important to note that the gerbils had an adequate vitamin A status and therefore future studies need to be conducted in vitamin A-depleted gerbils and humans to see if conversion factors differ by vitamin A status. Dark orange carrots may be an alternative source of provitamin A to typical carrots in areas of vitamin A deficiency. Moreover these studies showed that phenolics, including anthocyanins and phenolic acids, in purple carrot do not interfere with the bioavailability of β-carotene from purple carrots.

Another interesting outcome from the human and gerbil studies revolved around α-carotene. In the gerbils fed varying amounts of α-carotene and equal amounts of β-carotene from different colored carrots (*31*), the α-carotene concentration in liver increased dose dependently and did not contribute significantly to the vitamin A stores. Moreover in the human study comparing typical orange and dark orange carrots, the α-carotene serum concentration was identical in both treatments even though the concentration of α-carotene was more than two-times higher in the dark orange carrots. These studies are inconclusive concerning the vitamin A value of α-carotene during sufficient vitamin A status. To follow up this finding, α-carotene was isolated from carrots and 18.8 μg (35 nmol)/day was fed to vitamin A-depleted gerbils for 21 days (*32*). In the vitamin A-depleted gerbils, purified α-carotene maintained vitamin A status as well as β-carotene supplements when fed at twice the amount of β-carotene. Conversion factors were ~5.5 μg α-carotene or ~2.8 μg β-carotene to 1 μg retinol which are slightly higher than those proposed by the Institute of Medicine for oil-based supplements (*20*). Thus, α-carotene can support vitamin A status when needed by the host.

Color and Nutritional Value of Carrot Anthocyanins

Anthocyanins and Health

Anthocyanins are water-soluble red, blue, and purple pigments found in fruits, vegetables, and ornamental crops. Promising research has shown that dietary anthocyanins may serve an important role in promoting health.

Anthocyanins have been associated with reduced risk of atherosclerosis (*33, 34*) and cancer (*35 36*), reduction of inflammation (*37, 38*), and improved antioxidant status (*33, 34, 39*). Anthocyanins may be particularly beneficial to brain tissue and function. Rats fed blueberries had improved spatial learning and memory compared to control rats (*40*). Loren and colleagues recently reported that maternal supplementation with pomegranate juice protects the fetal brain against neonatal hypoxic-ischemic brain injury (*41*). Andres-Lacueva and colleagues (*40*) found anthocyanins in brain tissue of rats fed blueberries, suggesting that anthocyanins are able to cross the blood-brain barrier.

Purple Carrots as a Source of Anthocyanins

The six common anthocyanidin backbones are cyanidin, malvidin, delphinidin, peonidin, petunidin, and pelargonidin. These backbones can be glycosylated and form linkages with aromatic acids, aliphatic acids, and methyl ester derivatives (*42*). Anthocyanins found in *Daucus carota* L. (sometimes referred to as purple carrot or black carrot) are predominantly derivatives of cyanidin, though pelargonidin and peonidin glycosides have also been reported (*43*). The major forms have been identified as cyanidin-3-xylosyl-galactoside and cyanidin-3-xylosyl-glucosyl-galactoside. The latter can be acylated with ferulic, sinapic, or p-coumaric acid, and the acylated forms are predominant (*44*). This is in accord with the general finding that vegetable anthocyanins are more likely to be acylated than fruit anthocyanins (*45, 46*). Analysis of whole carrots showed the following anthocyanin derivatives: cyanidin-3-(2"-xylose-6-glucose-galactoside) (Cy3XGG), cyanidin-3-(2"-xylose-galactoside) (Cy3XG), cyanidin-3-(2"-xylose-6"-sinapoyl-glucose-galactoside) (Cy3XSGG), cyanidin-3-(2"-xylose-6"-feruloyl-glucose-galactoside) (Cy3XFGG), cyanidin-3-(2"-xylose-6"-(4-coumuroyl)glucose-galactoside) (Cy3XCGG) (*44*). Cultured purple carrot cells were also found to contain cyanidin-3-(2"-xylose-6"-(4-hydroxybenzoyl) glucose-galactose) (*47*).

Purple storage root flesh of USDA inbred B7262, which is a line with a purple exterior and an orange center, has been analyzed for anthocyanin content. One hundred grams of raw purple storage root material, after removal of the orange cores, contained on average 8.0 mg Cy3XGG, 8.3 mg Cy3XG, 99.8 mg Cy3XSGG, 47.6 mg Cy3XFGG, and 2.8 mg Cy3XCGG, for a total of 166 mg anthocyanin/100 g. Approximately 25% of orange taproot material was removed before analysis, thus the anthocyanin content of the whole taproot was approximately 133 mg anthocyanin/100 g. For comparison, anthocyanin contents of other sources include Bing cherries with on average 38 mg/100 g (*48*), pomegranate juice with approximately 14 mg/100 g (*49*), blueberries with on average 230 mg/100 g (*50*), blackberries with on average 179 mg/100 g (*50*), and black current with on average 207 mg/100 g (*50*). In the United States, anthocyanin consumption is estimated at about 215 mg/day during summer months and about 180 mg/day during winter months (*51*).

Carrot Anthocyanin Bioavailability in Humans

Due to their potential health-promoting effects, understanding factors affecting anthocyanin bioavailability has become important. Consideration of carotenoids, a well-studied class of phytonutrients, provides examples of factors that influence phytonutrient bioavailability, such as cooking, dose size, specific forms of the compounds ingested, and concomitant intake of other dietary components. For example, carotenoid absorption is improved when fat is consumed in conjunction with carotenoid-rich foods (*52*). Lycopene bioavailability is substantially higher from processed tomato paste compared to raw tomatoes (*53*). The carotenoid lutein appears to be significantly more bioavailable than β-carotene from kale in humans (*54*), thus showing different forms must be specifically studied. And lycopene absorption efficiency decreases with increasing dose size (*55*). Factors influencing bioavailability of anthocyanins are just beginning to be investigated. Our group has studied the bioavailability of purple carrot anthocyanins, including factors that influence their absorption.

Bioavailability studies have been performed with *Daucus carota* USDA inbred B7262 (*44*). In general, all anthocyanin forms found in the carrots except Cy3XCGG were identified in human blood and urine after carrot consumption, thus confirming that anthocyanins are bioavailable from purple carrots and can be absorbed intact. Previous studies have also shown absorption of intact glycosylated anthocyanins (*56-61*). Anthocyanins from carrots are quickly absorbed, appearing in plasma by 30 min after dosing, reaching peak plasma levels by 2 h after ingestion, then slowly decreasing, with anthocyanins still detectable in plasma at 8 h after dosing. Carrot anthocyanins can be detected in urine by 2 h after dosing, with the greatest rate of excretion occurring 4 h after dosing and with anthocyanins still detectable in the 16-24 h collection.

Recovery of carrot anthocyanins in blood and urine from human volunteers after carrot consumption was similar to recoveries of anthocyanins from other sources. Urinary recoveries of anthocyanins after ingestion of 250 g purple carrots were 0.014% for acylated anthocyanins, 0.19% for non-acylated anthocyanins, and 0.038% for total anthocyanins. Wu and co-workers (*57*) recovered 0.077% of anthocyanins in urine of volunteers after consumption of 12 g elderberry extract and 0.004% of anthocyanins in urine of volunteers after consumption of 189 g blueberries. Bub and colleagues (*62*) detected less than 0.03% of malvidin-3-galactoside in urine after a single ingestion of 500 mL of red wine, dealcoholized red wine, and grape juice. Others have found similar recoveries.

Since carrots are often served cooked, the effect of microwave cooking on anthocyanin bioavailability has been investigated. The microwave-cooked carrots showed a trend toward lower anthocyanin contents than raw carrots (23% decrease in total anthocyanin content), though this reduction did not reach statistical significance. Other studies have suggested that anthocyanins in juice or extracts are thermally unstable (*63, 64*), but studies have not previously considered thermal stability of anthocyanins in whole foods. When cooked and uncooked carrots were administered to study volunteers at matched carrot dose sizes, cooking did not appear to affect total anthocyanin mass in blood or urine after the dose. However, since the cooked carrots showed a trend (though not statistically significant) toward having slightly reduced anthocyanin contents, expressing the blood and urine appearance of anthocyanins as fractional recovery suggests cooking may be important. When anthocyanin appearance in urine is expressed as fractional recovery (anthocyanin mass in urine / anthocyanin mass in treatment), urinary recovery of anthocyanins from cooked carrots was found to be greater than that from the same size dose of raw carrots (*44*). Previous studies have shown that cooking and processing influence carotenoid bioavailability (*65, 66*). Thermal processing may disrupt cell walls, making compounds more accessible for absorption.

Since it has been suggested that anthocyanin absorption is mediated by a carrier in the gastrointestinal epithelium, absorption may be prone to saturation. To investigate this possibility, purple carrots were administered to study volunteers in two dose sizes. Plasma and urine appearances of anthocyanins were not different after ingestion of 250 g cooked carrots compared to 500 g (*44*). The 250 g carrot treatment contained approximately 350 μmol anthocyanins. This suggests that anthocyanin absorption may be saturated at levels of 350 μmol (or lower, which cannot be determined since our lowest dose was 350 μmol), though it should be noted that specific saturation levels would likely be compound specific. Saturation of an absorption mechanism has been observed previously for carotenoids (*55*). The apparent saturation supports carrier involvement in anthocyanin absorption. Additionally, the large size and the polarity of these compounds makes passive transport less likely since they would not partition into a lipid bilayer.

Carrots, like many other anthocyanin-rich vegetables, contain a substantial fraction of acylated anthocyanins. Both acylated and non-acylated anthocyanins have been recovered in blood and urine of volunteers after consumption of purple carrots (*44*). Other recent studies have also shown that acylated anthocyanins can cross the gastrointestinal tract intact (*67-69*). In human studies with purple carrots, acylated anthocyanins were recovered at much lower levels in blood and urine than non-acylated anthocyanins. Similarly, Wu et al (*69*) reported that recovery of the single acylated anthocyanin from freeze-dried marionberry powder was lower than that for the three non-acylated anthocyanins. And Mazza et al (*70*) were unable to detect acylated anthocyanins in serum of subjects after consumption of lowbush blueberries, while many non-acylated anthocyanins in serum were measurable. One possible explanation for this is that acylation of anthocyanins reduces bioavailability. Alternatively, it is possible that acyl groups are cleaved prior to appearance in blood and urine.

Consumer Evaluation and Promotion

Before growers embrace carrots of various colors, a market needs to be established. Consumer sensory evaluation showed that the high-β-carotene dark orange and white carrots were favored over the yellow, red and purple carrots in both blind and non-blind treatments (*26*). However, all the carrots were well-accepted by the consumer panel and therefore growers should be encouraged to cultivate specialty carrots to provide dietary sources of vitamin A and phytochemicals. Outreach activities with community gardens resulted in positive attitudes toward all the carrot colors. Seed companies and growers ranging from small-scale to large are beginning to direct their attention and resources to this wide array of carrot colors as this small niche market begins to grow.

Future Directions for Research

Genetic research has provided a useful basis for improving the nutrient composition of carrot, but many questions remain. Will use of molecular markers improve selection efficiency for desired pigments? Will changes in pigment composition alter color to reduce visual appeal for consumers? Can we select for one specific pigment, say α - carotene, to the exclusion of other carotene isomers? Can we breed for higher bioavailability of all pigments? Can we breed crops for those consumers who have specific nutrient needs? Can we determine the molecular basis of the mutations that resulted in high pigment accumulation? Can we take that knowledge and apply it to genetic improvement of nutrient composition in other crops? How far can we apply naturally-occurring crop variation, and what more could transgenic approaches bring?

More research is needed on vegetables' contribution to alleviating vitamin A deficiency globally. The carrot could play a major role in this regard as it can be grown in many areas of the world. Determining whether smaller amounts of dark-orange carrot with higher carotenoid concentrations can result in similar vitamin A status to a larger amount of typical orange carrot is an important next step in both human and animal models. Studies that compare the lycopene-rich red carrot to tomato in disease prevention models are also needed. The influence of anthocyanin presence in the purple carrot on carotenoid bioavailability is also needed in humans to ascertain phytochemical interactions from this unique vegetable.

Many interesting questions also remain to be answered with respect to nutritional aspects of anthocyanins and how carrots can provide anthocyanin-related health benefits. The mechanisms for anthocyanin absorption across the gastrointestinal tract are currently unknown. The effect of increasing dose size on anthocyanin absorption efficiency will be important for determining effective intake patterns. Dietary factors that increase or decrease absorption must be understood for the development of dietary recommendations. And the active forms of the compounds (parent anthocyanins or metabolites) must be elucidated. As the importance of anthocyanins to health continues to unfold, it is expected that consumer demand for anthocyanins will increase, and carrots will likely play an important role in providing dietary anthocyanins.

Carrot color has generated much interest in the past. As crop specialists collaborate with nutritionists to study the pigments that confer these distinctive colors, a broad new array of research questions arise, and consumers will ultimately benefit from those collaborations.

References

1. Mackevic, V. I. *Bull. Appl. Bot. Genet. Plant Breed.* **1929**, *20*, 517-562.
2. Banga, O. *Main types of the western carotene carrot and their origin;* W.E.J. Tjeenk Willink: Zwolle, The Netherlands, 1963, pp 13-26.
3. Simon, P. W. *Plant Breed. Rev.* **2000**, *19*, 157-190.
4. Bartley, G. E.; Scolnik, P. A. *Plant Cell* **1995**, *7*, 1027-1038.
5. Cunningham, F. X.; Gantt, E. *Ann. Rev. Plant. Phys. Plant Molec. Biol.* **1998**, *49*, 557-583.
6. Hirschberg, J. *Current Opinion in Plant Bio.* **2001**, *4*, 210-218.
7. Buckner, B.; Kelson, T. L.; Robertson, D. S. *Plant Cell* **1990**, *2*, 867- 876.
8. Buckner, B.; Miguel, P. S.; Janick-Buckner, D.; Bennetzen, J. L. *Genetics* **1996**, *143*, 479-488.
9. Thorup, T. A.; Tanyolac, B.; Livingstone, K. D.; Popovsky, S.; Paran, I.; Jahn, M. *Proc. Natl. Acad. Sci. U.S.A.* **2000**, *97*, 11192-11197.
10. Neuffer, M. G.; Coe, E. H.; Wessler, S. R. *Mutants of Maize*; Cold Spring Harbor Press: Plainview, NY. 1997; pp 304-307.
11. Chaim, B.; Paran, A.; Grube, I.; Jahn, R.; van Wijk, M.; Peleman, R. *Theor. Appl. Genet.* **2001**, *102*, 1016-1028.
12. Li, L.; Paolillo, D. J.; Prthasarathy, M. V.; DiMuzio, E. M.; Garvin, D. F. *Plant J.* **2004**, *26*, 59-67.

13. Liu, Y. S.; Gur, A.; Ronen, G.; Causse, M.; Damidaux, R.; Buret, M.; Hirschberg, J.; Zamir, D. *Plant Biotech. J.* **2004**, *1*, 195-207.
14. Santos, C. A. F.; Simon, P. W. *Mol. Genet. Genomics* **2002**, *268*, 122-129.
15. Simon, P. W. Genetic improvement of vegetable carotene content. In *Proc. Third International Symposium Biotech. and Nutrition;* Bills, D. D.; Kung, S.-D. Eds.; Butterworth-Heinemann: Boston, MA, 1992; pp 291-300.
16. Just, B. J.; Santos, C. A. F.; Fonseca, M. E. N.; Boiteux, L. S.; Oloizia, B. B.; Simon, P. W. *Theor. Appl. Genet.* **2007**, *114*, 693-704.
17. Tanumihardjo, S. A. *Int. J. Vitam. Nutr. Res.* **2002**, *72*, 40-45.
18. Tanumihardjo, S. A.; Yang, Z. Carotenoids: Epidemiology of Health Effects. In *Encyclopedia of Human Nutrition, 2nd ed.;* Caballero, B; Allen, L., Prentice, A, Eds.; Elsevier Ltd: Oxford, UK, 2005; pp 339-345.
19. *Dietary Reference Intakes for Vitamin C, Vitamin E, Selenium, and Carotenoids;* Institute of Medicine, Food and Nutrition Board; National Academy Press: Washington, DC, 2000; pp 325-382.
20. *Dietary Reference Intakes for Vitamin A, Vitamin K, Arsenic, Boron, Chromium, Copper, Iodine, Iron, Molybdenum, Nickel, Silicon, Vanadium, and Zinc;* Institute of Medicine, Food and Nutrition Board; National Academy Press: Washington, DC, 2001; pp 65-126.
21. Zhou, J. R.; Gugger, E. T.; Erdman, J. W. Jr. *J. Amer. Coll. Nutr.* **1996**, *15*, 84-91.
22. Brown, E. D.; Micozzi, M. S.; Craft, N. E.. *Am. J. Clin. Nutr.* **1989**, *49*, 1258-1265.
23. Horvitz, M. A.; Simon, P. W.; Tanumihardjo, S. A. *Eur. J. Clin. Nutr.* **2004**, *58*, 803-811.
24. Molldrem, K. L.; Li, J.; Simon, P. W.; Tanumihardjo, S. A. *Am. J. Clin. Nutr.* **2004**, *80*, 131-136.
25. Tanumihardjo, S. A.; Li, J.; Porter Dosti, M. *J. Am. Diet. Assn.* **2005**, *105*, 114-118.
26. Surles, R. L.; Weng, N.; Simon, P. W.; Tanumihardjo, S. A. *J. Agric. Food. Chem.* **2004**, *52*, 3417-3421.
27. Tanumihardjo, S. A.; Horvitz, M. A.; Simon, P. W. The Difference in Bioavailability of Beta-Carotene From Three Types of Carrots. In *Report of the XXI International Vitamin A Consultative Group Meeting, Marakech, Morocco*; International Life Sciences Research Foundation: Washington, DC, 2004; p 72.
28. Faulks, R. M.; Southon, S. *Biochim. Biophys. Acta* **2005**, *1740*, 95-100.
29. Molldrem, K. L.; Tanumihardjo, S. A. *Int. J. Vitam. Nutr. Res.* **2004**, *74*, 153-160.
30. Escaron, A. L.; Tanumihardjo, S. A. *Int. J. Vitam. Nutr. Res.* (in press).
31. Porter Dosti M. J. P.; Simon, P. W.; Tanumihardjo, S. A. *Br. J. Nutr.* (in press).

32. Tanumihardjo, S. A.; Howe, J. A. *J. Nutr.* **2005**, *135*, 2622-2626.
33. Ghiselli, A.; Nardini, M.; Baldi, A.; Scaccini, C. *J. Agric. Food Chem.* **1998**, *46*, 361-367.
34. Kaplan, M.; Hayek, T.; Raz, A.; Coleman, R.; Dornfeld, L.; Vaya, J.; Aviram, M. *J. Nutr.* **2001**, *131*, 2082-2089.
35. Katsube, N.; Iwashita, K.; Tsushida, T.; Yamaki, K.; Kobori, M. *J. Agric. Food Chem.* **2003**, *51*, 68-75.
36. Kang, S. Y.; Seeram, N. P.; Nair, M. G.; Bourquin, L. D. *Cancer Lett.* **2003**, *194*, 13-19.
37. Youdim, K. A.; McDonald, J.; Kalt, W.; Joseph, J. A. *J. Nutr. Biochem.* **2002**, *13*, 282-288.
38. Tsuda, T.; Horio, F.; Osawa, T. *J. Nutr. Sci. Vitaminol. (Tokyo)* **2002**, *48*, 305-310.
39. Youdim, K. A.; Shukitt-Hale, B.; MacKinnon, S.; Kalt, W.; Joseph, J. A. *Biochim. Biophys. Acta* **2000**, *1523*, 117-122.
40. Andres-Lacueva, C.; Shukitt-Hale, B.; Galli, R. L.; Jauregui, O.; Lamuela-Raventos, R. M.; Joseph, J. A. *Nutr. Neurosci.* **2005**, *8*, 111-120.
41. Loren, D. J.; Seeram, N. P.; Schulman, R. N.; Holtzman, D. M. *Pediatr. Res.* **2005**, *57*, 858-864.
42. Clifford, M. N. *J. Sci. Food Agric.* **2000**, *80*, 1063-1072.
43. Kammerer, D.; Carle, R.; Schieber, A. *Rapid Commun. Mass Spectrom.* **2003**, *17*, 2407-2412.
44. Kurilich, A. C.; Clevidence, B. A.; Britz, S. J.; Simon, P. W.; Novotny, J. A. *J. Agric. Food Chem.* **2005**, *53*, 6537-6542.
45. Wu, X.; Prior, R. L. *J. Agric. Food Chem.* **2005**, *53*, 3101-3113.
46. Wu, X.; Prior, R. L. *J. Agric. Food Chem.* **2005**, *53*, 2589-2599.
47. Glabgen, W. E.; Wray, V.; Strack, D.; Metzger, J. W.; Seitz, H. U. *Phytochemistry* **1992**, *31*, 1593-1601.
48. Jacob, R. A.; Spinozzi, G. M.; Simon, V. A.; Kelley, D. S.; Prior, R. L.; Hess-Pierce, B.; Kader, A. A. *J. Nutr.* **2003**, *133*, 1826-9.
49. Perez-Vicente, A.; Gil-Izquierdo, A.; Garcia-Viguera, C. *J. Agric. Food Chem.* **2002**, *50*, 2308-2312.
50. Moyer, R. A.; Hummer, K. E.; Finn, C. E.; Frei, B.; Wrolstad, R. E. *J. Agric. Food Chem.* **2002**, *50*, 519-525.
51. Kuhnau, J. *World Rev. Nutr. Diet.* **1976**, *24*, 117-191.
52. van Het Hof, K. H.; West, C. E.; Weststrate, J. A.; Hautvast, J. G. *J. Nutr.* **2000**, *130*, 503-506.
53. Gartner, C.; Stahl, W.; Sies, H. *Am. J. Clin. Nutr.* **1997**, *66*, 116-122.
54. Novotny, J. A.; Kurilich, A. C.; Britz, S. J.; Clevidence, B. A. *J. Lipid Res.* **2005**, *46*, 1896-1903.
55. Diwadkar-Navsariwala, V.; Novotny, J. A.; Gustin, D. M.; Sosman, J. A.; Rodvold, K. A.; Crowell, J. A.; Stacewicz-Sapuntzakis, M.; Bowen, P. E. *J. Lipid Res.* **2003**, *44*, 1927-1939.

56. Cao, G.; Muccitelli, H. U.; Sanchez-Moreno, C.; Prior, R. L. *Am. J. Clin. Nutr.* **2001**, *73*, 920-926.
57. Wu, X.; Cao, G.; Prior, R. L. *J. Nutr.* **2002**, *132*, 1865-1871.
58. Mulleder, U.; Murkovic, M.; Pfannhauser, W. *J. Biochem. Biophys. Methods* **2002**, *53*, 61-66.
59. Felgines, C.; Talavera, S.; Gonthier, M. P.; Texier, O.; Scalbert, A.; Lamaison; J. L., Remesy, C. *J. Nutr.* **2003**, *133*, 1296-1301.
60. Milbury, P. E.; Cao, G.; Prior, R. L.; Blumberg, J. *Mech. Aging Dev.* **2002**, *123*, 997-1006.
61. McGhie, T. K.; Ainge, G. D.; Barnett, L. E.; Cooney, J. M.; Jensen, D. J. *J. Agric. Food Chem.* **2003**, *51*, 4539-4548.
62. Bub, A.; Watzl, B.; Heeb, D.; Rechkemmer, G.; Briviba, K. *Eur. J. Nutr.* **2001**, *40*, 113-120.
63. Kasparaviciene, G.; Briedis, V. *Medicina (Kaunas)* **2003**, *39 Suppl 2*, 65-69.
64. Suhl, H. J.; Noh, D. O.; Kang, C. S.; Kim, J. M.; Lee, S. W. *Nahrung* **2003**, *47*, 132-135.
65. Rock, C. L.; Lovalvo, J. L.; Emenhiser, C.; Ruffin, M. T.; Flatt; S. W., Schwartz, S. J. *J. Nutr.* **1998**, *128*, 913-916.
66. Gartner, C.; Stahl, W.; Sies, H. *Am. J. Clin. Nutr.* **1997**, *66*, 116-122.
67. Harada, K.; Kano, M.; Takayanagi, T.; Yamakawa, O.; Ishikawa, F. *Biosci. Biotechnol. Biochem.* **2004**, *68*, 1500-1507.
68. Suda, I.; Oki, T.; Masuda, M.; Nishiba, Y.; Furuta, S.; Matsugano, K.; Sugita, K.; Terahara, N. *Food Chem.* **2002**, *50*, 1672-1676.
69. Wu, X.; Pittman, H. E., 3rd, Prior, R. L. *J. Nutr.* **2004**, *134*, 2603-2610.
70. Mazza, G.; Kay, C. D.; Cottrell, T.; Holub, B. J. *J. Agric. Food Chem.* **2002**, *50*, 7731-7737.

Beverages

Chapter 13

The Chemistry of Red Wine Color

James A. Kennedy

Department of Food Science and Technology, Oregon State University, Corvallis, OR 97331–6602

The chemistry of red wine color is a complex area of wine chemistry that begins during grape production and continues to the point of wine consumption. In the vineyard, variety, climate and management practices influence the amount and composition of anthocyanins produced in the grape. During wine production, time, temperature and cap management practices ultimately determine anthocyanin extraction as well as the extraction and production of other components that influence anthocyanin reactivity. Finally, the aging of wine ultimately transforms grape-derived anthocyanins into secondary pigments responsible for the long term color stability of red wine. This chapter summarizes the major factors that determine red wine color and the chemistry of reactions that anthocyanins undergo during aging.

Food color is an important aspect of its overall quality. In addition to it being the first aspect of food that is assessed, color can often indicate quality attributes beyond those observed with the eye. Red wine color is no exception. Understanding the chemistry of red wine color is therefore an area of interest to the practicing winemaker and wine scientists. Ribéreau-Gayon provided an excellent review of this topic in 1973 (*1*), and this chapter is an attempt to provide an update on our understanding of red wine color. This topic has also been reviewed recently by the American Chemical Society (*2*).

To understand red wine color, one needs to understand the chemistry of the anthocyanin compounds that are initially responsible for color as well as the other compounds found in wine. In addition, an understanding of red wine production practices are critical to understanding how management practices can be used to influence the quality and stability of wine color. Of particular interest is the degree to which the composition of the color changes with time.

Overview of Red Wine Production

The color of red wine is influenced by essentially all aspects of wine production, from the vineyard to aging in the bottle. Because there are so many factors that ultimately determine how a red wine appears at the time of consumption, it is important to have a general understanding of red wine production. Therefore, a short and simplified overview of red wine production is presented here (Figure 1). Beginning with the harvest, grapes are brought to the winery and are generally directed to a destemmer/crusher that, as the name implies, separates the grape berries from their stems and then crushes the fruit. The crushed berries are then directed to the fermentor and inoculated with yeast to conduct the fermentation. As the fermentation commences, fermentable sugars are converted to ethanol and carbon dioxide. As the carbon dioxide is liberated from the fermentor, the solid parts of the grape berry are buoyed to the top of the fermenting mass. Because anthocyanins and proanthocyanidins (in addition to flavor compounds) are concentrated in the solid parts of the grape berry (Figure 2), this "cap" must be mixed with the contents of the fermentor to aid in the extraction of these components. This is accomplished with a punchdown device that is used to physically mix the contents of the fermentor. Other means for mixing could be utilized including, for example, pumping the liquid over the top of the cap. Once the desired extraction is achieved, the new wine is separated from the solid parts of the grape berry and is then transferred to a storage vessel (e.g.: small oak barrels or stainless steel tanks) for aging. Aging practices are tremendously varied but overall, are designed to allow the wine time to undergo chemical and physical reactions which leads to a wine that is more stable and balanced. After aging the wine undergoes final finishing operations and is then bottled. Winemaking practices for most red wine produced in the world are such that the wine is ready for consumption shortly after bottling. Because of this most wine is consumed within five years of grape harvest. There are wines however that are designed to be much more age worthy (e.g.: first growth Bordeaux, Vintage Port, Napa Valley Cabernet Sauvignon), and the chemical reactions that occur in the bottle over 20-30 years and longer have intrigued scientists since the mid-19th century (*3*).

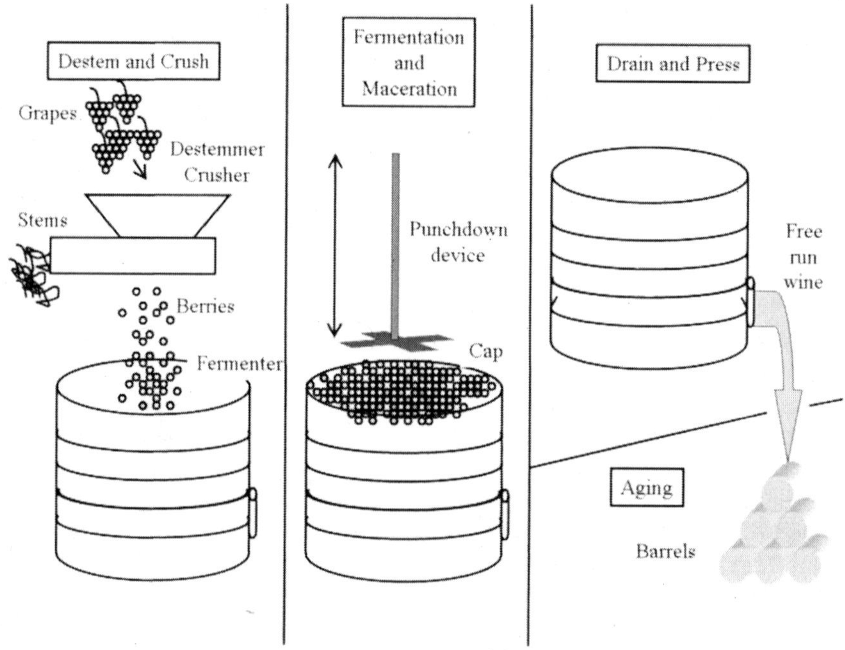

Figure 1. General overview of the major steps in red wine production. (See page 13 of color inserts.)

The above production process is very generalized but it gives the reader a general overview of the production process. It must be emphasized however that production methods are extremely varied and are constantly evolving.

Anthocyanin Chemistry

Although there are some 60 species of grape species known (*4*) essentially all red wines of commercial significance are derived from one species: *Vitis vinifera* L. The compounds initially responsible for the color of red wine are grape-derived anthocyanins, which in *Vitis vinifera* consist of an anthocyanidin

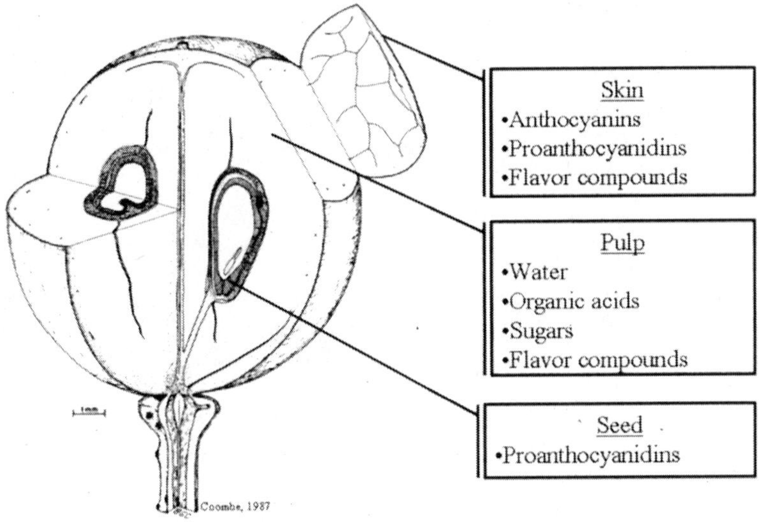

Figure 2. Diagram of a grape berry with the origin of compounds having importance in red wine indicated.

moiety glycosylated at C-3 with glucose (*5*, Figure 3). The anthocyanidin-3-*O*-monoglucosides can also be acylated at C-6 of the glucose. Anthocyanins are with few exceptions restricted to the skin of the grape berry.

The composition of anthocyanins in the grape varies considerably with variety (*6-8*). In addition to variations in the anthocyanidin portion of the molecule, anthocyanins also vary in the proportion of acylation. Generally speaking, malvidin-3-*O*-glucoside is the most abundant anthocyanin in commercially significant varieties. Acylation can vary from zero for cv. Pinot noir to substantial amounts for other varieties (e.g.: Tinta Cao).

Anthocyanins exist in various equilibrium forms (*9-11*; Figure 4). At wine pH (pH 3-4), two equilibrium forms dominate, the flavylium form and the hemiacetal form. The flavylium form is the desirable form from a winemaker's standpoint because it is the observed form (i.e.: the red form). The hemiacetal form in contrast, is colorless. The hemiacetal form comprises roughly 70% of the anthocyanins in wine, with the flavylium form present at about 20%.

In addition to the effect of pH, copigmentation affects the anthocyanin equilibrium in wine (*12, 13*). Copigmentation is characterized by inter or intramolecular non-covalent interactions between adjacent π orbitals resulting in the stabilization of the associated form (Figure 5). In the case of red wine color, the anthocyanin flavylium form is stabilized in the presence of other compounds (cofactors). Hydroxycinnamic acids, flavonols and flavan-3-ol monomers are examples of cofactors found in wine. Flavonols are known to be particularly strong cofactors and flavan-3-ol monomers are comparatively weak (*14*).

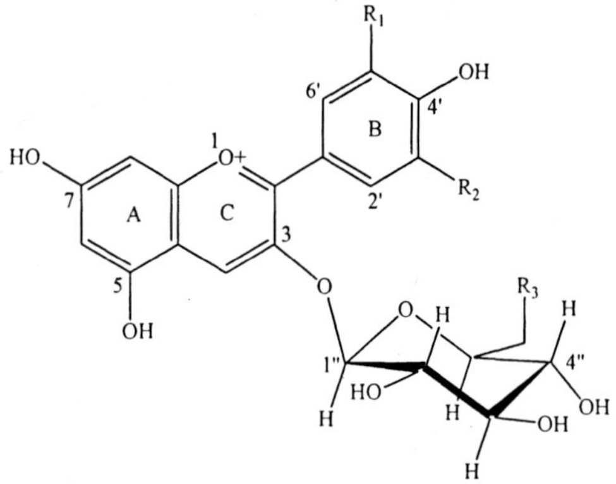

Anthocyanidin Moiety	R_1	R_2
cyanidin	OH	H
peonidin	OCH_3	H
delphinidin	OH	OH
petunidin	OCH_3	OH
malvidin	OCH_3	OCH_3

R3 substituents:

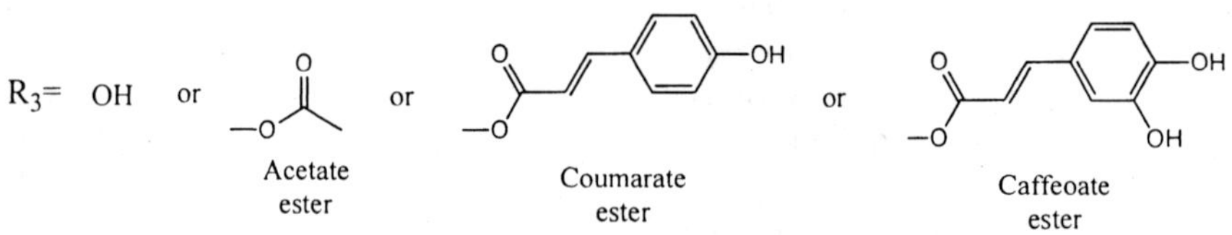

Figure 3. Anthocyanins found in Vitis vinifera sp.

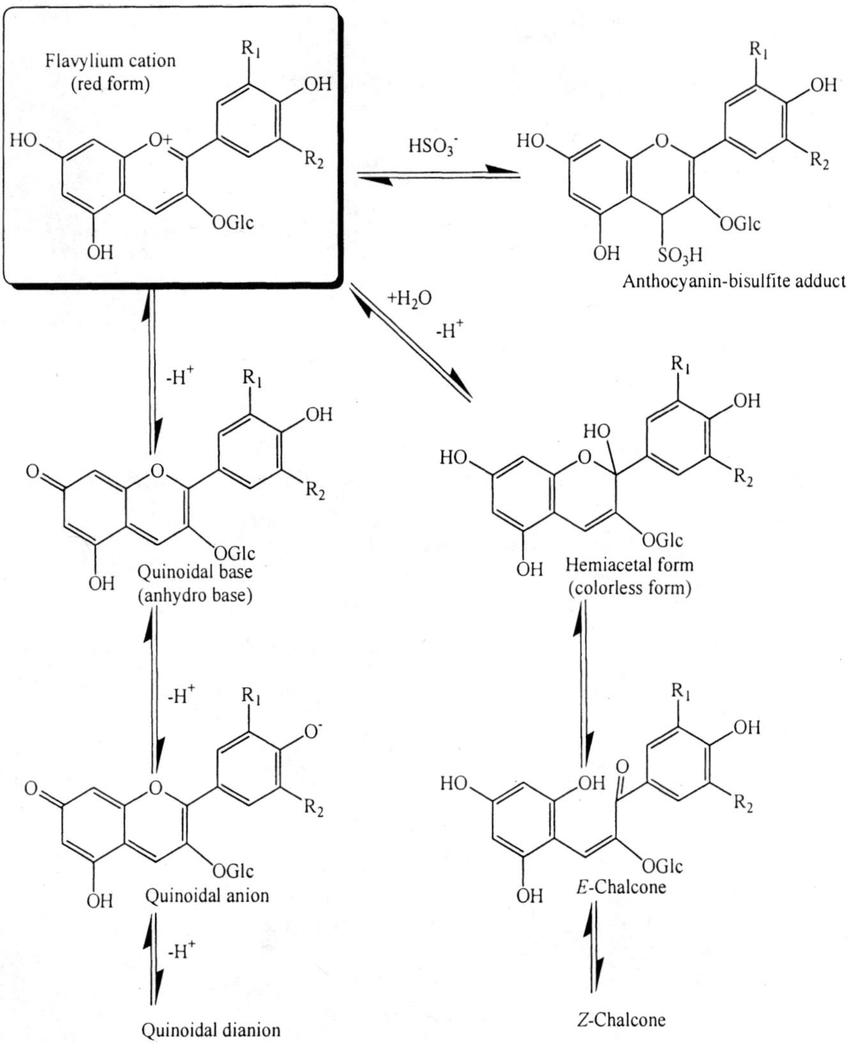

Figure 4. Various anthocyanin equilibrium forms observed in red wine.

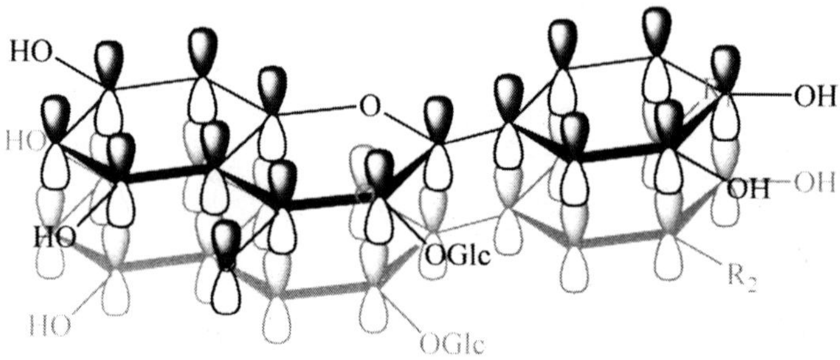

Figure 5. Copigmentation between an anthocyanin (gray) and a cofactor (black, flavonol) demonstrating noncovalent interaction between π orbitals. (See page 13 of color inserts.)

Factors Affecting Color in Wine

From a red wine color standpoint, and from the wine production overview described above, the major factors that determine the color properties of wine at the time of consumption can be divided into three broad areas (Figure 6). These are: the grape, fermentation/maceration and aging.

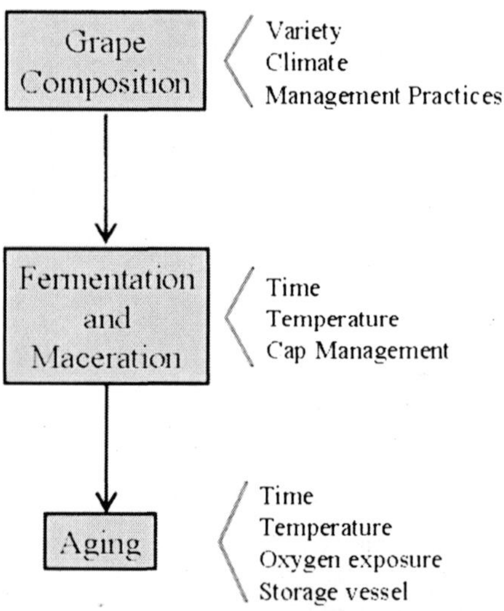

Figure 6. Major areas that influence red wine color and some considerations within each area.

The Grape

The grape is a large variable in determining anthocyanin amount in wine. As mentioned above, there is a large variation in anthocyanin composition and amount based solely upon variety (*6-8*). In addition to this, climate and management practices influence the composition and amount of anthocyanins present (*15-22*). From a management standpoint, and given that wine volume is related to berry volume, the proportion of skin tissue to berry volume influences the eventual amount of color present in the wine (*23*). Understanding this, and given that wine color density has been correlated with consumer preference (*24*), minimizing berry volume is a management concern. Managing berry volume is usually accomplished using irrigation and fruit load management (*25*).

Fermentation and Maceration

Because anthocyanins are localized in the skin tissue of most grape varieties, fermentation and maceration have a profound effect on the amount of anthocyanin present in the final wine. An extreme example of this would be the separation of the solid parts of the grape berry from the juice with little or no maceration resulting in a wine with little or no red color. Sparkling wines made from cv. Pinot noir grapes are example of how little color can be extracted.

Clearly, because of the localization of anthocyanins within plant cell vacuoles, diffusion needs to be considered. Diffusion is simply the process by which a compound moves from a region of high concentration toward a region of lower concentration (i.e.: from the plant cell into the wine). Considering the generally observed extraction curves for anthocyanins (*26, 27*) and the effects that wine processing variables have on the rate of the compounds' extraction, the overall process is generally consistent with diffusion. Diffusion is dependent upon the following:

1. Temperature
2. Molecular weight/size and type of molecule
3. Concentration gradient
4. Cell permeability
5. Surface area over the concentration gradient
6. Composition of extraction medium (such as ethanol concentration).

In all instances but molecular size, these variables contribute positively to the rate of diffusion. Time, of course, is an important variable for overall extraction.

Past research tells us that in any given fermentation, time and temperature are the two critical variables in determining the ultimate amount of anthocyanin present in wine (*28-35*). These are easily the most studied variables. For anthocyanins, higher temperatures reduce the time to maximum concentration, and increase the maximal amount. Besides the most obvious variables of time and temperature, the other variables undoubtedly also play a role in anthocyanin extraction. Determining their relative importance however is difficult because of the compound nature of their affects.

Aging

Once a red wine is pressed, the concentration of anthocyanins in the wine is maximal. During wine aging the concentration of anthocyanins declines fairly quickly (*36*), yet the color remains. It is this aspect of red wine color that has fascinated wine scientists for well over a century. To understand the transformations that occur in red wine, it is important to understand the reactivity of anthocyanins since it is generally accepted that the red color in older red wines is largely due to the presence of anthocyanins that have become modified.

From above, there are two predominant anthocyanin species in red wine, the flavylium form and the hemiacetal forms. From a reactivity standpoint these equilibrium forms differ in their reactivity, with the flavylium form being electrophilic and the hemiacetal form nucleophilic (Figure 7). It is expected then that the anthocyanins would combine with compounds that have contrasting reactivity. The bleaching of anthocyanins with bisulfite anion is one example of this type of reaction (*37*, Figure 4). Additional examples of species that would be expected to combine with anthocyanins are shown in Figure 7. It has been speculated therefore that products combining anthocyanins with these products would be a component of stable red wine color (*38-41*). Although until recently, these products were only speculated.

It is generally recognized that as a wine ages, its color becomes modified (*38*). As wine color transitions from grape-based anthocyanin to modified anthocyanin, its appearance changes from blue-red to brick red. One example of a modified anthocyanin combines the yeast metabolite pyruvic acid with anthocyanins to form vitisin A (*42, 43*; Figure 8). Vitisin A forms early after fermentation has completed (*44*). The formed vitisin A has an absorption maxima at about 509 nm which would appear like the familiar brick-red color of an older wine as opposed to a new wine which is dominated by absorption at 520 nm. In addition, the equilibrium proportion of the flavylium form of vitisin A increases relative to the initial anthocyanin.

As wine ages, it is exposed to oxygen and the effects of oxidation. A product of wine oxidation is ethanal, which is an oxidation product of ethanol oxidation (*45*). It has been observed for some time that ethanal formation leads to a modification of red wine color (*46-48*). One of the major findings in red wine color chemistry was the identification of the ethanal adduct of malvidin-3-

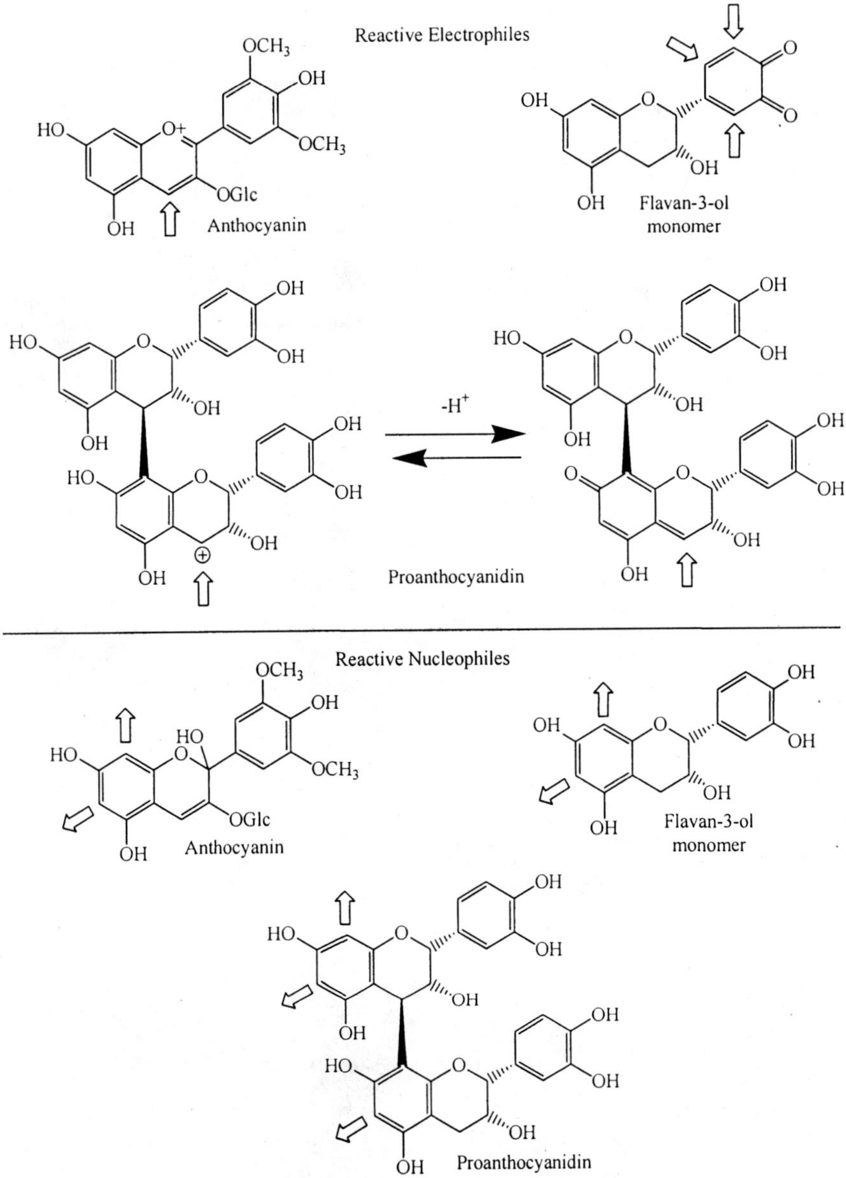

Figure 7. Electrophilic and nucleophilic reactive centers for anthocyanins and other common phenolics found in red wine with the reactive centers designated with arrows.

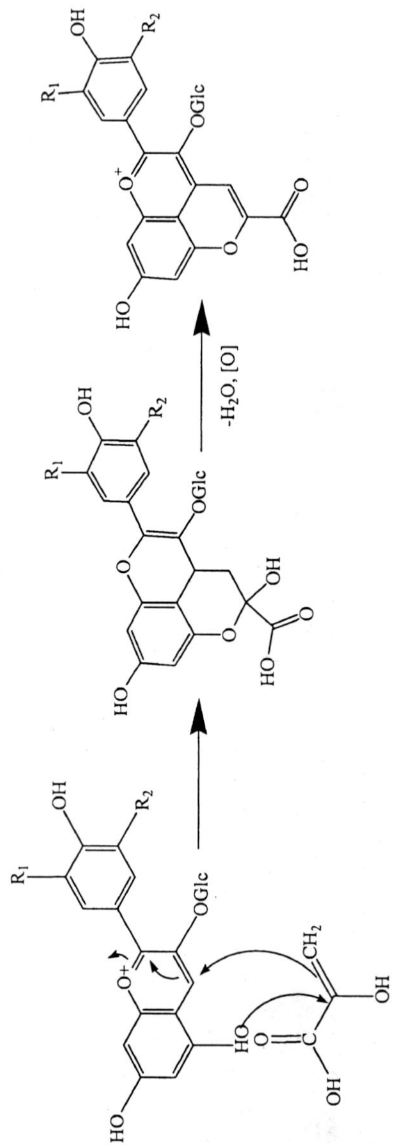

Figure 8. Reaction of an anthocyanin with pyruvic acid to produce vitisin A.

O-glucoside in wine (*42*). The formation of the vitisin pyranoanthocyanins in wine has subsequently led to the identification of other reactive nucleophiles including hydroxycinnamic acids and ethanal-flavan-3-ol reaction products (*49-58*; Figure 9). Given recent strides in mass spectrometry, there is little doubt that additional compounds will be identified that are relatively low in molecular weight (*59*) and which will contribute to the classic brick-red color of aged red wine.

Additional chemistry which is likely to lead to the formation of compounds which give red wine its brick-red color are the xanthylium-type compounds (*60*). These compounds have been known for sometime (*61, 62*), but advances in their chemistry have recently occurred. The products formed are purple-red or yellow in color. The formation of yellow compounds in wine coupled with the red color of anthocyanins would generate a wine color that is consistent with aged wines.

The biggest challenge for wine scientists is characterizing the large molecular weight pigmented material present in red wine. Some time ago, structures were proposed for pigmented polymers and because of the prevalence of proanthocyanidins, these were speculated to involve anthocyanins and proanthocyanidins (*38-41*). The accuracy of this speculation has been confirmed (*63-66*, Figure 10), and this chemistry has been extended into higher molecular weight material (*67*). It is interesting to note that pigmented polymers begin formation early in a wine's life (*68*).

The chemistry of red wine color has advanced tremendously and the identification of modified anthocyanins indicates that the color of wine is of interest for other sensory reasons, namely astringency. Given that astringency is an important attribute of red wine, modification of proanthocyanidins with anthocyanins are likely to be important for other reasons beyond simple color stabilization. Current evidence suggests that the mechanism for astringency perception is driven by hydrophobic interactions between salivary proteins and proanthocyanidins leading to the precipitation of the lubricating proteins (*69*). A reduction in the hydrophobicity of proanthocyanidins by modification with anthocyanins would likely reduce the interaction between salivary proteins and proanthocyanidins and therefore reduce their astringency. Therefore, while our understanding of red wine color has become clearer, it is likely that sensory studies will confirm that red wine color has importance for reasons beyond appearance. This then underlies the importance of red wine color beyond its appearance.

References

1. Ribéreau-Gayon, P. In *Chemistry of Winemaking*, Dinsmoor Webb, A. (Ed.) American Chemical Society: Washington, DC, **1973**, pp 50-87.

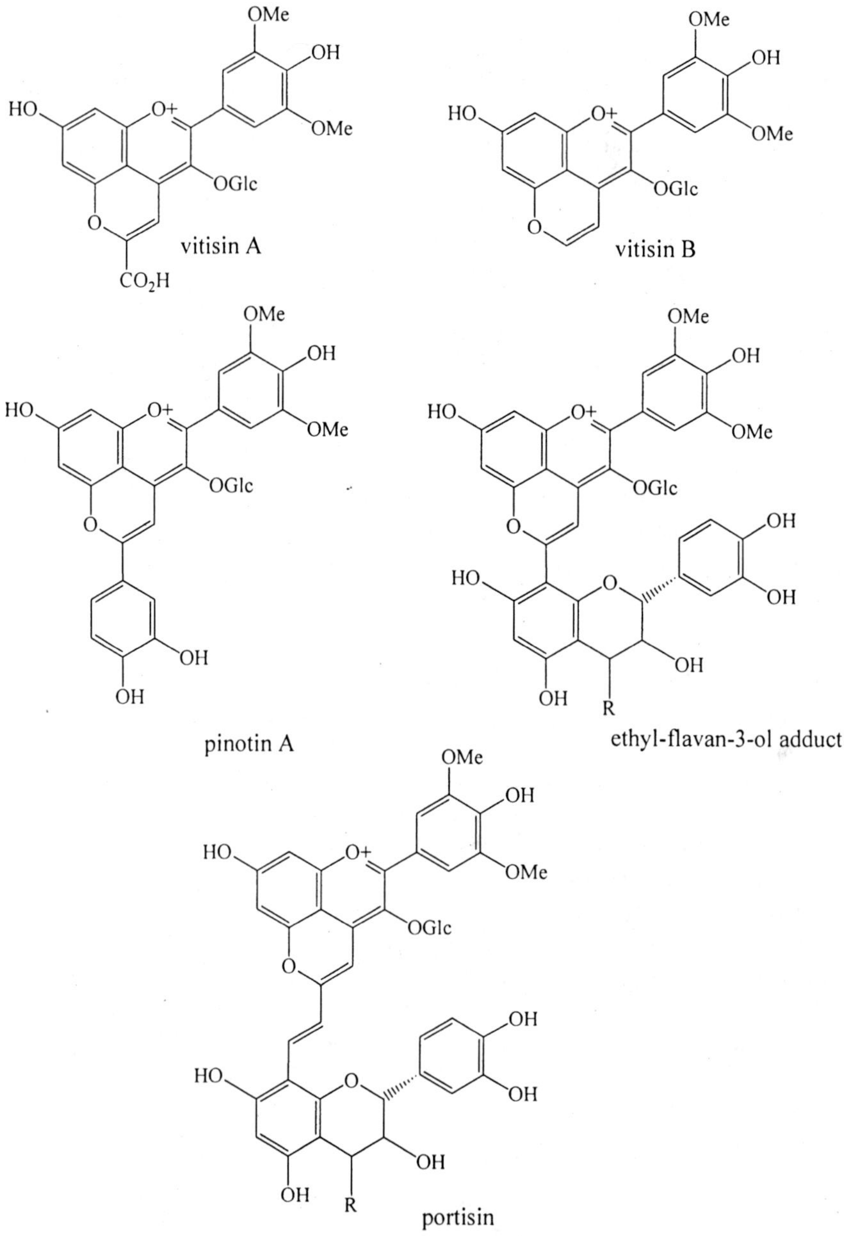

Figure 9. Examples of malvidin-3-O-glucoside-based pyranoanthocyanins that have been identified in red wine.

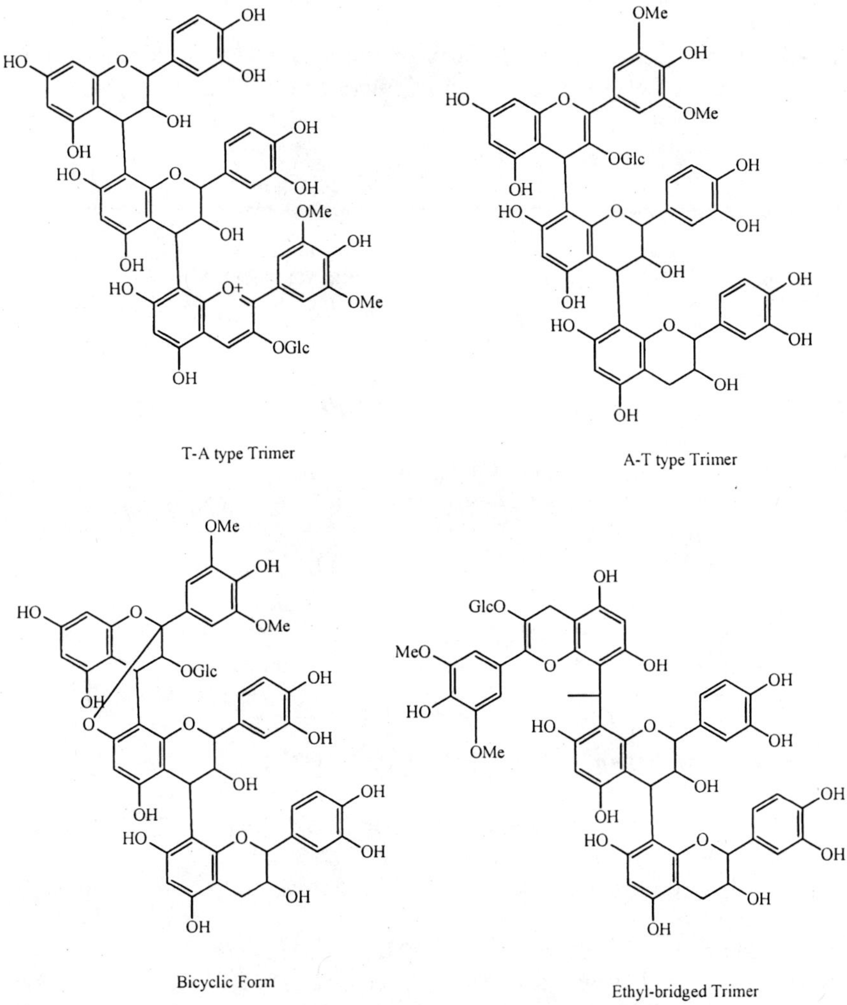

Figure 10. Examples of malvidin-3-O-glucoside proanthocyanidin derivatives.

2. *Red Wine Color: Revealing the Mysteries*; ACS Symposium series 866, Waterhouse, A. L.; Kennedy, J. A. (Ed.), American Chemical Society: Washington, DC, **2004**.
3. Pasteur, L. Etude sur le vin, ses maladies, causes qui les provoquent, procédés nouveaux pour le conserver et le vieillir. Paris Imp. Impériale **1866**.
4. *Biology of the Grapevine*, Biology of Horticultural Crops, Mullins, M. G.; Bouquet, A.; Williams, L. E. (Ed.) Cambridge University Press, New York, NY **1992**, pp 17-36.
5. Ribéreau-Gayon, P. *Ann. Physiol. Vég.*, **1964**, *6,* 211–242.
6. Mazza, G. *Crit. Rev. Food Sci. Nutr.* **1995**, *35*, 341-71.
7. Francis, F. J. *Crit. Rev. Food Sci. Nutr.* **1989**, *28*, 273-314.
8. Wenzel, K.; Dittrich, H.; Heimfarth, M. *Vitis* **1987**, *26*, 65-78.
9. Pina, F. *J. Chem. Soc., Faraday Trans.* **1998**, *94*, 2109-2116.
10. Brouillard, R.; Delaporte, B. *J. Am. Chem. Soc.* **1977**, *99*, 8461-8468.
11. Asenstorfer, R. E.; Jones, G. P. *Analytical Biochemistry* **2003**, *318*, 291-299.
12. Asen, S.; Stewart, R. N.; Norris, K. H. *Phytochemistry* **1972**, *11*, 1139-1144.
13. Boulton, R. B. *Am. J. Enol. Vitic.* **2002**, *52*, 67-87.
14. Mistry, T. V.; Cai, Y.; Lilley, T. H.; Haslam, E. *J. Chem. Soc. P2* **1991**, 1287-1296.
15. Ribéreau-Gayon, P. *Connaiss. Vigne Vin,* **1971**, *5*, 247-261.
16. Ribéreau-Gayon, P. *Connaiss. Vigne Vin*, **1972**, *6*, 161-175.
17. Dokoozlian, N. K.; Kliewer, W. M. *J. Am. Soc. Hort. Sci.* **1996**, *121*, 869-874.
18. Downey, M. O.; Harvey, J. S.; Robinson, S. P. *Austral. J. Grape Wine Res.* **2004**, *10*, 55-73.
19. Kliewer, W. M. *Am. J. Enol. Vit.* **1977**, *28*, 71-77.
20. Kliewer, W. M.; Torres, R. E. *Am. J. Enol. Vit.* **1972**, *23*, 71-77.
21. Mori, K.; Sugaya, S.; Gemma, H. *Scientia Horticulturae* **2005**, *105*, 319-330.
22. Spayd, S. E.; Tarara, J. M.; Mee, D. L.; Ferguson, J. J. *Am. J. Enol. Vit.* **2002**, *53*, 171-181.
23. Singleton, V. L. *Am. J. Enol. Vitic.* **1972**, *23*, 106.
24. Somers, T. C.; Evans, M. E. *J. Sci. Food Agric.* **1974**, *25*, 1369-1379.
25. Kennedy, J. A.; Matthews, M. A.; Waterhouse, A. L. *Am. J. Enol. Vit.* **2002**, *53*, 268-274.
26. Ribéreau-Gayon, P.; Milhé, J. C. *Conn. Vigne Vin* **1970**, *4*, 63-74.
27. Ribéreau-Gayon, P.; Sudraud, P.; Milhé, J. C.; Canbas, A. *Conn. Vigne Vin* **1970**, *4*, 133-144.
28. Aubert, S.; Poux, C. *Ann. Technol. Agric.* **1969**, *18*, 111-127.
29. Berg, H. W.; Akiyoshi, M. *Am. J. Enol. Vitic.* **1956**, *7*, 84-90.

30. Mayen, M.; Merida, J.; Medina, M. *Am J. Enol. Vitic.* **1994**, *45*, 161-166.
31. Ough, C. S.; Amerine, M. A. *Am. J. Enol. Vitic.* **1961**, *12*, 9-19.
32. Scudamore-Smith, P. D.; Hooper, R. L.; McLaran, E. D. *Am. J. Enol. Vitic.* **1990**, *41*, 57-67.
33. Timberlake, C. F.; Bridle, P. *Vitis* **1976**, *15*, 37-49.
34. Wightman, J. D.; Price, S. F.; Watson, B. T.; Wrolstad, R. E. *Am. J. Enol. Vitic.* **1997**, *48*, 39-48.
35. Zimman, A.; Joslin, W. S.; Lyon, M. L.; Meier, J.; Waterhouse, A. L. *Am. J. Enol. Vitic.* **2002**, *53*, 93-98.
36. Nagel, C. W.; Wulf, L. W. *Am. J. Enol. Vitic.* **1979**, *30*, 111-116.
37. Berké, B.; Chèze, C.; Vercauteren, J.; Deffieux, G. *Tetrahedron Lett.* **1998**, *39*, 5771-5774.
38. Somers T. C. *Phytochemistry* **1971**, *10*, 2175-2186.
39. Jurd, L. *Am. J. Enol. Vitic.* **1969**, *20*, 191-195.
40. Haslam, E. *Phytochemistry* **1980**, *19*, 2577-2582.
41. Ribéreau-Gayon, P.; Pontallier, P.; Glories, Y. *J. Sci. Food Agric.* **1983**, *34*, 505-516.
42. Bakker, J.; Timberlake, C. F. *J. Agric. Food Chem.* **1997**, *45*, 35-43.
43. Fulcrand, H.; Benabdeljalil, C.; Rigaud, J.; Cheynier, V.; Moutounet, M. *Phytochemistry* **1998**, *47*, 1401-1407.
44. Asenstorfer, R. E.; Markides, A. J.; Iland, P. G.; Jones, G. P. *Austral. J. Grape Wine Res.* **2003**, *9*, 40-46.
45. Wildenradt, H. L.; Singleton, V. L. *Am. J. Enol. Vitic.* **1974**, *25*, 119-126.
46. Trillat, A. *C.R. Ac. Sci.* **1907**, *24*, Juin.
47. Trillat, A. *Ann. De l'Institut Pasteur* **1908**, *22*, 704-719, 753-762,876-895.
48. Timberlake, C. F.; Bridle, P. *Am. J. Enol. Vitic.* **1976**, *27*, 97-105.
49. Mateus, N.; Carvalho, E.; Carvalho, A. R. F.; Melo, A.; González-Paramás, A. M.; Santos-Buelga, C.; Silva, A. M. S.; de Frietas, V. *J. Agric. Food Chem.* **2003**, *51*, 277-282.
50. Benabdeljalil, C.; Cheynier, V.; Fulcrand, H.; Hakiki, A.; Mosaddak, M.; Moutounet, M. *Sci. Aliment.* **2000**, *20*, 203-220.
51. Atanasova, V.; Fulcrand, H.; Cheynier, V.; Moutounet, M. *Anal. Chim. Acta*, **2002**, *458*, 15-27.
52. Atanasova, V.; Fulcrand, H.; Le Guerneve, C.; Cheynier, V.; Moutounet, M. *Tetrahedron Lett.* **2002**, *43*, 6151-6153.
53. Mateus, N.; Silva, A.M.S.; Santos-Buelga, C.; Rivas-Gonzalo, J.C.; de Frietas, V. *J. Agric. Food Chem.* **2002**, *50*, 2110-2116.
54. Mateus, N.; Carvalho, E.; Carvalho, A.R.F.; Melo, A.; González-Paramás, A.M.; Santos-Buelga, C.; Silva, A.M.S.; de Freitas, V. *J. Agric. Food Chem.* **2003**, *51*, 277-282.
55. Mateus, N.; Silva, A.M.S.; Rivas-Gonzalo, J.C.; Santos-Buelga, C.; de Freitas, V. *J. Agric. Food Chem.* **2003**, *51*, 1919-1923.

56. Mateus, N.; Oliveira, J.; Santos-Buelga, C.; Silva, A. M. S.; de Freitas, V. *Tetrahedron Lett.* **2004**, *45*, 3455-3457.
57. Schwarz, M.; Jerz, G.; Winterhalter, P. *Vitis* **2003**, *42*, 105-106.
58. Schwarz, M.; Wabnitz, T. C.; Winterhalter, P. *J. Agric. Food Chem.* **2003**, *51*, 3682-3687.
59. Hayasaka, H.; Asenstorfer, R. E. *J. Agric. Food Chem.* **2002**, *50*, 756-761.
60. Es-Safi, N.; Cheynier, V. In: *Red Wine Color: Revealing the Mysteries*; ACS Symposium series 866, Waterhouse, A.L.; Kennedy, J.A. (Ed.), American Chemical Society: Washington, DC, **2004**, 143-159.
61. Jurd, L.; Somers, T. C. *Phytochemistry* **1970**, *9*, 419-427.
62. Hrazdina, G.; Borzell, A.J. *Phytochemistry* **1971**, *10*, 2211-2213
63. Remy, S.; Fulcrand, H.; Labarbe, B.; Cheynier, V.; Moutounet, M. *J. Sci. Food Agric.* **2000**, *80*, 745-751.
64. Remy-Tanneau, S.; Le Guernevě, C.; Meudec, E.; Cheynier, V. *J. Agric. Food Chem.* **2003**, *51*, 3592-2597.
65. Dallas, C.; Ricardo-da-Silva, J. M.; Laureano, O. *J. Agric. Food Chem.* **1996**, *44*, 2402-2407.
66. Fulcrand, H.; dos Santos, P.; Sarni-Manchado, P.; Cheynier, V.; Favre-Bonvin, J. *J. Chem. Soc., P1* **1996**, 735-739.
67. Hayasaka, Y.; Kennedy, J. A. *Austral. J. Grape Wine Res.* **2003**, *9*, 210-220.
68. Eglington, J.; Griesser, M.; Henschke, P.; Kwiatkowski, M.; Parker, M.; Herderich, M. In: *Red Wine Color: Revealing the Mysteries*; ACS Symposium series 866, Waterhouse, A.L.; Kennedy, J.A. (Ed.), American Chemical Society: Washington, DC, **2004**, 7-21.
69. Haslam, E.; Lilley, T. H. *CRC Rev. Food Sci. Nutr.* **1989**, *27*, 1-40.

Chapter 14

Measuring Color in Wine: One Laboratory's Approach to Introducing a Color Measuring System

Steven F. Price

ETS Laboratories, 899 Adams Street, Suite A, St. Helena, CA 94574

Wine Industry standard measures for describing color are simple absorbance values at 420 and 520 nm. Color space measurement systems such as CIELab offer a more precise tool for color management and can be applied to describe the effects of a wide range of wine making processes. The use of graphic presentations and the conversion of the a* and b* coordinates to Hue° and Chroma has helped greatly in introduction of the color space concept to new users.

Introduction

Wine is a natural product and color variation from grape composition and wine making processes are common. Vintage variation in color is expected and at least partly accepted. Despite this acceptance, color has a pervasive effect on quality evaluation, influencing the perception of a range of quality attributes well beyond color (*1*). Most wineries would like to improve color consistency and many are aware of specific consumer color preferences for specific products. As wineries strive to control color variation they are finding a need for more informative color evaluation methods.

Standard red wine color analysis has been limited to absorbance at 420 and 520 nm measured on undiluted wine in a 1 mm cuvette (*2*). The sum of A_{420} +

A_{520} is traditionally referred to as "color intensity" and the ratio of A_{420}/A_{520} as "hue". Color analysis of white wines is often limited to A_{420} measured in a 10 mm cuvette. Tristimulus color measurements have been discussed (*3*) but have not been widely used in the industry. The primary limitations preventing use of more complex color measurement systems have been a lack of suitable equipment, and confusion about multi-component/multi-dimensional results.

ETS Laboratories offers a full range of analytical tools for the wine industry. The limitations of the standard color intensity and hue measures used in the industry became apparent when ratios of A_{420}:A_{520} were unable to accurately reflect perceived differences in wine hue. Client concerns about changes in color with wine age and processing effects on hue were not being adequately addressed. The 1976 CIEL*a*b* color space system was selected as an alternative tool for color analysis based on its linear relationship to perceived color, its widespread use in other food industries and apparent interest in the wine industry (*3*). An analytical method using spectrophotometric data was set up and a graphical report format developed.

Analysis

Samples are minimally handled prior to analysis as the intent of the analysis is to present the color of the sample as submitted. Samples with dissolved CO_2 must be degassed by sonication and turbid samples are filtered through a glass fiber syringe filter (Millex AP, Millipore, Billerica, MA). A range of filtration products were tested but all the available 0.45 μm filters removed tannin or highly hydrophobic compounds with potential color effects. Spectral data is collected at 2 nm intervals from 380 to 780 nm on an Agilent diode array spectrophotometer (Agilent GL03A, Agilent, Santa Clara, CA) using a 1 mm quartz flow cell for red wine and a 10 mm quartz cuvette for white wine. Spectral data is imported to an Excel spreadsheet via an in-house data extraction program. The spectral data is converted to the 1976 CIEL*a*b* values from standard tables. The 10° observer was selected as wine evaluation is not a spot measurement. The D55 illuminant provided better separation of wines than the warmer illuminants, particularly wines with purple and blue hues.

Data Presentation

A single page report is developed automatically for each analysis. The report is designed to present results from a single sample or two related samples. Having two samples in a report has been very useful for visually demonstrating the discriminating power of the CIELab information. As digital records of samples are maintained by the lab, a client can use a previously analyzed sample as a reference. For example, a wine can be compared to an earlier analysis of the same wine, to a wine from an earlier vintage or a control in an earlier experiment. An average color of a group can also be used as a reference to evaluate variation in a group such as bottle to bottle variation.

The report contains:

- a chart of the spectral data
- a table of A_{420}, A_{520} and their sum and ratio
- a table of CIEL*a*b* values plus hue angle (H), chroma (C) and difference values for two sample reports: ΔE, ΔL*, Δa* and Δb*
- a chart showing L* in grey scale and a* vs. b* with a colored outline representing hue

An example report is shown below for two Pinot Noir wines. The two wines were made by the same winery from two different vineyard blocks from the same vintage. For clarity, the report is shown broken into components in the tables and figures below but in the ETS Report the components all fit on one page.

Wine absorbance is the basis for perceived color. Having the raw spectral data on the report emphasizes this point (Figure 1). A_{520} and A_{420} are important reference points but the complete spectrum shows a more detailed approach is justified. Few wines have any absorbance above 730 nm. ETS analysts use this as a warning of potential turbidity problems. Wines with absorbance at 750 nm above 0.01 AU are refiltered or centrifuged. Any wines that receive treatment other than the standard filtration and/or sonication are noted in a comments section on the report. No report is generated for wines that can not be clarified as the light scattering from turbidity invalidates the transmitted spectral information.

The table of absorbance information and CIELab coordinates (Table I) presents the two measurement systems in an easily comparable format. The terms "color intensity" for $A_{420} + A_{520}$ and "hue" for A_{420}/A_{520} were deliberately avoided in the spectral table to encourage clients to use the CIELab hue angle value. The L* and b* values do not have equivalents in the standard absorbance system. Differences in a* values correlate with A_{520} in most red wines. Hue angle differences are more pronounced than differences in the A_{420}/A_{520} ratio and the chroma differences more pronounced than the $A_{420} + A_{520}$ differences. ΔE has been useful to clients as a ΔE value greater than 1 approximates a significant visual difference. The three dimensional aspects of ΔE are not explained to clients in the standard report (see below).

Hue angle has presented an interesting problem. Hue angle is equal to 0° when a* is greater than zero and b* is equal to zero. The hue angle then increases counter clockwise to 360° (or back to 0°). This works fine in the ETS

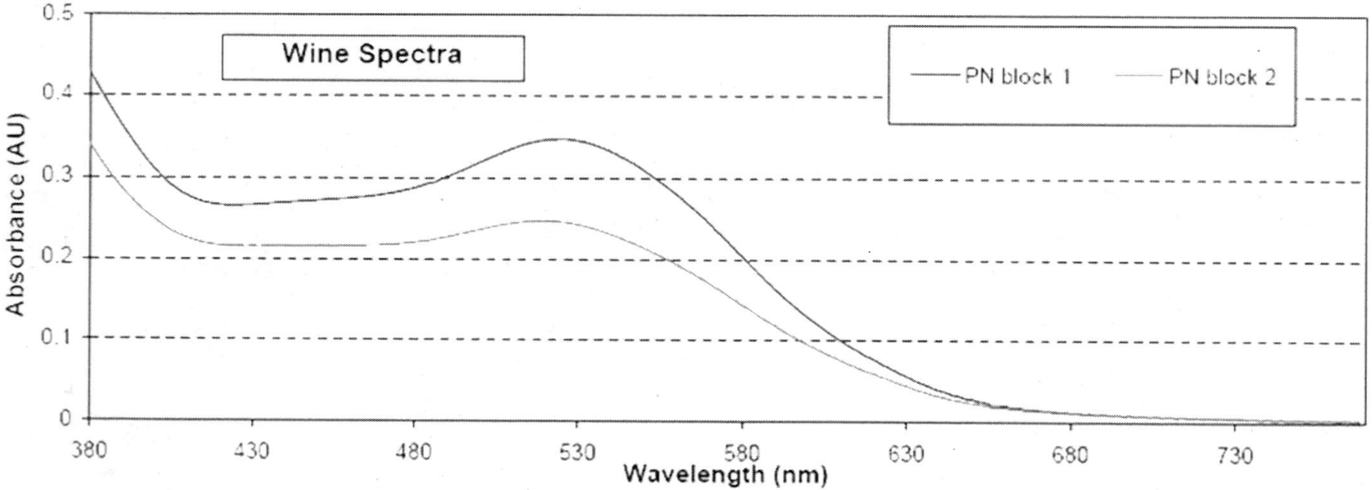

Figure 1. Spectral plot of two Pinot Noir wines. (See page 14 of color inserts.)

Table I. Spectral Data and CIELab Coordinates for Two Pinot Noir Wines

	PN Block 1	*PN Block 2*
Spectral Data		
420 nm	0.265	0.215
520 nm	0.346	0.247
420 + 520	0.512	0.462
420/520	0.767	0.875
CIELab Coordinates illuminant D650, Std Obs 10°		
L*	80.2	85.3
a*	20.4	14.1
b*	4.0	5.3
Hue Angle °	11.2	20.7
Chroma	29.3	20.8
ΔE	5.0	
ΔL*	6.3	
Δa*	1.3	
Δb*	8.2	

Report as long as b* is positive but in some cases young wines may have a slightly negative b* value. It is difficult to explain to a client that a hue angle of 1° is very similar to a hue angle of 359°. While we present the data in this format on the chart (Figure 2), we use negative values in the table so that a hue angle of 359° is presented as -1°. The discrepancy in discussed in an annotated report available to all ETS clients.

The graphs of L* and a* vs. b* (Figure 2) have been helpful in explaining lightness, hue and chroma to clients. The L* graph is very straight forward and there are seldom client questions about its meaning. The a* vs. b* plot is more complicated. We have chosen to emphasize hue and chroma over a* and b* as it is a more intuitive presentation. By coloring the outside of the graph to roughly represent hue it is quickly apparent to clients that samples with a higher hue angle are more orange and ones with a lower value more blue. In the example shown here, PN block 2 is considerably more orange than PN block 1 and this is readily apparent on the graph. We explain chroma as the intensity of a given hue. In the most basic terms, we describe sample hue as which color a sample is and chroma as how much of that color is present.

We do not attempt to explain the three dimensional aspect of the CIELab color space. Although the L* graph could be considered the third axis of the a*

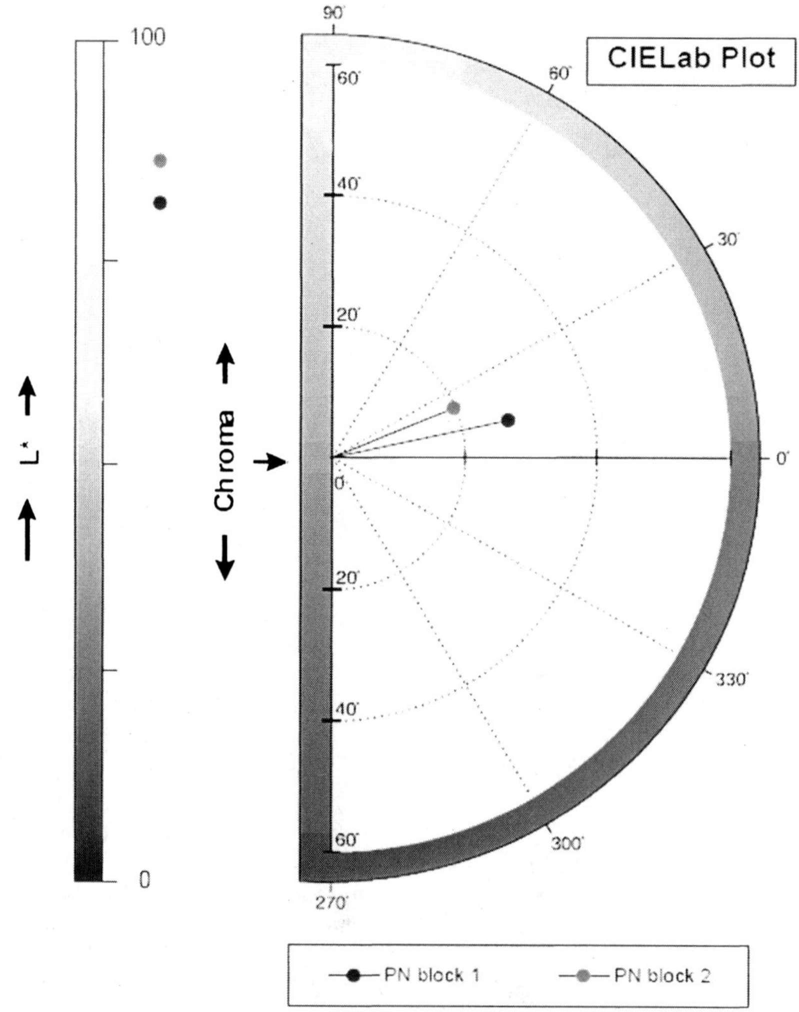

Figure 2. Plots of CIELab data L and a* vs. b*.*
(See page 15 of color inserts.)

vs. b* plot we have found three dimensional presentations too confusing for many clients.

White wines are presented with a similar format but the a* vs.b* graph has been modified. Most white wines have hue angles between 30° to 120° so the plot has been modified to fit those dimensions. The outside hue band thus goes from orange into the green.

ETS Laboratory clients routinely use the color report for research, quality control and process management. The extended set of information contained in the report addresses a range of client skill levels and the presentation of visual and numeric information appeals to a broad clientele. Although CIELab has still not been widely adapted by the wine industry clients come to ETS for color analysis when an exact color description is required.

References

1. Parr, W.V.; White, G.; Heatherbell, D.A. *J. Wine Research* **2003**, *14*, 79-101.
2. Zoecklein B.W.; Fugelsang K.C.; Gump B.H.; Nury F.S. *Wine Analysis and Production;* Chapman and Hall: New York, NY, 1995; pp 382-383.
3. Pérez-Caballero, V.; Ayala, F.; Echávarri, J.F.; Negueruela, A.I. *Am. J. Enol. Vitic.* **2003**, *54*, 59-62.

Chapter 15

Assessing Color Quality of Beer

Thomas H. Shellhammer[1] and Charles W. Bamforth[2]

[1]Department of Food Science and Technology, Oregon State University, 100 Wiegand Hall, Corvallis, OR 97331
[2]Department of Food Science and Technology, University of California at Davis, 127 Cruess Hall, Davis, CA 95616

The visual quality of beer depends on color, clarity and foam characteristics and how these support or negate consumer expectations regarding the particular style and brand. Beer color originates principally with Maillard reaction products formed during malting within the grain used to prepare the wort for fermentation. Kilning malted barley creates the greatest input to beer color while boiling wort can add substantial color in lighter colored worts, along with the oxidation of grain and hop polyphenols. In some instances, beer color is modified post-fermentation using caramel color or roasted malt extracts. The conventional method for measuring beer color throughout much of the world examines the absorbance at 430 nm. While this works well for light colored lagers it lacks the ability to measure nuances of darker, redder beer or cloudy beer color. Tristimulus measurement improves on A_{430} measurement in this case.

The Appearance of Beer

The aesthetics of beer as judged by the consumer and the quality as interpreted by the brewing scientist lies in large part with its appearance. One's first impression of whether the beer is true to type is dependent on the product's color. The presence or absence of haze can be viewed either as a defect or characteristic feature depending upon the style. For instance, world-wide pale lager beer is expected be brilliantly clear and the presence of a haze represents a defective product to both the consumer and scientist alike. Similarly, beers with the purposeful inclusion of yeast, such as German and North American hefeweizens, are meant to be cloudy, and one measure of quality is the extent and duration of the biological haze present in this style.

While the absolute color may be key to a consumer's perception of beer quality, brewers recognize that reproducible and recognizable color may be of greater importance. Consumers' expectation of flavor can hinge on the color of the product they are consuming. Experiments in which the color of a pale lager beer is modified illustrate this impact. A collection of British beer tasters, some fully trained as sensory panelists, were given two beers to evaluate: a pale colored lager beer and the same beer to which caramel coloring had been added to increase the color value by 9 °EBC (European Brewing Convention) color units thereby giving it a color of a typical British pale ale. When asked to rate each beer on a scale of 1 (most lager-like) to 10 (most ale like) nine out of ten panelists rated the colored beer higher (more ale like) than the uncolored lager beer. In fact, four of the panelists offered aromatic descriptors of the colored lager which one would normally associate with an ale, i.e., "lacks dimethyl sulfide", "more bitter", "full", "heavy", and "malty" (*1*). In a separate experiment a leading brand of American commercial light colored lager beer was colored using flavorless food dyes at four different levels up to a color that matched the color of a leading British commercial ale. When American beer drinking consumers were asked to rank the five samples in increasing order of quality (low to high), the lager beer with the greater amount of color was ranked as being higher in quality and the degree of quality improvement was correlated with the extent of color increase (*2*). Clearly, the color of the beer has a significant impact on the perceived, or expected, flavor and quality regardless of the beer type.

Standard Methods for Measuring Beer Color

The color of beer ranges from the palest yellow for lightly flavored lager beer through brown, with reddish and amber hues, for many ales and porters to deep black for stouts. Stylistic differentiation is due in large part to the grist composition which impact both flavor and color. Grist color is due principally to Maillard reaction products formed during kilning and in some instances to caramelization of sugars in specialty malts. Transmission spectra for commercial beer display significant differences from 400 – 550 nm, violet through green, and slightly lower differences from 550 – 700 nm (Figure 1).

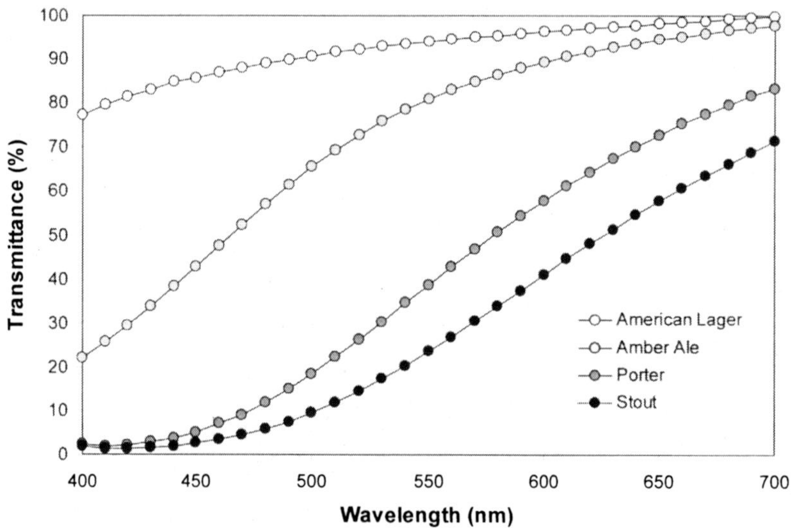

Figure 1. Transmission spectra for four different colored commercial beer styles.

Historically, beer color was determined by visual comparisons against a set of color standards developed by Joseph Lovibond in the late 19th century as well as against solutions of potassium chromate in the early 20th century. Color blindness in the human comparator, roughly 10% in men and 1% in women (*3*), inconsistent illumination spectra, and aging of the color comparison standards were key flaws in this method. These issues prompted the need for precise and reproducible measurements and led color measurement to evolve into one of spectrophotometric absorbance and/or transmission measurements.

The standard method of the American Society of Brewing Chemists (ASBC) and the European Brewery Convention (EBC) relies on absorption at 430 nm (*4, 5*). In the case of beer, selection of the proper wavelength of absorption is complicated by the fact that beer does not possess a wavelength of maximum absorbance (Figure 2). The single wavelength selected for color measurement is often one that is associated with the product's complimentary color. Since beer color typically varies from yellow to brown, and within some cases slight reddish hues (Table I), absorbance in the blue – indigo region (430 nm) is reasonable.

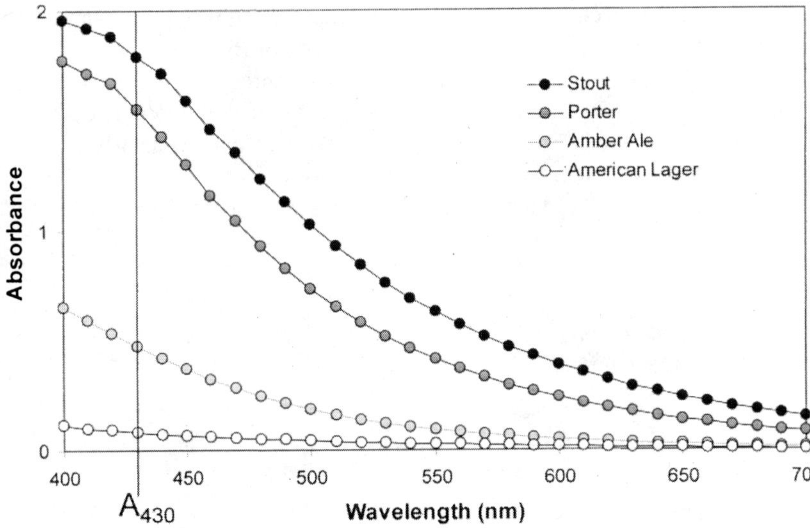

Figure 2. Absorbance spectra for four different colored commercial beer styles. Standard methods of beer analysis utilize absorbance at 430 nm.

The ASBC standard was developed in the 1950's using a half-inch cell. Absorbance readings were adjusted such that they were close in value to the standard at the time, °Lovibond (Equation 1).

$$\text{ASBC color} = A_{430}\ (\tfrac{1}{2}\ \text{inch cell}) \times 10 \tag{1}$$

Since most modern cuvettes have a path length of 10 mm, color measurements using a 10 mm cuvette are multiplied by 1.27 to accommodate for the shorter path length. The ASBC method is often referred to as being the (U.S.) Standard Reference Method (SRM), and color values may be expressed as °SRM to denote the method. Color expressed as °SRM agrees closely with color in °Lovibond and the two can be used interchangeably.

Table I. Beer color across a range of beer styles

Style	*Color*	*Color units*	
		SRM	*EBC*
American/European light lager	Yellow, straw, golden	2 - 4	4 - 8
British Pale Ale	Amber	10 - 15	20 - 30
American Porter	Dark brown	20 – 30	40 - 60
Irish Stout	Black	35 - 70	70 - 140

The caveat to spectrophotometric color measurement is that the sample must be visually clear. Thus, the ASBC color analysis is also performed at 700 nm. Should the $A700 \leq 0.039 \times A430$, the beer is considered "free of turbidity" and the color of beer can be determined from its absorbance at 430 nm. If the sample is not "free of turbidity" it requires further clarification by centrifugation or filtration followed by a repeat absorbance measurement at 430 nm. The EBC also uses absorbance at 430 nm but with a slight modification (Equation 2).

$$\text{EBC color} = A_{430}\ (10\ \text{mm cell}) \times 25 \qquad (2)$$

Thus for the same color EBC units will be approximately twice (1.97 times) as large as SRM color units.

Origins of Color in Beer

Four key ingredients serve as the basis for all beer: water, malted barley, hops and yeast. Of these, malted barley represents the single greatest impact on beer color because it is the ingredient that is used in the largest quantity and can be produced in a wide range of colors and flavors. Hops have a negligible direct contribution to beer color, but they can potentially impact the color of very pale beer via oxidation of their polyphenols. Yeast does not contribute to color unless it remains present in the final product, as in a hefeweizen, where it contributes to turbidity that is critical in the visual appearance of this particular style. Yeast can result in color loss via adsorption of colored materials to their cell wall.

The color of malted grain develops as a result of the biochemical and thermal processes during malting. Briefly, malting is controlled germination of grain with the three key steps being steeping, germination, and kilning. In the production of standard malt, with wort colors ranging from 2 – 10 °SRM, the precursors to color are reducing sugars and free amino nitrogen (FAN) that will participate in the Maillard reaction during kilning. The final amounts of these reactants are determined by the moisture content of the grain as it leaves the steeping operation and the extent to which the maltster allows the grain to germinate. Higher steep-out moisture, in the range of 45 – 50% w/w, leads to more rapid malting with greater amounts of reducing sugars and FAN at the end of germination. Similarly, lower steep-out moisture (40 – 43%) and cooler malting conditions lead to lower levels of these Maillard reaction precursors.

The biochemical sequence of events during germination of barley is initiated by water entering the kernel via the micropyle, the release of gibberellic acids via the migration of moisture across the scutellum and activation and/or release of enzymes in the aleurone layer at the periphery of the endosperm (Figure 3). Sequentially, beta glucanases, proteases, and amylases are produced within the

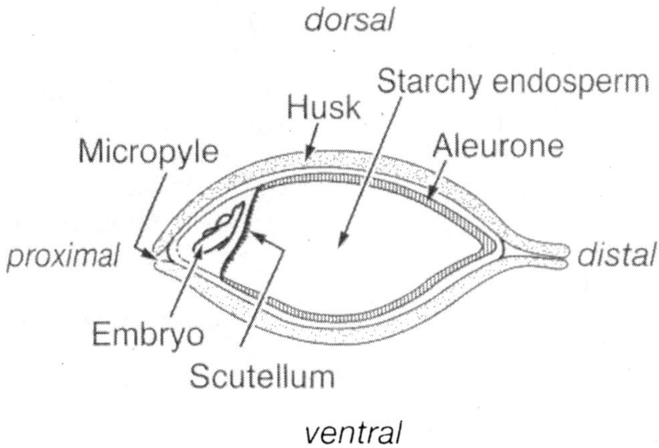

Figure 3. Cross-section of a barley kernel (Reproduced from (6). Copyright 2006 American Society of Brewing Chemists)

aleurone and slowly migrate through the endosperm toward the center of the kernel. In order to grow, the embryo must ultimately utilize amino acids produced via the action of exo- and endo-proteases on storage protein plus sugars produced from the amylolytic action on the starch granules. Enzymatic attack on these internal structures is only possibly once the endosperm cell walls, which are comprised of (1→3),(1→4)-β-D-glucan (~75%), arabinoxylan (~20%), and protein (~5%), have been sufficiently degraded to allow entry of this large molecular weight machinery. However, it is the maltster's objective to achieve near complete destruction of the grain's endosperm cell walls, and hence minimize the amount of beta glucans, without undue loss of starch via the amylases.

The final step in malting is the kilning operation, one that has two distinct phases – drying followed by curing. In the former phase, surface water and free moisture is removed under relatively cool drying conditions in order to preserve enzyme activity. In this first phase the air-on temperature is between 50-60°C. Once the surface moisture has been removed, the moisture content of the barley is approximately 12%, and the second phase of kilning begins by increasing the temperature to 80-110°C and reducing the air flow (*7*). The objective of this second phase is the formation of color and flavor via the Maillard reaction. The timing of the second phase, the final curing temperature, and the length of time at this curing temperature dictate the type and extent of color formation. Color from this type of malt is due to high molecular weight melanoidins which are yellow, orange and red initially and turn to brown as the Maillard reaction is allowed to proceed (*8*).

In the production of specialty malts, such as Caramel malt, the kilning operation is replaced with a combination of stewing and roasting in a roasting drum. Green malt, having just completed germination (45% moisture), is placed in a drum roaster and heated to a temperature which is optimal for the newly created and released amylases, 65-75°C. During the stewing phase all of the starch is converted into fermentable reducing sugars such as maltose, maltotriose, and to minor degree glucose, and nonfermentable dextrins, a process not unlike mashing which occurs in the brewery. Once this conversion is complete, the temperature is increased substantially to 80-145°C and the large concentration of reducing sugars results in the creation of substantial amounts of Maillard reaction products. The high temperatures results in the formation of nitrogen heterocyclic products that have strong toffee and nutty flavors, and in extreme cases the formation of pyrroles and pyrazines bring burnt and bitter flavors (*9*). The high concentration of maltose and high temperatures also result in caramelization reactions occurring. Caramel malts tend to be amber with red hues and are often used in the production of "red" beers.

Another class of specialty malts is produced by roasting green or kilned lager malt in a drum roaster. These roasted malts are subjected to very high

Table II. Appearance and flavor of malted barley

Malt type	*Color (SRM)*	*Appearance*	*Beer type*	*Flavor*
Standard Malts				
Wheat	1	Pale straw	Weizen	Malty
Pale Lager	2	Pale yellow	Light lagers	Cereal, DMS
Pale Ale	3	Yellow, golden	Ales	Biscuit, toasted
Vienna	4	Amber	Dark lagers	Nutty Toffee
Color/Caramel Malts				
Munich	10–20	Amber, brown	Amber beer	Intense malty
Cara Pils	5-15	Pale	Lagers	Sweet, biscuit
Caramel/ Crystal	20–120	Amber, brown, red	Ales & lagers	Toffee, nutty, burnt
Roasted Malts				
Chocolate	350	Brown, black	Porters & stouts	Coffee
Black	400-600	Black	Porters & stouts	Neutral
Roasted barley	300-800	Black	Irish stout	Bitter, burnt

temperatures, as high as 230°C which is within 20°C of the combustion temperature (248°C) for malt (*9*). The flavors and colors from these malts are caused by pyrroles and pyrazines as well as thermal degradation of carbohydrates and phenolic acids. The former produce coffee and burnt aromas while the later can result in smoky and clove-like aromas. Roasted malts offer colors which range from brown to black. Unique to this class of malts are black malts which act essentially as coloring agents offering little in flavor. High molecular weight extracts of this type can be used in small amounts to trim color of pale beer without impacting flavor just prior to packaging. Conversely, low molecular weight extracts can have the complimentary effect of adding flavor with minimal color.

Processing Impacts on Wort and Beer Color

Several steps exist throughout beer production that have a measurable impact on the color of the final product. First and foremost is the mashing step. At this point the grist composition will determine the color and flavor of the final product. Utilizing the unique color and flavor impacts of various malts (Table II), the brewer blends these to achieve a specific target. Color intensity is determined in part by wort concentration; the values listed in Table II are obtained from worts produced via a Congress mash at roughly 8°P (% wt/wt).

Wort boiling potentially has the second largest impact on wort, and hence beer, color. Boiling typically lasts 1 – 2 hours at temperatures of 93 – 110°C, with lower temperatures occurring at altitude and higher temperatures as a result of pressure boiling. The extent of melanoidin production that occurs during wort boiling is a function of temperature, time, wort pH, and concentrations of FAN and sugars. While the majority of color input to the wort/beer system comes during the kilning phase of malting, color increase due to boiling is measurable albeit minor in many cases. Nevertheless, in pale worts (2 SRM) with high FAN, it is possible to significantly increase the wort's color with a long boil. Remaining operations downstream, namely fermentation and filtration, result in minor color depletion due to adsorption on yeast cell walls and filtration media, such as cellulose fibers in pad filtration.

Oxidation of barley and hop polyphenols during beer storage will result in increased reddish, brown color. Color changes as a result of polyphenol oxidation are most apparent in pale lager beers following extended storage post-packaging. Such oxidation will be promoted by high levels of dissolved oxygen in the packaged beer as well as the presence of soluble iron imparted by brewhouse equipment, brewing water, or diatomaceous earth filtration media. Fining with polyphenol adsorbents such as polyvinylpolypyrrolidone (PVPP), prior to packaging helps mitigate oxidative browning by reducing levels of potential browning polyphenolic precursors. In darker beers, oxidative browning is masked by the colors from the colored & roasted malts.

Coloring agents such as malt extracts and caramel coloring can be added post-fermentation as a means of adjusting the beer's final color. These products are intensely colored with colors ranging from 250 – 3,500 °SRM for malt extracts and 5,000 – 30,000 °SRM for caramels (*10*). At the levels used for beer coloring applications they present a negligible impact on flavor. Caramel as a coloring agent is discussed in depth in Chapter 20 of this monograph.

Comparison of Conventional Color Measurement Techniques

The brewing industry retains A_{430} readings as the conventional standard for online and offline color measurements. Single wavelength measurement is rapid, straightforward, and easily transferable to automated online color monitoring. While the use of a single wavelength color measurement is satisfactory in representing color for some beers, such as pale yellow lager beer, it becomes less satisfactory for darker and reddish beers.

The retina of the human eye is sensitive to wavelengths ranging from 400 to 700 nm (Figures 1 & 2); and the perception of color relies on the assessment of electromagnetic absorption across the entire visible spectrum. Cone cells at the center of the retina respond individually to red, green, or blue light and together they are interpreted as color (*11*). Color measurement using tristimulus data representing lightness, red-green quality, and yellow-blue quality (CIE L*a*b*, respectively) better characterize how the human eye perceives color. The ASBC approved tristimulus color measurement of beer using the CIE L*a*b* color space as a standard method in 2002(*12*). Transmission data can be collected using a precision spectrophotometer at 5 nm intervals over the visible range and converted to chromacity values (X, Y, Z) using the spectral power distributions and color matching functions at each wavelength for the standard illuminant C. The chromacity values in turn are used to create L*, a* and b* values. Data collection software on many spectrophotometers has this function built in. The 9th edition of the ASBC Methods of Analysis includes a tristimulus color calculator written in an Excel spreadsheet format whereby the user pastes the transmission spectra into a preprogrammed spreadsheet to obtain ASBC color and tristimulus color (L*a*b* and L*C*h).

Beer color determination using tristimulus colorimetry is typically more accurate, allows for better comparisons of samples, and gives information about measured color shifts which are not available from single-wavelength methods. An example illustrating this was presented by Stephen Smedley at the Brewing Research Foundation International (*13*) in which pairs of beers having equal A_{430} readings had different L*a*b* values. Six sets of beer pairs ranging in color from 3.5 to 17 °SRM were found statistically to be visually different by a panel of human judges. Across these six sets, the net difference in L*a*b* values defined as ΔE in Equation 3 ranged from 0.43 to 7.40. The darkest pair (49.4 SRM) was not judged to be visually different despite a ΔE of 1.39. Transmission spectra for an analogous case are presented in Figure 4.

$$\Delta E = \sqrt{(\Delta L*)^2 + (\Delta a*)^2 + (\Delta b*)^2} \qquad (3)$$

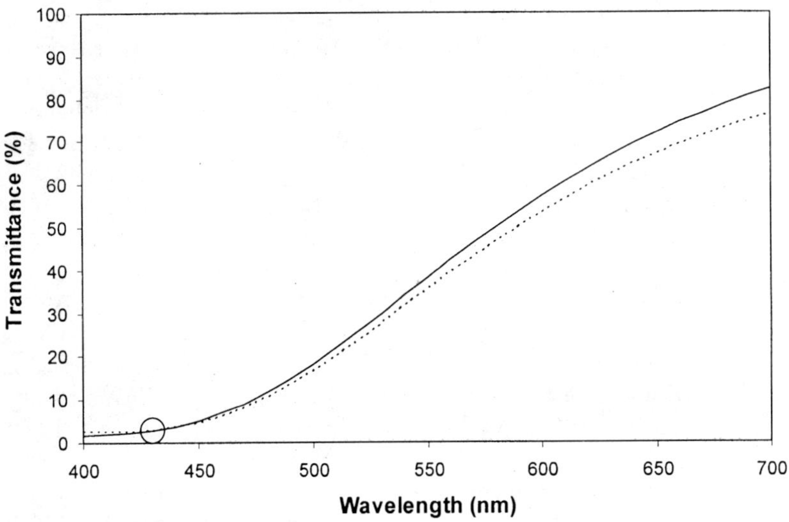

Figure 4. Transmission spectra for two beers that have different tristimulus color but have identical absorbance at 430 nm. Color = 29.7 °SRM, 38.8 °EBC. ΔE = 1.98

Beer color is determined primarily by the type and concentration of Maillard reaction products created during the malting of barley or other brewing grains, such as malted wheat. Secondary inputs to color arise from boiling wort and/or oxidation of polyphenols during beer aging. Both of these situations are more apparent in lightly colored wort and beer. Post-fermentation color adjustments are often made with caramel color or intensely colored malt extracts. Color measurement in the brewing industry is primarily based on a single measurement at 430 nm despite its apparent shortcomings. Although tristimulus color measurement has been approved by the ASBC as a standard method for quantifying beer color, it is still relegated to the laboratory. Nevertheless, equipment for rapid online tristimulus data collection using tristimulus filters, diode array detectors, or narrow bandwidth filters, is available to the brewer.

Incorporation of this technology and the use of a critical ΔE will allow the brewer to produce a visually consistent color over time.

References

1. Bamforth, C. W.; Butcher, K. N.; Cope, R. *Ferment* **1989**, *2*, 54-58.
2. Traina, C. D. The sources and measurement of beer color and its impact on consumer perception of quality. M.S., University of California, Davis, 2004.
3. Hough, J.; Briggs, D.; Stevens, R.; Young, T., *Malting and Brewing Science*. 2 ed.; Chapman and Hall: London, 1982; Vol. 2, p 914.
4. American Society of Brewing Chemists, Beer-10A. Spectrophotometric color method. In *Methods of Analysis*, 9th ed.; American Society of Brewing Chemists: St. Paul, MN, 2004.
5. European Brewery Convention, Section 9 Beer Method 9.6 Colour of Beer. Spectrophotometric Method: Instrumental Method. In *Analytica-EBC*, Verlag Hans Carl Getranke Fachverlag: Nurnburg, Germany, 2004.
6. Bamforth, C. W. *Scientific Principles of Malting and Brewing*. American Society of Brewing Chemists: St. Paul, MN, 2006; p 246.
7. Bamforth, C. W.; Barclay, A. H. P. Malting Technology and the Uses of Malt In *Barley: Chemistry and Technology*, MacGregor, A. W.; Bhatty, R. S., Eds. American Association of Cereal Chemists: St. Paul, MN, 1993; pp 297-354.
8. Nursten, H. The Maillard Reaction In *Chemistry, Biochemistry and Implications*. The Royal Society of Chemistry: Cambridge, UK, 2005; p 214.
9. Gretenhart, K. E. *MBAA Technical Quarterly* **1997**, *34*, 102-106.
10. Smedley, S. M. *Brewers' Guardian* **1995**, *124*, 42-45.
11. Smedley, S. M., *Jounal of the Institute of Brewing* **1992**, *98*, 497-504.
12. American Society of Brewing Chemists, Beer-10A. Tristimulus Analysis (Colorimetric or Spectrophotometric). In *Methods of Analysis*, 9th ed.; American Society of Brewing Chemists: St. Paul, MN, 2004.
13. Smedley, S. *Brewers' Guardian* **1995**, *124*, 44-47.

Chapter 16

Pigment Composition and Stability in Berry Juices and Wines

Maarit J. Rein[1,2] and Marina Heinonen[1]

[1]Department of Applied Chemistry and Microbiology, Food Chemistry Division, University of Helsinki, Helsinki FIN00014, Finland
[2]Current address: Nestlé Research Center, Vers-Chez-Les-Blanc, 1000 Lausanne, Switzerland

The color of berry juices and wines was enhanced and stabilized by the addition of different plant extracts and phenolic acids. This enhancement was established also with pure anthocyanins in model solutions. The color quality of black currant wine was improved by the addition of crowberry juice and grape skin extract. Strawberry and raspberry juice color was improved by the addition of black carrot, grape skin, and rosemary extracts. These commercial color enhancers immediately increased the color intensity of these juices, but during storage these improvements were not very stable. Phenolic acids improved and stabilized the colors of strawberry, raspberry, lingonberry, and cranberry juices during storage. Sinapic and ferulic acids enhanced the color of strawberry and raspberry juices the most. Rosmarinic acid intensified and stabilized the color of lingonberry and cranberry juices the most. Novel anthocyanin derivatives were formed between the phenolic acids and juice anthocyanins during storage.

Attractive color is one of the most important sensory characteristics and a significant quality parameter of fruit and berry products. The color of red berry products is, however, unstable and easily susceptible to degradation due to the reactivity of anthocyanins, the chromophore present in these products. The stability of anthocyanins is affected by pH, light, storage temperature, presence of enzymes, oxygen, structure and concentration of the anthocyanins, and the presence of other compounds such as minerals and proteins. The stability of anthocyanin color can be improved by copigmentation, where the anthocyanin molecule reacts with other natural plant components directly or through weak interactions, resulting in an enhanced and stabilized color (*1, 2*). Copigmentation is known to be responsible for the abundant color variability of bluish flowers and for stable wine colors (*3-9*), through which the phenomenon was first investigated.

Anthocyanins are natural pigments widely distributed in nature. Anthocyanin color molecules belong to the group of flavonoids having the typical $C_6C_3C_6$ skeleton, which is polyhydroxylated and polymethoxylated (*10*). In nature the aglycone skeleton is glycosylated, and the sugar moiety can also be acylated. Anthocyanins are responsible for the reds, purples, and blues in many flowers, fruits and vegetables. They are found in the petals of petunia, stems of rhubarb, and roots of red radish, for example. Fruits and berries are the most ample sources of anthocyanins in nature. In fruits and berries, anthocyanins are mainly located in the peel, like in apples and grapes, but they are also found in the pulp, as in the case of cherries or blueberries.

Anthocyanins are considered to contribute to the protective effect of fruits and berries for their antioxidant, anti-carcinogenic, anti-inflammatory, and anti-angiogenic properties, for example (*11-13*). The positive effect of fruit and berry intake on human health has been reported in several studies (*14-18*). Anthocyanins can also improve the nutritional value of processed foods by preventing oxidation of lipids and proteins in food products (*19-21*). The stability of anthocyanins becomes most significant in the case of nutritional value, as well as in the case of color quality.

Color enhancement of Berry Wine and Juices

In wine and juice making the anthocyanin content and color quality of the product depends mostly on raw materials and processing methods used. Maintaining a strong and stable color in berry wines and juices is problematic during processing and storage. In the study of black currant wine, it was shown that the anthocyanin color can be enhanced by grape skin extract or by crowberry juice addition (*22*). Black currant wine enriched with crowberry juice had the strongest color, but the most stable color appeared in black currant wine enhanced with grape skin extract. During storage the anthocyanin content declined in all the wines. However, the decline of color intensity did not take place in the same ratio with the decrease of anthocyanin content. This is presumably due to copigmentation reactions where anthocyanins condense with each other and with other organic molecules to form new pigments.

Natural plant extracts, black carrot, grape skin, and rosemary extracts, enhanced the color of strawberry and raspberry juices immediately after their addition, but also during a storage period of 100 days. This enhancement is most probably due to the overall increment of anthocyanin content in the berry juices when black carrot and grape skin extracts were added, since these products contain high amounts of anthocyanins. This is also applicable to the case of black currant wine. However, other substances in the plant extracts, such as phenolic acids and other flavonoids, can also take part in the color enhancement reactions.

The color of berry juices was also successfully enhanced by the addition of pure phenolic acids (*23*). The color of four berry juices (strawberry, raspberry, cranberry, and lingonberry) faded quickly during storage, and their anthocyanin content diminished likewise (Figure 1). Phenolic acids (ferulic, sinapic, and rosmarinic acids) improved the color of the juices by stabilizing and enhancing their color.

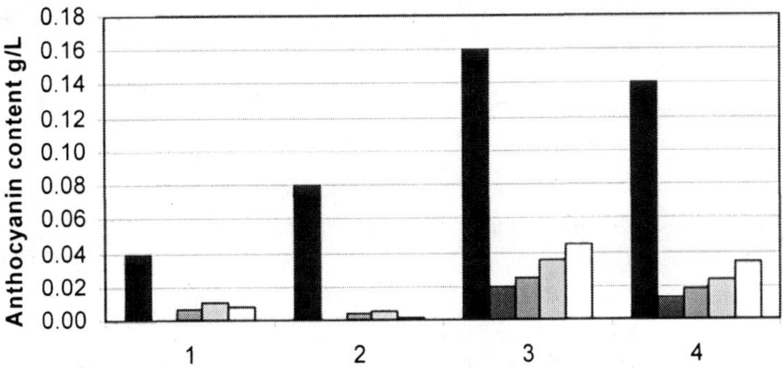

Figure 1. Changes in the anthocyanin content of berry juices during storage. 1) Strawberry juice, 2) Raspberry juice, 3) Lingonberry juice, and 4) Cranberry juice. The black bars present the initial anthocyanin concentration. Dark gray bars present the anthocyanin content of non-enhanced juices after 100 days of storage; Medium gray bars present juice enhanced with ferulic acid after storage; Light gray bars juice+sinapic acid; White bars juice+rosmarinic acid.

In strawberry juice, sinapic acid enhanced and stabilized the anthocyanin color significantly; by the end of storage the juice color intensity was 104% of the original intensity of the non-enhanced juice. Ferulic and rosmarinic acid had similar effects on strawberry juice, although not as vigorous. In raspberry juice the simple cinnamic acids, ferulic and sinapic acids, improved the juice color the most. At the end of the storage period the color intensity in raspberry juice enhanced with ferulic acid was 35%, and with sinapic acid 30%, more intense than in the non-enhanced raspberry juice at the same time point. In raspberry juice rosmarinic acid enhancement started off strong but was not stable throughout the storage period (Figure 2).

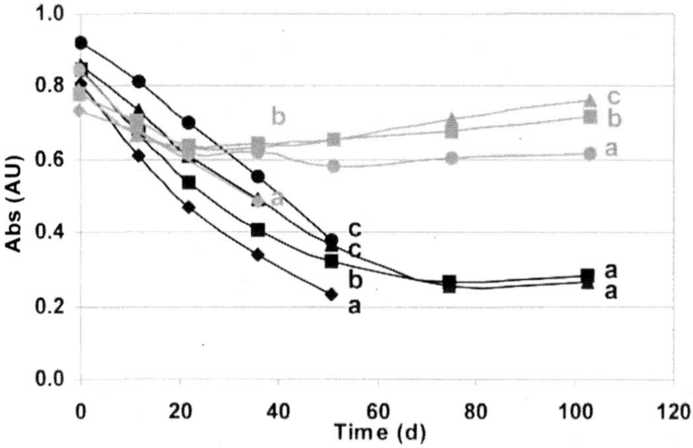

Figure 2. The color enhancement of berry juices by phenolic acids during storage detected as a change in the absorbance of λ_{max}. Black lines: raspberry juice; Gray lines: strawberry juice. Plain non-enhanced juice –◆–; enhanced with ferulic acid, –■–; enhanced with sinapic acid –▲–; enhanced with rosmarinic acid, –●–. Within the specific time point values marked by the same letter are not significantly different.

Rosmarinic acid enhanced the color of cranberry and lingonberry juices the most. The color intensity of cranberry juice enhanced with rosmarinic acid at the end of storage was 110% more than the intensity of the non-enhanced juice in the same time point. In lingonberry juice rosmarinic acid intensified the color by 50% compared to the non-enhanced juice at the end of storage. Sinapic and ferulic acids also enhanced the color of these two juices but more moderately (Figure 3).

The reactions observed within the four berry juices with added natural phenolic acids differed significantly by their mechanisms and manifestations. Intermolecular copigmentation reactions are most likely responsible for the

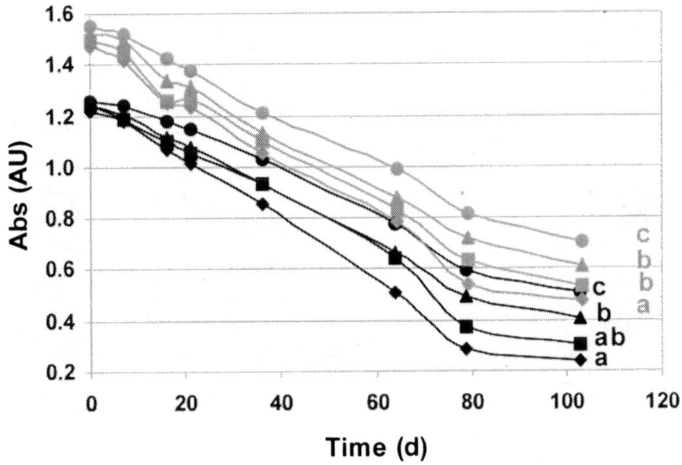

Figure 3. The color enhancement of berry juices by phenolic acids during storage detected as a change in the absorbance of λ_{max}. Black lines: cranberry juice; Gray lines: lingonberry juice. Plain non-enhanced juice –◆–; enhanced with ferulic acid,–■–; enhanced with sinapic acid –▲–; enhanced with rosmarinic acid, –●–. At the end of storage values marked by the same letter are not significantly different within a juice.

color enhancement by the conjugated cinnamic acid, rosmarinic acid, which protected lingonberry and cranberry juice anthocyanins. In raspberry and strawberry juices sinapic and ferulic acids formed new anthocyanin derived pigment molecules, pyranoanthocyanins, which possessed more stable and stronger colors compared to the color of the intact berry juices. The new anthocyanin derivatives found in the juices were 4-vinylguaicol and 4-vinylsyringol adducts of pelargonidin and cyanidin depending on the used cinnamic acid, linked to 4- carbon and 5-hydroxyl positions of the anthocyanin. This was the first time pelargonidin 3-glucoside based vinylphenol pyranoanthocyanins, but also pyranoanthocyanin of cyanidin aglycone with more complex glycosyl residues were found. This was also the first time these new derivatives were detected in non-fermented strawberry and raspberry juices (*24*).

Color Enhancement of Pure Anthocyanins

The color of pure anthocyanin molecules was intensified by the addition of phenolic acids (*25*). The immediate copigmentation effect, i.e. the color enhancement perceived directly after preparation, differed from the phenomena observed during storage. The copigmentation reactions were studied with five different anthocyanins and five different phenolic acids, and the phenomenon was observed both as a bathochromic shift and as a hyperchromic effect (figure 4).

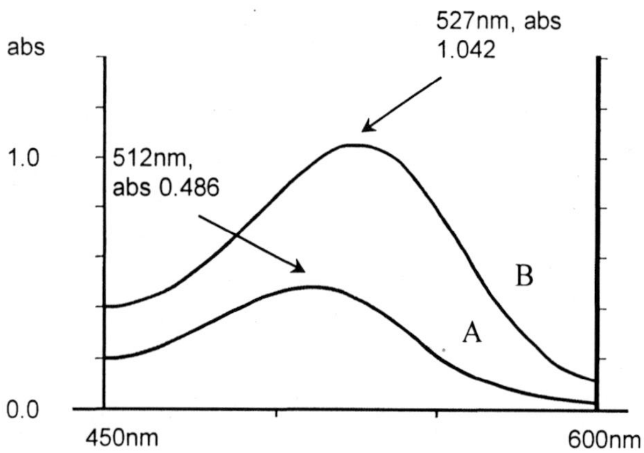

Figure 4. Change in absorption maximum wavelength (bathochromic shift) and in the color intensity (hyperchromic effect) of an anthocyanin copigmented with a phenolic acid. A) Cyanidin 3-glucoside; B) Cyanidin 3-glucoside + rosmarinic acid

Ferulic and rosmarinic acids induced the strongest color enhancement to the monoglucosidic anthocyanins, malvidin 3-glucoside, pelargonidin 3-glucoside, and cyanidin 3-glucoside immediately after solution preparation. Caffeic acid and chlorogenic acid appeared as moderate color enhancers and gallic acid was the weakest copigment on the day of preparation (Figure 5).

During storage the most efficient color enhancement was observed with ferulic and caffeic acids, especially in pelargonidin 3-glucoside solution, where the color intensity was stabilized and intensified 220% by the former and 190% by the latter of the original intensity, respectively (Figure 6). The copigmentation effect was not significant with acylated and trisaccharidic anthocyanins. Most of the copigments reduced the acylated anthocyanin color during storage.

It is most probable that the same new pigments, pyranoanthocyanins, which were detected in berry juices, were generated in the course of time in the model solutions of the pure compounds, but since they were not monitored, no distinct elucidation is available to support this assumption.

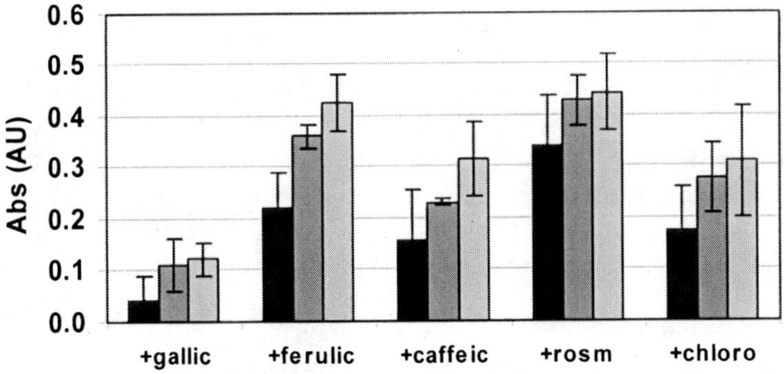

Figure 5. Increment of color intensity of anthocyanin monoglucosides due to the addition of different phenolic acids measured immediately after solution preparation. Black bar: Pelargonidin 3-glucoside; Dark gray bar: Cyanidin 3-glucoside; Light gray bar: Malvidin 3-glucoside. Zero level represents the absorption of the plain non-enhanced anthocyanin.

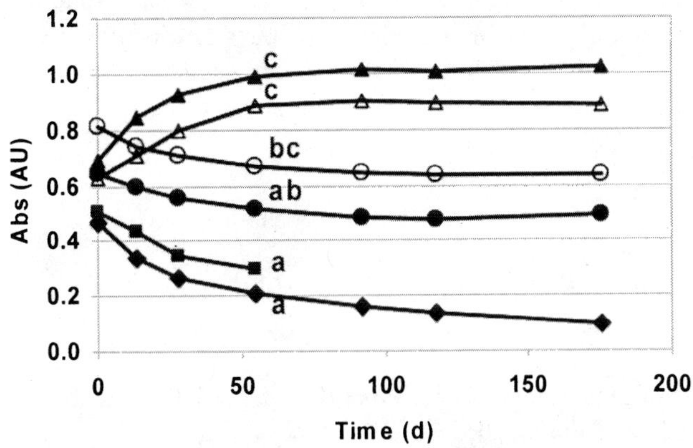

Figure 6. Copigmentation effect of pelargonidin 3-glucoside with phenolic acids during storage, detected as a change in the absorbance of λmax. Plain anthocyanin –◆–; anthocyanin +gallic acid,–■–; anthocyanin +chlorogenic acid, –●–; anthocyanin +rosmarinic acid, –○–; anthocyanin +caffeic acid, –△–; anthocyanin +ferulic acid –▲–. Within the specific time point values marked by the same letter are not significantly different.

Conclusion

Anthocyanin colors can be stabilized and enhanced both with different plant extracts and pure phenolic acids. With pure compounds and in berry juices the observed copigmentation reactions differed significantly. Conjugation of a cinnamic acid affected the way the copigment interacted with anthocyanins. Conjugation results in a weaker color stability and enhancement than what was observed with non-conjugated cinnamic acids. The simple cinnamic acids produced stronger and more stable colors and formed new anthocyanin derived molecules, whereas conjugated cinnamic acids did not.

The substitution pattern of each anthocyanin affects the chemical behavior of the pigment molecule. It was observed that methoxylation increased the color enhancement reactions and hydroxylation decreased them. Since the attached glycosyl units and acyl groups of an anthocyanin also have a strong effect on the pigment properties, they of course also affected the color enhancement reactions. Monoglucosidic anthocyanins were more susceptible to color enhancement than trisaccharide and acylated anthocyanins. It is most likely that the acylated and trisaccharidic anthocyanins are sterically too compact for further copigmentation reactions to take place.

The current results may be of use in improving the color quality of berry products and in the development of foods with anthocyanin-rich ingredients. To fully benefit from these research findings, the physico-chemical characteristics of the new pyranoanthocyanins should be investigated in respect to their other qualitative and sensory aspects for the development of berry food products.

References

1. Darias-Martin, J. ; Martin-Luis, B. ; Carrillo-Lopez, M.; Lamuela-Raventos, R.; Diaz-Romero, C.; Boulton, R. *J. Agric. Food Chem.* **2002**, *50*, 2062-2067.
2. Talcott, S. T.; Brenes, C. H; Pires, D. M.; Del Pozo-Insfran, D. *J. Agric. Food Chem.* **2003**, *51*, 957-963.
3. Asen, S.; Stewart, R. N.; Norris, K. H. *Phytochemistry* **1972**, *11*, 1139-1144.
4. Asen, S.; Stewart, R. N.; Norris, K. H. *Phytochemistry* **1975**, *14*, 2677-2682.
5. Brouillard, R. *Phytochemistry* **1983**, *22*, 1311-1323.
6. Liao, H.; Cai, Y.; Haslam, E. *J. Sci. Food Agric.* **1992**, *59*, 299-305.
7. Brouillard, R.; Dangles, O. *Food Chem.* **1994**, *51*, 365-371.
8. Yabuya, T.; Nakamura, M.; Iwashina, T.; Yamaguchi, M.; Takehara, T. *Euphytica* **1997**, *98*, 163-167.

9. Bloor, S.J.; Falshaw, R. *Phytochemistry* **2000**, *53*, 575-579.
10. Brouillard, R. In *Anthocyanins as Food Colors*; Markakis, P., Ed.; Academic Press Inc: New York, NY, 1982; pp 1-38.
11. Clifford, M. N. *J. Sci. Food Agri.c* **2000**, *80*, 1063-1072.
12. Kong, J.; Chia, L.; Goh, N.; Chia, T.; Brouillard, R. *Phytochemistry* **2003**, *64*, 923-933.
13. Rossi, A.; Serraino, I.; Dugo, P.; Di Paola, R.; Mondillo, L.; Genovese, T.; Morabito, D.; Dugo, G.; Sautebin, L.; Caputi, A. P.; Cuzzocrea, S. *Free Radic. Res.* **2003**, *37*, 891-900.
14. Hollman, P. C. H.; Hertog, M. G. L.; Katan, M. B. *Biochem. Soc. Trans.* **1996**a, *24*, 785-789.
15. Hollman, P. C. H.; Hertog, M. G. L.; Katan, M. B. *Food Chem.* **1996**b, *57*, 43-46.
16. Youdim, K. A.; Spencer, J. P. E.; Schroeter, H.; Rice-Evans, C. *Biol. Chem.* **2002**, *38,* 503-519.
17. Knekt, P.; Kumpulainen, J.; Jarvinen, R.; Rissanen, H.; Heliovaara, M.; Reunanen, A.; Hakulinen, T.; Aromaa, A. *Am. J. Clin. Nutr.* **2002**, *76*, 560-568.
18. Rissanen, T. H.; Voutilainen, S.; Virtanen, J. K.; Venho, B.; Vanharanta, M.; Mursu, J.; Salonen, J. T. *.J Nutr.* **2003**, *133*, 199-204.
19. Kähkönen, M. P.; Heinämäki, J.; Ollilainen, V.; Heinonen, M. *J. Sci. Food Agric.* **2003**, *83*, 1403-1411.
20. Kähkönen, M. P.; Hopia, A. I.; Heinonen, M. *J. Agric. Food Chem.* **2001**, *49*, 4076-4082.
21. Viljanen, K.; Kivikari, R.; Heinonen, M. *J. Agric. Food Chem.* **2004**, *52*, 1104-1111.
22. Eiro, M.; Hopia, A.; Kaukovirta-Norja, A.; Lehtinen, P.; Heinonen, M. *Vitic. Enol. Sci.* **2000**, *55*, 3-6.
23. Rein, M. J.; Heinonen, M. *J. Agric. Food Chem.* **2004**, *52*, 3106-3114.
24. Rein, M. J.; Ollilainen, V.; Vahermo, M.; Yli-Kauhaluoma, J.; Heinonen, M. *Eur. Food Res. Technol.* **2005**, *220*, 239-244.
25. Rein, M. J.; Heinonen, M. *J. Agric. Food Chem.* **2002**, *50*, 7461-7466.

Chapter 17

Non-Enzymatic Browning in Orange Juice and Mango Puree

Robert S. Greenberg[1], Catherine A. Culver[1], Nicholas S. Kretchman[1], and Jennifer DiCicco[2]

[1]Pepsi-Cola Company, R&D Center, 100 Stevens Avenue, Valhalla, NY 10595
[2]Sensory Consultant, Beavercreek, OH 45432

The quality of Not-From-Concentrate Orange Juice (NFC) and Totapuri mango puree deteriorates during storage because of changes caused by non-enzymatic browning. Juice and puree quality is typically determined by trained sensory panels. This study investigates the relationship between dissolved and headspace oxygen levels, ascorbic acid levels, juice color, and sensory scores for NFC orange juice stored under various conditions. The color and sensory scores for mango puree were evaluated as a function of storage time, temperature, and package type. These studies showed that juice quality can be rapidly evaluated by measuring the hue angle, an indicator of juice browning. The role of tin packaging in preserving mango puree color remains unclear.

The quality of fruit juices depends on the preservation of juice color, flavor, and aroma during processing and distribution. Quality losses are typically due to browning and the development of off-flavors (*1*). Juice browning develops from carbonyl-amine reactions between amino acids, sugars, and ascorbic acid breakdown products (*2, 3*). Browning degradation products isolated from citrus model systems include aroma active organic acids, furans, furanones, pyranones, and pyrroles (*1*).

The kinetics of juice browning and ascorbic acid degradation have been reported as either zero order or first order (*3, 4*). These kinetics may actually describe multiple reactions, depending on temperature and oxygen levels. The initial rapid loss of ascorbic acid may reflect oxidation by dissolved and headspace oxygen, followed by slower thermal degradation. For aseptically packaged orange juice stored six months at ambient temperature, approximately one third of ascorbic acid loss is attributed to oxidative and two thirds to non-oxidative pathways (*5*). The oxidative reaction proceeds approximately 10-1000 times more rapidly than non-oxidative processes (*2*).

Browning in juices other than citrus, apple, and pear has received relatively little attention. Processors have long known that non-enzymatic browning of pineapple and mango products can be greatly reduced by use of tin-lined steel cans (*6, 7, 8*). The chemistry behind this protection is not understood. It has been suggested that tin alters the redox potential of the juice system.

The current work explores the quantitative relationship between juice browning and flavor loss as a function of time and storage temperature for aseptically packaged not-from-concentrate (NFC) orange juice. It also examines the effect of packaging on the color and flavor of mango puree under similar storage conditions.

Procedure

Samples of Florida Not-From-Concentrate (NFC) orange juice were aseptically cold-filled into 14 oz. PET bottles. To reduce headspace oxygen, the headspace was flushed with nitrogen prior to capping. Samples were stored at 35°F, 45°F, 90°F and 110°F (2°C, 7°C, 32°C and 43°C) for five weeks. Samples were analyzed weekly for ascorbic acid content (*9*), color, dissolved oxygen, and headspace oxygen. L*a*b* values were measured with a Gretag Macbeth Color Eye 3100 colorimeter. The hue angle was calculated as $\tan^{-1}\{b^*/a^*\}$. Dissolved oxygen was measured with a YSI Model 52CE dissolved oxygen meter with a YSI 5905 BOD probe. Headspace oxygen was measured with a Mocon HS-750 headspace analyzer. At four weeks, the flavor was profiled by descriptive analysis by a ten member trained panel using a fifteen point anchored universal scale. Samples were presented in a balanced complete block design spread across two sittings.

Commercial samples of Totapuri mango puree were obtained form the Bangalore region of southern India. Product was supplied in A12 (5 kg) tin-lined steel cans and 5 L aseptic laminate foil bags. Samples were approximately four weeks old at time of receipt. Samples were stored at 40°, 90° and 110°F (4°C, 32°C, and 43°C) for seven months, or until puree color became unacceptable. Samples were analyzed monthly for ascorbic acid (*10*), beta-carotene (*11*), color, and flavor. L*a*b* values were measured with a Hunter ColorQuest XE colorimeter. Product flavor, aroma and appearance were evaluated by the Tropicana Product Development team in Valhalla, NY on a 5-point hedonic scale.

Results

Not-From-Concentrate (NFC) Orange Juice

Ascorbic Acid and Color of Stored Juice

Many studies have investigated the relationship between ascorbic acid loss and orange juice browning as a function of storage time and temperature. Comparison of test results is difficult because of differences in oxygen levels, solids content, juice thermal history, and test methods.

At the beginning of this storage study, NFC orange juice contained 1.2 ppm dissolved oxygen and 9.7% headspace oxygen. These levels decreased to approximately 0.2 ppm and 7%, respectively, after 25 days. The influence of temperature on the rate of oxygen loss is shown in Figure 1 (dissolved) and Figure 2 (headspace). The container manufacturer estimates the rate of oxygen migration through the bottle to be approximately 0.033 mL per day at room temperature. This rate has little impact on dissolved or headspace oxygen levels, but may become more significant at elevated temperatures. The initial rapid drop in dissolved oxygen is greater than the rate of ascorbic acid loss, presumably due to oxidation of other juice constituents, such as citrus oil terpenes. The headspace oxygen level does not decrease at the same rate, suggesting that the rate of oxygen diffusion into juice is slow.

Ascorbic acid loss in NFC orange juice as a function of storage time and temperature is shown in Figure 3. This loss is a key indicator of juice quality; it corresponds to loss of fresh-fruit flavor and the appearance of "old", "stale", or "oxidized" off-flavors (see below).

Aseptically cold-filled NFC orange juice is typically distributed and sold as a refrigerated product with a shelf life of 90 days. Studies of ascorbic acid

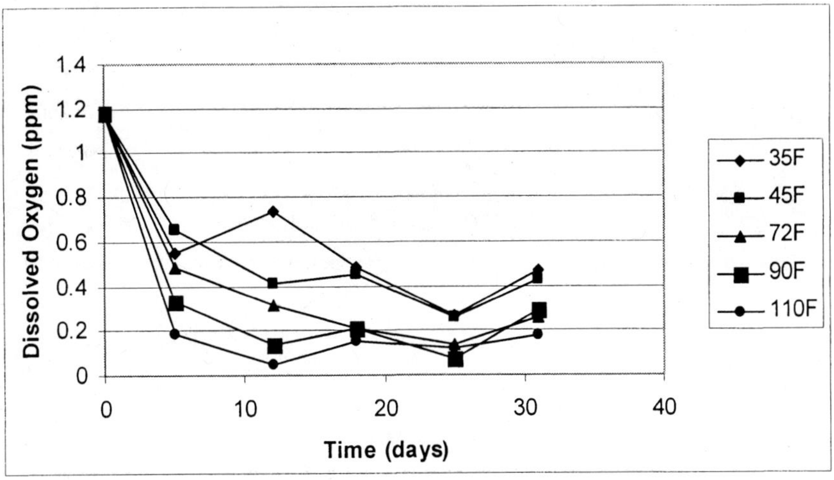

Figure 1. Dissolved oxygen as a function of time and storage temperature

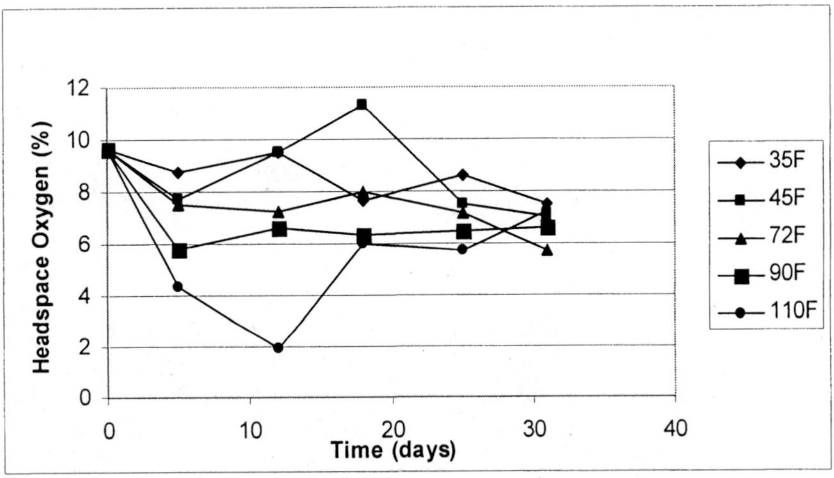

Figure 2. Headspace oxygen as a function of time and storage temperature

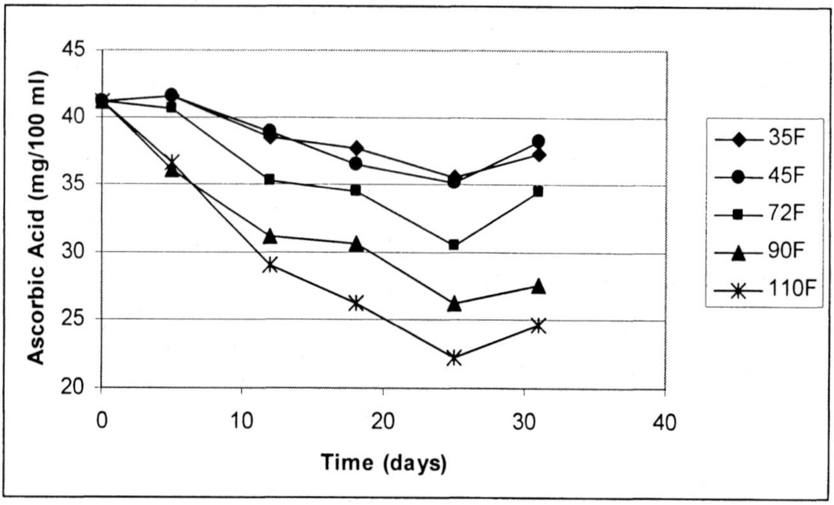

Figure 3. Ascorbic acid loss as a function of storage temperature

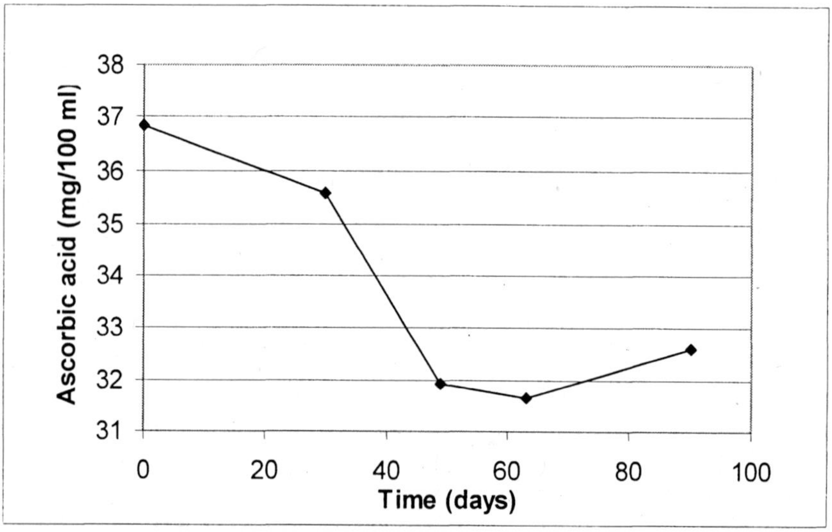

Figure 4. Ascorbic acid loss in NFC orange juice at 45°F (7°C)

losses in chilled juice were extended for an additional two months to follow the product to the end of shelf life. Ascorbic acid levels, shown in Figure 4, decreased from approximately 37 mg/100 mL to 32 mg/100 mL This 14% decrease is within the acceptable range. This data reinforces the citrus industry's current practice of distributing and selling NFC orange juice at refrigerated temperatures with a limited shelf-life.

Orange juice browning can be measured at 420 nm for clarified juice or by monitoring changes in L*a*b* values as the color shifts from yellow-orange to brown. The rate and extent of browning should increase with the storage temperature. The effect of storage time and temperature on the color of NFC orange juice is shown in Figures 5 (L*) and 6 (hue angle). As expected, juice color changed rapidly at elevated storage temperatures, indicating a substantial loss of juice quality. The hue angle is a more sensitive indicator of juice browning. L* values will eventually decrease as the juice color darkens. Studies on chilled juice were extended to the end of its typical 90 day shelf life. Changes in juice color as a function of storage time are shown in Figure 7 (L*) and Figure 8 (hue angle). Typically, industry standards consider changes of more than one L* unit or one hue angle degree to be unacceptable. The samples stored at refrigeration temperatures fell within acceptable ranges for both L* and hue angle. Samples stored at room temperature were acceptable, but with some loss of color quality. Samples kept at elevated temperatures browned rapidly, and were considered unacceptable after less than a month of storage.

Sensory Evaluation of Stored Juice

The quality and acceptability of processed orange juice is routinely assessed by trained panelists. "Raw orange" is a sensory descriptor associated with the fresh fruit character of a Florida Valencia orange. "Cooked orange" and "cooked vegetable" are descriptors associated with undesirable flavors from excessive thermal exposure during processing or storage. The effect of storage temperature on the sensory attributes of NFC orange juice is shown in Figure 9. Samples in this study were evaluated after thirty days at the designated temperature. As expected, raw orange flavor decreased and cooked flavor increased with both storage time and temperature. NFC juice stored at 35°F and 45°F received acceptable sensory scores. Juice stored at 72°F was considered acceptable, but of lower quality, while juice stored at 90°F and 110°F was unacceptable. These flavor changes are similar to those seen with other measures of juice quality, such as changes in ascorbic acid (Figure 3), L* value (Figure 5) and hue angle (Figure 6).

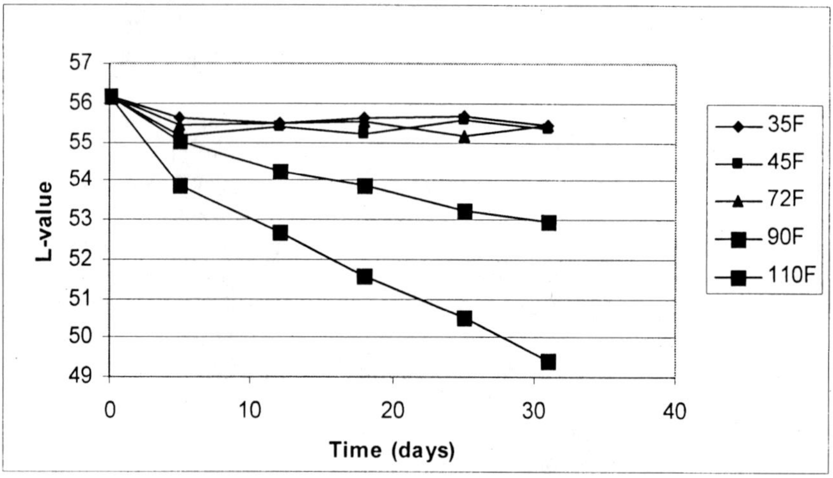

Figure 5. L value for NFC orange juice as a function of time and temperature*

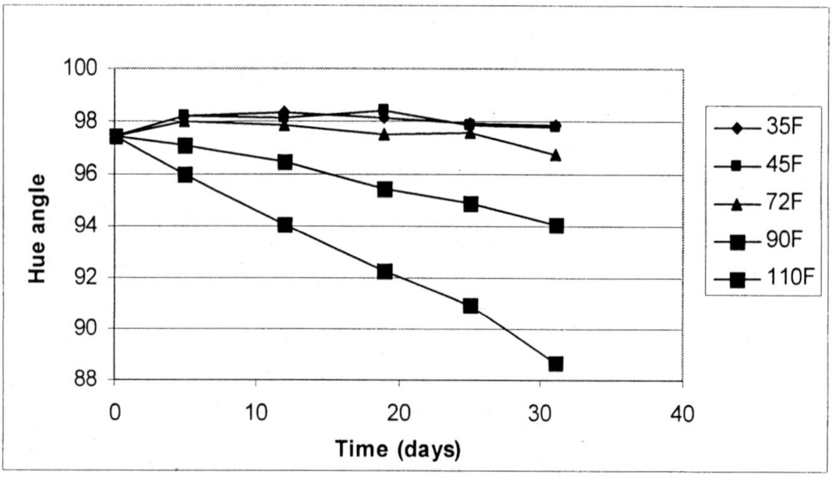

Figure 6. Hue angle for NFC orange juice as a function of time and temperature

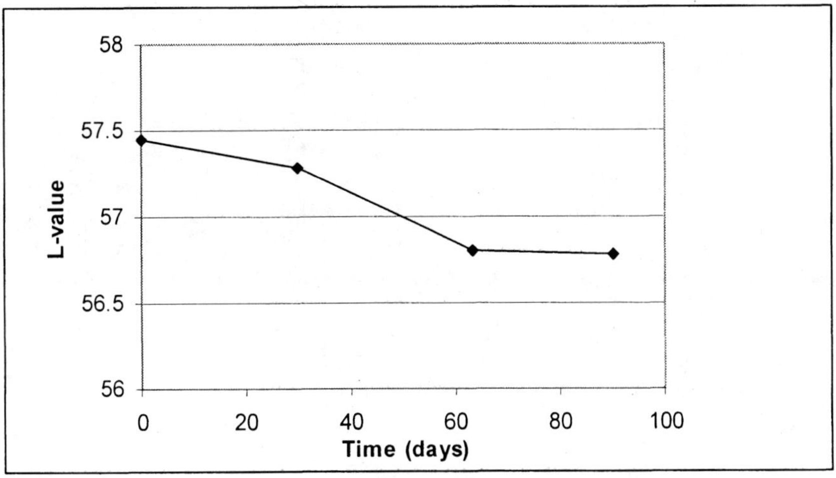

Figure 7. L value for NFC orange juice as a function of storage at 45°F (7°C)*

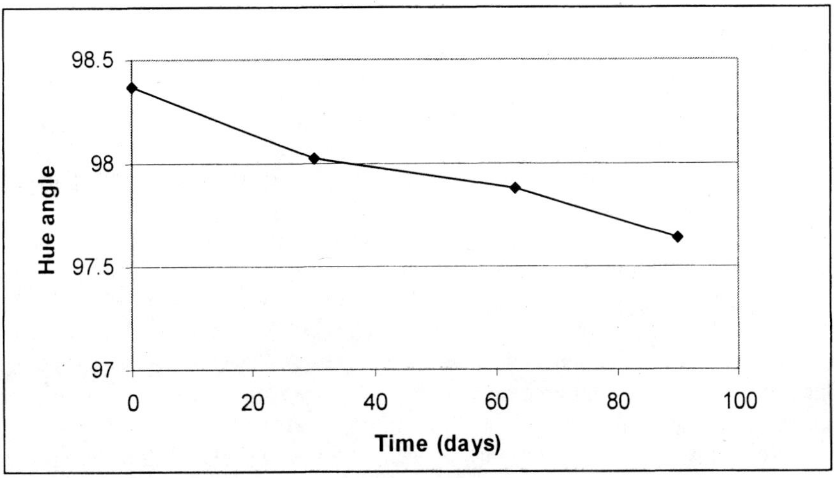

Figure 8. Hue angle for NFC orange juice as a function of storage at 45°F (7°C)

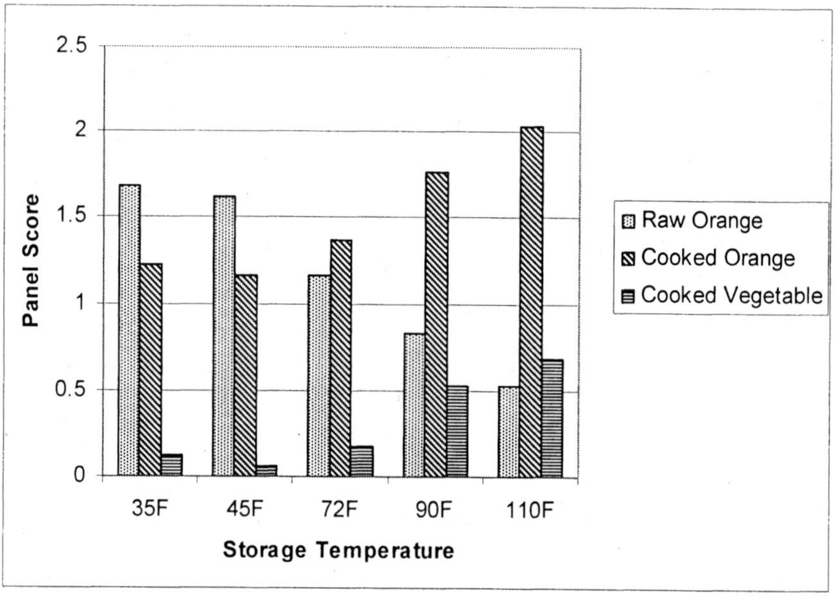

Figure 9. NFC orange juice flavor as a function of storage temperature

Totapuri Mango Puree

Color of Stored Puree

Mango products are extremely susceptible to browning if not packaged in tin-lined cans. Unlike citrus juices, processed Totapuri mango puree contains insignificant levels of ascorbic acid (less than 1 ppm), so browning is not linked to ascorbic acid degradation. Tests also showed that the beta-carotene content remained essentially unchanged over the first three months of storage even though the puree showed significant levels of browning. Analysis was discontinued after this period because of variability in can-to-can results and the lack of correlation with puree browning. Similar results have been reported for mango juice stored for 8 months at 37°C in tin-lined cans and glass jars (*6*).

Color proved to be a reliable measure of storage-related defects, such as browning and flavor changes. The change in mango puree color as a function of storage time and temperature is shown in Figures 10 (L*) and 11 (hue angle) for product in tin-lined cans, and in Figures 12 (L*) and 13 (hue angle) for product in aseptic laminate bags. Higher initial L* values and hue angles indicate lighter purees with lower levels of browning. Mango puree stored in tin-lined cans was

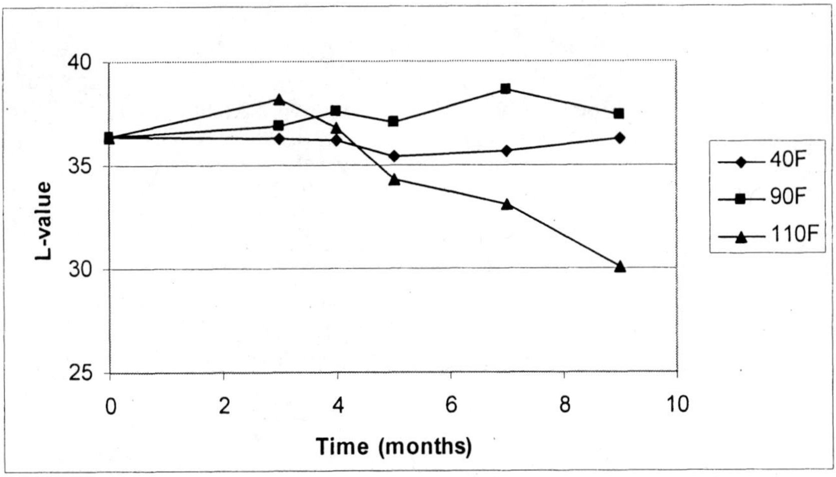

Figure 10. L value of mango puree (cans) as a function of storage time*

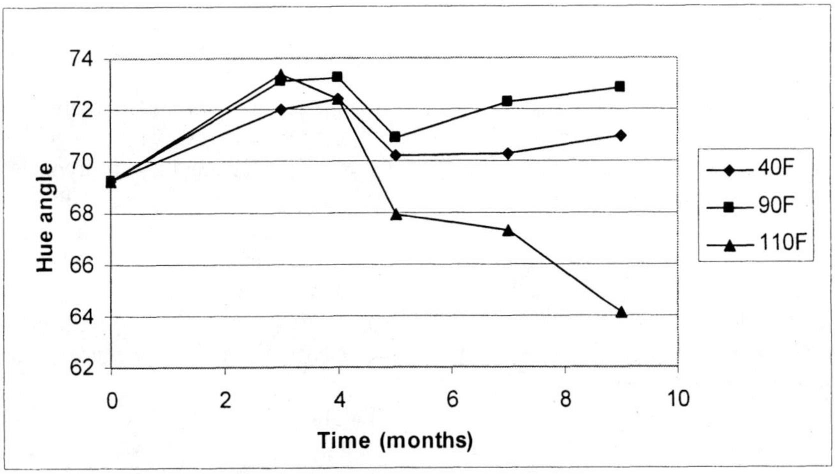

Figure 11. Hue angle of mango puree (cans) as a function of storage time

an acceptable color after seven months of storage, even at elevated temperatures. Puree stored for nine months at 110°F was considered unacceptably brown.

Mango puree packaged in aseptic bags was darker and browner than canned puree at the beginning of the storage study, presumably due to quality changes that occurred during the four week shipping period from India to the US. The color of this mango puree was acceptable at the beginning of the storage study, but dropped to unacceptable levels after two months. Mango puree in aseptic bags was much darker than canned samples at the end of the study.

Sensory Evaluation of Stored Puree

Tropicana's Valhalla-based Product Development team determined the overall acceptability of stored mango puree by evaluating its flavor, aroma, and visual appearance. Positive attributes included fresh mango flavor, a light and fresh fruity aroma, and yellow color. Negative attributes were defined as a cooked or oxidized flavor, cooked or caramelized aroma, and brown color. The overall acceptability of stored mango puree is shown in Figures 14 and 15. These results correlate well with puree color. Mango puree stored in cans was considered acceptable, except for puree stored for nine months at 110°F. The flavor of puree stored in aseptic bags was considered unacceptable after two months of storage at all temperatures.

Conclusions

Deterioration of NFC orange juice quality is due to a combination of reactions leading to browning, loss of ascorbic acid and fresh fruit flavor, and the formation of undesirable cooked flavors. Juice quality is typically determined by sensory panels. The strong linkage between juice color and flavor makes hue angle measurements an excellent indicator of quality. The challenge for juice processors is to use this information to gain a better understanding of the factors driving the chemistry of quality loss and how to control these factors. Refrigerated distribution of beverages is not feasible in many markets. Better control of juice chemistry will lead to increased juice distribution and quality.

Hue angle was also a good predictor of the quality of mango puree. Product stored in tin-lined cans maintained its color and quality over extended storage periods, even at elevated temperatures. The color and flavor of puree deteriorated rapidly when stored in aseptic bags. The effect of tin is not understood, but may be due to a change in the redox potential. Alternate systems producing a similar effect would allow the industry to use lighter and more flexible packaging.

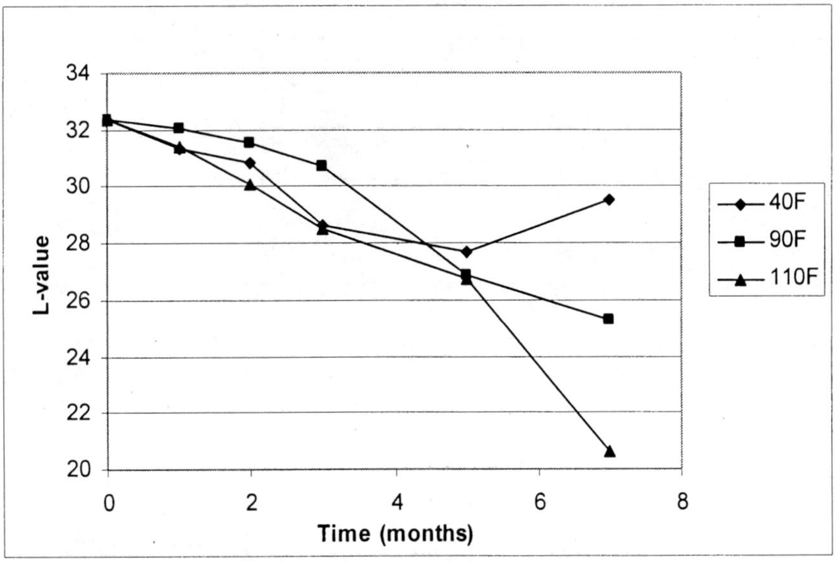

Figure 12. L value of mango puree (bags) as a function of storage time*

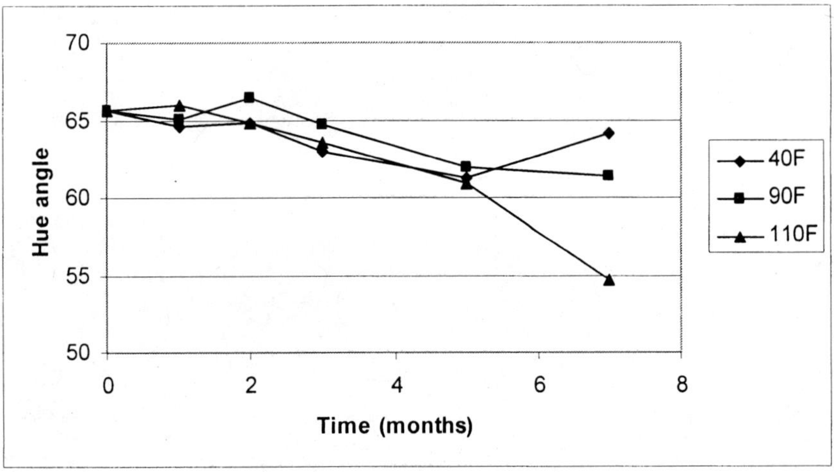

Figure 13. Hue angle of mango puree (bags) as a function of storage time

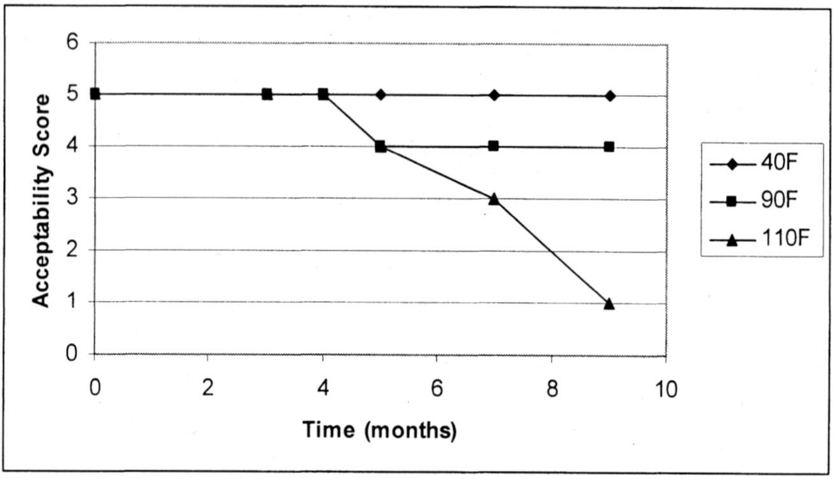

Figure 14. Acceptability of mango puree (cans) as a function of storage time

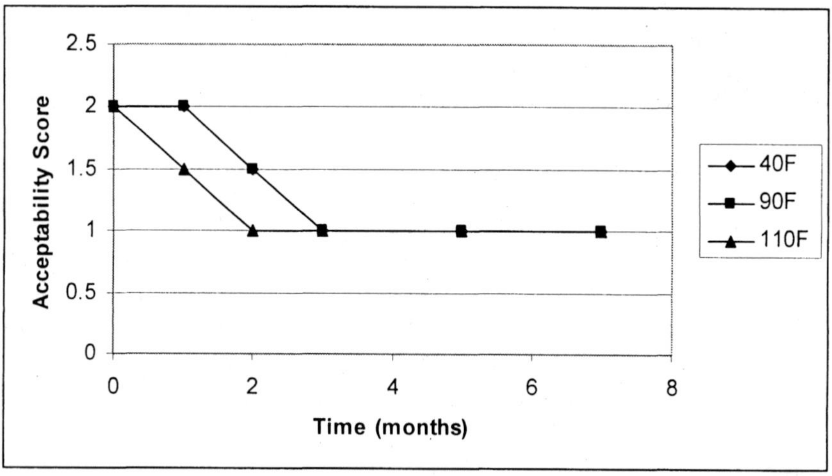

Figure 15. Acceptability of mango puree (bags) as a function of storage time

References

1. Nagy, S.; Rouseff, R.L.; Lee, H.S. Thermally Degraded Flavors in Citrus Juice Products in *Thermal Generation of Aromas;* Parliament, T. H., McGorrin, R. J., Ho, C. T., Eds., American Chemical Society: Washington, DC, 1989; p 331.
2. Johnson, J.R.; Braddock, R.J.; Chen, C.S. *J. Food Sci.* **1995**, *72*, 502-505.
3. Roig, M.G.; Bello, J.F.; Rivera, Z.S.; Kennedy, J.F. *Food Res. Intl.* **1999**, *32*, 609-619.
4. Koca, N.; Selen Burdurlu, H.; Karadenlz, F. *Turkish J. Agric.* **2003,** *27*, 353-360.
5. Sizer, C.E.; Waugh, P.L. *Food Technol.* **1988**, *42*, 152-159.
6. Mahadeviah, M.; Sheshadri, K.S.; Gowramma, R.V.; Crown, J.K. *Indian Food Packer* **1994**, *July-August*, 23-28.
7. Wu, J.S.B.; Hsueh-er,C.; Fang,T. Mango Juice in *Fruit Processing Technology;* Nagy, S., Chin, S.C., Shaw, P.E., Eds.; AgScience Inc.: Auburndale, FL, 1993; pp 620-655.
8. Nagy, S.; Barros, S.; Carter, R.; Chen, S.C. *Proc. Florida State Hort. Soc.* **1990**, *103*, 277-279.
9. *Official Methods of Analysis of AOAC International (18th Ed.);* Method 986.13; AOAC International, Gaithersburg, MD, 2005.
10. *Official Methods of Analysis of AOAC International (18th Ed.);* Method 967.22; AOAC International, Gaithersburg, MD, 2005.
11. *Official Methods of Analysis of AOAC International (18th Ed.);* Method 941.15; AOAC International, Gaithersburg, MD, 2005.

Chapter 18

Global Color Quality of Beverages Utilizing Caramel Color

Hilary A. Sepe[1], Owen D. Parker[2], Alexander R. Nixon[1], and William E. Kamuf[2]

[1]FB3 Development, LLC, 624 East Market Street, Louisville, KY 40202
[2]D. D. Williamson Company, Inc., 100 South Spring Street, Louisville, KY 40206

Caramel color, from the palest yellow to the deepest brown, has a simple job to do - create visual appeal. Caramel, accounts for more than 90% by weight of all the colors, and is produced by the controlled heat treatment of carbohydrates. Over 70% of world caramel color consumption is in soft drinks, where liquid caramel colors are almost exclusively used. High demand for caramel color in the soft drink industry lead to trials of alternative applications for caramel, primarily alcoholic and non-alcoholic beverages, such as blended whiskey, beer, and fruit juices. Combining natural and caramel colors allows for the creation of vivid beverages, such as elderberry juice concentrates that resemble grape juice and red soft drinks that develop depth as you view them. The flexibility of caramel allows for the creation of a wide variety of beverages that cater to all the senses of the consumer.

As members of the food industry know, the color of foods and beverages is of the utmost importance. Since color is often the only characteristic consumers can easily evaluate before making a purchase, they rely heavily on the color to be an indicator of appeal, quality, flavor, and consistency.

Food producers are very aware of this mindset and know how important it is for a product to look just right. For this reason, a wide variety of approved food colors are added to foods and beverages. Addition of color can help to make up for color lost in processing, add consumer appeal to a "colorless" product, or help the color of the product match the flavor. If a grape-flavored beverage appeared orange, many consumers would be unsure of the product. Their senses would be confused as what they taste would be nothing like they anticipate from its appearance. Colors added to food products can help protect light sensitive components, provide consistency to the product, so that it appears the same no matter where in the world it is purchased, and can help to add color where a matured product doesn't gain enough. When a blended whiskey has "aged" fully, it is removed from a barrel and bottled. If the beverage hasn't gained enough color from the wooden barrel during the aging process, caramel color may be added to help provide a product that is more consistent with other batches of whiskey.

Food Color Consumption

Dr. Fergus Clydesdale from the University of Massachusetts, USA aptly stated the importance of color in food products. He indicates that taste thresholds, sweetness perception, food preference, pleasantness and acceptability are all greatly influenced by the visual aspects, particularly the color, of a food (*1*). Due to the significance of color, it is important to understand the opportunities and challenges that colors pose during their incorporation into products. Generally producers struggle with determining what color a consumer would expect the product to be and finding a colorant that will provide the necessary hue while easily functioning as a component in the production system. Suppliers, on the other hand, are challenged by the fact that quite often, color is one of the last parameters considered in the research and development process. This poses an extreme challenge as replacing the color is not always a straightforward process.

Annual consumption of caramel color exceeds 200,000 tons on an annual basis and accounts for more than 80% (by weight) of all colorants added to the foods we eat and drink. It is estimated that every day over 1 billion servings of caramel color are consumed world-wide. Caramel color is an important component in many products. Major uses include pet foods, baked goods, sauces, soups, and seasonings. However, the majority of caramel color is used in the beverage industry (Figure 1).

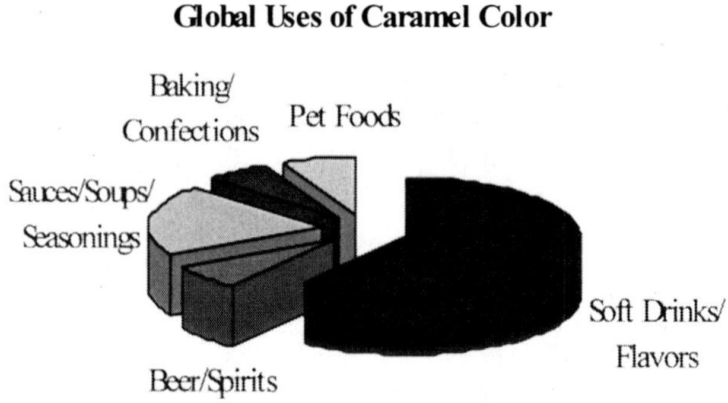

Figure 1. Global uses of caramel color, by weight.

Browning Reactions

Caramel color is a dark-brown to black liquid or solid having an odor of burnt sugar and a pleasant, somewhat bitter taste. It is totally miscible with water and contains colloidal aggregates that account for most of its coloring properties, and its characteristic behavior toward acids, electrolytes and tannins.

There are two types of caramelization reactions in food products; enzymatic browning, illustrated when damaged or cut fruit darkens at the exposed surface (Figure 2), and non-enzymatic browning which occurs when food products such as coffee beans, meat, bread or sugars are heated (Figure 3)(*2, 3*).

Non-enzymatic browning in foods proceeds in several ways with two of the most important being 1) the well-known Maillard reaction in which sugars, aldehydes and ketones, react with naturally occurring nitrogen containing compounds such as amines and proteins, to form brown pigments known as melanins and 2) caramelization reactions in which sugars are heated in the absence of nitrogen containing compounds. In the latter reaction the sugars initially undergo dehydration and then condensation or polymerization into complex molecules of varying molecular weight. Lightly colored, pleasant tasting caramel flavors are produced in the initial stages but as the reaction continues more high-molecular-weight color bodies are produced and the flavor characteristics become more bitter.

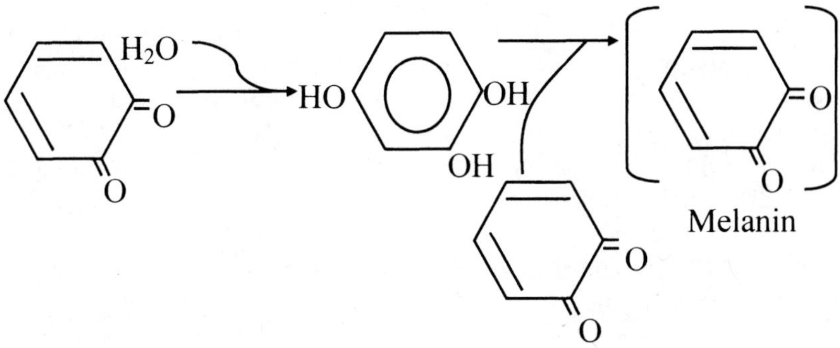

Figure 2. Unstable Quinones proceed to Melanin, producing brown color on cut or damaged surfaces of fruits during enzymatic browning.

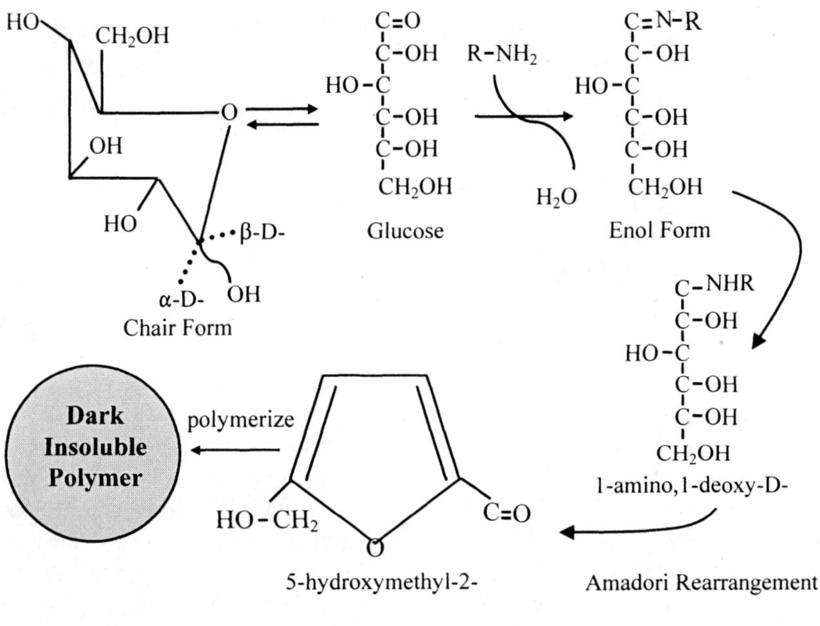

Figure 3. The Maillard reaction is the most closely related to the reaction which produces caramel color.

Caramel Color Chemistry

Caramel color first gained commercial importance as an additive in brewery products--porter, stout, dark beers and ales and as a colorant for brandy. In 1858, the first known published technical study of caramel color was authored by the French chemist M. A. Gelis (*4, 5*). Gelis' work indicated that caramelized sucrose contained three main products; a dehydration product, Caramelan $C_{12}H_{18}O_9$, and two polymers, Caramelen $C_{36}H_{50}O_{25}$ and Caramelin $C_{96}H_{102}O_{51}$. Greenshields indicated that it is common for both Maillard and caramelization reactions to give aldehydes and dicarbonyl compounds but the former type incorporates nitrogen containing components (*6*). For this case Hodge and Greenshields grouped the reaction mechanisms as follows (*7*):

1. Starting reactions
    a. sugar-amino condensation
    b. Amadori or Heyns rearrangement
2. Degradative reactions causing the formation of colorless or yellow products with strong ultraviolet absorbance and the release of carbon dioxide
    a. Sugar dehydration
    b. Ring splitting (Strecker degradation).
3. Polymerizing or condensing reactions forming the strongly colored components of relatively high molecular weight
    a. Aldol condensations
    b. Aldehyde/amino polymerization and formation of hetero-cyclic nitrogen compounds.

Production Standards for Caramel Color

Caramel color is prepared by the controlled heat treatment of carbohydrates. The carbohydrate raw materials are commercially available food-grade nutritive sweeteners which are the monomers, glucose and fructose, and/or polymers thereof (e.g., glucose syrups, sucrose and/or invert sugar, and dextrose). To promote caramelization, food grade acids, alkalis and salts may be employed in amounts consistent with Good Manufacturing Practices (GMP) (*8*).

The ammonium compounds that are employed are ammonium hydroxide, carbonate, bicarbonate, phosphate, sulfate, sulfite and bisulfite. The sulfite compounds are sumptuous acid, and potassium, sodium and ammonium sulfites and bisulphates. The compounds that can be used for all four types or caramel color are sulfuric and citric acid, and sodium, potassium and calcium hydroxide. Food grade polyglycerol esters of fatty acids may be used as processing aids (antifoam) in amounts not greater than that required to produce the intended effect. There are four types of caramel color which are of commercial importance and which have distinctive applications in foods and beverages. Each type of caramel color has specific functional properties which ensure compatibility with a product and eliminate undesirable effects such as haze, flocculation and separation. The four types of caramel color are Class I (also known as Plain or Spirit caramel), Class II (Caustic Sulfite caramel), Class III (Ammonia or Beer caramel), and Class IV (known as Sulfite-Ammonia, Soft Drink, or Acid-Proof caramel).

Caramel Standards

Caramel colors have been used for so long and in such a wide variety of food products that consumers tend to think of them as a single substance when in reality they are a family of similar materials with slightly differing properties. There are, in fact, four distinct types of caramel color to satisfy the requirements of differing food and beverage systems.

When the FAO/WHO Joint Expert Committee on Food Additives (JECFA) first examined caramel colors in 1919, they were thought to be a large number of ill-defined and complex products for which no adequate specifications existed. All that was known at that time was that they were formed from various carbohydrates when heated with a range of acids, bases and salts. In addition, an interim report by the United Kingdom Food Additives and Contaminants Committee on the Review of Coloring Matter in Food Regulation 1973, published in 1979, reported caramels to be "a multiplicity of ill-defined products geared to meet the special needs of particular users" (*9, 10, 11*)

Many studies of caramel color were undertaken thereafter with the largest and most comprehensive being initiated by the International Technical Caramel Association (ITCA). ITCA is the industry group composed of major users and manufacturers throughout the world formed to sponsor studies to further assure consumers and government agencies of the safety and suitability of caramel color ah a food and beverage colorant. ITCA commissioned the Ontario Research Foundation (ORF) of Mississauga, Ontario, Canada to do characterization work on caramel color to prove its homogeneity. The project began in November 1979 and was completed in May 1985. The main objective of the program was to develop an analytical procedure for the characterization of caramel color that was quantitative, reproducible and comprehensive. Caramel color was separated into size fractions using ultra-filtration, which were in turn resolved into subfractions by techniques based on charge, polarity, solubility and functional groups. The subfractions were then profiled using chromatographic procedures along with other physical and chemical techniques. These studies defined the four distinct types of caramel color and showed that, while each of the four types gave differing chemical profiles, the profiles of colors varying in color intensity within a type were essentially the same. This data was submitted to JECFA and other interested regulatory agencies throughout the world in June 1987 (*12*).

Caramel Color Classification

At this time, ITCA also submitted the proposed specifications and classification scheme which were based on the ORF database and on consultations with regulatory agencies and industry groups worldwide, including the European Technical Caramel Association, the British Caramel Manufacturers Association, the Canadian Caramel Association and the Japanese Caramel Industrial Association. The specifications and classification which are under review by JECFA and have been accepted as tentative by the European Economic Community (EEC) are described below (*13*).

Class I caramel colors can be produced from a variety of carbohydrate sources including glucose and sucrose. Recent research has led to Class I caramels from apple, onion and garlic sources as well as the development of Organic caramel color for use in Organic foods and beverages. Ammonium and sulfite compounds cannot be used as reactants for Class I caramel colors.

Class I caramels possess a slightly negative colloidal charge. They have a high Hue Index, indicating that they impart a yellow color to products and are stable in solutions containing up to 70% alcohol. Additionally, Class I's are known to contain high levels of furfuryl and furfural alcohol, which can bring out important flavors when incorporated into foods and beverages. For this reason, Class I caramel colors are widely used in alcoholic beverages and coffee products.

For the production of Class II caramel colors, sulfite compounds must be used and ammonium compounds cannot be used as reactants for Class II caramel color. These colors have a negative colloidal charge and are stable in alcohol up to 70%. These properties give Class II's a unique property; stability in high proof alcohols containing tannins, such as Cognac. Like charges on tannins and Class II caramel colors will repel, helping to keep the product in solution. This characteristic is also advantageous in other high proof spirits containing vegetable extracts.

Ammonium compounds must be used and sulfite compounds cannot be used as reactants for Class III caramel color production. Class III caramel colors have a positive colloidal charge, making them stable in beers which contain positively charged proteins. The same principle of like charges repelling one another applies with Class III caramel colors and beer. These caramels impart a red-brown to yellow-brown to products due to their Hue Index. Also, Class III's help to maintain foaming properties, making them the appropriate choice for malt based beverages such as malta and beers.

Class IV caramel colors must be produced using both ammonium and sulfite compounds as reactants. Class IV caramel color accounts for over 70% of all caramel color and will provide color which can vary from red-brown to gray-brown in food products. These caramels have foaming properties which are beneficial in root beer and cola-type beverages. Class IV caramels have a negative colloidal charge and are the most versatile caramel colors as they have good alcohol, acid, and salt stability.

Additionally, with the increase in popularity of diet colas, double strength Class IV caramel colors were developed. These products have comparable color intensity, yet contain less residual sugars. This helps to insure that the caloric content of these diet beverages is kept low.

Consistency of Color

One challenge faced by food ingredient suppliers is to provide consistent food color and insure product quality around the world when so many different raw materials and manufacturing facilities are utilized. Quality control becomes a very important aspect for these suppliers and insuring that test methodologies are standardized across locations can play a large role in the quality of product manufactured.

When producing a color additive, insuring consistent color is of the utmost importance. Color intensity is compared to standardized color references and measured using a documented (Food Chemicals Codex) method using a spectrophotometer.

Of additional importance is product pH. Measuring the acidity or alkalinity of a product and insuring that a consistent pH is maintained helps the customer be sure that caramel color will act in the same fashion every time it is formulated into that product.

Hue index (HI) is another useful tool to assist with color determination in a final product. The HI is calculated as

$$\log \frac{(Absorbance)_{510nm}}{(Absorbance)_{610nm}} \times 10 \tag{1}$$

and indicates varying degrees of redness in a finished product. A low HI indicates a more brown product, whereas a higher HI indicates a more yellow product. For example, a chocolate of coffee liquor would have a lower HI than beer. Different classes of caramel color will provide a different HI to a final product (Figures 4 & 5) (*14*).

Hue Index	Class I	Class II	Class III	Class IV
Calculated Value	> 7.0 – 6.3	5.5 – < 4.5	6.3 – 5.0	5.5 – < 4.5
Descripton	Pale to bright yellow	Amber to dark brown	Golden yellow to reddish brown	Reddish brown to dark brown

Figure 4. Hue Index of Different Classes of Caramel Color

Another important characteristic of caramel color is how it functions over the entire spectrum of visible light. While most other colors will have a maximum absorbance (λ_{max}) and "peak" somewhere between 400 and 700 nm, caramel color does not have such a peak.

The absorbance values of each class of caramel color vary as the product is scanned across the wavelengths, yet the shape of the curve will be similar for all classes (Figure 5). This is an important characteristic of caramel color which provides a quick indication of the sample having been blended with a different class of caramel color or adulterated by other materials, such as synthetic food colors.

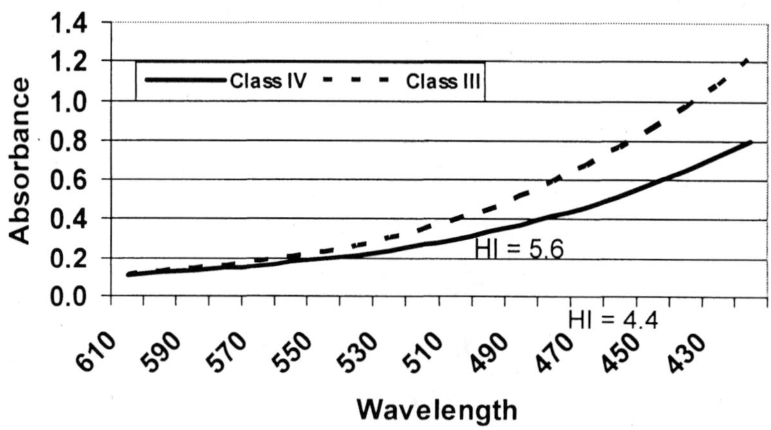

Figure 5. Absorbance and Hue Index (HI) of Class III and Class IV caramel color when scanned from 420 nm to 610 nm using a spectrophotometer.

Stability

When working with caramel color, it helps to first understand how caramel color behaves on its own, as this can provide indications as to how caramel color will react when incorporated into various food products.

It is recommended to store caramel color inside, in closed containers, under ambient conditions (<22°C and <60% relative humidity). Shelf-life studies have been performed to determine how various storage conditions can change the characteristics of caramel color. Storage time, temperature and exposure to light can have dramatic effects on caramel color. Different classes of caramel color will react differently to changing conditions.

Class IV Single Strength Caramel Colors have been shown to increase in color for approximately 30 days after manufacture. Residual sugars, in combination with high levels of base allow the reaction to continue beyond the caramelization stage. This generally yields a 6% increase in color during the first 30 days (Figure 6). Following this time, the sugars are used up and the reaction ceases, maintaining a stable color for up to two years.

Class IV Double Strength Caramel Colors will also gain color during the course of storage, but much more gradually. Assuming ambient storage conditions, a Class IV Double Strength will gain approximately 0.6% color per month (Figure 7). This generally occurs over the first 6 months of storage and results in an approximately 3% color gain during that time.

As shown above, caramel color in its concentrated form will gain color. However, when it is diluted or used in a finished product, it too is susceptible to ultraviolet light (Figure 8). To help prevent fading of the caramel color and damage to other product components, beers are sold in colored glass and blended whiskeys are sold in dark boxes or bags to prevent exposure to light.

Functionality

Caramel color has also shown to add complexity to food systems, as it has many components which can impact the flavor of finished products. Using a gas chromatograph/mass spectrometer (GC/MS) it is possible to isolate various flavor components (Figure 9). Pyrazines, for example, provide a "nutty" or "cracker-like" flavor. Furfural is a compound which can give a "bread, almond, or sweet" note, similar to the flavor provided by the charring of oak barrels for bourbon (*15*). "Heated sugar", "cotton candy", and "maple" flavors are imparted by components like 5-methyl-2 (5H)-furanone and maple lactone (*16*). Each class of caramel color has a unique flavor profile which can impact a finished beverage. Class I compounds are high in furfural and furfuryl alcohol, which are

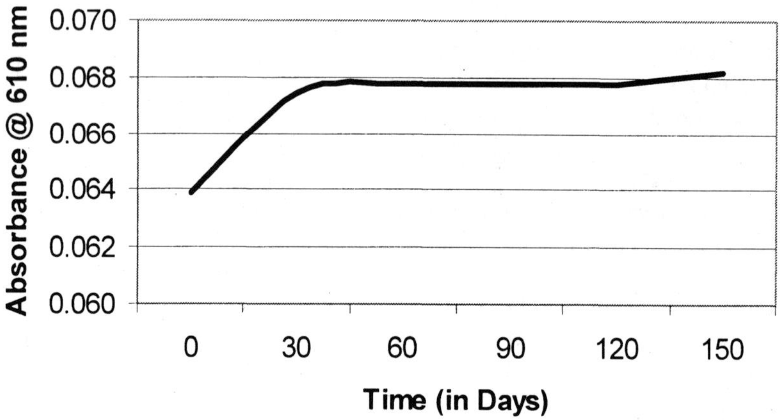

Figure 6. Effect of time on color of Class IV Single Strength Caramel Color.

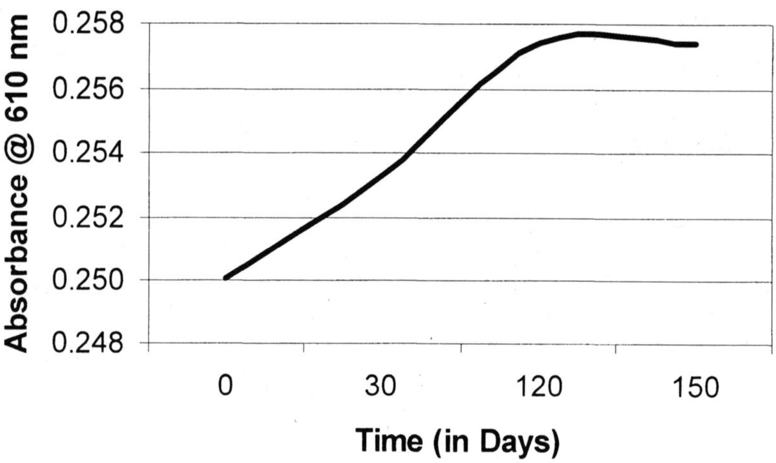

Figure 7. Effect of time on color of Class IV Double Strength Caramel Color.

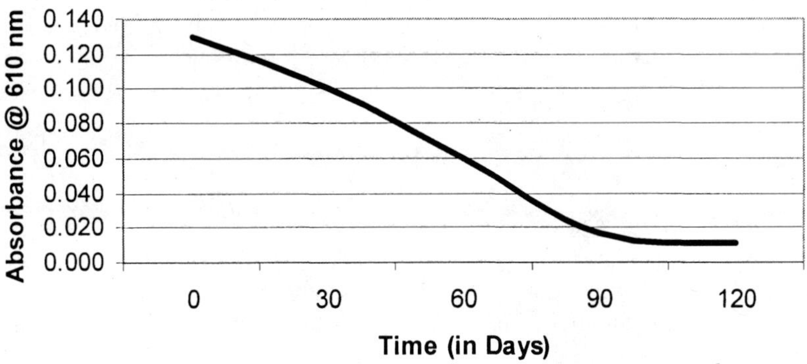

Figure 8. Light stability of Class III Caramel Color.

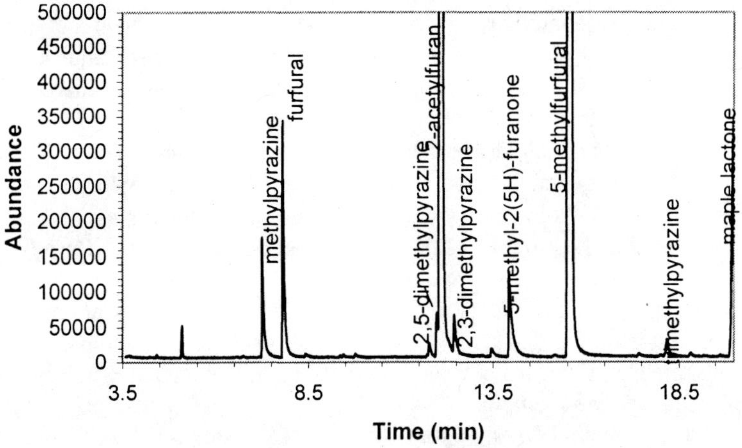

Figure 9. Flavor profile of a Class IV Caramel Color as determined by GC/MS.

common in coffee. Addition of a Class I Caramel Color to a coffee-type beverage will help enhance the flavor (*17*).

A significant discovery was made in 1971 with respect to the interactions of caramel color in finished beverages. Anwar and Calderon discovered that caramel color has the ability to act as an emulsifier in the presence of flavors. Therefore, if caramel color is the only component removed from a beverage system, it is highly unlikely for the same taste to be reproduced (*18*).

Applications

Caramel color is useful in a wide variety of applications. Although the sauce, soup, seasoning, baking, and pet food industries are large consumers of caramel color and have found great success in enhancing their products through incorporation of caramel, the beverage industry remains the largest user of caramel color by far.

The low pH of soft drinks and the necessity for an added emulsifier make Class IV Caramel Colors a typical recommendation for sodas. Addition of caramel can produce a wide variety of soft drinks. It can add the classic "cola" color that we have come to expect, can be used in minute quantities to provide a slight caramel color to ginger ale, or can be incorporated into fruit drinks to give the beverage depth and provide a more "natural" hue. This functionality is evident when caramel color is incorporated into the beverage and works well when liquid caramel is added to a finished product or when a powdered caramel is added to a powdered drink mix.

Beer and other malted drinks require Class III Caramel Colors (*19*). The proteins in these beverages interact with the positively charged caramel color and prevent either from falling out of solution, allowing for evenly dispersed color. Due to malt proteins, Malta, a common beverage in the Caribbean and Latin America, is one that also requires a Class III Caramel.

Most spirits and liqueurs require Class I or Class IV Caramels. These products are stable in 70% and 40% alcohol, respectively. Some possibilities of products manufactured with these caramels include blended whiskeys, cream liqueurs, brandy and tequila.

A Class I caramel color powder can very effectively serve as a coffee flavor extender. The furfural and furfuryl alcohol present in the caramel helps to enhance the flavor and add red tones which provide a more realistic color to instant coffee. Generally 5 to 10% addition of caramel color to coffee is required to achieve significant flavor impact. Other Class I's can function as adhesives in granola bars and rice cakes.

Color Blends

Caramels are a very diverse group of colors. The versatility of caramel is evident in the fact that it can be used to replace a combination of three certified colors. For certain applications, such as pet foods, baked goods, and instant coffee or cocoa, FD&C Red #40, FD&C Yellow #5 and FD&C Blue #1 are blended to make a brown color. By substituting caramel color for these three synthetic colors, the ingredient label on a finished product will be much "cleaner" or more consumer friendly.

Additionally, caramel color helps to add versatility when used in conjunction with natural colors. A blend of turmeric and caramel color can produce a shelf-stable bright yellow, which is a challenge with the extreme light sensitivity of turmeric. A combination of carmine and caramel color yields a heat-stable tomato red color. Blends using caramel color can help to create the color that is "just right" for a diverse group of products.

Summary

Caramel, is a flexible and versatile food and beverage color. Whether the application calls for a red hue, increased foaming properties, or stability in alcohol, acid or beer, there is a caramel color suitable to meet those needs. Caramel, like all colors, has a simple job to do – to create visual appeal in any food or beverage. Making the right decision about what color to use is an art in itself. Yet, making sure the color you choose will get the job done consistently and reliably in a wide range of applications is all science.

References

1. Clydesdale, M. F. Color as a Factor in Food Choice. In *CriticalReviews in Food Science and Nutrition;* Clydesdale, M. F., Ed.; CRC Press: Cleveland, OH, 1993; Vol. 33; p 81-101.
2. Mathewson, P. *Enzymes in Food Processing.* Food Technology Resource Group: Salt Lake City, UT, 1999.
3. *Food Chemistry: Carbohydrates;* Fennema, O. R., Ed.; 2nd edition; Marcel Dekker: New York, NY, 1985; pp 50, 98, 133, 144, 448.
4. Gelis, M. A. *Ann. Chim. Phys.* **1858**, *3*, 352.
5. Fetzer, W. R. *Ind. Eng. Chem.* **1938**, 10.
6. Greenshields, R.N. *Proc. Biochem.* **1973**, *April.*
7. Hodge, J. E. *J. Agric. Food Chem.* **1953**, *1*, 928-934.

8. National Academy of Science. *Food Chemicals Codex*, 4th edition; National Academy Press: Washington, DC, 1996; pp 80-85.
9. Food and Agriculture Organization of the United Nations. Technical Report Series No. 617. World Health Organization, Rome, 1977.
10. Twenty-ninth report of the Joint FAO/WHO Expert Committee of Food Additives. World Health Organization, Geneva, 1986.
11. Official Journal of the European Communities, *Commission Directive 95/45/EC*, July 1995, pp 22-24.
12. JECFA. *Compedium of Food Additive Specifications.* Joint Expert Committee on Food Aditives, FAO/WHO. Foreign National Press, 1992.
13. ITCA. *Specifications for caramel color.* International Technical Caramel Association, Washington DC, 1984.
14. Linner, R. T. *Proc. Soc. Soft Drink Tech*, Washington, DC, 1970.
15. 15. Acree, T.; Arn, H. Gas chromatography – olfactometry (GCO) of natural
16. products. www.flavornet.org (accessed Mar 11, 2007).
17. 16. McGorrin, R. J. Character-Impact Flavor Compounds. In *Sensory-*
18. *Directed Flavor Analysis*; Marsili, R., Ed.; Taylor and Francis:
19. New York, 2007; pp 223-267.
20. 17. Kroplein, U. *J. Chromatogr.* **1986**, *362*, 286-290.
21. 18. Anwar, M. H.; Calderon, M. Emulsions of flavoring oils and process for
22. making same. U.S. Patent 3,622,343, November 23, 1971.
23. 19. EBC method. *J. Inst. Brewing* **1950**, *56*, 373.

Meats and Seafood

Chapter 19

Color Quality of Salmon

Grete Skrede and Jens-Petter Wold

Matforsk AS, Norwegian Food Research Institute, Osloveien 1, N–1430 Aas, Norway

Farming of the pink-red colored Atlantic salmon (*Salmo salar* L.) is an expanding industry in many countries. The pigments of the salmon are carotenoids, mainly astaxanthin, which are supplied through the diet. In commercial farming, the fish are graded according to flesh color, an important price setting and consumer acceptance factor. As a consequence, a need for non-destructive color assessments appeared at an early stage in the development of the salmon farming industry. Instrumental color analysis based on the CIELAB color space proved well suited for assessing salmon flesh color. A color card was developed based on selection of suitable standards from the Natural Color System (NCS), combined with visual judgment of flesh color of representative salmon by a trained sensory panel. Further developments have resulted in the *Salmo*Fan™ which today is used world-wide for grading salmon color. On-line monitoring of salmon fillet color, using the visual spectra from a multispectral camera, is currently being developed.

Wild salmon is a traditional and highly valued food with an attractive pink-red color. The fish is also a popular prey among anglers. The fish is anadrome, i.e. hatching and the initial growth period occurs in fresh water. The second growth period is in salt water when the fish migrate to the sea. When the time comes, salmon return to fresh water, normally their hatching river, to spawn.

Wild salmon around the world belong to various species. In the Pacific there are five salmon species of commercial importance (*1*). These are the Chinook (*Onchorynchus tshawytscha*), the Chum (*O. keta*), the Coho (*O. kisutch*), the Sockeye (*O. nekra*) and the Pink salmon (*O. gorbuscha*). The various species vary widely in color. The flesh of Pink salmon is pink, of Chum salmon reddish-pink, of Coho salmon reddish-orange, and of Chinook salmon (springs) red. The Sockeye salmon has the reddest flesh among all wild salmon species. In the Atlantic, the Atlantic salmon (*Salmo salar* L.) is the predominant species (Figure 1). The Atlantic salmon has a medium color intensity compared with the Pacific salmon.

In the late nineteen sixties farming of Atlantic salmon (*Salmo salar* L.) was initiated in Norway and today salmon farming is an expanding industry in many countries. Total world production in 2005 amounted to 1,257,000 metric tons (*2*). Norway produced about 47% of the total, Chile 30% and the UK about 11% of the world production. The farmed fish is traded worldwide.

Pigmentation of Wild and Farmed Salmon

The pigments giving the pink-red color to salmon flesh are carotenoids, the main pigment being astaxanthin. There may also be minor levels of canthaxanthin. In the wild, astaxanthin accounts for more than 90% of the total carotenoid content found in salmon flesh (*3*). In addition to giving the characteristic salmon color, astaxanthin has been shown to have a role in the growth and health of the fish (*4*), possibly due to its antioxidant properties (*5*).

The salmon can not endogenously synthesize astaxanthin and thus rely on supply of pigments from the diet to obtain the characteristic salmon color (*6*). The pigments are absorbed during the digestive processes and deposited in the fish flesh. In the wild, the salmon obtain carotenoids from microscopic algae, which can synthesize the pigment, and from organisms having fed on these algae.

Pigments in Salmon Farming

During the initial stages of the commercial salmon farming, processing waste from the shrimp, crab and crayfish industry was used for pigmenting the fish (*7*). However, as the demand for pigments increased and the industry turned from semidry to dry feed, pure carotenoids were taken into use, first

Figure 1. Farmed Atlantic salmon (Salmo salar L.). (See page 16 of color inserts.)

canthaxanthin and then later astaxanthin. Today nature identical astaxanthin is the main source of pigment in diets for farmed salmon. The pigment is distributed as small water-dispersible beads of gelatin encapsulated astaxanthin. These beadlets are ready for inclusion into the feed mix during processing of the diets.

Astaxanthin obtained from microorganisms including red yeast (*Xanthophyllomyces dendrorhous* (formerly *Phaffia rhodozyma*)) is being tested as a future pigment source for fish feed (*8*).

Factors Affecting Pigment Deposition

There are a number of factors affecting flesh pigmentation of salmons. The various salmon species vary greatly in color, mainly because they have different abilities to deposit carotenoids in the flesh (*6*). Also within a certain species there are individual variations in pigment incorporation and different fish may have quite different colors even when fed identical diets (*9*). Further, there is significant temporal variation in the pigmentation of salmon as the fish flesh will lose pigments and become paler during spawning (*10*).

The pigment deposition in the salmon will increase with increasing pigment levels in the diet, up to a certain level (*11*). In general, however, the deposition rate of dietary astaxanthin in salmon is rather low (≈ 15%) (*6*). With Atlantic salmon, astaxanthin concentrations up to about 10 mg/kg may be obtained with excess astaxanthin level in the diet. In comparison, some Pacific species may have considerably higher carotenoid concentrations.

The level of carotenoids in salmon feed is regulated internationally. At present the EU sets a maximum level of 100 mg/kg feed for astaxanthin and 25 mg/kg feed for canthaxanthin (*12*). In practical farming today, levels of 50-60 mg astaxanthin/kg feed would be typical.

There is a great focus on optimizing the level of pigment in the feed, as pigment cost represents a substantial proportion of the total feed costs. In commercial salmon farming, the cost of pigment amounts to approximately 15%

of the total feed cost, which again amounts to about 50% of the total production costs at farm level (*6, 13, 14*).

Color Assessment of Salmon

Color of salmon may be evaluated in several ways: chemically, instrumentally and visually. Regardless of which methods are chosen, sampling of salmon flesh is a challenge. The salmon body is not evenly pigmented or colored. The pigment level varies along the fish body, being higher at the tail end and at the lower section of the fish body. Also muscle structure (binding tissue, fat) varies along the fish body or fillet and this variation may influence color achievement. To make sampling reproducible, standardized sampling areas like the Norwegian Quality Cut (NQC) has been established (*15*). The cut is preferably used when assessing color and other quality traits of salmon. Other standards like the Scottish Quality Cut (SQC), are also in use (*16*).

Chemical Carotenoid Analysis

For quantitative and qualitative carotenoid analysis, traditional chemical methods are used. The salmon flesh samples are extracted, followed by non-specific carotenoid analysis using a spectrophotometer (*17*) or by more specific methods using liquid chromatography (*18*).

The lipophilic character of the carotenoids requires the use of strong organic solvents for extraction. As the carotenoids are very susceptible to oxidation, an antioxidant is often added to the extraction medium. To prevent degradation by light, the analyses are often carried out in dim light or under dark cloth.

In spectrophotometric methods, concentrations are calculated by means of molar absorptivity at absorbance maximum for the dominant carotenoid. In liquid chromatography, external or internal standards are used and the concentrations of these are also often determined from their absorbtivity. Carotenoid levels in salmon are normally reported as mg/kg (ppm).

A typical extraction procedure includes extraction of minced muscle samples by a 1:1:3 (v/v/v) mixture of distilled water, methanol (containing 500 mg/L butylated hydroxytoluene (BHT)) and chloroform (*18, 19*). After centrifugation, aliquots of the chloroform containing the carotenoids are evaporated to near dryness on a water bath by a gentle N_2 flow. The samples are dissolved in 20% acetone in hexane and filtered (0.45 μm) directly into vials, sealed and injected into the HPLC the same day. The chromatographic system includes a Spherisorb S5-CN nitrile column (PhaseSep, Queensferry, Clywd, UK; 250 mm column length; 4.6 mm internal diameter; particle size 5 μm) using

20% acetone in hexane as the mobile phase under isocratic conditions (1.5 mL/min). These conditions also allow for analysis of various astaxanthin isomers that may occur in the fish muscle.

Instrumental Color Analysis

Quality grading in practical farming and international trading requires methods that are more rapid and nondestructive. Both instrumental and visual analysis is currently utilized in assessing color of salmon.

There are, however, challenges connected with color assessment of salmon, as both raw and cold smoked salmon flesh is translucent. A salmon fillet will vary in thickness and the skin may vary in color. Unless special precautions are taken, this will influence the results of both instrumental and visual color analyses (*20*). Conditions must be standardized with regard to sample size, sample background and light source. Although the optimal way would be to remove skin and cut out samples, the practical solution is most often to evaluate flesh color at the quality cut (NQC) or directly on the fillet.

Instrumental color analysis of salmon is based on the CIELAB color scales (*20*). With the red/yellowish color of salmon flesh, the color is presented by positive values for red (a^*) and yellow (b^*). Ranges obtained for color parameters CIE 1976 $L^*a^*b^*$ (CIELAB) of raw Atlantic salmon are L^* 32 - 51, a^* 1 - 14, b^* 4 – 16, resulting in hue (h_{ab}) values ($\tan^{-1}(b^*/a^*)$) between 16 and 86 (*11*).

The relationships between color parameters obtained from instrumental analysis and astaxanthin concentrations in the fish have been studied in detail. By applying linear regression analysis, significant correlations between astaxanthin concentration and lightness L^* ($r = -0.58$, $p < 0.05$), redness a^* ($r = 0.89$, $p < 0.001$), and hue h_{ab} ($r = -0.69$, $p < 0.01$) have been reported (*9*). By applying non-linear regression models and higher numbers of fish, even higher correlation coefficients have been obtained (*11*).

Several instrument models are used for instrumental color analysis, resulting in slightly different color values (*9, 21*). Due to the need for flexibility in the farming industry and the market place, hand-held models are the most popular.

Color of Processed Salmon

The most common ways of processing salmon are cold smoking and baking. Such processing influences color as can be described by instrumental analysis (*9, 17*). With cold smoking, lightness (L^*) increases while intensity in red (a^*) and yellow (b^*) as well as hue (h_{ab}) decrease (Table I). This means fillets turn lighter, less red and yellow and develop a more reddish hue during cold smoking. When baked, lightness (L^*), intensity in yellow (b^*) and hue (h_{ab}) increase and intensity in red (a^*) decreases. This corresponds to a more light and yellowish color of the baked flesh compared with the raw fillets.

There are significant correlations between the initial carotenoid concentration of the fillets and the color parameters of the final products (*9*). The initial color quality of the salmon is thus of major importance for the color characteristics of the final products.

Table I. Carotenoid Concentration and Instrumental Color Parameters (CIELAB) of Flesh from Farmed Raw, Cold Smoked and Baked Atlantic Salmon

Processing	*Lightness* (L^*)	*Redness* (a^*)	*Yellowness* (b^*)	*Hue* (h_{ab})
Raw	41.4 ± 1.8	10.5 ± 1.0	14.1 ± 1.0	53.1 ± 1.4
Cold smoked	46.3 ± 1.9	8.2 ± 1.4	8.9 ± 2.0	47.2 ± 5.1
Baked	73.2 ± 1.5	5.2 ± 1.1	18.3 ± 1.1	74.1 ± 2.7

NOTE: Carotenoid concentration of raw flesh averaged 6.4 mg/kg.

SOURCE: From Reference 9.

Visual Color Assessment

The human eye is excellent in differentiating and rating colors, especially when given the possibility of comparing colors physically placed close to each other (*20*). However, when left without any reference color to compare with, the ability of humans to rate colors is greatly reduced.

Trained sensory panels are frequently used for assessing color of salmon. By using specific samples for calibration of the panelists, letting them have reference samples or printed color standards to compare with, reproducible color assessments are obtained.

Color Standards for Atlantic Salmon

Developing a Color Standard for Raw Salmon

During the development of the international salmon farming industry it was early realized that some sort of a color standard would be useful in grading salmon in the marketplace. In the late eighties Matforsk was hired by the main astaxanthin producer at that time, F. Hoffmann-La Roche Ltd., to contribute in the development of a color card for salmon (*22*). The color card should be used by the salmon farming and processing industry in evaluating pigmentation of salmon.

The development should be based on a representative selection of farmed Atlantic salmon with the widest possible range in astaxanthin available at that time. The fish should be visually matched with proper color standards selected from a color system by using a trained sensory panel. The color card should be developed from the selected color standards.

Based on carotenoid analyses, 51 salmon samples with astaxanthin content ranging from 3.0 to 9.5 mg/kg were selected. The Natural Color System (NCS) was used for selecting the standardized colors in the experiment. The system has the notations of blackness (s), chromaticity (c) and hue, all ranging from 0 to 100% in intensity. With salmon color, hue describes the percentage of red in yellow. Thus, a color notation of 2050-Y70R corresponds to blackness s = 20, chromaticity c = 50, and 70% red (R) in yellow (Y).

In a preliminary test with salmon of high and low astaxanthin concentrations, a number of presumably suitable NCS standards were selected. In the cases where the gap between color standards of the NCS seemed too wide, additional standards with more close notations were ordered. The main experiment included 51 different NCS color standards obtained as painted papers with the NCS notations printed on the back. All standards had 60% gloss.

In the main experiments, flesh samples from each fish were circulated among the sensory panelists who compared each sample with each of the color standards to select the best match. For each salmon, average NCS notations (blackness, chromaticity and hue) were calculated and systematic trends in the NCS notations were studied.

The results revealed that there was no significant correlation ($r = 0.19$) between NCS blackness and NCS chromaticity. Thus, the variation seen in blackness was not related to color intensity. Between average NCS hue and NCS chromaticity, however, there was a significant linear relationship ($r = 0.87$, $p < 0.001$). This means that the color of Atlantic salmon followed a linear trend from low color intensity and a yellow-red hue to higher color intensity and a more red hue. Along this trend regularly spaced chromaticity and hue notations could be chosen for a color card. With the lack of correlation between blackness and hue, blackness was not specifically defined and should be presented at more than one level.

The results served as the basis for the Color Card for Salmonid produced by F. Hoffmann-La Roche and used by the salmon farming industry for many years. The card had 8 different colors, each at two lightness (blackness) levels. Instructions both regarding sampling position and lightning conditions were supplied with the card.

*Figure 2. Salmo*Fan^{TM} *used for evaluating color of salmon world wide. (See page 16 of color inserts.)*

Color Standard in Use

Since 1997 a modified edition, the Roche *Salmo*Fan™, has been used worldwide as an important tool for the industry (Figure 2). One of the major changes compared to the initial color card is the wider range of colors, allowing more highly pigmented salmon species to be evaluated as well. The fan has 15 blades numbered from 20 to 34 corresponding to increasing salmon color. The color scale is also produced as a *Salmo*Fan™ Lineal with the 15 color standards (Figure 3), and as a mini card with 10 color standards.

Significant linear relationships between astaxanthin concentration of salmon and SalmoFan^{TM} scores have been reported (*23*). Little has been published on the correlation between score of the *Salmo*Fan^{TM} and CIELAB color values. One study reported redness a^* values to correspond roughly to *Salmo*Fan^{TM} scores, but found that instrumental analysis was preferable to differentiate between samples (*24*).

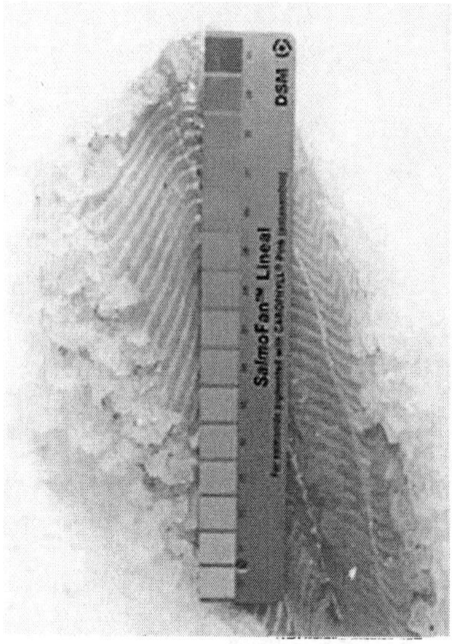

Photo: Annica Thomsson

*Figure 3. Salmo*Fan™ *Lineal applied in evaluating salmon fillet color. (See page 17 of color inserts.)*

Consumer Preferences

In commercial salmon farming, successful pigmentation is vital for consumer preference, wherever the fish is sold. In a focus group study performed in Seattle and Boston, consumers who reportedly ate salmon on average once a month, reported color along with rich taste, nice texture and versatility, among their immediate associations for salmon quality (*3*). The color was equated to that of red meat. When the focus groups were shown pictures representing different color levels, two-thirds preferred a red color (33 on the *Salmo*Fan™). The consumers felt they received salmon corresponding to a score of 27 on the *Salmo*Fan™ in the market while they preferred a color of 33.

In a Norwegian stated-choice-study, consumers were presented pictures of salmon fillets with increasing intensity in color (*Salmo*Fan™ 21 - 32) (*13*). The color of salmon sold in Norway normally ranges from 23 to 30 on this scale (most commonly 25 – 27). The consumers were asked for their willingness-to-pay (WTP) within the color range presented. The results revealed a WTP with increasing redness of the fillets above 23 on the *Salmo*Fan™. Salmon fillets paler than 23 would be difficult to sell at any price. When informed about the origin of the salmon color, the consumers maintained their WTP for the pale and normal red salmon, while they lowered their WTP for salmon with color above

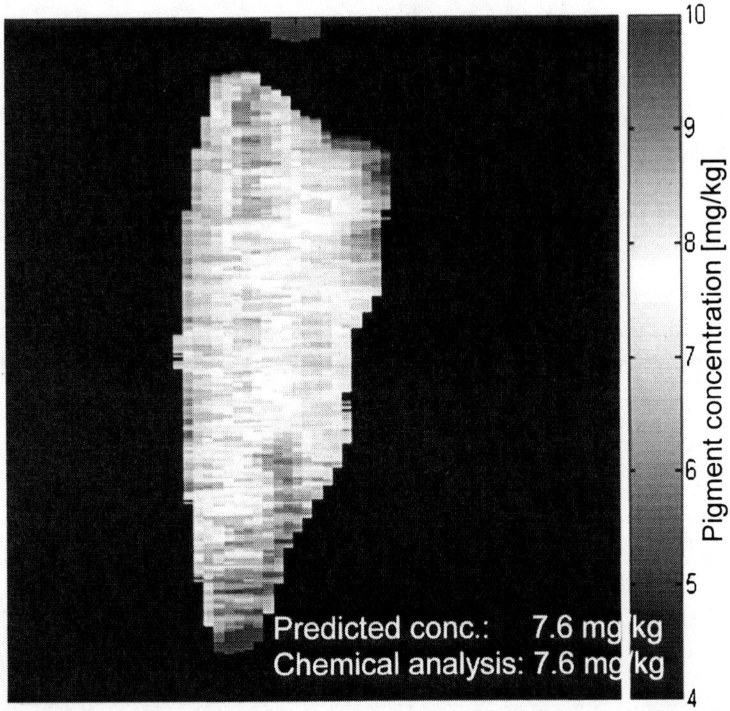

Figure 4. Output from the on-line color scanner: Pigment distribution and average concentration in each fillet. (See page 17 of color inserts.)

what they considered as normal red. Thus, the consumer wanted pink-red salmon also when they knew that the fish had been fed nature identical astaxanthin.

Multispectral Modeling of Visual Color and Pigment Concentration

A different approach for color assessment of salmon fillets is currently being developed at Matforsk, in collaboration with SINTEF IKT (The Foundation for Scientific and Industrial Research at the Norwegian Institute of Technology) and a commercial company (QVision AS) (*25*). By use of a line scanner mounted above a conveyor belt, a multispectral image of the sample is obtained in a fraction of a second. The camera is set up to scan the near infrared spectrum and the visible spectrum in each pixel of the sample image. Based on previous analysis of standards, the spectra are analyzed and used for prediction of pigment concentration in each pixel (Figure 4). Finally, an average pigment concentration across all pixels is calculated. Similar predictions are also obtained for *Salmo*FanTM scores and the fat level of the fillet. As the spectrometer is mounted above a processing line, the method has a large potential as a rapid, non-destructive tool for assessing color quality and fat level of salmon during processing.

References

1. Agriculture and Agri-Food Canada, *Fact Sheets.* Fish and Seafood. http://atn-riae.agr.ca/seafood/factsheet-e.htm (accessed Mar 11, 2007)
2. Norwegian Seafood Federation, *Numbers and Facts 2005*; Appendix to Annual Report 2005, http://www.fhl.no (accessed Mar 11, 2007)
3. Anderson, S. *Proceeding IIFET2000;* International Institute of Fisheries Economics and Trade, Oregon State University: Corvallis, OR, 2000; 3 pp.
4. Christiansen, R.; Lie, Ø.; Torrissen, O. J. *Aquacult. Nutr.* **1995**, *1*, 189-198.
5. Bell, J. G.; McEvoy, J., Tocher, D. R.; Sargent, J. R. *J. Nutr.* **2000**, *130,* 1800-1808.
6. Buttle, L.; Crampton, V.; Williams P. *Aquacult. Res.* **2001**, *32,* 103-111.
7. Hardy, R. W.; Barrows, F. T. In *Fish Nutrition*, 3rd ed.; Halver, J. E.; Hardy, R. W., Eds.; Academic Press: San Diego, CA, 2002; pp 505-600.
8. Storebakken, T.; Sørensen, M.; Bjerkeng, B.; Hiu, S. *Aquaculture* **2004**, *236*, 391-403.
9. Skrede, G.; Storebakken, T. *Aquaculture* **1986**, *53*, 279-286.
10. Gobantes, I.; Coubert, G.; Milicua, J. C.; Gómez, R. *J. Agric. Food Chem.* **1998**, *46*, 383-387.
11. Christiansen, R.; Struksnæs, G.; Estermann, R., Torrissen, O. J. *Aquacult. Res.* **1995**, *25,* 311-321.
12. List of the Authorised Additive in Feedstuffs Published in Application of Articl 9t (b) of Council Directive 70/524/EEC Concerning Additives in Feedingstuffs. *Off. J. Euro. Union* **2004**, *C50*, 1-144.
13. Steine, G.; Alfnes, F.; Rørå, M. B. *Marine Resource Econ.* **2005**, *20*, 211-219.
14. Guttormsen, A. *Marine Resource Econ.* **2002**, *17*, 91-102.
15. Standards Norway, *Reference Sampling for Quality Assessment*, NS 9401.E, http://www.standard.no (accessed Mar 11, 2007)
16. Morris, P. C.; Beattie, C.; Elder, B.; Finlay, J., Gallimore, P.; Jewison, W.; Lee, D.; Mackenzie, K.; McKinney, R.; Sinnott, R.; Smart, A.; Weir, M. *Aquaculture* **2005**, *244*, 187-201.
17. Skrede, G.; Storebakken, T. *J. Food Sci.* **1986**, *51*, 804-808.
18. Bjerkeng, B.; Følling, M.; Lagocki, S.; Storebakken, T., Olli, J. J.; Alsted, N. *Aquaculture* **1997**, *157*, 63-82.

19. Birkeland, S.; Bjerkeng, B. *Int. J. Food Sci. Technol.* **2005**, *40*, 963-976.
20. Hunter, R. S.; Harold, R. W. *The Measurement of Appearance,* 2nd ed.; John Wiley & Sons: New York, NY; 1987.
21. Cardinal, M.; Gunnlaugsdottir, H.; Bjoernevik, M.; Ouisse, A.; Vallet, J. L.; Leroi, F. *Food Res. Int.* **2004**, *37*, 181-193.
22. Skrede, G.; Risvik, E.; Huber, M.; Enersen, G.; Blümlein, L. J. Food Sci. **1990**, *55*, 361-363.
23. Johnston, I. A.; Alderson, R.; Sandham, C.; Dingwall, A., Mitchell, D., Selkirk, C., Nickell, D.; Baker, R.; Robertson, B.; Whyte, D.; Springate, J. *Aquaculture* **2000**, *189*, 335-349.
24. Morris, P. C.; Beattie, C.; Elder, B.; Finlay, J.; Gallimore, P.; Jewison, W.; Lee, D.; Mackenzie, K.; Sinnott, R.; Smart, A.; Weir, M. *Aquaculture* **2003**, *225*, 41-65.
25. Wold, J. P.; Johansen, I. R.; Haugholt, K. H.; Tschudi, J; Thielemann, J; Segtnan, V. H.; Narum, B.; Wold, E. *J. Near Infrared Spectrosc.* **2006**, *14*, 59-66.

Chapter 20

Coloring Technology for Surimi Seafood

Jae W. Park

Seafood Research and Education Center, Oregon State University, Astoria, OR 97103

Surimi seafood was invented to mimic primarily Alaska king crab in Japan in the mid 1970s. Over 30 years of the production history, various surimi seafood products have been developed beyond shellfish (crab, shrimp, and lobster) and finfish (salmon and anchovy) substitutes. Now surimi seafood stands strong with its own identity, not as a shellfish substitute. Their color ranges from dark-red to bright blood-red, and dark brown-red to orange-red. Currently the worldwide industry achieves the desired surface color (hue) by blending various colorants such as carmine, monascus, paprika oleoresin, caramel, canthaxanthin, lycopene, and emulsifiers. All these colorants are derived from natural sources except canthaxanthin and beta-carotene which are chemically synthesized. In the body of white meat except salmon and anchovy style products, whitening agents such as vegetable oil, calcium carbonate, and titanium carbonate are used. This chapter covers colorants used in surimi seafood, color application to crabstick, application problems (flaking and bleeding/transfer), quality measurement, and labeling requirements.

Surimi is concentrated fish protein manufactured after extensive washing and dewatering of fish mince, primarily from finfish. Frozen surimi is a fish protein concentrate containg primarily myofibrillar proteins stabilized with cryoprotectants. It typically contains 74-76% water, 16-17% protein, 9-10% cryoprotectants (4-5% sorbitol, 4% sugar, 0.2-0.3% sodium polyphosphate), and less than 0.5% fat. The quality of frozen surimi is determined by gel strength, color, impurities, and moisture content. Among these variables, gel strength affecting texture and gel color are the most important. Even when frozen surimi is stabilized with cryoprotectants and kept well in frozen storage, gel strength and color diminishes gradually as frozen storage extends. Commercial surimi is marketed with a 2-year frozen shelf life.

Surimi seafoods are developed in several styles, but the majority is crabmeat style. There are two categories of crabmeat style, depending on the manufacturing methods: filament meat and solid meat (*1*). In this chapter, primarily filament style crabmeat and its color application will be covered.

Manufacturing of Filament Style Crabmeat (Crabstick)

Once frozen surimi is chopped with salt in a silent cutter and salt-soluble proteins are extracted through chopping, water/ice and other ingredients (starch, egg white, flavors, and others) are uniformly mixed while the temperature is maintained properly depending on the fish species. For example, when Alaska pollock surimi is manufactured, the temperature of chopping and holding should be maintained as cold as possible. When warm water fish species are used, warmer chopping temperatures 20-30 °C would be preferred (*2*). Blended surimi paste is extruded onto a cooking belt or drum as a continuous sheet. The surimi sheet is cooked rapidly under steam and/or gas and then cooled off before going through a slitting step. Slitted sheets are then bundled, wrapped, cut, packaged, pasterurized, chilled, and/or frozen. A detailed processing diagram is shown in Figure 1.

Color application

Color application using a plastic wrap film (Figure 2) often causes color flaking when the finished product is cut into a different shape. Therefore color co-extrusion was initiated in the United States in the late 1980s. Colored paste is co-extruded on the edge of a sheet of uncolored surimi paste prior to the initial cooking step. This obviates the need for an intermediate wrapping step and subsequent cooking for color setting and unwrapping (Figure 1). Co-extrusion color, however, can sometimes bleed or transfer into the white portion of the product when vacuum-packed products are pasteurized at high temperatures (*1*).

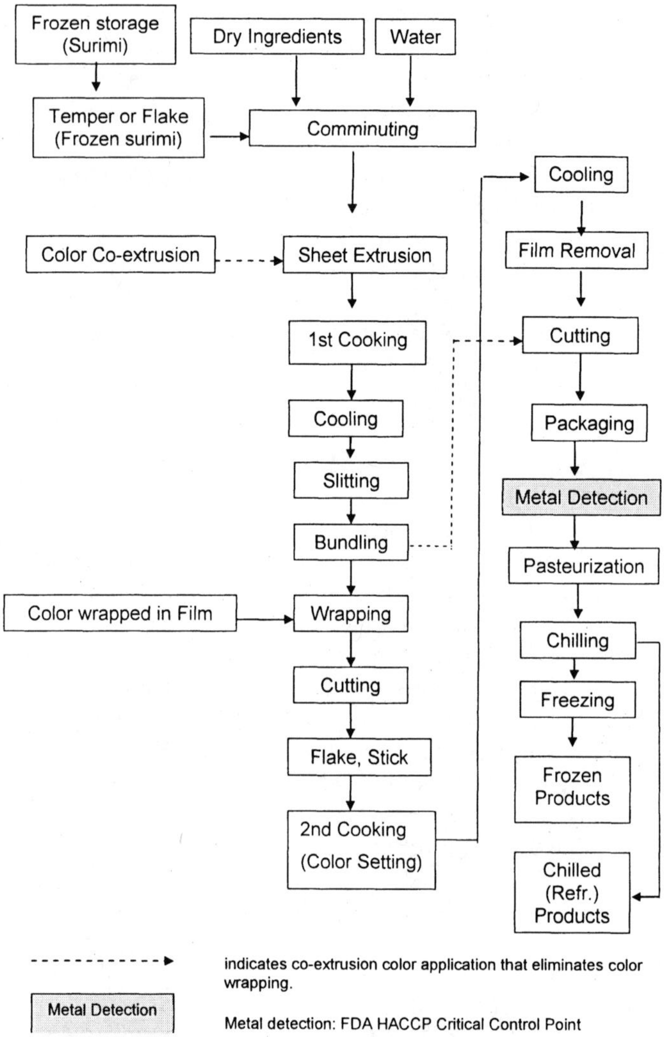

Figure 1. Flow chart of crabstick manufacture including color applications

1. *Regulatory Guidelines Concerning Food Additives;* Bureau of Food and Drug, Ministry of Health: 1984, http://www.doh.gov.ph/bfad2/ao_88-a_1984.html

(Reproduced with permission from reference 1. Copyright 2005 Taylor & Francis.)

Colored meat preparation

The surimi seafood industry prepares colored meat using two different processes. One is to prepare colored meat by mixing surimi paste, already prepared for extrusion of body meat sheet, with the desired colorants. The other method is to prepare colored meat by chopping frozen surimi with other ingredients and finally mixing it with the desired colorants. The latter process makes it much easier to control color variation compared to the former. It is important that the temperature of colored meat must be controlled during mixing and its holding temperature (0-3 °C) must also be cold to prevent possible setting. If paste is made with warm water species of surimi, it may not be necessary to keep it that cold. However, a colder temperature is still better to prevent setting. Otherwise, gelled particles may block extrusion nozzles resulting in non-uniform color application. It is also not uncommon to add sodium citrate or citric acid (less than 1%) to prevent setting by controlling the pH of colored meat. Especially for colored meat to be ued for the wrapping method, a lecithin solution is often added to facilitate removal of the wrapping film after cooking.

Chopping always introduces a large number of air bubbles. For consistent color application, the use of vacuum during chopping is recommended. Unused colored meat can be rapidly frozen and reused with proper cautions. Frozen colored meat must be chopped finely in a silent cutter and possibly blended with new batches. Otherwise, aggregated colored particles will affect the shade and uniformity of applied color.

Colorants used in surimi crabstick

Surimi seafood manufacturers have two groups of color additives to choose from when selecting colorants: certified color additives and color additives exempt from certification (exempt colorants). The first group currently contains seven synthetic colorants, referred to the industry as FD & C (Food, Drug, and Cosmetic). Each batch of these colorants is required by US law to be analyzed and certified by the Food and Drug Administration to comply with purity specifications. The second group currently contains 26 colorants that do not require FDA analysis and certification. Within in this group are several colorants that are referred to as "natural" in other countries (*3*). In the commercial production of crabstick, almost all colorants used are exempt colorants.

The ever-increasing consumer demand for "all natural" products favor natural colorants because they are perceived as being more compatible with the image of surimi seafood. Although there exists a demand for "natural" from both the consumer and the manufacturer, there is confusion as to what is a "natural" color. The industrial world appears to accept as "natural" any colorant

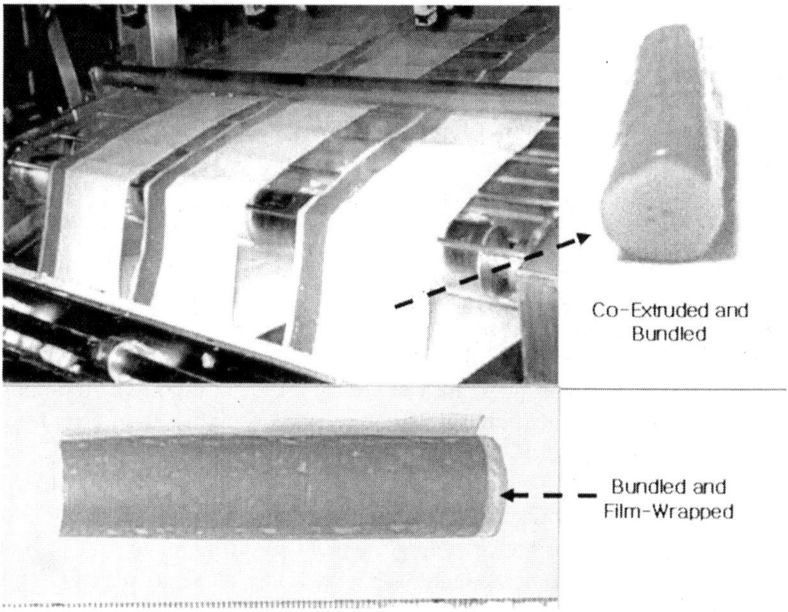

Figure 2. Two different methods of color applications. (See page 18 of color inserts.)

that is extracted from agricultural, biological, and at times, mineral sources. The acceptance can be very broad or narrow, depending on the individual agenda. The consumer, on the other hand, believes that natural colors are from fruits and vegetables, and by inference, also healthy. With all of the interest in "natural", one might expect the government regulators to define "natural", but instead the US government has opted to not recognize the term "natural". Therefore, in a legal sense, "natural colors" do not exist (*3*). In other words, unless the colorant is an extract from crab itself, the surimi industry will have difficulties in persuading FDA to describe surimi crabsticks as "naturally colored".

In lieu of an artificial/natural classification, the FDA created two categories of colorants: those that require certification and those that do not require certification. The latter category is often incorrectly referred to as the "natural colors", and indeed, includes those colorants which are conventionally called natural, i.e. carmine, paprika, annatto, turmeric, beet juice extract, and grape skin extract. (*4*)

Within FDA's exempt colorant listing (21CFR 73) are also colorants that are synthetically derived, much like the certified colors, which behave chemically identical to the similar colorants extracted from natural sources.

Caramel is also on FDA's exempt color listing, and some have argued that it should not be accepted as a "natural colorant" as it is a reaction end product. Mineral colorants, such as titanium dioxide and calcium carbonate, have also been characterized as natural by some seeking to satisfy an "all natural" claim. In addition, there are other colorants on FDA's list of "colorants not requiring certification" which may or may not be accepted as natural depending again on one's point of view (*3*).

Carmine

Carmine is a vivid purple-red powder offered in "lake" form (a lake, by definition, is oil dispersible (water insoluble) and functions as a pigment) (*3*). This colorant is a primary red colorant used in the manufacture of crabstick. It is a purified pigment from the belly of a female insect cochineal residing on a cactus found in Peru, Chile, Canary Islands, and Mexico. About 10-20% of the dry weight is carminic acid, which is the coloring principle (*3*). Liquid carmine is usually offered with a 3 to 7% carminic acid content and is alkalized with either ammonium or potassium hydroxide. The pH can very from 8 to 11. Carmine solution has a strong affinity for proteins and thus works very well in surimi seafood. It is known to be stable to light, heat, oxidation, and storage. It bleeds if not properly adjusted. It is often used with paprika oleoresin and a stabilizing emulsifier (*3*)

Paprika oleoresin

Paprika, as used in color applications, is an oleoresin. It is made by percolating any one of a combination of approved solvents (such as hexane, acetone, various alcohols, etc.) through dried, ground or pelleted sweet red pepper (free of seeds). Solvent is removed by vacuum distillation. Oleoresin can also be obtained by supercritical CO_2 extraction (*3*).

The oleoresin is a viscous dark reddish-brown colored liquid, pourable at room temperature, and containing a mixture of some 10 or more carotenoids (such as xanthophylls, capsanthins, and capsorubins) (*5*, *6*). Capsorubins, the red fraction, can be as high as 40-59% of the oleoresin. The exact hue of the oleoresin can vary depending on growing, harvesting, drying, storage and extraction conditions. It is standardized for color strength (not hue) with vegetable oil to a level referred to as 40,000 and 80,000 CU (although higher concentrations are available). CU values are directly proportional to the optical density of a solution of oleoresin in acetone. Some suppliers prefer to use ASTA values to designate strength. A 1000 ASTA is equivalent to 40,000 CU (*3*).

Canthaxanthin

Canthaxanthin exists widely in nature. It was first discovered in edible mushrooms (*Cantharellus cinnabarinus*) (*7*). Today, canthaxanthin that is commercially available as a colorant is produced by chemical synthesis. It is insoluble in water and only very slightly soluble in oil. Several commercial forms are available and have been used as a substitute for paprika in carmine-based blends for surimi crabstick. As with all carotenoids, it is degraded by light, heat and oxidation (*3*).

Monascus

Monascus colorant (Angkak or Chinese red rice) has been used for over a hundred years as a food color in Malaysia and China (*3*). It was the color of choice in Japan from the inception of surimi-based crabstick production. This colorant is produced by solid or liquid mold fermentation of bread crumbs or rice grains. The colorant is made up of a mixture of red, yellow, orange and purple pigments (*8*). Like carmine, monascus colorant has a strong affinity for proteins but its poor light stability limits its usefulness in seafood manufacture. When monascus colored surimi products are placed in supermarket showcases, the color fades within a week of exposure to fluorescent lighting. Monascus is not allowed as a food colorant in either the US or Europe simply because no GRAS (generally recognized as safe) applications have been submitted..

Lycopene

Lycopene is a bright red carotenoid pigment, a phytochemical found in tomatoes and other red fruits. Lycopene is the most common cartenoid in the human body and is one of the most potent cartenoid antioxidants. Its name is derived from the tomato's species, *Solanum lycopersicum*. Lycopene can be found in other fruits such as watermelon, papaya, and pink grapefruit. Unlike other fruits and vegetables, where nutritional content such as vitamin C is diminished upon cooking, processing of tomatoes increases the bioavailablity of lycopene. Thus processed tomato such as tomato juice and ketchup contain the highest concentration of lycopene. There is evidence that frequent intake of such products is associated with reduced risk of cardiovascular disease and cancer, especially prostate cancer (*9*). Processing through physical and thermal treatments has the potential to enhance absorption by disruption of the tomato tissue matrix, dissociation of cartenoid-protein complexes, enhanced surface area, and increases in solubility (*10*). With health-consious trends of consumer's preference, some surimi seafood is manufactured with lycopene properly stablized with emulsifiers.

Other naturally derived colorants used in crabmeat body

Some surimi seafood manufacturers use various whitening colorants derived from natural compounds (Figure 3). They are all water-insoluble and make the surimi meat white when the fine particles are homogeneously distributed through comminution. Since its whitening effect is derived from uniform distribution, the smaller particles show stronger effects.

Titanium dioxide

Titanium dioxide is an approved colorant with an end use restriction of 1 % (w/w) maximum in the finished product. Although TiO_2 exists in nature, usually in three crystalline forms, only synthetically prepared pigment can be used as a color additive. It is extremely effective in whitening surimi. However, the usage concentration must be carefully controlled. At a concentration >0.05%, the pH of the meat can decrease, negatively affecting the texture. In addition, when its usage level exceeds 0.08%, the surimi meat color appears to be an artificial chalky white.

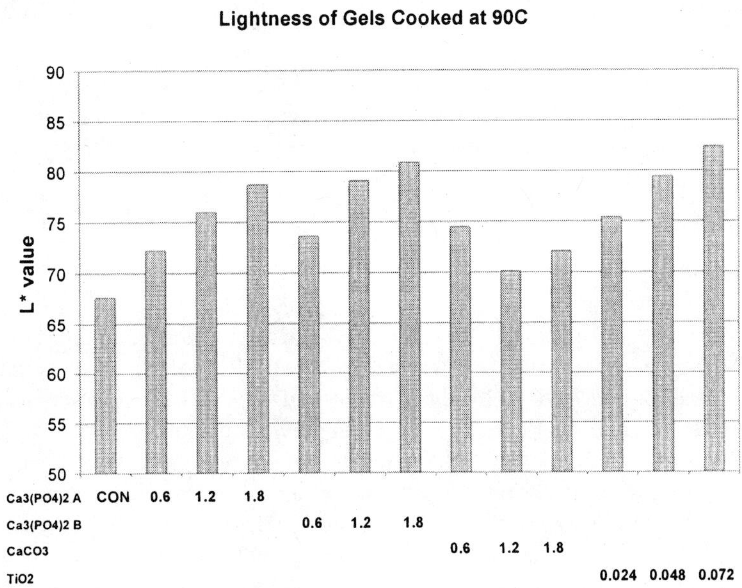

Figure 3. Whitening effects of calcium phosphates, calcium carbonate, and titanium dioxide on surimi gels. Numbers on the botton indicate the percentage of ingredient tested. Con: control without any addition.

Calcium carbonate

Calcium carbonate is also a synthetically prepared powder composed primarily of precipitated $CaCO_3$. This pigment has considerably less whitening strength compared to TiO_2 (Figure 3). There is always the risk that carbon dioxide gas could be formed in the presence of phosphate (a cryoprotectant from surimi) and acid (a salad ingredient).

Vegetable oil

Vegetable oil has been used to improve the perceived whiteness of surimi paste. With the mechanical action of cutting of oil drops and surimi proteins in the presence of water, emulsion-like paste is perceived as white through the light scattering affect.

Problems in Color Application

Color flaking

Color paste for meat is applied directly on the wrapping film and the film surrounds the cooked rope before cooking in conventional crabstick production. Due to poor interaction between color paste and the wrapping film, it is common to see the coloring flakes appearing as specks. This problem has been completely solved by co-extruding color paste at the same time as the body meat paste is extruded onto the cooking belt.

Bleeding and transferring

Color flaking is solved by color co-extrusion, but color bleeding or transferring becomes a new problem when co-extruded color meat is vacuum – packed and pasteurized. However, this problem can be overcome using a effective emulsifier like PGPR (polyglycerol polyricinoleate).

The surimi industry has experimented with a novel process for controlling bleed; this process is a "hot" topic among surimi seafood processors. The process involves the formation of a physical barrier to bleeding rather than chemically altering the colorant or surimi system. The most effective physical barrier between carmine and the wet surimi paste is obtained by the use of water in oil emulsifiers. The process requires that the emulsion hold until gelling is complete. Once physically trapped, the carmine is not likely to bleed. The most functional emulsifier is polyglycerol ester of ricinoleic acid (PGPR), a major component (at times, as high as 80 - 90%) of castor oil (*3*).

Even though this emulsifier has been used on and off by the color manufacturers since the inception of the crabstick industry in the mid-1970s in Japan and early 1980s in the US, no GRAS (generally recognized as safe) application of PGPR in the seafood manufacturing has been submitted to the US FDA. However, the US chocolate industry has successfully used PGPR at 0.3% as emulsifier based on Quest International's GRAS application. The FDA issued no questions on Quest's conclusion that PGPR is GRAS under the intended conditions of use (*11*). The agency emphasized that it is the manufacturer's continuing responsibility to ensure that food ingredients marketed are safe (*11*).

Recently the FDA has accepted a GRAS petition made by Stepan Company for the use of PGPR as an emulsifier in margarines, low fat margarines, spreads, creamers, and dairy analogs at levels no greater than 1.0 percent by weight (*12*). FDA replied to the petitioner with no further question and requested for continuous safety (*12*). Even though it is allowed to be used for crabstick coloring in Japan and other countries, this PGPR is not permitted in the US crabstick simply because no GRAS application has been submitted on behalf of the US crabstick industry.

Other studies indicate that bleeding may be suppressed by the addition of calcium chloride (*13*) to the surimi paste. The effect of calcium ion addition is shown in Figure 3. The use of calcium salt, however, can speed up the rate of gel setting (*14*). The rate of setting, on the other hand, can be slowed by lowering the processing temperature. Satisfactory results can therefore be obtained by balancing setting time against bleed.

Color quality measurement of surimi gels

Color, like texture and flavor, is an important quality aspect of surimi and and its related products. The world surimi industry uses three analytical instruments: Minolta, Hunter, Nippon Denshoku. All are tristimulus filter colorimeters based on the same technological principles. Lightness (L*) and yellow hue (+b*) often determine the color quality of surimi gel. Park (*15*) evaluated the effects of various compositional and physical conditions during preparation and color measurement of gels from Pacific whiting and Alaska pollock. Water addition, sample size, species, cooking and testing temperatures, and freeze/thaw cycling affected lightness and yellowness values, while outside light condition at measurement did not affect colors. The green hue (-a*) was less affected by all conditions.

Park (*16*) compared two whiteness indices to each other when the effect of protein additives on surimi gels was investigated (Figure 4). This study reported the effects of protein additives on surimi gels were more clearly demonstrated using whiteness index II than whiteness index I considering visual differences between samples. These indices are defined as:

- Whiteness I
  $= 100 - [(100 - L^*)^2 + a^{*2} + b^{*2}]^{1/2}$

- Whiteness II
  $= L^* - 3b^*$

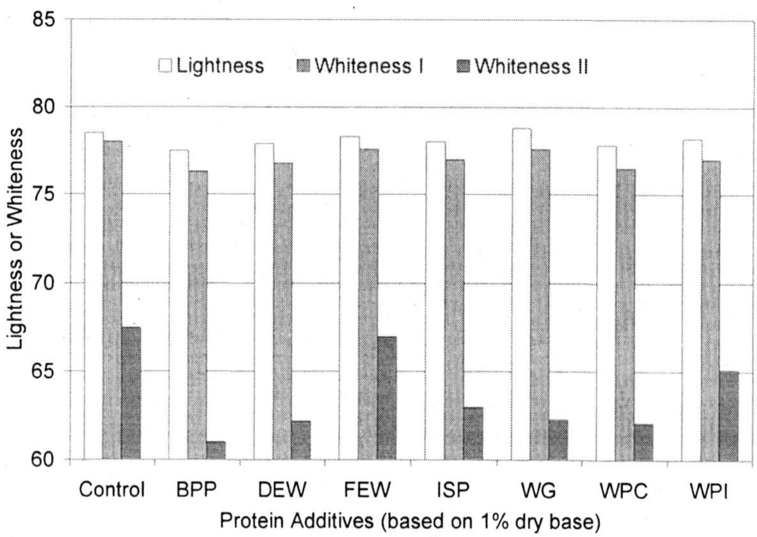

Figure 4. Whiteness of surimi gels as affected by various protein additives. Adapted from Ref. (16). BPP: beef plasma protein; DEW: dried egg white; FEW: frozen egg white; ISP: isolated soy proein; WG: wheat gluten; WPC: whey protein concentrate; WPI: whey protein isolate.

Color labeling

Labeling laws are often amended either by court action, legislative action, or regulatory agency action and therefore this topic may soon become out-dated. It is recommended that legal counsel be consulted before proceeding with labeling involving the declaration of natural color(s) in an ingredient statement (*3*).

The Guide to US Food Labeling Law (August 1991, Tab 400, Pg.19) reported that "on January 3, 1978, former FDA Commissioner Donald Kennedy wrote to the 100 largest manufacturers and distributors of packaged foods requesting that they voluntarily label all the specific colors used in their foods. There remains no legal requirement, however, for labeling individual uncertified colors in food." The FDA still encourages firms to voluntarily declare colors, spices and flavorings when they are added to foods and "supports declaration of ingredients on food labels to the extent that it is practical." (56 FR 28592, Proposed Rule B.2). In this Rule, FDA indicated it would prefer as part of the ingredient list, "terms such as 'colored with ________' and '__________(color)', "as an example of appropriate terms for the voluntary declaration of exempt colors when the blank is completed with the utilized colorants listed in 21 CFR, Part 73 [20] (*3*).

The Nutrition Labeling and Education Act of 1990 did not change the FDA's position, but it did allow for the natural colorants to be listed generically, i.e. without naming the specific colorant (21 CFR, Part 101.22(k)(2)) in place of individual listing. This section states "color additives not subject to certification may be declared as "Artificial Color", "Artificial Color Added", or "Color Added" (or by an equally informative term that makes clear that a color additive has been used in the food). Alternatively, such color additives may be declared as "Colored with __________" or " _____________color", the blanks to be filled in with the color additive listed in the applicable regulation in Part 73. Almost all the US surimi seafood manufactuers currently use "color added" for the use of colorants (carmine, paprika, canthoxanthin, and others), which are exempted from certification

References

1. Park, J.W. In Surimi and Surimi Seafood; Park, J.W., Ed., Second Ed.; CRC Press: Boca Raton, FL, 2005; pp 749-801.
2. Esturk, O.; Park, J.W.; Thawornchinsombut, S. *J. Food Sci.* **2004**, *69*, E412-6.
3. Lauro, G.J.; Inami, O.; Johnson, C. In Surimi and Surimi Seafood; Park, J.W., Ed., Second Ed.; CRC Press: Boca Raton, FL, 2005; pp 375-434.
4. Lauro, G. J. *Cereal Food World* **1991**, *36*, 949-953.
5. Minguez-Mosquera, M. I.; Hornero-Mendez, D. *J Agri.c Chem.* **1994**, *42*, 38-44.
6. Minguez-Mosquera, M. I.; Jaren-Galan, M.; Garrido-Fernandez, J. *J. Agric. Chem.* **1994**, *40*, 2384-2388.
7. Haxo, F. *Botan. Gaz.* **1950**, *112*, 228-232.
8. Black, P. J.; Loret, M. O.; Santerre, A. L.; Pareilleux, A..; Prome, D.; Prome, J. C.; Lausac, J. P.; Goma, G. *J. Food Sci.* **1994**, *59*, 862-865.
9. Bowen, P.; Chen, L.; Stacewicz-Sapuntzakis, M.; Duncan, C.; Sharifi, R.; Ghosh, L.; Kim, H. S.; Christove-Tzelkov, K.; van Breemen, R. *Exp. Biol. Med.* **2002**, *227*, 886-893.

10. Schwarz, S. J. *J. Nutr.* **2005**, *135*, 2040S-2041S.
11. US FDA. Agency response letter GRAS notice No. GRN 000009, 1999. http://vm.cfsan.fda.gov/~rdb/opa-g009.html (accessed Mar 11, 2007)
12. US FDA. Agency response letter GRAS notice No. GRN 000179, 2006. http://www.cfsan.fda.gov/~rdb/opa-g179.html (accessed Mar 11, 2007)
13. Kanmuri, Y.; Hashino, M. ; Yokota, N. Japan Patent H06-61237. **1994**.
14. Saeki, H.; Shoji. J.; Hirata, F.; Nonaka. M.; Arai, K. *Nippon Suisan Gakkaishi.* **1992**, *58*, 2137-2146.
15. Park, J. W. *J. Food Sci.* **1995**, *60*, 15-18.
16. Park, J. W. *J. Food Sci.* **1994**, *59*, 525-527.

Chapter 21

Color Quality in Meat

Donald H. Kropf

Department of Animal Sciences and Industry, Kansas State University, Manhattan, KS 66506–0201

Chilled, frozen, cured and cooked color, also color of fat and - bones all influence the quality image of meat. These can be affected by animal genetics, feeding, handling, carcass chilling, time postmortem, cutting, packaging system, display condition and cooking. Time post-cutting affects myoglobin chemical state and must be identified. Illuminant, observer angle, and aperture size affect instrument readings and must be identified. Modified atmosphere packaging, could be high or ultra-low oxygen, or carbon monoxide, are useful in case-ready distribution and marketing. Each represents technical challenges regarding color and color measurement. Unique challenges to traditional reflectance readings are lack of sample uniformity, but scanning technology and color transformation to cyan, magenta, yellow are useful techniques in quantifying discoloration. Image analysis of layers of myoglobin chemical forms offers promise in understanding color changes.

Introduction

Meat color and color stability can be affected by numerous decisions and circumstances that begin with live animal selection and breeding, and include nutrition, environment, handling, holding conditions before slaughter, stun/bleed variables, chilling protocol, aging, holding time and conditions (especially temperature), fabrication conditions, time from fabrication to packaging, successful packaging, type of packaging; storage, distribution and handling of product, display conditions and handling by customers. The above apply to fresh, chilled, frozen and to cured products. Enhancements to chilled non-cured meat can markedly influence color stability. Cured meat is additionally affected by added ingredients, physical manipulation conditions, time and temperature protocols for smoking and heat processing and post-cook chilling and packaging. Those who troubleshoot color stability problems tend to concentrate on the most recent events but color could be affected in many places in the livestock to processor to distributor to consumer chain.

This chapter will briefly discuss the specifics of many of the above circumstances, how they influence meat color and how the meat-food industry attempts to deal with them. A brief discussion will deal with cooked color, premature browning and persistent red and pinking; all affect consumer safety and perceptions of safety.

The chapter will deal with basic meat characteristics that are responsible for meat color including chemistry of myoglobin and other pigments, but also meat structural influences on appearance traits, primarily color. It will also deal with methods of color evaluation including visual or subjective judgments and numerous instrumental techniques for assessing color.

Importance of Meat Color

Appearance is rapidly assessed and directed into a decision to buy or not buy, to eat or not eat (*1*). Another researcher (*2*) stated that color has both a psychological and a real effect. The psychological effect results in an almost immediate positive or negative response; the real effect is a more reasoned assessment of quality and factors affecting "wholesomeness" and "freshness" such as perceived time product has been held, handling effects, and temperature. Another report (*3*) stated that color of meat impacted two consumer visual judgments: an immediate one of meeting or not meeting acceptance standard and secondly, one reflecting perceived palatability. In summary, safety, perceived freshness and wholesomeness and expected sensory satisfaction make up meat quality judgments.

Regardless of how these judgments are made, and who makes them, i.e., retail or food service personnel, or the retail purchaser, the collective influence results in price discounts for nearly 15% of retail beef packages (*4*).

Color stability implies the duration of acceptable, saleable color, which is relatively short-lived. Surface discoloration is inevitable with increasing time and is interpreted by customers to indicate unwholesomeness or lack of acceptable freshness, even though color usually deteriorates earlier than microbial or serious oxidative consequences occur. Meat discoloration is defined as divergence from the consumer-defined ideal to something less desirable, for example, cherry red to brown for beef or bright pink to brown, green, or gray for pork.

Strategy for maximizing acceptable color (and color life) involves delaying pigment oxidation and/or enhancing reduction of oxidized pigment. No single factor is responsible for meat discoloration and in a complex food system like meat, no factor acts independently.

Discoloration of meat cuts can result in a large monetary loss to the retailer, and thus to other segments of the industry. Retail loss estimates are scarce and difficult to obtain, but one study (*5*) reported 3.1 and 4.6% case pulls for all beef retail packages in two stores and cite the following disposition of retail packages in three stores studied: converted (probably ground) and repackaged 4.1%, trimmed and repackaged 2.9%, rewrapped 2.9%, price reduced for quick sale 0.6% and discarded 0.2%. Another study (*6*) cites beef shrink, defined as losses due to reduced value to be 6% compared to 4% for overall meat market operations but losses are higher, up to 23%, for some high value beef cuts. Nineteen retail beef cuts exceeded 10% loss in this survey of ten large supermarkets in the USA. Losses up to 10% were stated by retailer executives at a symposium (*7*).

While we usually think of muscle color, meat color quality also includes fat color and bone color. Slight to intense yellow fat can have consequences in marketing beef and pork. Bone discoloration to gray or green or darkening (almost to black in some cases) can pose a problem for pork, beef and frozen chicken.

Myoglobin Chemistry

Many sources present the structure of myoglobin which is involved in controlling and transferring oxygen to its critical use sites, notably to mitochondria. A concise description of its structure and major reactions was summarized by Mancini and Hunt for presentation at the 2005 International Congress of Meat Science and Technology (*8*).

This sarcoplasmic protein has the major role in imparting meat color with lesser role attributed to hemoglobin and cytochrome C. This water soluble protein is made up of 8 α-helices linked by smaller non-helical sections with a prosthetic group within its hydrophobic pocket. This heme ring contains a centrally located iron atom with the ability to form six bonds, four with pyrrole nitrogens, one to coordinate with the proximal histidine-13 and one site to reversibly bind ligands. This ligand and whether the iron is ferrous or ferric provide chromophoric identification which is responsible for meat color, although some alteration results from such physical influences as light scatter because of free water, intramuscular fat and muscle structure.

The meat color triangle (Figure 1) has frequently been used to describe relationships of the three different major chemical states of myoglobin in fresh meat. These are sometimes referred to in the literature as myoglobin species. A recent presentation (*8*) has humorously described this triangle pattern of most common myoglobin reactions in fresh meat as the "ferrous-ferric-wheel" with the notable components of deoxymyoglobin, oxymyoglobin and metmyoglobin.

Deoxymyoglobin

Deoxymyoglobin has heme iron in the ferrous (reduced, Fe^{2+}) state with no ligand at the 6th site. This state results because of the postmortem cessation of oxygen transport to muscle cells and to myoglobin, creating an anaerobic environment. This purple-red or purplish-pink color state is apparent briefly in freshly cut muscle and is more noticeable in meat with greater pigment content. It also is formed when oxygen is removed from contact with the meat surface by effective vacuum packaging or by ultra-low oxygen modified atmosphere packaging. Very low oxygen partial pressure ($\leq$1.4 mm Hg) is required to maintain this deoxymyoglobin state (*9*).

Introduction of oxygen into these very low oxygen in-package environments can occur by ineffective initial evacuation of air, by more than very low oxygen ingress because of unsatisfactory package seals or lack of sufficient film oxygen barrier property or by inadvertent introduction of oxygen by the vehicle of meat which has been allowed too much oxygen uptake prior to anoxic packaging. These create problems that shorten shelf life of meat because it discolors too soon. Certain process protocols like passing meat pieces through a cuber (mechanical tenderizer) or preparing meat cubes with a high ratio of meat surface to meat mass will result in a high degree of oxygen diffusion into the meat cuts and after vacuum packaging contributes to a too-high partial oxygen pressure within packages. Consequences of exceeding the very low partial oxygen pressure levels needed to maintain deoxymyoglobin will be discussed in the discoloration discussion.

In many meat marketing systems, it is desirable to keep meat in the deoxymyoglobin state, even though this form is vulnerable to undesirable oxidation unless conditions are very carefully controlled. Unnecessary and

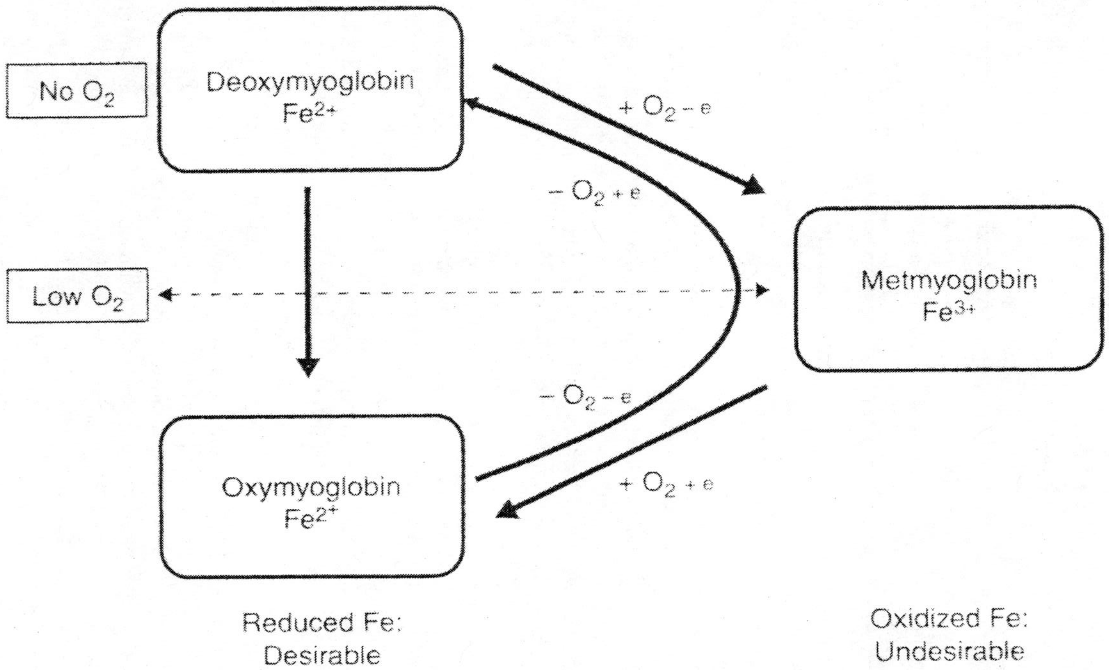

Figure 1. Fresh meat color triangle.

uncontrolled pigment changes may be costly in terms of lessening the ability of other myoglobin forms to return to the deoxy form and will likely shorten color life of meat. This will be discussed later in this chapter.

Oxymyoglobin

Most research studies involving display life or color stability usually require a relatively stable starting point at which visual or instrumental measures of oxygenation have already undergone the rapid changes of blooming, instead of a moving base line. If studies require a deoxymyoglobin initial value, color evaluation must be done immediately after or before exposing muscle to air. If the initial or zero time value is one for oxymyoglobin, preliminary studies should be used to establish a point of maximum or stable oxygenation before any discoloration occurs. This could be for visual evaluation or for selected instrumental color values.

Several studies have dealt with bloom time (time of exposure to air) on muscle color traits. The beef industry relies strongly on carcass quality grades to establish value and muscle color is one factor that influences carcass maturity and quality grade. Carcasses are ribbed to expose the longissimus muscle surface at the 12th vertebral location and then are conveyed past a grading station for evaluation as rapidly as appropriate color traits have reached an acceptable bloom.

One study (*10*) was conducted to determine effects of bloom time from 0 to 93 minutes evaluated every 3 minutes (on 10 randomly chosen carcasses) and study other measurement conditions using Minolta Chroma Meter CR-310 reflectance readings. L* (lightness) values stabilized after about 30 minutes bloom time, while a* (redness) and b* (yellowness) took about 78 minutes to stabilize, however relative differences among carcasses did not change after 3 to 12 minutes bloom time. These instrumental readings were not affected by days postmortem (1, 3 and 4 days), anterior versus posterior cut surface, and medial or lateral sampling location within muscle surface.

Another study (*11*) used 12 beef loins with vacuum packaged portions held at -2.2, 1.7 or 4.4°C. These were unwrapped at 2, 14 and 26 days postmortem and bloom rate measured with a Hunterlab Miniscan XE Plus, 3 cm aperture size, Illuminant A for CIE L*, a*, b*, hue angle and saturation index determinations. Spectral % reflectance every 10 nm from 400 to 700 nm was measured and % oxymyoglobin calculated. All color measurements were taken at 0 time and 5, 10, 15, 20, 30, 40, 50 and 60 minutes post exposure to air. Temperature and days postmortem showed an interaction as at 2 days, -2.2°C was fastest and 4.4°C slowest for bloom rate. At 14 days, -2.2°C caused slowest bloom and 4.4°C was fastest. At 26 days, bloom rates were similar at the 3 temperatures. At 2 days, enzyme competition for oxygen at the warmest (4.4°C) temperatures likely slowed bloom. With longer aging, decreased enzyme activity competed less for oxygen. Authors stated that a minimum of 20 minutes was needed for adequate bloom, especially in plants with good chill systems. At -2.2°C, 65% of myoglobin was calculated to be in oxy form in 20 minutes, and 75% in 30 minutes.

A comprehensive study (*12*) done with five pork muscles, using both Hunterlab and Minolta instruments and 3 illuminants essentially found L* to be not affected by bloom time, while hue angle stabilized in 5 minutes, a* and b* in 10 minutes and chroma in 20 minutes.

Blooming was stated to be more efficient under conditions which increase oxygen solubility and discourage oxygen using enzyme activity, notably colder temperature and lower pH (*13*). Several studies have reported that meat first exposed to air at 5 to 10 days postmortem or removed from vacuum at longer times postmortem, bloomed more rapidly (*14*, *15*).

Colleagues in industry have noted this increase in bloom rate with longer time postmortem, a process they refer to as "ripening". Another study (*16*) using observation of beef samples placed in plexiglass cubes to observe changes in myoglobin chemical state, noted faster bloom shown by deeper oxymyoglobin depth for samples first exposed to air at 3 and 5 days postmortem compared to shorter times. Lamb muscle observations noted slower ability to bloom than beef at several times post-slaughter and explained this by a higher oxygen consumption rate (*15*). Limited data suggests that turkey muscle has less ability to change from deoxy to oxy form of myoglobin (*13*).

Time to bloom as well as intensity of bloom to an acceptable bright red color is an important consideration for meat cuts that are distributed in a low or ultra-low oxygen system (*17*). When individual retail cut packages are removed from the master pack or the outer oxygen-barrier film is removed, rapid oxygenation is expected, so these individual retail packages, which are also wrapped in an oxygen permeable film, will quickly reach an acceptable bright color so they may be placed in display. The same is also required of Cryovac® Peelable which has individual retail packages with an inner oxygen permeable packaging and an outer oxygen barrier packaging which is removed (peeled off) shortly before placing in retail display. This system and also the Pactiv system produce more acceptable bloom when the oxygen level is very low during distribution. Pactiv utilizes an added oxygen scavenger attached to the inside of its outer barrier bag, which requires an injected activator. Another similar technology uses a polymeric oxygen scavenging system which absorbs oxygen after activation inside an oxygen barrier bag. The oxygen scavenging layer consists of three components, including an oxidizable polymer that binds oxygen molecules, a photoinitiator and a catalyst. After the package is sealed it passes under an ultraviolet light, whereupon the photoinitiator provides energy to start the reaction and the catalyst speeds up the scavenging. This system is very convenient as no fluid needs to be injected. However, the scavenging may be slow for some situations. Another approach to scavenging is the marketing of a soaker pad which is placed under meat to absorb fluid but also serves a second function as an oxygen scavenger.

The scavenging ability of muscle also should contribute to lessening the amount of in-package oxygen to the very low levels desired. These meat packages are frequently kept very cold, but scavenging is more rapid at somewhat warmer temperatures so a post-packaging hold at 3 to 4°C for several hours rather than at -1 to 0°C could facilitate increase oxygen scavenging.

The above systems function most satisfactorily when newly fabricated cuts are immediately packaged to minimize exposure to air so as to lessen amount of oxygen conveyed into the package from the meat cut. A similar meat packaging system also has perforations in its inner oxygen permeable film, presumably to speed up blooming.

A comprehensive study (*18*) investigated the influence of pH, muscle and storage time on the ability of vacuum packaged pork to bloom. A very high final muscle pH (over 6.0) was created by epinephrine injection. Low pH (5.2 to 5.5) and medium pH (5.6 to 6.0) conditions were also studied. All pH values were based on 24 hour postmortem readings. The authors stated that L* was not a good indicator of blooming as it changed very little in 30 minutes. Both a* and b* increased during the 30 minute bloom appraisal although authors stated that only those changes within 20 minutes would usually be perceptible. Longer storage, up to 28 days at 4°C and lower pH resulted in more rapid blooming for longissimus, gluteus medius, and semimembranosus muscles, the three muscles studied.

Relatively fresh muscle, that with a good supply of reducing capability, will have two pigment layers, oxymyoglobin on the surface and deoxymoglobin at deeper locations (*13*, *16*, *19*, *20*). For intact muscle, oxygen diffuses more deeply with increasing time after initial exposure to air. This increasingly deeper penetration of oxymyoglobin may continue for several days, depending on the mini-environment surrounding the muscle and the supply of reducing capability of the muscle, but ultimately the oxymyoglobin front will begin to move back closer to the surface.

Several techniques have been used to observe the movement of the oxymyoglobin front into the muscle, a change that can be observed during blooming by cutting perpendicular to the muscle surface. The edge of the oxymyoglobin front was observed to be 84% as deep into beef muscle as the detectable oxygen penetration (*19*). A steady state of oxymyoglobin penetration was observed in seven hours, although this could be highly influenced by extrinsic and intrinsic variables. The oxymyoglobin layer depth change was less at warmer temperatures with the following data: 0°, 2.2 mm; 5°, 1.6 mm; and 15°, 0.9 mm.

The location of the exact front of this pigment state is not clear-cut and somewhat difficult to identify although in our lab (*16*) a color transformation to the cyan, yellow, magenta and black color mode was used. This made the layer boundary sharper and more discernable. Even so, there is a location of varying proportions of both myoglobin chemical forms.

Figure 2 shows the relationship of partial pressure (or %) of oxygen to the proportion of the three myoglobin forms. At this point, if we focus only on deoxy and oxy forms, the need for extremely low pO_2 for deoxy and of high pO_2 for oxymyoglobin is clearly shown.

Air at sea level with approximately 21% oxygen (or 152 mm Hg of pO_2) obviously promotes oxymyoglobin formation. Use of greater oxygen % as used in a Hi Oxygen Modified Atmosphere package (MAP) was reported (*21*). The following % oxygen and % CO_2 in-package mixtures gave these reported depths of oxygenated layer at 1, 2 and 5 days: 80, 20 gave 11, 11 and 14 mm; 60, 40 gave 11, 9 and 7 mm and 40, 20 with 40% nitrogen gave 14 and 8 with discoloration apparent at 5 days for the 40% oxygen. Use of Hi Oxygen MAP is widely practiced today for distribution and marketing of case-ready meat products. The deeper oxygenated layer provides the longer time before meat discoloration. This is essential for long-distance distribution. Applying high-Ox using hyperbaric oxygen is immediately effective in oxygenation of myoglobin such as total oxygenation of a one-inch thick beef steak, but is not suited for commercial use.

Enhancement of meat marketed with the high-Ox system is used to stabilize color and control development of oxidative flavor and microbial growth. The system has the additional benefit of increased yield is widely used today.

Oxidation

When reducing mechanisms of muscle approach depletion and pO_2 in deeper muscle locations are no longer favorable to either deoxy- or oxy-myoglobin formation or retention, a third layer of pigment, brown metmyoglobin, forms between the oxygmyoglobin and deoxymyoglobin layers. The oxymyoglobin layer becomes thinner and the metmyoglobin layer becomes thicker and moves closer to the surface and both visual observers and reflectance instruments begin to "see" the brown discoloration.

The metmyoglobin front is seldom in a straight line as it approaches the meat surface, therefore discoloration as viewed from the surface often is visually spotty and not uniform. Visual evaluation of the meat surface is influenced not only by pigment state at the surface but also by subsurface pigment. The depth to which myoglobin chemical form and meat structure influence a surface

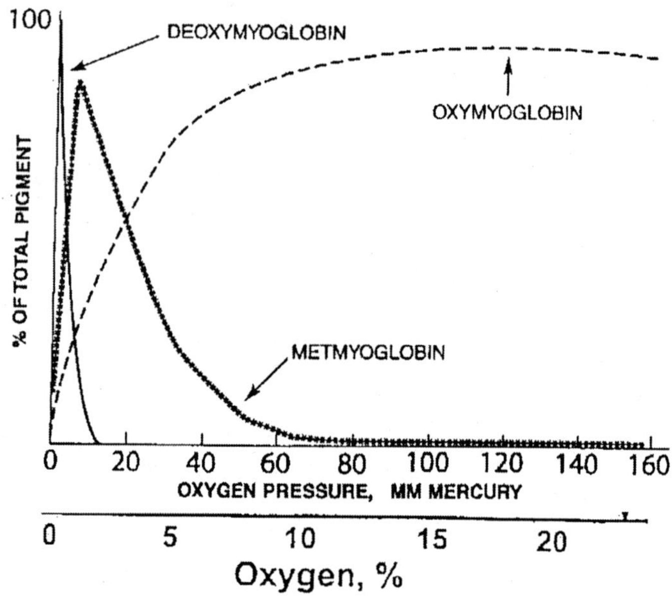

Figure 2. Effect of pO_2 and % O_2 on fresh meat myoglobin form.

appearance is limited by optical infinity (22), the maximum depth to which at least some light can penetrate and be reflected to the meat surface.

Observation of all three layers has been frequently reported as early as 1929 (*19*), but distance from surface and thickness of layers has been difficult to quantitate.

Figure 2 indicates that low partial oxygen pressure (pO_2), with maximum effect at 4 mm Hg and at as low as 1.4 mm Hg is favorable to metmyoglobin formation. Considering this information, lessening oxygen level to "just low" by either vacuum or low oxygen MAP is simply not good enough to avoid oxidation to metmyoglobin at too rapid a pace and to limit meat shelf (display) life.

Consumer discrimination begins at 20% metmyoglobin, a distinctly brown color is evident at 40% metmyoglobin and a brown to grey-greenish color appears when 60% metmyoglobin is detected by reflectance instrumentation (*20*). When pigment in the meat surface is about 20% metmyoglobin, ratio of sales of discolored versus bright beef was reported to be 1 to 1. A linear increase in consumer discrimination was reported as percentage metmyoglobin in fresh beef increased.

Green discoloration in fresh meat has been attributed to an altered heme structure, with two forms responsible. Choleglobin results from the interaction of myoglobin (ferrous or ferric) and hydrogen peroxide, with the most favorable pH at 4.5 to 6.0 (*13*). The source of hydrogen peroxide may be microbial.

Sulfmyoglobin, formed by reaction of sulfide and oxygen with reduced myoglobin also results in green discoloration. Sodium sulfite has been used as a release agent for hair or wool and if not quickly neutralized or removed will cause green color.

Extended oxidation may break down the porphyrin structure and results in yellow or colorless appearance.

Cured Color

Another myoglobin form of large economic importance is formed by nitrite which when reduced to nitric oxide reacts with myoglobin forming nitric oxide myoglobin. Upon heating, this is changed to nitric oxide hemochrome which forms a pink color in the curing reaction. Nitrite use has had the additional benefits of being an antioxidant and antimicrobial which is especially useful against *Clostridium botulinum*.

The curing reaction is most effective for producing cured color when meat is fresh and myoglobin is in the reduced state. Use of added reductant i.e., erythorbate (iso-ascorbate) and sufficient heating time and temperature at color fixing temperature is essential.

Retention of cured color and avoidance of fading is maximized by maintaining low oxygen (0.5% or less) in the package, by low-temperature and by reducing light exposure to lessen photochemical oxidation.

Oxidation and Reducing Reactions

Relatively stable and desirable color in meat results from myoglobin being in a reductive state, but this is not a static unchanging situation. The reaction from the deoxy to the oxy state of myoglobin is quite straight forward, but the direct reversal of this reaction is minimal and may not happen. Most likely deoxymyoglobin is formed from oxymyoglobin in a two step reaction which proceeds through oxidation to metmyoglobin where oxygen is removed and an electron is lost and then further proceeds to deoxy under conditions of very low oxygen and sufficient metmyoglobin reducing ability.

This reaction of proceeding through the myoglobin triangle seems inevitable but should be minimized. It is expensive in terms of loss of reducing capability and use of the NADH pool, which are limited in postmortem muscle. It is beneficial to slow these reactions by cold temperature and lessening oxidation promoting local environments. Controlling reactions by limiting myoglobin exposure to oxygen until the point where product is displayed would conserve reducing capability until its expenditure is warranted. This would be a very useful strategy for extending color life of fresh meat. Both deoxy and oxy forms are liable to undergo oxidation, therefore such driving forces as higher temperature and intense light exposure should be minimized. The reduction of metmyoglobin is critical to maintaining a saleable color life. Scavenging of oxygen is a normal occurrence in muscle and drives myoglobin reaction toward metmyoglobin, unless oxygen level is very low, at 1.4 mm partial oxygen pressure or less in which case deoxymyoglobin is formed if sufficient reducing capability of muscle is available. However, when this metmyoglobin reducing capacity (MRA) is depleted, myoglobin will persist in the oxidized state.

Excellent review discussions regarding myoglobin chemistry (*8*, *13*, *23*) are available and space in this chapter is too limited for this discussion.

Oxidation Promoting Forces

Oxidation promoting forces within muscle are affected by the composition, concentrations and reactivity of oxidation susceptible substrates, by pro-oxidants which catalyze oxidation and by antioxidants which minimize these reactions (*24*). Freezing, thawing, temperature fluctuation, cutting and grinding may affect meat membrane structure, thus enhance oxidation (*23*).

Some muscles are more oxidative in nature, thus oxidize faster. This could be because of lipid traits such as a greater proportion of phospholipids, more unsaturation of lipids, more free fatty acids, a greater content of transition metals (Fe, Cu), or possibly proteins that are more vulnerable to oxidation.

In trouble shooting discoloration problems for either fresh or cured meat, the tendency is to concentrate on late oxidation promoting events in the whole sequence of animal production, processing, to retail chain of events. However, numerous contributions to oxidative potential can occur at any time in the sequence. These are frequently additive but the consequences are not visible until some occurrence, such as exposure to display lighting appears to create the discoloration. However, a large contribution to oxidative potential may have occurred at some earlier time in the sequence.

Antioxidants

Antioxidants include lipid soluble α-tocopherol which plays a major role in stabilizing membranes. Other endogenous antioxidants include lipid soluble carotenoids and water soluble ascorbic acid,. At high levels ascorbic acid may also act as a pro-oxidant.

Cytosolic antioxidants include carnosine, anserine, glutathione, polyamines, uric acid and superoxide dismutase and enzymic ones include catalase and glutathione peroxidase (*23*, *24*).

The mechanisms of myoglobin chemistry have been widely studied, especially lipid-myoglobin relationships. Some of these can be altered by minimizing stress before slaughter, by nutritional supplements, controlled animal handling and slaughter procedures and appropriate chill.

Cooked Color Variables

With the cooking of fresh meat, the expected progression of cooked color with higher temperature or longer cooking time usually proceeds from red to lighter red, pink, lighter pink, gray, and brown. This progression might be appropriate, especially considering ground beef, for product with a normal ultimate pH of 5.6 to 5.8 from carcasses up to 42 months animal age (USDA quality grades A and B) and these colors might relate quite well to final cooked temperature. However, many product samples are exceptions so in view of consumer safety, cooked meat color, juice color and product texture should not be used to predict degree of doneness (*25*). The only reliable indicator is temperature determined by a fast response small sensing location device.

Premature Browning

Brown cooked color, giving an appearance of doneness, has been reported to occur at temperatures as low as 55°C (131°F) which is not hot enough to impact on safety from pathogens. Incidence is highly dependent of redox state of ground beef (*25*). Both oxy- and met- states favor premature browning. Deoxymyoglobin shows a normal pattern of color change with heating time and temperature.

Persistent Red Cooked Color

This problem (*25*), also known as pinking especially in poultry meat, is a contrast to premature browning, since it is the persistence or development of red or pink at temperatures where brown color and myoglobin denaturation should occur. It occurs in restructured meat product where ground or chunk meat is mixed with ingredients and shaped; then cooked usually with some time delay, but also occurs in some solid muscle like turkey or chicken breast meat.

One cause is contamination with nitrite, possibly by exposing fresh meat to equipment or surfaces previously exposed to nitrite containing meat. Other sources could be added water containing nitrite. Addition of ingredients containing nitrite ranging from vegetables (onion, celery, etc.) where pieces often leave tracks, i.e., colored areas the shape of these pieces; to soy material dried in nitrogen. Very low levels of nitrite can result in pinking: namely 14, 4, 2 and 1 ppm for beef round, pork shoulder, turkey breast and chicken breast (*26*). Nitrite, not nitrate, was the causative agent in direct dried soy isolate.

A large portion of persistent red or pinking problems in cooked beef and other meats is due to the formation of reduced hemochrome pigment. Beef can be cooked high enough (to 71.1°C) to cause a color change to gray-brown, but this color can revert to red-pink under reducing conditions.

Another closely related cause occurs when such products are allowed to be stored, such as overnight, before cooking. This allows a gel formation which provides some protection against heat denaturation during subsequent cooking and results in a more pink or red color. This can be counteracted by slight acidification with citric acid or by addition of certain dairy proteins (*27*). Negative aspects of these are the lessening of product yields by citric acid and the conflict with some religious requirements by combining meat and dairy proteins.

A very important cause of persistent red color and one that is easy to check is the use of beef with a pH of 6.0 or higher. The high pH protects myoglobin from heat denaturation so that the bright red pigment (oxymyoglobin) or purple-red (deoxymyoglobin) may be present at higher cooked temperatures, especially in the middle of hamburger patties, even when the outside surface is charred. High pH beef frequently is derived from sources such as cow, bull or dark cutting carcasses. This problem is more extensive if patties are not cooked immediately after forming, but are held under refrigeration before cooking. Remixing and reforming can break up the formed gel and counteract this protection, but represents a further processing cost.

Ground beef patties (113 g, 20% fat) made from muscle of pH 5.5, 5.6, 5.7, 5.8, 5.9, 6.1, 6.2, 6.3, and 6.4 were cooked to end-point temperatures of 66, 71, 77, and 82°C and were evaluated for internal cooked color for both patties that had primarily oxymyoglobin internally before cooking and also deoxymyoglobin internally (*25*). Myoglobin denaturation also was measured. Oxymyoglobin patties became sharply more resistant to thermal denaturation at pH 6.2 or higher while deoxymyoglobin patties had a more linear decrease in thermal denaturation with increased pH. Ingredients that would elevate pH, e.g., phosphates, could enhance the persistent red color.

Dirty gas jets or a poorly adjusted gas fired oven which may encourage incomplete combustion may also cause this problem with nitrogen dioxide, not carbon monoxide, being identified as the cause. This can react with meat pigments and will produce a heat stable red color on the surface of the meat to a depth of 1/8 to 1/4 inch, but not in the product interior. Recent work (*28*) found surface pinking in both cooked turkey and beef was not observed with up to 149 ppm carbon monoxide or 5 ppm nitric oxide (NO), but as little as 0.4 ppm nitrogen dioxide (NO_2) caused pinking.

Persistent red may also be caused by exposing ground or other meat to ammonia leaks or may result from treatment with high levels of ionizing radiation. An additional cause, more likely in poultry breast meat, is the effect of preslaughter stress leading to elevated muscle cytochrome content and formation of reduced nicotinamide-denatured globin hemochrome during cooking.

Reflectance spectra of the red areas or absorbance spectra of extracts from these areas can provide useful clues about the chemical state of myoglobin, which may be helpful in identifying the cause of a persistent red color problem. Nitrosohemochrome, the heated cured meat pigment, and nicotinamide hemochrome pigment, either of which could contribute to the problem, have unique reflectance or absorbance characteristics at specific wavelengths (*29*).

The use of carbon monoxide in MAP has caused researchers to investigate cooked color of meat containing carboxymyoglobin. One study (*8*) reported that carboxymyoglobin was more heat stable compared to oxymyoglobin and another (*8*) reported 58% denaturation for deoxymyoglobin at 71.1°C and 49% for carboxymyoglobin. Carbon monoxide in low-oxygen packages may result in more persistent pinking, possibly caused by heat-denatured CO-hemochrome and not by undenatured carboxymyoglobin.

Meat Color Evaluation

An excellent guide for meat color evaluation is available in a somewhat less accessible publication as "Guidelines for Meat Color Evaluation" written by a committee of the American Meat Science Association chaired by my colleague from Kansas State University, M. C. Hunt (*30*). This guide is very thorough and uses almost a step-by-step procedure for both visual and instrumental methods as used at that time, 1991. These procedures will be followed closely in the following section, but with some modifications in sample preparation, visual appraisal and instrumental appraisal.

Sample Preparation

Color measurement has a fundamental problem of obtaining a representative sample and also doing color measurement at a step in the protocol that answers the specific questions needing study (*30*).

Care must be taken in sample selection regardless of the means to be used for examining and evaluating the sample. Sample preparation for color measurement requires standardized procedures that are both repeatable (by the same person in the same laboratory) and reproducible (by different people in different laboratories at different times). All samples must be handled in exactly the same manner to prevent artifacts. This is of particular importance when live animal treatments are being evaluated for their effects on meat color. Muscle foods are generally opaque and will absorb, reflect or scatter incident light but generally do not transmit much light. Factors for which standardization is especially important include (unless the factor is an experimental variable) animal nutritional regiment, carcass chill rate, muscle, sample locations within a muscle, fiber orientation, muscle pH, postmortem time, temperatures of samples, time and temperature of storage, simulated or actual distribution or display, specific details of display, marbling content and distribution, surface wetness and gloss and packaging.

Some disagreement exists between researchers on this point, but the author believes samples should have approximately the same mass as those cuts for which study results will be applied; i.e., use 1 inch thick steaks instead of small sample cores. This avoids unusual sample temperature fluctuation during display case defrost.

For an initial fresh meat color evaluation that answers specific questions about effects of animal selection and breeding, nutritional and handling impacts, or influence of stunning, slaughter protocol or chilling, careful attention must be paid to bloom time, the time the fresh muscle surface to be evaluated is exposed to air. Effects of bloom time on oxygenation have been discussed under oxygenation.

Visual Appraisal

Visual appraisals of color in many research studies are closely related to consumer evaluations. They are not easy to conduct with either trained or consumer panels, since human judgments may not be repeatable from day to day and are influenced by personal preference, lighting, visual deficiencies of the eye and appearance factors other than color. Various scoring scales have been utilized for panel evaluations. Many of these are descriptive and imply averaging the color over the entire meat surface area. Others utilize a "worst-point" color score for a single or cumulative discolored area of at least 2 cm^2 in diameter, whereas still other systems are various combination of these two methods.

Some scales with hedonic terms have been misused for descriptive panels. Hedonic terms are a "like", "dislike" evaluation and are strongly affected by the varying preferences of panelists. Their use should be limited to consumer panels which require a minimum of no less than 50 panelists and preferably 100.

Pictorial color standards or colored chips with built in textural traits are useful to support specific color scales. Appropriate scales and standards for specific experiments are dependent upon project objectives, as is the decision of whether trained (descriptive) panels, consumer (acceptance) panels or both are required. The most suitable scales usually are best constructed through preliminary studies in which the product is treated in a similar manner as will be used in the actual study and ideally should cover the full range of color anticipated in the study. These can not only be used for panel training but may also serve as reference standards during the course of a study. Correctly structured scoring scales and appropriate standards substantially improve panel consistency and validity.

Steps in Visual Appraisal

1. Select panelists with normal color vision. Screen them using the Farnsworth Munsell 100 Hue Test.
2. Standardize conditions and sample variables, except for the ones to be studied.
3. Develop pre-trial samples to cover the range of color expected in the actual study.
4. Have panelists score these samples, perhaps 3 or 4 different ones at a time. Discuss after each of these scoring sessions.
5. Try to include some unusual sample problems in these pre-trial sessions. Try to address such issues as how to arrive at an average color score or a worst-point color score. Decide if discoloration would include sample darkening or brown (metmyoglobin) color. Such evaluation criteria should not be drastically changed during a study, but sometimes require fine tuning. Separate evaluation for areas of "normal" color and discoloration may be appropriate.
6. Develop scoring scale terms that are appropriate to the study and have equal differences between terms in a logical way to fit sample changes during the study.
7. When possible, use color prints, tiles or chips for both panel training and for the daily (or other time period) re-orientation or standardization of panelist evaluation. Colored standards should be stored in the dark when not in use to avoid color fading.
8. Conduct appraisals under conditions (i.e., retail cases, lighting, temperature, defrost cycling, case fill, packaging) that simulate the conditions under which consumers make their selections. Avoid evaluation during time of case defrost. Our cases are set for defrost twice daily at times when sample evaluation will not be done.

9. Use samples at least 12 to 15 mm thick; stack wafer-type samples and evaluate against a white background. Most backing boards or plastic foam trays meet this requirement.
10. Overwrap with the type of film most commonly used in merchandising the specific product or for special packaging conditions required for specific research studies.
11. Keep panelist viewing angle constant relative to the light source, about 45° to view mainly diffuse reflectance.
12. Rotate packages daily from front to back and side to side in the display area to help minimize variations in temperature and air movement and light intensity in display cases.
13. Standardize the type of lighting used for color evaluations and display studies. If fluorescent, state the specific lamp used, i.e., the wattage and color temperature such as 3000° Kelvin. Currently we are using 3000° Kelvin lamps. One recommendation (*22*) in the literature suggests lighting ≥90 Color Rendering Index (CRI), but CRI is evaluated at 8 color points and is not always an appropriate description for meat display. Lamps used for display studies should be similar to those accepted by retailers.

Application of Visual Appraisal

Visual appraisal of color of the longissimus of ribbed beef carcasses is a part of beef carcass quality grading since it is considered in determining carcass maturity.

Visual color evaluation is sometimes used as a criterion for "accept" or "reject". An example is the evaluation of pork meat quality for the export to Japan trade. They will frequently have their quality evaluators in pork plants in the U.S.A. for this purpose.

Another example of "accept" or "reject" is the decision by a retail customer or sometimes by a food service manager. The retail shopper's decision may be influenced by the color variation of meat cuts in the display case. Slightly discolored cuts have a lesser chance of acceptance when they are in competition with cuts of acceptable color.

Shoppers have "learned" through family instruction that the bright cherry red color of oxymyoglobin is desired and that darker red, the color of beef primarily in the deoxymyoglobin form, is less desirable. This is unfortunate because the highly oxygen permeable packaged product often leads to shorter shelf life. Currently used high-oxygen modified atmosphere packaging with injected enhancement solutions has extended distribution life to possibly 10 days with up to 5 more days for display.

A 1986 study (*31*) composed of questionnaires (and brief educational statements presented to about half) of 1750 grocery shoppers indicated that those not receiving the educational statements had a strong preference for bright cherry red colored ground beef. Those receiving this material were as likely to choose purple-red as bright red. However, retailer efforts to sell beef in the deoxymyoblobin state have not been successful.

Color Evaluation Challenges

Some color abnormalities are not measurable by instrumentation, these should be evaluated visually.

One of these is iridescence, sometimes called rainbow or mother-of-pearl color effect which may range from green to yellow to orange to a weird red. This condition is caused by diffraction of light rays by the distance between some microstructural units of muscle and is greatest in cuts made at a 90° angle to the muscle fiber direction. By definition, during rotation of sample, iridescence will appear and disappear, or become stronger and weaker. Therefore, sample orientation during evaluation can influence both the intensity and the proportion of sample affected. We solved this problem by mounting samples on our cooler wall so the most prominent iridescence faced the floor and samples were positioned so all panelists had to look up.

Striping, the appearance of intermittently dark and light color, is another unusual pattern that varies mostly in the difference in color intensity.

Instrumental Color Measurement

A serious problem in sample evaluation is that of uneven color or variable discoloration. Specific areas on a meat product may be severely discolored, whereas other areas are normal in color. When such a problem occurs, it may be useful to determine the proportion of the surface each section represents and to measure them separately. This process or the determination of what surface area to sample make the process more subjective rather than strictly objective. Use of a larger aperture enables a more representative reading or averaging. Use of the largest aperture size that will fit all samples in a study is recommended. For different aperture size on the same sample different reflectance readings were found (*35*), especially for longer wavelengths. Some new scanning procedures (to be discussed later) may overcome this problem.

Samples should be thick enough to be opaque or a white opaque backing must be used. Orientation of muscle fibers within the sample may influence readings. For samples with this problem, one reading should be taken, and then the sample should be rotated 90° and another reading should be taken; instruments are now capable of averaging multiple readings. The type of film used to overwrap the sample can influence readings, and the same film should be used for any research study, unless the study deals with film comparisons. Reflectance instruments can be standardized for film differences. Samples should have a flat surface that covers the entire aperture. A unique problem is presented by gas packed samples or modified atmosphere packages which have a space between sample and top film layer. Such packages can be placed upside down over the aperture to bring sample surface in contact with the film to eliminate the space effect, but this procedure might leave some moisture or residue on the inside layer of the film.

Several options are available for instrumental color reading of meat samples. Extraction and determination of absorption or transmittance is suitable for total heme pigment or myoglobin or for nitrosohemochrome determination for cured product. It is not suitable for quantitation of myoglobin forms or for determining spectra of deoxy- and oxymyoglobin containing samples as these quickly change upon exposure to air. Any such procedure entails the critical decision of how deep from the surface to sample. Estimation of metmyoglobin by extraction of sample surface by a buffered extract solution is an accepted procedure when done in cold conditions, but the same procedure results in interconversion between deoxy and oxy forms of myoglobin (*33*).

Reflectance Measurements

Such measurements can be closely related to what the eye and brain perceive. Over the course of a display- or shelf-life study, repeated readings can be made at the same sample location.

Important criteria to be indicated for color measurement studies or for color specifications include:

1. What color scale used
2. What observer angle (2° or 10°)
3. Which Illuminant (A, C, D^{65}, F)
4. Instrument aperture size
5. Instrument geometry
6. Specular included (yes or no)
7. Procedure details such as number of readings and sample orientation for each temperature, time interval into a study, humidity conditions.

The color scale could be Hunterlab L, a, b or more commmonly today is the CIE (1976) L*, a*, b*. Some advocate tristimulus scales (red, green, blue) or the Munsell scale.

While the earlier 2° observer angle was widely used, the more recent CIE addition of 10° viewer angle is more representative of what we see and is in more widespread use today (*22*). The same publication presents comparisons of use of several viewing angles, illuminant and aperture size, with too much detail to be discussed here.

Illuminant is an important factor to specify. We have frequently used Illuminant A (2854° Kelvin) which represents incandescent light in spectral energy balance and is richer in the red end of the visible spectrum. Illuminant C was the original synthetic light to represent daylight, but has been widely replaced by D^{65} (6500°K) based on a "cool phase of white Planckian daylight typical of overcast northern light including the effect of ultraviolet and occluded sunlight" (*22*). Illuminant effect calculations are made by software and apply to the color scale measurements such as CIE L*, a* and b* and their mathematical derivatives.

The effect of instrument aperture size has seemingly received little recognition except for a brief report in 1989 (*34*) at the International Congress of Meat Science and Technology and some data in 1994 (*22*). The 1989 report indicated a lesser percent reflectance in the longer red wavelengths for a 10 mm compared to 50 mm aperture.

A recent Kansas State University study (*35*, Table I) compared the effect of 0.64, 1.27, 2.54, and 4.45 cm aperture sizes for readings on beef longissimus muscle, with readings centered on the same location of visually uniform samples. The beef samples showed little difference in % reflectance at shorter wavelengths but lower readings for the 0.64 cm size at 10 nm intervals from 610 to 700 nm and lower readings for the 1.27 cm aperture at 680, 690 and 700 nm compared to the 2 larger aperture sizes. For both Illuminant A and D^{65}, this had a minor effect on L*. For Illuminant A, each smaller aperture size resulted in lesser a* and the 0.64 size resulted in lower b* and saturation index (chroma) and higher hue angle. For Illuminant C and D^{65}, the smallest aperture resulted in lower a*, b* and saturation index and a greater hue angle value.

Many meat color studies done recently use the CIE color system. L* deals with lightness, up to 100, to dark, down to 0. The a* value is usually stated as redness with increasing positive a* interpreted as redder samples. The b* value is often defined as yellowness, although negative values would mean blueness. One study (*36*) has indicated an a* change of 0.6 to be the smallest perceptible visual change for beef and chicken samples ranging from 15 to 28 a*.

Hue angle is calculated as $\tan^{-1}$ (b*/a*) and a higher value is frequently interpreted to mean greater discoloration. The smallest visually perceivable change was found to be 0.9 for beef and chicken samples in the 34 to 55 range.

Saturation index (chroma) is calculated as $\sqrt{(a^*)^2 + (b^*)^2}$ and a low value indicates a low color intensty where grey may more easily dominate.

The above values do not always relate well to meat discoloration, thus a* divided by b* has also been considered in some studies.

An interesting approach (*37*) to achieve instrumental values that relate well to discoloration used both Minolta CR 300 readings and color gauge measurements to produce Hewlett-Packard 4400 C images for larger sample areas to determine "trajectories" or progressive numerical changes that relate well to discoloration. Spoilage was most closely related to (a* minus b*) or a new coordinate system which rotated the a*b* axis by +45 degrees for reporting results.

Image analysis in another study (*38*) used photographs of total beef muscle surface of loin strip steaks for which RGB (red, green, blue) images were

Table I. Mean values of C I E L*, a*, b*, hue angle, and saturation index of beef longissimus under illuminant A, C, and D^{65} for four aperture sizes

		Aperture Size (cm)			
		4.45	2.54	1.27	0.64
Illuminant A	L*	38.45^{a}	38.34^{a}	37.85^{ab}	35.23^{b}
	a*	30.92^{a}	30.84^{ab}	29.28^{b}	23.75^{c}
	b*	25.37^{a}	25.49^{a}	24.38^{a}	20.87^{b}
	Hue Angle	39.27^{b}	39.47^{b}	39.68^{b}	41.14^{a}
	Sat. Index	40.00^{a}	40.02^{a}	38.11^{a}	31.63^{b}
Illuminant C	L*	34.73	34.59	34.32	32.45
	a*	21.48^{a}	21.46^{a}	20.01^{a}	14.60^{b}
	b*	22.38^{a}	22.63	21.90^{a}	19.74^{b}
	Hue Angle	46.21^{c}	46.60^{bc}	47.70^{b}	53.92^{a}
	Sat. Index	31.05^{a}	31.22^{a}	29.70^{a}	24.61^{b}
Illuminant D^{65}	L*	34.69	34.56	34.30	32.45
	a*	22.20^{a}	22.22^{a}	20.78^{a}	15.31^{b}
	b*	21.89^{a}	22.14^{a}	21.42^{a}	19.29^{b}
	Hue Angle	44.64^{b}	44.97^{b}	45.97^{b}	51.97^{a}
	Sat. Index	31.20^{a}	31.39^{a}	29.87^{a}	24.68^{b}

NOTE: [a,b,c] Means, within a row, with different supercripts differ ($P < 0.05$)

converted to a cyan magenta yellow (CMY) image, which enabled a sharper perception both visually and to instrument scanning. With a color calibration, this enabled scanning to result in a very quantitative determination of % discolored pixels. The above two studies are a good illustration of fresh, new, more accurate assessments of discoloration. Some similar and creative approaches were summarized in a review paper (*8*).

Reflectance to Estimate Myoglobin Chemical Form

Deoxy, oxy and metmyoglobin can be estimated from reflectance at certain wavelengths and even though these are accurate only to 6 or 7%, this is useful information (*30*). This is based on reflectance readings at wavelengths that are isobestic (equal) for two or three myoglobin forms. The 525 nm value is isobestic for all three, 474 is isobestic for oxy and met, thus useful for estimating deoxy. The same reasoning is the basis for estimating metmyoglobin at 572 nm and oxymyoglobin at 630 nm and at 610 nm (*32*).

The Kubelka-Munk equation, used to convert data to K/S (absorption/scattering) makes for a more linear relationship, to facilitate calculation when 0 and 100% of appropriate myoglobin form is established as a base for this calculation.

Efforts to put color reflectance values into meaningful numbers can be affected by sample characteristics such as surface wetness, % intramuscular fat (marbling) and myoglobin concentration. Wetness and % fat can influence light scatter of the sample and restrict the depth to which light wavelengths penetrate samples, come back and are detected and measured. Optical infinity is a term to describe this characteristic and can vary from sample to sample.

Some ways to correct for such influences is to use ratios or differences of % reflectance at two different wavelengths, such as 630 nm/580 nm or 630 nm minus 580 nm readings, presumably because this factors out the structural effects. Reflectance at 730 nm has been suggested as a corrector for structural influences and the Kubelka-Monk equation discussed earlier also accomplished this.

Limits on the target length of this chapter do not allow discussion of effects of pre- and post-harvest factors that influence color of meat. A very thorough discussion is available (*8*) and covers effects of genetics, environment, nutrition, animal handling and chilling. Genetics has a major impact especially on pork color as it influences stress susceptibility which impacts on glycolytic potential and rate of postmortem muscle pH decline, thus pale, soft, exudative muscle can result with a resulting shorter color life.

The tendency for such effects can be partially overcome by accelerated chill of carcasses.

Trouble shooting color problems involves application of knowledge about meat color and myoglobin chemistry and how it may be impacted by decisions and protocol along the entire live animal pre-harvest to post-harvest chain of events.

References

1. Mackinney, G; Little, A.C.; Briner, L. *Food Technol.* **1966**, *20*, 1300.
2. Hiner, R.L. *Proc. Recip Meat Conf.* **1954**, 7, 66-70.
3. Naumann, H.D.; McBee, J.L. Jr.; Brady, D.E. *Food Technol.* **1957**, *11*, 31 Abst.

4. Smith, G.C.; Belk, K.E.; Sofos, J.N.; Tatum, J.D.; Williams, S.N., Eds. Decker, E.A.; Faustman, C.; Lopez-Bote, C.J. *Antioxidants in Muscle Foods*, Wiley Intersciences, New York 2000; pp 397-426.
5. Doordan, M.; Anderson, D.L.; Naumannn, H.D.; Stringer, W.C. *Agric. Res Serv. Bull.* **1969**, *53*, 33.
6. Anonymous. National Live Stock & Meat Board 1990.
7. Jansen, C. Symp. *Influence of Vitamin E on Meat Quality* **1991**; p 29.
8. Mancini, R.A.; Hunt, M.C. *Meat Sci.* **2005**, *71S*, 100-121.
9. Brooks, *J. Proc. Royal Soc. London Series B, Biol. Sci.* **1935**, *118*, 560-577.
10. Wulf, D.M.; Wise, J.W. *J. Anim. Sci.* **1999**, *77*, 2418-2427.
11. Trater, C.; Hunt, M.C. Personal Comm. 2002.
12. Brewer, M.S.; Zhu, L.G.; Bidner, B.; Meisinger, D.J.; McKeith, F.K. *Meat Sci.* **2001**, *57*, 169-176.
13. Ledward, D.A. Chem. *Muscle Based Foods*. Ed. Johnston, D.E.,; Knight, M.K.; Ledward, D.A. Spec. Publ. 106, *Royal Soc. Chem*. 1992.
14. O'Keefe, M.; Hood, D.E. *Meat Sci.* **1982**, *7*, 209.
15. Atkinson, J.L.; Follet, M.J. *J. Food Technol.* **1973**, *8*, 51.
16. Limsupavanich, R.; Kropf, D.H.; Hunt, M.C.; Boyle, E.A.E.; Boyle, D.L.; Loughan, T.M. *Congr. Meat Sci. Technol.* **2005**, Paper 2, *27*, 93 Abst.
17. Kropf, D.H. In *Encyclopedia of Meat Sciences*; Ed. Jensen, W.K.; Devine, C.; Dikeman, M.; Elsevier Academic Press, 2004; Vol. 3, pp 962-969.
18. Zhu, L.G.; Bidner, B.; Brewer, M.S. *J. Food Sci.* **2001**, *66*, 1230-1235.
19. Brooks, J. *Biochm J.* **1929**, *23*, 1391-1400.
20. Kropf, D.H. *Meat Focus Int.* **1993**, *June*, pp 269-275.
21. Taylor, A.A. *Int. Symp. Meat Sci. Technol.* Ed. Franklin, K.R.; Cross, H.R. Nat'l Live Stock and Meat Board 1982; pp 353-366.
22. Macdougall, D.B. In *Adv. Meat Research*, Ed. Pearson, A.M.; Dutson, T.R., Blackie Academic and Professional 1994; Vol. 9, pp 79-93.
23. Faustman, C.; Cassens, R.G. J. *Muscle Foods* **1990**, *1*, 217-243.
24. Bertelsen, G.; Jakobsen, M.; Juncher, D.; Moller, J.; Kroger-Ohlsen, M.; Weber, C.; Skibsted, L.H. *Proc. Int. Symp. Meat Sci. Technol.* **2000**, *46*, 516-524.
25. Kropf, D.H.; Hunt, M.C. *Proc. Recip. Meat Conf.* **1998**, 144-148.
26. Heaton, K.M.; Cornforth, D.P.; Moiseev, I.V.; Egbert, W.R.; Carpenter, C.E. *Meat Sci.* **2000**, *55*, 321-329.
27. Slesinski, A.J.; Claus, J.R.; Anderson-Cook, C.M.; Eigel, W.E.; Graham, P.P.; Lenz, G.E.; Nobel, R.B. *J. Food Sci.* **2000**, *65*, 417-420.
28. Cornforth, D.P.; Rabovister, J.K.; Akuja, S.; Wagner, J.C.; Hanson, R.; Cummings, D.; Chudnovsky, Y. *J. Agric. Food Chem.* **1998**, *46*, 255-261.
29. Cornforth, D.P.; Calkins, C.R.; Faustman, C. *Proc. Recip. Meat Conf.* **1991**, *51*, 53-58.
30. Am. Meat Sci. Assn. Comm. Guidelines for Meat Color Evaluation; Chair, Hunt, M.C.; *Proc. Recip. Meat Conf.* Addendum **1991**, pp 1-17.

31. Lynch, N.M.; Kastner, C.L.; Kropf, D.H. *J. Food Sci.* **1986**, *51*, 253-255, 272.
32. Mancini, R.A.; Hunt, M.C.; Kropf, D.H. *J. Muscle Foods* **2003**, *64*, 157-162.
33. Kryzwicki, K. *Meat Sci.* **1979**, *3*, 1-10.
34. Sterrenburg, P. *Proc. Int. Congr. Meat Sci. Technol.* **1989**, *35(II)*, 3:21, pp 610-613.
35. Yancey, J. Personal communication, 2005.
36. Zhu, L.G.; Brewer, M.S. *J. Muscle Foods* **1999**, *10*, 131-146.
37. Cusick, R.S.; LaBudde, R.A. *Proc. Recip. Meat Conf.* **2003**, *56*, 117, Abst.
38. Ringkob, T.P. *Proc. Recip. Meat Conf.* **2001**, *54*, 97-98.

NOTE: All Proc. Recip. Meat Conf. available to members at website: www.meatscience.org.

Oils, Emulsions, Cereals, and Dairy Products

Chapter 22

Color Quality in Olive Products

M. I. Mínguez-Mosquera, B. Gandul-Rojas, L. Gallardo-Guerrero, M. Roca, and D. Hornero-Méndez

Research Group of Chemistry and Biochemistry of Pigements, Department of Food Biotechnology, Instituto de la Grasa, Consejo Superior de Investigaciones Científicas, Sevilla, Spain

Detailed analysis of the pigments responsible for the color in the fruit of olive (*Olea europaea* L.) has allowed establishing indices for quality traceability of virgin olive oil and table olives. Studies on chlorophyll and carotenoid metabolism during development and ripening of different Spanish varieties have shown the existence of intervarietal metabolic differences. This knowledge allows use of these pigments as biomarkers for an integral system of traceability of the virgin olive oil, which identifies all the steps of production, i.e. cultivar, geographical origin and the extraction in the case of olive oil. Particularly interesting in the sector of the table olive has been the clarification of the kinetics and mechanisms involved in the pigment transformation that occurs during fermentation. Based on this we have discovered the multifactor system leading to the insertion of endogenous copper fruit in the chlorophyll molecule, visualized as a superficial alteration known as *green-staining*.

The olive tree (*Olea europaea* L.) is the only member of the Oleaceae family producing edible fruits. The Oleaceae family contains about 22 genera and 500 species. Olive is by far the most economically important member of the family. The genus *Olea* contains about 20 species.

The olive has been cultivated by man since ancient times (6000 years ago), with its origins located in the eastern Mediterranean area. The olive was spread west on both sides of the Mediterranean basin, reaching Spain with the Roman Empire, and later cultivated largely by the Arabs, and Spain soon becoming the main producing country in the world. Today, the industry remains largely confined to Mediterranean countries of Europe, the Middle East, and North Africa, where it began thousands of years ago.

There are about 850 million olive trees in the world over an extension of about 8.7 million ha of which 95% are concentrated in countries of the Mediterranean basin. Spain, Italy and Greece are the main producers with 170, 125 and 120 million trees respectively, and other important producing countries are Turkey, Tunisia, Portugal and Morocco. During recent years, the average annual world production is estimated to be around 10 millions tons, of which 10% is used for table olive production and the rest for olive oil extraction (*1*). For updated information about current and past olive world production, visit the website of the International Olive Oil Council (IOOC) at www.internationaloliveoil.org.

Olive oil is an important component of the Mediterranean diet. Several studies has demonstrated that those eating the Mediterranean diet (rich in olive oil, fruits, vegetables, and fish) are known to have lower rates of colon, breast, and skin cancer, and coronary heart disease (*2-4*). The active principals in olive oil are thought to be monounsaturated fats (primarily oleic acid), squalene, and phenolic compounds that function as antioxidants in the body. Extra virgin oils are higher in these protective compounds than processed oils. Olive oil may act by reducing the LDL and raising the HDL forms of cholesterol in the blood (*5*).

Color Quality

As for many food products, color is one of the most important quality attributes. Consumers evaluate the quality of food throughout the external appearance in first instance, using the color as a primary tool (*6*). The external color aided by our own experience, education and some innate knowledge will even give some information about other quality attributes such as taste, odor and flavor, and in some way may inform us about the nutritional value and hygienic conditions of the food. Therefore color can be considered as a quality index, but only with a deep study of the chemical components responsible for these attributes, the pigments, will we be able to adequately evaluate color and the factors determining its quality. In conclusion, color as a quality index and the quality of that color should be inseparable concepts (*7*). In this way, and after characterizing the pigment profile, qualitative and quantitative, in raw and processed olive products it is to assess the effect of the standard process and therefore to establish the genuineness of the product, the occurrence of some alteration and even the detection of some bad practices (i.e. adulteration of extra virgin olive oil).

In the case of olive products, virgin olive oils and table olives, the color of these products is due to the presence of two main families of natural pigments, chlorophylls and carotenoids. In the case of ripe (black) olive fruits the color is mainly due to the presence of anthocyanins biosynthesized during fruit ripening. The color of olive products, olive oils and table olives may vary depending on several factors such as cultivar, stage of fruit ripening, climate and agronomical conditions, growing area, processing techniques and storage conditions. Among these factors ripening stage and cultivars are the most important (*8-15*). Figure 1 shows olive fruits at the stage where they are green (unripe), turning color and black (ripe). During this physiological process the change in color is mainly due to the partial or total disappearance of chlorophylls accompanied by a concomitant biosynthesis of anthocyanins that are responsible for the dark color of the ripe fruit. In the ripe fruit there are also carotenoids and chlorophylls, although they remain masked by the anthocyanins. Later, when these fruits are used for olive oil production, only chlorophylls and carotenoids will be transferred to the oil, whereas the hydrophilic anthocyanins will go with the aqueous phase. In Spain, table olives are produced from green fruits (Spanish-style green table olives) while extra virgin olive oil is obtained from ripe fruits that are richer in oil and with finest flavor.

In the Mediterranean basin each country has its own autochthonous cultivars, with about 1500 cultivars in the world that can be classified into three categories according to the final use given to the fruits: table olive processing, oil extraction, and dual use cultivars (*16*). In Spain there are 262 identified cultivars, out of which 24 are considered as main cultivars regarding the occupied geographical area. In the case of olive trees cultivated for olive oil production, the four main cultivars are *Picual*, *Cornicabra*, *Hojiblanca* and *Lechín de Sevilla*, producing together 60% of Spanish olive oil. Other important varieties are *Arbequina* and *Blanqueta*. In the case of cultivars used for table olive production, the two main varieties in Spain are *Gordal Sevillana* and *Manzanilla de Sevilla*, the last one considered one of the best by most table olive processors and internationally most widespread because of its high productivity and quality. The *Hojiblanca* variety is also used in some regions (Córdoba and Málaga) for table olives, so this can be considered as a dual-use cultivar.

In respect to the color, the qualitative pigment composition is basically the same in all fruits regardless the cultivar, and this composition is not modified

Figure 1. Changes in the color of olive fruits during ripening. A: green (unripe) stage, B: turning color stage, C: black (ripe) stage.
(See page 19 of color inserts.)

during ripening although quantitative changes will take place with different extension depending on the cultivar. Chlorophylls are represented by chlorophyll *a* and *b*, while the carotenoid fraction is composed of lutein (the major one), followed by β-carotene, β-cryptoxanthin, violaxanthin, neoxanthin and antheraxanthin, which can be referred to as the minor carotenoid fraction. Figures 2 and 3 show the chemical structures of these pigments.

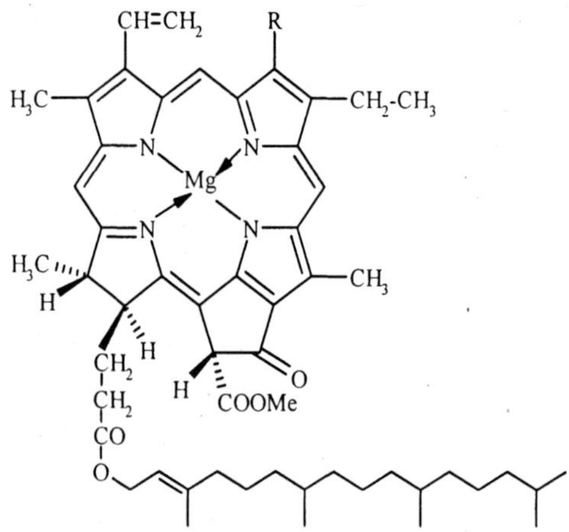

Figure 2. Chemical structures of chlorophyll a (R= $-CH_3$) and chlorophyll b (R= -CHO).

In some cultivars, such as *Blanqueta* and *Arbequina*, some exclusive pigments have been found that can be used for authentication of single-variety extra virgin olive oil produced from these varieties. In the case of *Arbequina* fruits the occurrence of xanthophylls (neoxanthin and violaxanthin) esterified with fatty acids have been described for the first time and for the moment only in this variety (*17*). In *Blanqueta* fruits have been found the existence of some hydroxy chlorophyll derivatives as a consequence of a particular chlorophyll catabolism mechanism characteristic of this variety. In summary, cultivars may be differentiated by the total pigment content (both chlorophylls and carotenoids) and by analyzing the individual carotenoid composition (*17, 18*). Figure 4 shows the changes in chlorophyll and carotenoid pigments during ripening for *Picual* as a representative cultivar. In the case of the chlorophyll fraction a continuous decrease is observed with the progress of ripening from intense green (IG) to black (B) fruits. In all ripening stages chlorophyll *a* is the major pigment being 2-3 times higher in concentration that chlorophyll *b*. In some varieties such as *Blanqueta*, chlorophyll almost totally disappears at the ripe stage and in consequence fruits have a pale yellowish color giving way to a pale colored olive oil (*19*). In respect to the carotenoid fraction a decrease in concentration is also observed throughout ripening, lutein being the major pigment at all ripening stages. Lutein will be the major pigment in the corresponding olive oil or table olives produced from these fruits. In the case of *Arbequina* fruit in which esterified xanthophylls have been found, the decrease in carotenoid concentration is less pronounced due to the compensation of catabolism by the carotenogenic process leading to the biosynthesis of the esters.

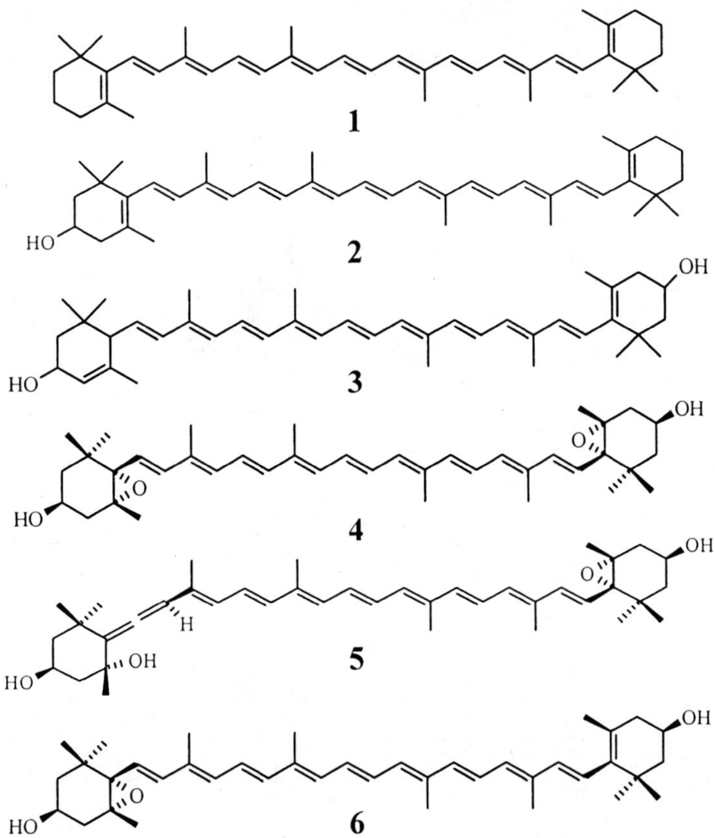

Figure 3. Chemical structures of carotenoids present olive products.
1. β-carotene, 2. β-cryptoxanthin, 3. lutein, 4. violaxanthin, 5. neoxanthin,
6. antheraxanthin .

Processing of Spanish-Style Green Olives

Today table olives can be considered the most important fermented vegetable, with Spain as the main producer and exporter in the world (*20, 21*). Table olives are defined by the Unified Qualitative Standard Applying to table olives in International Trade as "the sound fruits of specific varieties of the cultivated olive tree (*O. europaea sativa* Hoffm Lin) harvested at the proper stage of ripeness and whose quality is such that, when they are suitably processed, produce an edible product and ensure its good preservation as marketable goods. Such processing may include the addition of various products or spices of good table quality" (*22*).

The main purpose of table olive processing is the removal of the natural bitterness of the fruit to make it acceptable as a food. In the Spanish-style

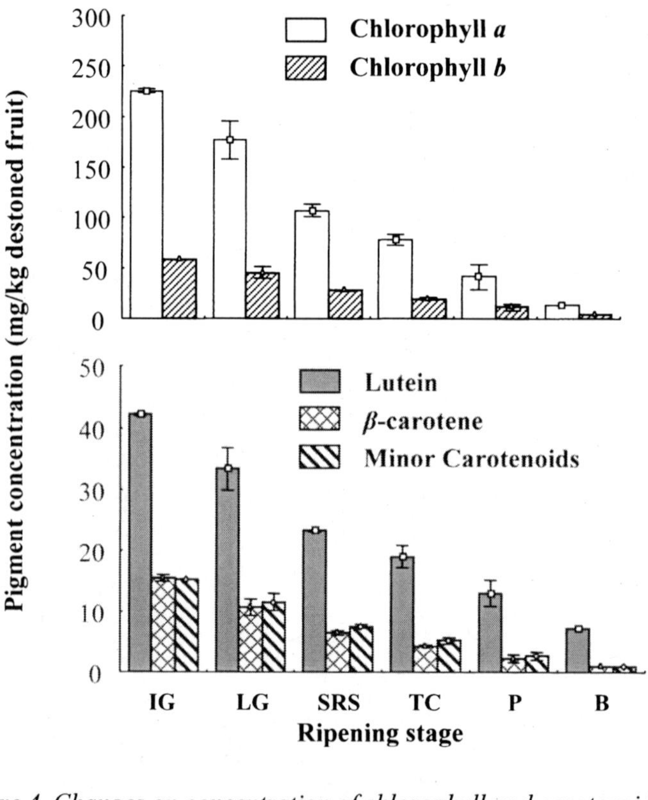

Figure 4. Changes on concentration of chlorophyll and carotenoid pigments during ripening of Picual variety olive fruits. IG: intense green; LG: light green; SRS: small reddish spots; TC: turning color, P: purple; B: black.

process the fruits are harvested at a degree of ripening characterized by a yellowish-green color and the bitterness is removed by means of alkaline hydrolysis. Fermentation by lactic acid bacteria confers to the Spanish-style green table olives their unique and highly valued organoleptic characteristics.

Before processing the fruits have to be harvested in their optimum stage, characterized by the external yellowish-green color of the fruit, which is due to the presence of both chlorophylls and carotenoids. In general, fruits are harvested manually to avoid damage. Figure 5 shows a simplified scheme for the Spanish-style process with four main steps: alkaline treatment, washing, brining and fermentation.

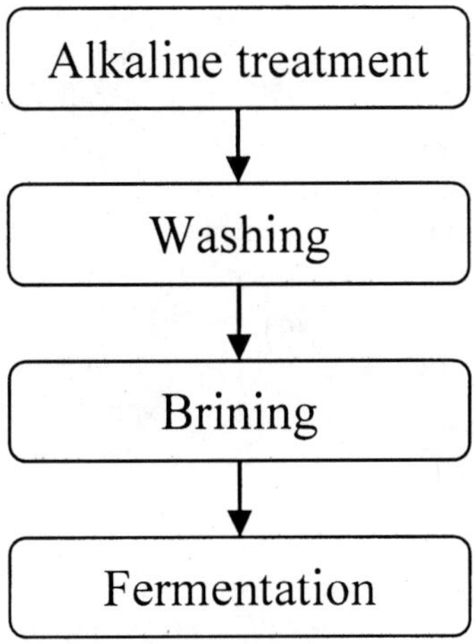

Figure 5. Simplified scheme for the Spanish-style green table olive process.

The alkaline treatment is performed with a diluted sodium hydroxide solution (commonly referred as "lye solution") as a result of which the glucoside oleuropein is hydrolyzed. However, the action of this process is much more complex and affects other components of the olive, enabling the development of a suitable culture medium for lactic fermentation during brining. Depending on the characteristics of the cultivar fruits, this treatment is adjusted in time and concentration to allow penetration of the lye into the pulp at 2/3 or 3/4 of the distance from skin to pit. In a normal procedure 9-10 h are necessary to accomplish the treatment. After this, alkali in the fruits is removed by two consecutives washings with water for a period of 2-3 h and 12-15 h respectively. Finally, the fruits are transferred to a fermentor and 10% sodium chloride solution (brine) is added. Within a few days this solution equilibrates with the inner part of the fruit, and compounds such as sugars, organics acids, vitamins and amino acids pass into the brine by osmosis, converting the brine into a rich culture medium suitable for microorganism growth. The succession of different microorganism species reduce the pH values from 10 to 4 or less, helping to develop the proper conditions for the abundant growth of *Lactobacillus plantarum,* which is the responsible for the fermentation of the Spanish-style green table olives. After these bacteria have consumed all the fermentable matter, the production of lactic acid is stopped, making it necessary to increase the salt concentration to 9% for preventing the development of unwanted bacteria and in that way ensuring the proper preservation of the product. Figure 6 shows typical Spanish-style green table olives.

During Spanish-style processing, minor components may undergo some changes, producing very relevant and important modifications in the organoleptic characteristics of the final product, among which color changes stand out. In this way, and promoted by the alkaline treatment and the acidity generated during fermentation, the chlorophylls and carotenoids responsible for the yellowish-green color of the fresh fruit undergo certain structural transformations. Chlorophylls *a* and *b* initially present in the fresh fruit are totally degraded by various mechanisms. In traditional table olive processing, two coexisting mechanisms are involved (*23*). One of the mechanisms is governed by the action of the chlorophyllase enzyme that is activated under alkaline conditions; the other one is chemical and governed by the acid medium promoted during fermentation. The deesterification of the phytol ester by the action of chlorophyllase does not affect the chromophoric properties of the resulting chlorophyllides in respect to the original chlorophylls. However, the acidic fermentation medium causes drastic color changes as a consequence of the replacement of Mg^{2+} by H^{+} in the porphyrin ring of chlorophylls and chlorophyllides. This reaction is known as pheophytinization, and is responsible for the color change from green (chlorophylls and chlorophyllides) to grey brown (pheophytins and pheophorbides).

Within the carotenoid pigment fraction, only those members with a molecular structure sensitive to an acidic medium, that is the epoxidated xanthophylls, violaxanthin, neoxanthin and antheraxanthin, will undergo isomerization giving way to auroxanthin, neochrome and mutatoxanthin respectively, and producing a decrease in the intensity of yellow coloring.

In general all these transformations are normal and desirable, as they are responsible for the highly valuable yellow-golden color of Spanish-style green table olives. However during the last decade several innovations have been introduced in the traditional processing system (reuse of brines and alkaline solutions, elimination of the short washing, etc), which have modified among other things the described degradation mechanism for chlorophylls, with the appearance of oxidative reactions affecting the chlorophyll isocyclic ring and yielding allomerized chlorophylls (*24*). These reactions are also involved in the

Figure 6. Typical Spanish-style green table olives (See page 18 of color inserts.)

development of a color surface alteration that occasionally appears in fermented fruit of the *Gordal* cultivar called *green-staining alteration* (*25*). The research work carried out by the authors of the present chapter have demonstrated that the visible manifestation of this alteration in the processed fruit, bluish-green spots, is due to the formation and oxidation of copper-chlorophyll derivatives (*17, 26*). In addition, it has been found that the copper ions chelated by the chlorophyll derivatives are endogenous to the fruits (*25*). The authors concluded that the appearance of copper-chlorophyll complexes in the fermented fruits of the *Gordal* cultivar is mediated by the industrial process, leading to aggressive cell and chloroplast disintegration, allowing oxidation of chlorophylls and subsequent chelation with copper ions within the fruits. Figure 7 shows fruits of Gordal variety with *green-staining alteration.*

Processing of extra virgin olive oil

Among the vegetable oils there is no doubt that virgin olive oil is the oldest known and the only one that can be consumed as obtained, preserving intact the nutritive and functional properties of its main components (*27*). Olive oil is clearly characterized by its fine and balanced aroma and flavor, and for its characteristic long shelf-life (Figure 8). Moreover, in the last decades increasing evidence has been found in relation to its biological and positive effects on health (*28*). Today, olive oil is considered an exceptional food ingredient having a great demand and high profitability. The IOOC estimated the production for the 2002-2003 season around 2,500,000 tons, being 98% produced in the

Figure 7. Spanish-style green table olives (Gordal variety) affected by the green-staining alteration. (See page 20 of color inserts.)

Figure 8. Bottle extra virgin olive oil with its characteristic greenish color. (See page 20 of color inserts.)

Mediterranean Basin countries, with Spain (36%), Italy (21%) and Greece (17%) as the main producers (*29*). According to the IOOC criteria, olive oil can be categorized into various grades (*22*), namely virgin olive oil, refined olive oil and pure olive oil. Virgin olive oil, the one with highest quality, is obtained from good quality olive fruits by mechanical procedures, which include milling, beating, centrifugation and decantation. However the final quality is affected not only by processing but also by the agronomic techniques, seasonal conditions, ripening stage of fruits, cultivar, harvesting and transportation systems, storage, etc (*11, 13*). Figure 9 shows a general scheme of olive processing to obtain virgin olive oil (*27*). The fruits should be harvested at the optimum degree of ripeness according to the maximum oil content and other characteristics related to the organoleptic properties of the resulting oil. Harvesting procedures producing damage to the fruits should be avoided. In general, fruits are manually collected from trees, although mechanical harvesting is now being used. The first operation, after fruits arrive at the mill, is leaf removal and washing to prevent, among other things, contamination from impurities that may affect the organoleptic properties of the oil. Subsequently, a paste is prepared by crushing the fruits, allowing the liberalization of oil droplets. The olive paste obtained after crushing is mixed slowly with the objective of gathering together the large size drops of oil and combining into a continuous oily phase. During this operation, the temperature should not exceed 25-30°C to prevent loss of aromatic compounds and the increase of oxidative processes. The next step is the liquid-solid separation which constitutes the fundamental part of olive oil processing with the aim of separating the oil and residual water (liquid phase) from the solid phase (known as *olive-pomace*) consisting of skin, pulp and stone particles. This separation maybe performed in three ways: selective filtration, extraction by pressure and extraction by centrifugation, the latter being the preferred method nowadays. The obtained oil contains some residual water and solid debris, and therefore to purify the oil a liquid-liquid separation is performed after a prior sieving by natural decantation, centrifugation, or a combination of both. The resulting virgin olive oil is stored for a limited period of one season or part of the following one. During this period is extremely important to avoid contact with light and maintain temperature between 15-18°C.

As in the case of table olives, the color of the virgin olive oil is due to two families of pigments, chlorophylls and carotenoids, the first group responsible for the green hues and the second one responsible for the yellow colors. During extraction these pigments are transferred to the oils due to their lipid-soluble nature, and at the same time they undergo some structural modifications affecting the final color of the product. The pigment composition of virgin olive oil may be quite variable depending on factors such as cultivar, degree of ripeness of fruits, geographical and environmental conditions, processing techniques and storage conditions (*13, 17, 30*). However, in qualitative terms

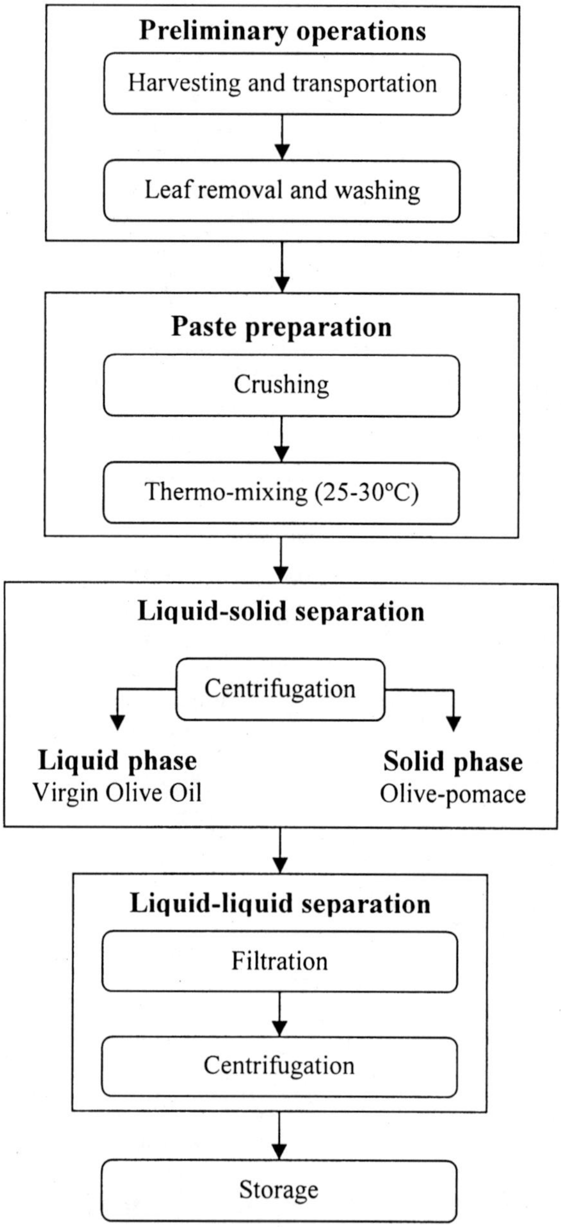

Figure 9. Scheme of olive processing to obtain virgin olive oil.

the pigment composition profile is nearly the same in all cultivars and is not changed during fruit ripening. Therefore the color intensity of the olive oil is directly related to the chlorophyll and carotenoid concentration, while the ratio of these two pigment fractions will moderate the tone from green to yellow. Moreover during fruit ripening the ratio between chlorophylls and carotenoids decreases due to the higher disappearance rate of chlorophyll, and this is the reason why the oil obtained from fruits at early ripening stages (green or mottled olives) has a greenish color compared to the oil obtained with more ripe fruits, with less intense color due to the decrease in the total pigment content in the fruit, but more yellowish as a result of an increase in the relative proportion of carotenoids (*18, 31*). In general terms olive fruit used for olive oil production are harvested at the ripening stage at which the chlorophyll-to-carotenoid ratio is about 3, being independent of the cultivar, nevertheless this ratio changed to 1 as a result of processing oils (*32*). This modification in the value of the ratio occurs as a result of a differential transfer of chlorophylls and carotenoids, so that only about 20% of the chlorophyll content of the fruit is transferred to the oil, while the carotenoid fraction is 50% (*10, 33*).

As a result of processing most of the olive components, including pigments, come into intimate contact with enzymes and other released components, promoting some important changes that may affect the quality attributes of the olive oil, including color. The modification of the pigments present in olive oil is influenced by the release of acid compounds. In the case of chlorophylls the pheophytinization reaction is the most important, producing magnesium-free chlorophyll (pheophytins). In the carotenoid fractions, as mentioned during the fermentation of table olives, xanthophylls containing 5,6-epoxide groups such as violaxanthin, neoxanthin and antheraxanthin are transformed to luteoxanthin, auroxanthin, neochrome and mutatoxanthin. The other carotenoids, β-carotene, β-cryptoxanthin and lutein remain unchanged (*10*). The detailed analysis of the pigments responsible for the color of olive oil has allowed the use of the chlorophyll and carotenoid profile as a quality parameter since the presence of pigments others than the normal ones or the detection of an unusual level of pigment transformation is indicative of bad or fraudulent practices (*34*).

Regarding quality traceability of olive oil, two parameters, based on chlorophyll and carotenoid composition, have been proposed for single-variety Spanish virgin olive oils (*18*). As mentioned before it has been demonstrated that the ratio chlorophylls-to-carotenoids shows a constant value (around 1) regardless of the variety and the ripeness stage. The second quality index has been defined with respect to the carotenoid fraction, so that the ratio between lutein and the rest of the minor carotenoids has values around 0.5. An exception to this is the oil obtained from *Arbequina* cultivar, in which this value is greater than 1, allowing not only authenticaton of the quality of the virgin olive oil but also its varietal origin. In addition, the percentage of violaxanthin and lutein, together with the total pigment concentration, allows distinguishing between single-variety Spanish virgin olive oils (*18*).

Although these parameters remain at constant values for up to one year after production when stored at 15°C in the dark (*15*), some modifications affecting the pigment may change the pigment profile compared to a recently extracted virgin olive oil. The color change of olive oils during storage is mainly due to the acid catalyzed transformation of chlorophylls into pheophytins, and in minor extension to the isomerization of 5,6-epoxide xanthophylls. During this period a small amount of pyropheophytin *a* has been detected in the stored oils, although this pigment is not present in recently extracted oils., the concentration levels are always lower than 3% of the the total chlorophyll content, and the ratio of pheophytin a to pyropheophytin a is higher than 20. Therefore, the presence of this pigment can be used as an indicative tool that oil has been stored for a time. Moreover, inadequate storage conditions, with variations in temperature, cause an increase in the concentration of this pigment (*35*), so that the presence of this pigment at unusual levels may suggest bad practices during storage that might compromise the preservation of the quality attributes of the original extra virgin olive oil. In this way, the relative amount (%) of pyropheophytin a to the total amount of pheophytins (%Pyrophy) seems to be the best quality index for assessing the goodness of the storage conditions. At 15°C, %Pyrophy has been found to be lower than 5% after 12 months, while at ambient temperature (25-35°C), %Pyrophy reached an average value of 14% after 12 months, but the quality parameters of the olive oil no longer correspond to an extra virgin grade. Finally, taking into consideration that pyropheophytinization is the prevalent reaction in thermally treated oils (such as deodorized oils), high values of %Pyrophy can be indicative of refining of oils or even could be used for detecting fraudulent additions of deodorized oils to extra virgin olive oils (*36*).

References

1. IOOC. Catálogo mundial de variedades de Olivo. Madrid: International Olive Oil Council, 2000.
2. Pérez-Jiménez, F.; de Cienfuegos, G.A.; Badimon, L.; Barja, G.; Battino, M.; Blanco, A.; Bonanome, A.; Colomer, R.; Corella-Piquer, D.; Covas, I.; Chamorro-Quiros, J.; Escrich, E.; Gaforio, J.J.; Luna, P.P.G.; Hidalgo, L.; Kafatos, A.; Kris-Etherton, P.M.; Lairon, D.; Lamuela-Raventos, R.; López-Miranda, J.; López-Segura, F.; Martínez-González, M.A.; Mata, P.; Mataix, J.; Ordovas, J.; Osada, J.; Pacheco-Reyes, R.; Perucho, M.; Pineda-Priego, M.; Quiles, J.L.; Ramírez-Tortosa, M.C.; Ruiz-Gutiérrez, V., Sánchez-Rovira, P., Solfrizzi, V., Soriguer-Escofet, F., de la Torre-Fornell, R., Trichopoulos, A., Villalba-Montoro, J.M.; Villar-Ortiz, J.R. *Eur. J. Clin. Invest.* **2005**, *35*, 421-424.

3. Carluccio, M.A.; Siculella, L.; Ancora, M.A.; Massaro, M.; Scoditti, E.; Storelli, C.; Visioli, F.; Distante, A.; De Caterina, R. *Arterioscler. Thromb. Vasc. Biol.* **2003**, *23*, 622-629.
4. Bartsch, H.; Nair, J. and Owen, R.W. *Carcinogenesis* **1999**, *20*, 2209-2218.
5. Pérez-Jiménez, F.; López-Miranda, J.; Mata, P. *Atherosclerosis* **2002**, *163*, 385-398.
6. Artés, F.; Mínguez-Mosquera, M.I.; Hornero-Méndez, D. In *Colour in food: improving quality*. MacDougall, D., Ed.; Woodhead Puclishing Ltd.: Cambridge, UK, 2002; pp 248-282.
7. Mínguez-Mosquera, M.I.; Jarén-Galán, M.; Gandul-Rojas, B; Hornero-Méndez, D.; Garrido-Fernández, J.; Gallardo-Guerrero, L. In *Clorofilas y carotenoides en tecnología de alimentos*. Mínguez-Mosquera, M.I., Ed.; Servicio de Publicaciones de la Universidad de Sevilla: Sevilla, Spain, 1997.
8. Mínguez-Mosquera, M.I.; Garrido-Fernández, J. *J. Agric. Food Chem.* **1989**, *37*, 1-7.
9. Mínguez-Mosquera, M.I.; Garrido-Fernández, J.; Gandul-Rojas, B. *J. Agric. Food Chem.* **1989**, *37*, 8-11.
10. Mínguez-Mosquera, M.I.; Garrido-Fernández J.; Gandul-Rojas, B. *J. Agric. Food Chem.* **1990**, *38*, 1662-1666.
11. Mínguez-Mosquera, M.I.; Gandul-Rojas, B.; Garrido-Fernández, J.; Gallardo-Guerrero, L. *J. Am. Oil Chem. Soc.* **1990**, *67*, 192-196.
12. Mínguez-Mosquera, M.I.; Gallardo-Guerrero, L. *J. Sci. Food Agric.* **1995**, *69*, 1-6.
13. Gandul-Rojas, B.; Mínguez-Mosquera, M.I. *J. Sci. Food Agric.* **1996**, *72*, 31-39.
14. Gallardo-Guerrero, L.; Roca, M.; Mínguez-Mosquera, M.I. *J. Am. Oil Chem. Soc.* **2002**, *79*, 105-109.
15. Roca, M.; Gandul-Rojas, B.; Gallardo-Guerrero, L.; Mínguez-Mosquera, M.I. J. Am. Oil. Chem. Soc. **2003**, *80*, 1237-1240.
16. Barranco, D. In *El cultivo del Olivo*; Barranco, D.; Fernández-Escolar, R.; Rallo, L., Eds.; Junta de Andalucía and Ediciones Mundi-Prensa: Madrid, Spain, 1997; pp 59-60.
17. Gandul-Rojas, B.; Roca, M.; Mínguez-Mosquera, M.I. *J. Agric. Food Chem.* **1999**, *47*, 2207-2212.
18. Gandul-Rojas, B.; Roca, M.; Minguez-Mosquera, M.I. *J. Am. Oil Chem. Soc.* **2000**, *77*, 853-858.
19. Roca, M.; Mínguez-Mosquera, M.I. *J. Agric. Food Chem.* **2001**, *49*, 832-839.
20. Fernández-Díez, M.J.; Castro Ramos, R.; Garrido Fernández, A.; González Cancho, F.; González Pellissó, F.; Nosti Vega, M.; Heredia Moreno, A.; Mínguez Mosquera, M.I.; Rejano Navarro, L.; Durán Quintana, M.C.; Sánchez Roldán, F.; García García, P.; Castro Gómez-Millán, A. 1985

Biotecnología de la aceituna de mesa. Consejo Superior de Investigaciones Científica-Instituto de la Grasa: Madrid-Sevilla, Spain, 1985.

21. *Table olives. Production and Processing.* Garrido-Fernández, A.; Fernández-Díez, M.J.; Adams, M.R., Eds.; Chapman & Hall: London, 1997.
22. IOOC. Unified Qualitative Standard Applying to Table Olives in International trade. Madrid: International Olive Oil Council, 1980.
23. Mínguez-Mosquera, M.I.; Garrido-Fernández, J.; Gandul-Rojas, B. *J. Agric. Food Chem.* **1989**, *37*, 8-11.
24. Mínguez-Mosquera, M.I.; Gallardo-Guerrero, M.L. *J. Food Prot.* **1995**, *58*, 1241-1248.
25. Mínguez-Mosquera, M.I.; Gallardo-Guerrero, M.L.; Hornero-Méndez, D.; Garrido-Fernández, J. *J. Food Prot.* **1995**, *58*, 567-569.
26. Gallardo-Guerrero, M.L.; Gandul-Rojas, B.; Mínguez-Mosquera, M.I. *J. Food Prot.* **1999**, *62*, 1167-1171.
27. Boskou, D. *Olive Oil. Chemistry and Technology.* Boskou, D., Ed.; AOCS: Champaign, IL, 1996.
28. Owen, R.W.; Giacosa, A.; Hull, W.E.; Haubner, R.; Würtele, G.; Spiegelhalder, B.; Bartsch, H. *Lancet Oncol.* **2000**, *1*, 107-112.
29. IOOC. *Olivae* **2003**, *99*, 42-45.
30. Morello, J.R.; Motilva, M.J.; Ramo, T.; Romero, M.P. *Food Chem.* **2003**, *81*, 547-553.
31. Mínguez-Mosquera, M.I.; Rejano-Navarro, L.; Gandul-Rojas, B.; Sánchez-Gómez, A.H.; Garrido-Fernández, J. *J. Am. Oil Chem. Soc.* **1991**, *68*, 332-336.
32. Roca, M.; Mínguez-Mosquera, M.I. *J. Am. Oil Chem. Soc.* **1991**, *78*, 133-138.
33. Gallardo-Guerrero, L.; Roca, M.; Mínguez-Mosquera, M.I. *J. Am. Oil Chem. Soc.* **2002**, *79*, 105-109.
34. Gandul-Rojas, B.; Roca, M.; Mínguez-Mosquera, M.I. In *Proceeding of 1st International Congress on Pigments in Food Technology.* Edited by Mínguez-Mosquera M.I.; Jarén-Galán, M. and Hornero-Méndez, D., Eds.; PFT99: Sevilla, Spain, 1999; pp 381-386.
35. Gallardo-Guerrero, L.; Roca, M.; Gandul-Rojas, B.; Mínguez-Mosquera, M.I. In *Proceeding of 3st International Congress on Pigments in Food, more than colors...* Dufossé, L., Ed.; PFT04: Quimper, Francia, 2004; pp 85-87.
36. Hornero-Méndez, D.; Gandul-Rojas, B.; Mínguez-Mosquera, M.I. *Food Res. Int.* **2005**, *38*, 1067-1072.

Chapter 23

Color Quality in Red Pepper (*Capsicum annuum*, L.) and Derived Products

M. I. Mínguez-Mosquera, A. Pérez-Gálvez, and D. Hornero-Méndez

Research Group of Chemistry and Biochemistry of Pigements, Department of Food Biotechnology, Instituto de la Grasa, Consejo Superior de Investigaciones Científicas, Sevilla, Spain

The red pepper fruit (*Capsicum annuum*, L.) has been used since ancient times as a spice and source of pigments to change the color and flavor of foodstuffs, making them more attractive and acceptable for the consumer. Pepper used as food colorant has traditionally been in the form of paprika (ground powder), although today concentrated oleoresins are more widely used in the food industry. The fruits of *C. annuum* owe their intense red color to carotenoid pigments that are synthesized in very high amounts during ripening. This chapter discussed various aspects regarding color quality and carotenoid composition in red pepper and its derived products.

The red fruits of pepper (*Capsicum annuum,* L.) have attracted man for a long time due to their characteristic color, taste and aroma, being frequently added to foodstuffs, making them more attractive and acceptable for the consumer. The main traditional use of red pepper has been as a food colorant, mainly as a dry ground powder (paprika), and more recently as concentrated oleoresins. The intense red color of the ripe fruits of *C. annuum* is due to presence of carotenoid pigments that are synthesized massively during ripening. Identities and properties of these pigments will be discussed later.

All peppers belong to the genus *Capsicum* of the family *Solanaceae*, which includes other species such as tomatoes, potatoes, and tobacco. According to the Russian botanist N.I. Vavilov, the main center of diversity and/or origin of the genus *Capsicum* is South America (particularly Brazil, Mexico, Bolivia and Peru), from where it was brought to Spain following the discovery of America by Columbus, and promptly distributed throughout Europe as an alternative hot spice to black pepper. Today, it is commercially grown in the United States, Brazil, India, Taiwan, South Africa, Zimbabwe, and through all Europe, Spain and Hungary being the major paprika producers. During the period 1995-2004 the world area dedicated for growing pepper for paprika production has increased 25%, from about 1.200.000 to 1.500.000 ha, and in the same period the annual world production has continuously increased from 14.000.000 to 23.000.000 tons (*1*). In the case of oleoresin, India and Spain are the main producers with 96.600 and 72.800 kg respectively during 1998, with the world production of about 178.400 for the same season.

It is generally accepted that genus *Capsicum* comprises 26 species, of which only 11 have been used by man; the rest are considered wild species (*2*). Among the former are five domesticated species: *C. annuum*, *C. baccatum*, *C. chinense*, *C. frutescens*, and *C. pubescens*. The species *C. annuum* and *C. frutescens* include most of the peppers commonly found in the market and used in foodstuffs. Traditionally, growers have selected pepper cultivars for the properties and characteristics that were most popular and/or most profitable agriculturally. The result is a great number of very different cultivars. Table I summarizes the main subdivisions of the species *C. annuum*, each comprising a wide range of different cultivars.

C. frutescens is a short-lived perennial plant with woody stems that reach a height of two meters, glabrous or pubescent leaves, has two or more greenish-white flowers per node, and extremely pungent fruit. This plant is cultivated in the tropics and warmer regions of the United States and Mexico, and is used mainly for Tabasco sauce manufacturing. *C. annuum* is an herbaceous annual plant that reaches a height of one meter and has glabrous or pubescent lanceolate leaves, white flowers, and fruits that vary in length, color, and pungency depending upon the cultivar. The non-pungent forms are used for direct consumption, fresh or dried, whole or ground. In both species, the fruits are berries, with different shapes (rounded or elongated) and sizes (few grams up to 250 grams) depending on the variety.

The numerous cultivars of pepper (*C. annuum*) show a wide range of morphological and organoleptic characteristics, including color, that determine their use, and which may be merely ornamental. There are about 59.000 recorded *Capsicum* accessions located in more than 140 germ-plasm banks (*3*), representing an invaluable source of potential genetic variation that can be used in breeding programs. Not all cultivars can be used to produce paprika; they must first meet a series of appropriate agronomic and industrial requirements (*4*). The most highly valued of these is high carotenoid content, as ultimately the commercial value of paprika depends on its coloring capacity, which depends directly on relative pigment richness. Other varietal characters of interest are low content in capsaicinoids (that is, reduced hotness or pungency); low moisture content and not very thick pericarp when ripe (to shorten the drying step of paprika processing, thereby reducing the production cost); simultaneous and grouped ripening (to help in mechanical harvesting); and, of course, high agronomic production and yield, together with resistance to factors such as disease, high/low temperatures, and salinity of soil and irrigation water.

Table I. Main taxonomic subdivisions of the *Capsicum annuum* L. specie.

C. annuum Subdivisions
Capsicum annuum var. abbreviatum
Capsicum annuum var. acuminatum
Capsicum annuum var. aviculare
Capsicum annuum var. cerasiforme
Capsicum annuum var. conoide
Capsicum annuum var. conicum
Capsicum annuum var. cordiforme
Capsicum annuum var. fasciculatum
Capsicum annuum var. glabriusculum
Capsicum annuum var. grossum
Capsicum annuum var. leucocarpum
Capsicum annuum var. longum
Capsicum annuum var. luteum
Capsicum annuum var. lycopersiciforme
Capsicum annuum var. minimum
Capsicum annuum var. minus
Capsicum annuum var. microcarpum
Capsicum annuum var. pyramidale
Capsicum annuum var. violaceum

In Spain, two main cultivars have been traditionally used for paprika production, namely *Agridulce* (*C. annuum* var. *longum*) and *Bola* (*C. annuum* var. *grossum*) (*5,6*). In the case of Hungary, the other important paprika producing country, *Szeged* and *Kalocsa* cultivars are the outstanding ones. As these cultivars have been subjected to a long and slow process of selection by agricultors, the local cultivars that have appeared are well adapted to the climate and very similar to each other genetically, and thus unlikely to be improved substantially. However, the industrialization of paprika production, the opening of new markets, and the introduction of the crop into countries with cheap labor have stimulated breeding programs for more-competitive cultivars, such as the recent example of *Jaranda* and *Jariza* new cultivars, and some other interesting cultivars obtained by classical selection and breeding (*7, 8, 9, 10*).

Color Nature in Pepper Fruits, Paprika and Oleoresins

Color of food products can be considered one of the most important quality attributes, by means of which consumers evaluate the quality of food (*11*). In the case of red pepper and derived products, mainly paprika and oleoresin, their color is due to the outstanding presence of natural pigments, the so called carotenoids. As in most fruits, the color of pepper fruits and derived products may vary depending on several factors such as cultivar, ripening stage of fruit, agronomic conditions, growing area, processing techniques and storage conditions. During ripening of *Capsicum* fruits there is a characteristic color change from green (unripe fruit) to red (ripe fruit) which is mainly due to the partial or total disappearance of chlorophylls accompanied by a massive *de novo* biosynthesis of carotenoids which are responsible for the characteristic red color of the ripe fruit. Figure 1 shows a pepper plant (cv. *Agridulce*) with fruits at different ripening stages, and Figure 2 shows the changes in total chlorophyll and carotenoid pigments during ripening of *Agridulce* fruits as a representative cultivar.

Carotenoids are important natural pigments found in all plants, algae, many bacteria and fungi, as well as in some animals (*12*). Carotenoid pigments are the compounds responsible for the attractive colors of fruits and flowers, having an important role in attracting animals that act as pollinators and seed dispersion vehicles, including in this process the consumption of food by humans. When carotenoids are ingested, they show important biological actions such as antioxidants, reducing risk of cancer, positive effect on the immune response, and in addition some of them (β-carotene, β-cryptoxanthin, etc) have provitamin A activity (*14, 14, 15*). From the point of view of color, the now very familiar *Capsicum* fruit, can be considered exotic and unique, since some of the

Figure 1. Spanish red pepper (C. annuum) cv. Agridulce cultivated for paprika production at La Vera region (Cáceres, Spain). (See page 21 of color inserts.)

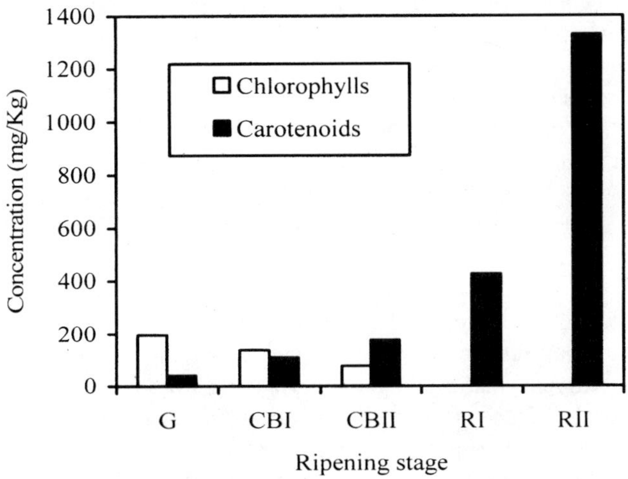

Figure 2. Evolution of the chlorophyll and carotenoid content during ripening of red pepper(C. annuum) cv. Agridulce. G: green; CBI: color break I; CBII: color break II; RI: red I; RII: red II.

carotenoid pigments responsible for the final red color of the fruits are almost exclusive to the genus *Capsicum,* namely capsanthin, capsorubin, and capsanthin 5,6-epoxide (*5, 16*).

The basic structure of carotenoids is essentially a C40 terpenoid compound formed by the condensation of eight isoprene units, their biosynthesis being an important branch of the complex isoprenoid pathway. The basic carotene structure (i.e. lycopene) undergoes several structural modifications namely cyclization, hydroxylation, epoxidation and ring rearrangement, giving way to the great variety of carotenoids in Nature (more than 650) (*17*). During ripening of the pepper, there is a spectacular *de novo* biosynthesis of carotenoid pigments. Figure 3 shows the structures of the main carotenoids found in red pepper fruits. All the carotenoid pigments present in the mature pepper contain 9 conjugated double bonds in the central polyene chain, although with different end groups (β, κ, 3-hydroxy-5,6-epoxide, 5-hydroxy-3,6-epoxide), which change the chromophore properties of each pigment, allowing them to be classified in two isochromic families: red (R) and yellow (Y). The red fraction contains capsanthin, capsanthin-5,6-epoxide, and capsorubin, while the yellow fraction comprises the rest of the pigments (zeaxanthin, violaxanthin, antheraxanthin, β-cryptoxanthin, β-carotene, and cucurbitaxanthin A), which act as precursors of the former (*5, 18*).

Earlier work on red pepper (*19, 20, 21, 22*) established that the total carotenoid content and R/Y ratio can be used to define the quality of the raw material (fruits) and derived products (paprika and oleoresins). Fluctuations in the total carotenoid content and the R/Y ratio with respect to those found in the source fruit can indicate inadequate processing, the existence of degradative reactions (acting preferentially on one of the isochromic fractions), or even possible color adulteration.

It is noteworthy to mention that during *de novo* biosynthesis triggered by the ripening process, the new formed xanthophylls become esterified with fatty acid (*23, 24*). The function of xanthophyll esterification, occurring in general in most ripening fruits, is still unknown although some functions or benefits may result from this process, for example the increase of liposolubility, facilitating the accessibility to biosynthetic enzymes or even enhancing the stability of these pigments towards enzymatic and non-enzymatic oxidative reactions. As this process is closely related to ripening, esterification has been proposed as a ripeness index (*25*). In the case of red pepper xanthophylls, yellow xanthophylls are mainly esterified with unsaturated fatty acids like oleic and linoleic, while saturated fatty acids esterify the red xanthophylls (*24, 26*), important fact that may govern in part the differential stability of these xanthophylls during paprika processing.

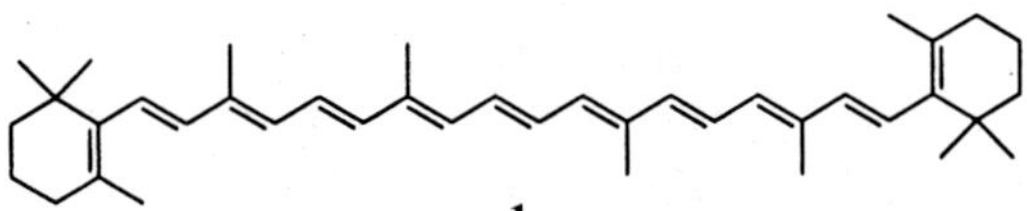

1

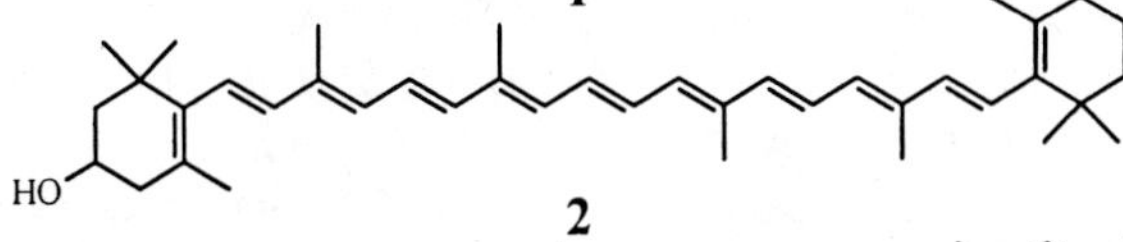

2

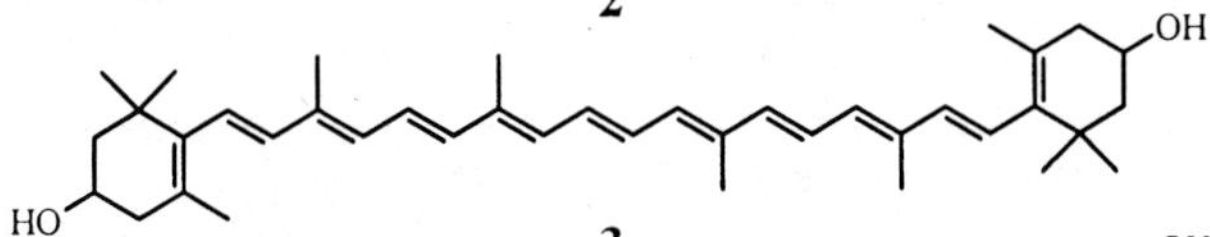

3

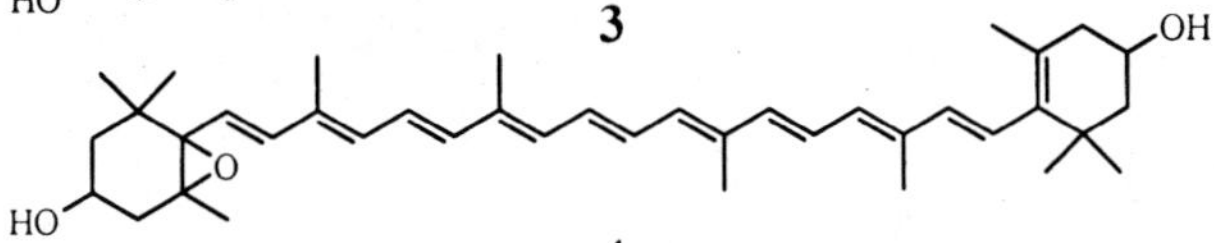

4

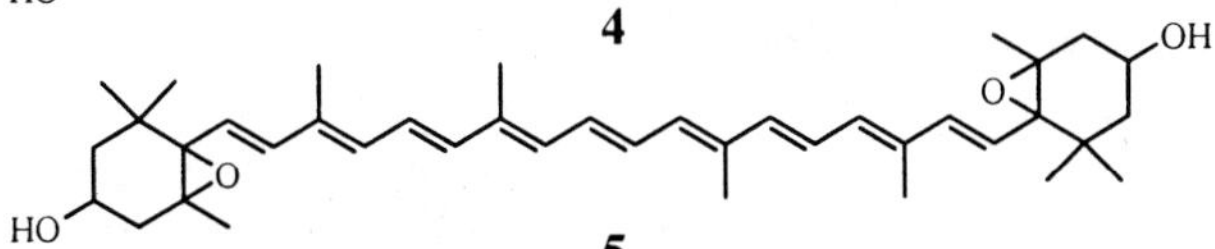

5

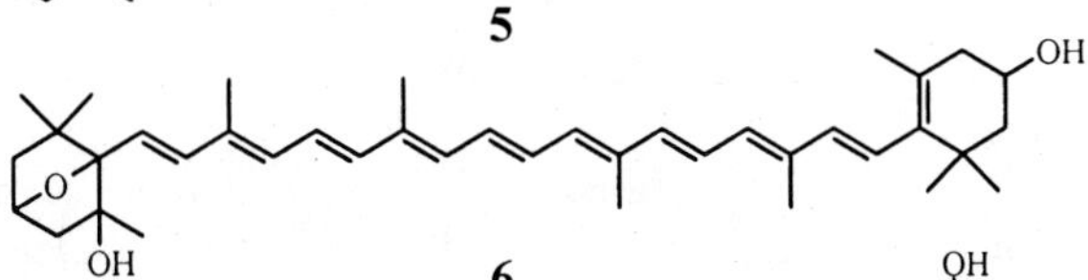

6

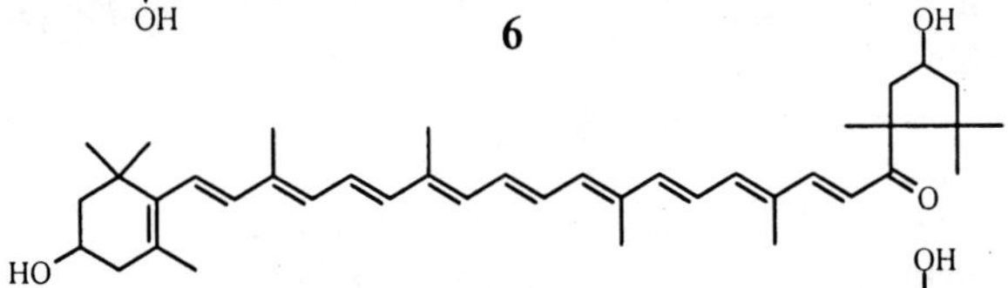

7

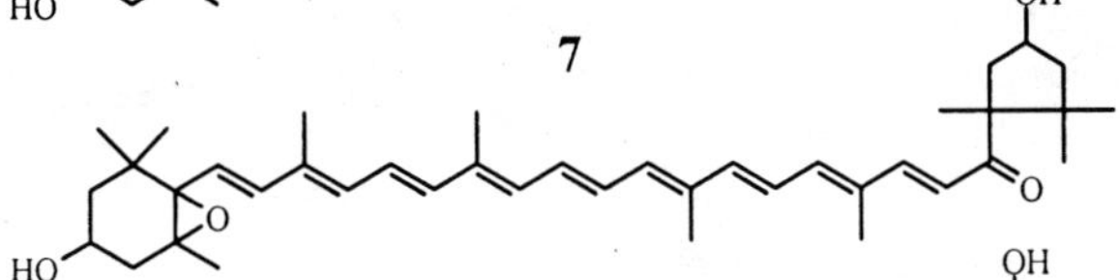

8

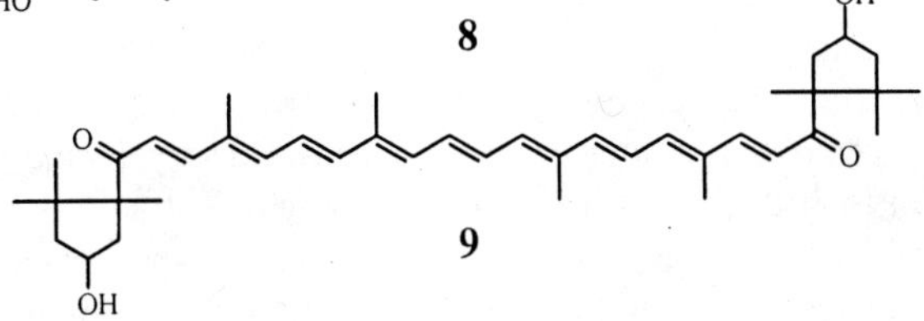

9

Figure 3. Chemical structures of the main carotenoids present red pepper fruits and derived products. 1. β-carotene, 2. β-cryptoxanthin, 3. zeaxanthin, 4. antheraxanthin,5 . violaxanthin,6. cucurbitaxanthin A, 7. capsanthin, 8. capsanthin 5,6-epoxide, 9. capsorubin.

Processing of Paprika

Figure 4 shows a simplified scheme of the main processing steps for paprika production. Apart from the normal preliminary operation such harvesting and transportation, the main operations are dehydration and milling. The dehydration of fruits is needed to extend the shelf-life of fruits from days to months, since the water content in the fresh fruit is around 70-85% which leads to spoilage if stored fresh. In addition, size reduction is done by milling or grinding the dry fruits, which is necessary to homogenize the color and to facilitate handling and transportation. The final product of this simple two-step processing is paprika, a fine, brilliant, highly colored powder. Paprika can be added directly to foods for improving their appearance and flavor, and for that reason the food industry uses paprika extensively in barbecue sauces, salad dressing, ketchup, cheese, snack food, sausages, flavored oils, etc. Figure 5 shows the intense and characteristic red color of Spanish paprika from the La Vera region (Cáceres, Spain).

Dehydration of pepper fruits can be carried out in different ways. Traditionally, two conventional dehydration techniques have been used, via open air exposure of fruits to sunlight, or in drying chambers where the heat source is the burning of tree logs. The first system has been extensively used in Murcia region (Southern Spain) and Turkey, but now it is being abandoned as the quality of dry peppers is affected by oxidative degradation promoted by sunlight (*27*), so that the industrial processors tend to replace this technique by emerging ones. The traditional method applied in La Vera region (Cáceres, Spain), is however still in use. The drying chambers consisted of a two floor construction where the oak logs are placed on the ground floor while the fruits are piled on the top one, and hot air circulates from the heat source through the mass of fruits. By using this technique, fresh fruits are dehydrated in 7-10 days (*27, 28*). The characteristic smoked flavor of this product is very highly appreciated by the consumer and food processors.

High scale paprika producers prefer industrial dryers because the dehydration process applied is more uniform and rapid than in the traditional systems. In general, the main advantages of the modern industrial processes are the reduction of residence time of fruits in the dryer to 4-6 hours, and the possibility of using continuous systems. Industrial dryers may reduce fruits moisture down to 3-4%, while the product obtained by traditional processes has moisture level of about 7-9% (*29, 30*). Hot tunnel dryers are often employed, where hot air at 60-80°C is insulated under counter-current conditions through a tunnel where the fresh fruits are intermittently introduced on trays (*31*).

Milling aims to reduce particle size and increase homogeneity of different dry fruit batches, improving the suitability of the resulting paprika for further processing such as oleoresin production or inclusion in food formulations and dosage. Dry fruits are generally milled by using hammer or ball mills, and adding a certain percentage of fruit seeds (30-35%), acting as a dilution factor

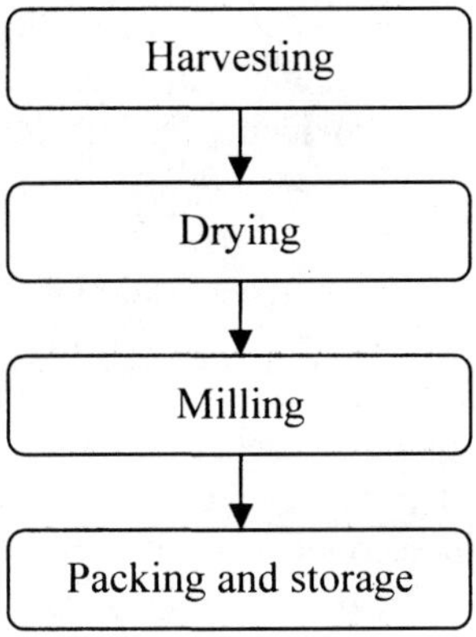

Figure 4. Simplified scheme for paprika production process.

Figure 5. Spanish paprika from La Vera region (Cáceres, Spain).
(See page 21 of color inserts.)

for the total carotenoid content (*32*). However, the addition of seeds has further implications in stabilizing carotenoid content during storage. A comparative study of paprika stability with different seed percentage (0%, 20%, 40% and 60%) showed that high addition of pepper seeds during milling might increase color stability with the drawback of high color dilution, while paprika without added seed showed the lower carotenoid stability (*32*). A possible explanation is the participation of natural antioxidants, mainly tocopherols, contained in the pepper seeds (*33*).

Dehydration conditions (temperature and drying time) as well as ripening stage of fruits and their moisture content determine the extent of changes in the fruit carotenoid composition responsible for quality of the resulting paprika (*6, 8*). The oxidation rate of carotenoid pigments increases as the drying temperature increases (*28, 30*), and in consequence stability of coloring power during processing and storage is dependent on drying conditions. In consequence, any dehydration systems should keep a compromise between drying temperature and residence time. This is achieved in both industrial and traditional drying processes by lowering residence time or temperature respectively, although in both systems the thermal stress will compromise the shelf life of paprika.

Mild temperature conditions (average 40°C) are applied in the traditional process carried out in the La Vera region (Cáceres, Spain), which is obviously characterized by a longer residence time (7-10 days), but with the advantage of maintaining constant the carotenoid content during processing, that is, color quality remains unchanged. The advantage of this process is emphasized if one considers the fact that over-ripening of fruits may take place during the dehydration process (*6, 8, 34, 35*). Peppers from *Agridulce*, *Bola*, *Jaranda* and *Jariza* cultivars subjected to the traditional La Vera dehydration process increase their initial carotenoid content about 15-20%. The ratio between the two isochromic pigment fractions (R/Y) increases due to the fact that pigments belonging to the yellow fraction are biosynthetically converted to red pigments. In the case of processing of *Bola* cultivar fruits, R/Y changes from 1.88 to 5.36 (*35*).

During milling, fruits are subjected to heating from friction generated by the mill hammers or balls, and this together with the reduction in particle size and the increase in active specific surface facilitates oxidative reactions. During operating time of milling, about 2-3 hours, carotenoid content may be affected (*6, 8*). Losses during the milling step are around 10-15% of the carotenoid concentration present in dry fruits. In general, the R/Y ratio in the resulting paprika is higher than that of the corresponding fresh fruit. This implies that the degradative processes occurring affect mainly to the yellow fraction.

During paprika storage the stability of carotenoids will be a result of previous processing steps (*34*). Over-drying or excessive milling promotes oxidation reactions that not only affect carotenoid concentration but also the rest of the fruit components, especially fatty acids that may undergo oxidation, and seriously affect the carotenoid profile (*36*). In general, air and light contact must be avoided during paprika storage.

Paprika Oleoresin

Paprika oleoresin is obtained from paprika powder, as raw material, by means of solvent extraction of the lipid soluble components, including the carotenoid pigments. One kilogram of oleoresin is equivalent to approximately 18 to 20 kilograms of paprika. The paprika oleoresin is a deep red oily liquid (Figure 6). Paprika oleoresins have important applications in the food industry, as trends in the use of natural colorants are more widespread to substitute for those of synthetic origin. The oily substrate, where the carotenoid fraction is dissolved, is the ideal vehicle to incorporate this product in multiple food formulations. Some applications include processed meats, pasta and pizza, dairy products, such as soups, sauces, and snacks to supply or improve the color of those foodstuffs. Figure 7 shows a scheme of the main steps for paprika oleoresin process by means of solvent extraction. This is not a very selective procedure, extracting not only carotenoids, but also the rest of components of the lipid fraction. All lipophillic solvents are in principle suitable for this step including both organic solvents and supercritical fluids.

Figure 6. Paprika oleoresin prepared by solvent extraction from Spanish paprika. (See page 22 of color inserts.)

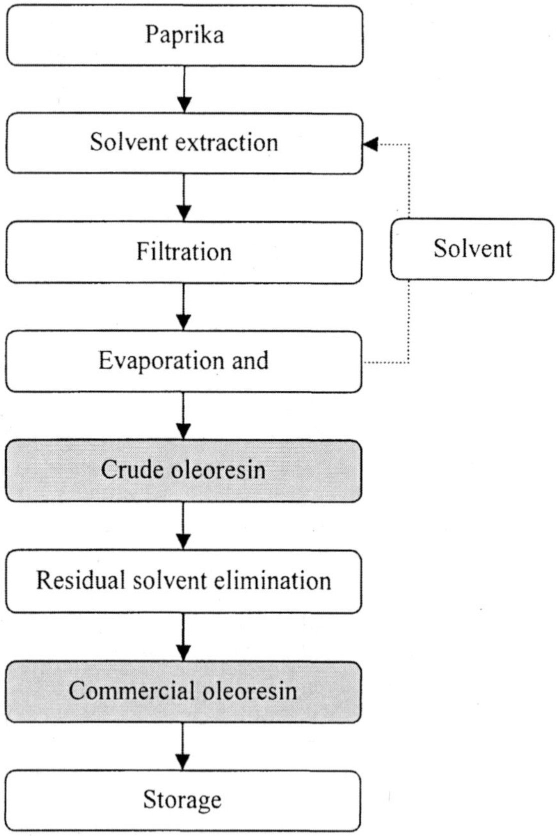

Figure 7. Scheme for paprika oleoresin production system by classical solvent extraction.

Classic extractions with organic solvents are usually carried out with hexane, methylene chloride and ethyl acetate. Depending on the country, some food legislation and regulations may restrict the use of particular solvents. The simplest extraction method is countercurrent extraction either in a batch or continuous process. Batch processes use a paprika to solvent ratio of 0.6 kg/L, and extraction times of up to 5 hours, whereas in continuous countercurrent extractors the ratio is 4 kg/L and processing time is 3-5 hours (*37*). The remaining color in the solid material at different times will determine the operating time. The solvent is usually removed by concentration of the extract in vertical evaporators under reduced pressure at about 50-80 °C, giving way to a crude oleoresin. The product is further treated to remove solvent residues by either a steam spray at 160 bars for 15-30 minutes or indirect heating with water at 100 °C. Selection of solvent and optimization of operating conditions at the extraction stage is very important to improve quality and stability indexes of the final product.

The increasing restrictions on the use of organic solvents and the lack of selectivity of this process represents a serious disadvantage for organic solvent-extraction procedures. Extraction with supercritical fluids is an interesting and powerful alternative that has been used for extraction of caffeine from tea and coffee, and essential oils from a wide range of spices like basil, oregano, turmeric and paprika (*38, 39, 40, 41*). Working with supercritical fluids as an extraction solvent has several advantages. This procedure includes at the same time both an extraction step and a fractionation one, and temperature conditions are reduced to 40 °C, gentler than in classical procedures, so thermal stress is considerably reduced. Additionally the necessity of solvent removal is practically eliminated because decompressing the extract will remove any trace of extraction fluid, which can be recovered and recycled. Carbon dioxide under supercritical conditions is the gas most frequently applied for extraction purposes because it is inert, non-toxic, highly available and inexpensive. As described by Govindarajan, paprika extraction at 120 bars and 40 °C produces a highly pungent and aromatic oleoresin, while a second extraction at 320 bars and 40 °C produces a highly colored oleoresin free of capsaicinoids and aroma compounds (*42*). This two-step procedure, with increasing operation pressure, has also been reproduced with sweet paprika as the raw material (*39*). In this case, a β-carotene rich oleoresin was first obtained, while a purified carotenoid extract resulted from the second step.

As stated before, processing is a continuous heating line, where the extract is subjected to thermal stress. High temperature conditions particularly affect the R isochromic fraction while the Y one is more stable, the opposite effect to that described for paprika. The change in the stability pattern is due to the existence of an isokinetic point that, as a function of the temperature, degrades preferentially one fraction over the other (*39*). Other stability factors are the molecular environment that covers the carotenoid and the oily matrix where the oleoresin can be diluted (*43, 44*).

Problems during storage of paprika oleoresins can arise from flocculation of gums and waxes that cause cloudiness and changes in density and refractive index of the end product. In addition, if this sludge is not removed, it could interfere in the clarity and solubility of the oleoresin in the food applications. Therefore, physical or chemical refining is often applied to avoid such problems during long storage periods. With respect to the stability of the carotenoid fraction, it will depend on conditions reached during processing, especially during the solvent removal step. The industrial sector has some predictive tools allowing them to calculate carotenoid content after a storage period, that is, the content of color in the sample (*39*).

Determination of Quality of Paprika and Oleoresin

In a product economically evaluated by its coloring capacity, measurement of this organoleptic quality must be the reference and standard procedure for ranking different qualities of paprika. Several standard methods have been developed to achieve the goal of obtaining rapid quantifiable data via easy measurement techniques. A few standard methods for obtaining a "color value" are commonly applied, such as the STANDARD and ASTA measurements (*45, 46*). Briefly, these methods consists of recording the absorbance value at a specific wavelength (around 460 nm) and application of dilution factors that give a color value as STANDARD or ASTA units.

These spectrophotometric methods have been adopted by industry as the routine and reference methods to control color quality as an alternative to HPLC methods for determination of carotenoids in paprika and paprika oleoresins (*47*). To take the advantages of spectrophotometric measurements and to improve the standard methods, some approaches have been developed to obtain a more approximated carotenoid composition of paprika. A method developed by the authors (*22*) allows a simultaneous quantification of the red (R) and yellow (Y) fractions with low determination error and lower analysis time in comparison with that required by chromatographic methods. The R/Y ratio has been proposed as a useful quality index for both the fruits and their processed products (*19, 50, 51*). It is known that yellow fraction is more prone to oxidation than the red one, so a high value for the ratio should indicate that degradative reactions have taken place. However, the determination of quality of a sample based just on a spectrophotometric measurement of color gives only an isolated static idea of it, without the possibility of predicting its change with time. Measurement of promoters of degradative reactions, such as peroxides, could help to predict the potential stability of paprika. Hornero-Méndez *et al.* (*48*) have developed a simple spectrophotometric method for the determination of the peroxide value in paprika samples that may help for that purpose. In addition, predictive models for assessing carotenoid stability during storage have also been proposed (*52*). However the amount of information we can get by using spectrophotometric methods is limited, and only the chromatographic analysis of individual carotenoid composition will allow us to characterize the sample quality, and perhaps make possible the determination of color adulteration with other natural sources of pigments (*49*).

References

1. Food and Agriculture Organization of the United Nations, FAOSTAT data. http://faostat.fao.org (accessed Mar 11, 2007)

2. Eshbaugh, W.H. In *New Crops*. Janick, J. and Simon, J.E. Eds. Wiley, New York. 1993; pp. 132-139.
3. Bioversity International Home Page. http://wwwbioversityinternational. org (accessed Mar 11, 2007)
4. Costa, J.C. In *Actas de las III Jornadas sobre la Mejora de Tomate y Pimiento*. Sociedad Española de Ciencias Hortícolas: Tenerife, Spain. 1980; pp 1-25.
5. Mínguez-Mosquera, M.I.; Hornero-Méndez, D. *J. Agric. Food Chem.* **1994**, *42*, 38-44.
6. Mínguez-Mosquera, M.I.; Hornero-Méndez, D. *J. Agric. Food Chem.* **1994**, *42*, 1555-1560.
7. Almela, L.; López-Roca, J.M.; Candela, M.E.; Alcázar, M.D. *J. Agric. Food Chem.* **1991**, *39*, 1606-1609.
8. Mínguez-Mosquera, M.I.; Pérez-Gálvez, A.; Garrido-Fernández, J. *J. Agric. Food Chem.* **2000**, *48*, 2972-2976.
9. Hornero-Méndez, D.; Gómez-Ladrón de Guevara, R.; Mínguez-Mosquera, M.I. *J.Agric. Food Chem.* **2000**, *48*, 3857-3864.
10. Hornero-Méndez, D.; Costa-García, J.; Mínguez-Mosquera, M.I. *J. Agric. Food Chem.* **2002**, *50*, 5711-5716.
11. Artés, F.; Mínguez-Mosquera, M.I.; Hornero-Méndez, D. In *Colour in food: improving quality*. MacDougall, D., Ed.; Woodhead Publishing Ltd.: Cambridge, UK, 2002; pp 248-282.
12. Weedon, B.C.L. In *Carotenoids*. Isler, O., Ed. Birkhäuser Verlag, Basel. Chapter 2. 1971. pp. 29-59.
13. Edge, R.; McGarvey, D.J.; Truscott, T.G. *J. Photochem. Photobiol. B* **1997**, *41*, 189-200.
14. Olson, J.A. *J. Nutr.* **1989**, *119*, 94-95.
15. Ziegler, R.G. *J. Nutr.* **1989**, *119*, 116-122.
16. Davies, B.H.; Matthews, S.; Kirk, J.T.O. *Phytochem.* **1970**, *9*, 797-805.
17. Britton, G. Overview of Carotenoid Biosynthesis. In *Carotenoids: Biosynthesis and Metabolism*; Britton, G.; Liaaen-Jensen, S. and Pfander, H. Eds.; Birkhäuser: Basel. 1998; Vol. 3, pp 13-147.
18. Hornero-Méndez, D; Mínguez-Mosquera, M.I. *J.Agric. Food Chem.* **1998**, *10*, 4087-4090.
19. Mínguez-Mosquera, M.I.; Garido-Fernández, J.; Pereda-Marín, *J. Grasas Aceites* **1984**, *35*, 4-10.
20. Mínguez-Mosquera, M.I.; Pérez-Gálvez, A. *J. Agric. Food Chem.* **1998**, *46*, 5124-5127.
21. Mínguez-Mosquera, M.I.; Jarén-Galán, M.; Garrido-Fernández, J. *J. Agric. Food Chem.* **1992**, *40*, 2384-2388.
22. Hornero-Méndez, D.; Mínguez-Mosquera, M.I. *J. Agric. Food Chem.* **2001**, *49*, 3584-3588.

23. Camara, B.; Monéger, R. *Phytochem.* **1978**, *17*, 91-93.
24. Mínguez-Mosquera, M.I. and Hornero-Méndez, D. *J. Agric. Food Chem.* **1994**, *42*, 640-644.
25. Hornero-Méndez, D.; Mínguez-Mosquera, M.I. *J. Agric. Food Chem.* **2000**, *48*, 1617-1622.
26. Biacs, P.A.; Daood, H.G.; Pavisa, A.; Hajda, F. *J. Agric. Food Chem.* **1989**, *37*, 350-353.
27. Mínguez-Mosquera, M.I.; Jarén-Galán, M.; Garrido-Fernández, J.; Hornero-Méndez, D. In *Carotenoides en el pimentón. Factores responsables de su degradación.* Mínguez-Mosquera, M.I.; Jarén-Galán, M.; Garrido-Fernández, J.; Hornero-Méndez, D. Eds. CSIC, Madrid. 1996.
28. Pérez-Gálvez, A.; Garrido-Fernández, J.; Lozano-Ruíz, M.; Montero-de-Espinosa, V.; Mínguez-Mosquera, M.I. *Grasas Aceites* **2001**, *52*, 311-316.
29. Chung, S.K.; Shin, J.C.; Choi, J.U. *Korean Soc. Food Nutr.* **1992**, *21*, 64-69.
30. Levy, A.; Harel, S.; Palevitch, D.; Akiri, B.; Menagem, E. and Kanner, J. *J. Agric. Food Chem.* **1995**, *43*, 362-366.
31. McGaw, D.R.; Commissiong, E.; Holder, R.; Seepaul, N.; Maxwell, A. In *Proceedings 1st Nordic Drying Conference - NDC'01.* Alves-Filho, O.; Eikevik, T.M. and Strømmen, I., Eds., Trondheim, Norway, 2001.
32. Pérez-Gálvez, A.; Garrido-Fernández, J.; Mínguez-Mosquera, M.I. *J. Am. Oil Chem. Soc.* **1999**, *76*, 1449-1454.
33. Okos, M.; Csorba, T.; Szabad, J. *Acta Aliment.* **1990**, *19*, 79-86.
34. Pérez-Gálvez, A.; Hornero-Méndez, D.; Mínguez-Mosquera, M.I. *J. Agric. Food Chem.* **2004**, *52*, 518-522.
35. Mínguez-Mosquera, M.I.; Jarén-Galán, M.; Garrido-Fernández, J. *J. Agric. Food Chem.* **1994**, *42*, 1190-1193.
36. Malchev, E.; Chenov, N.S.; Ioncheva, I.; Tanchev, S.S.; Kalpakchieva, K.K. *Nahrung* **1982**, *26*, 415-418.
37. Cofield, E.P. *Chem. Eng.* **1951**, *58*, 127-140.
38. Sugiyama, K.; Saito, M.; Hondo, T.; Senda, M. *J. Chromatogr.* **1985**, *332*, 107-116.
39. Jarén-Galan, M., Nienaber, U.; Schwartz, S.J. *J. Agric. Food Chem.* **1999**, *47*, 3558-3564.
40. Marr, R.; Gamse, T. *Chem. Eng. Proc.* **2000**, *39*, 19-28.
41. Uquiche, E.; del Valle, J.M.; Ortiz, J. *J. Food Eng.* **2004**, *65*, 55-66.
42. Govindarajan, V.S. *Crit. Rev. Food Sci. Nutr.* **1986**, *23*, 207-288.
43. Pérez-Gálvez, A.; Garrido-Fernández, J.; Mínguez-Mosquera, M.I. *J. Am. Oil Chem. Soc.* **2000**, *77*, 79-83.
44. Pérez-Gálvez, A.; Mínguez-Mosquera, M.I. *Biochim. Biophys. Acta.* **2002**, *1569*, 31-34.
45. Guenther, E. In *The essential oils.* Van Nostrand: New York, Vol I, 1948.

46. ASTA. In *Official analytical methods of the American Spice Trade Association.* ASTA Ed., Englewood Cliffs, NJ, 1986.
47. Mínguez-Mosquera, M.I.; Hornero-Méndez, D. *J. Agric. Food Chem.* **1993**, *41*, 1616-1620.
48. Hornero-Méndez, D.; Pérez-Gálvez, A.; Mínguez-Mosquera, M.I. *J. Am. Oil Chem. Soc.* **2001**, *78*, 1151-1155.
49. Mínguez-Mosquera, M.I.; Hornero-Méndez, D. and Garrido-Fernández, J. *J. AOAC Int.* **1995**, *78*, 491-496.
50. Mínguez-Mosquera, M.I.; Fernández-Díez, M.J. *Grasas Aceites* **1981**, *32*, 293-298.
51. Mínguez-Mosquera, M.I.; Jarén-Galán, M.; Garrido-Fernández, J. *J. Agric. Food Chem.* **1992**, *40*, 2384-2388.
52. Jarén-Galán, M.; Pérez-Gálvez, A.; Mínguez-Mosquera, M.I. *J. Agric. Food Chem.* **1999**, *47*, 945-951.

Chapter 24

Avocado Oil: The Color of Quality

M. Wong[1], O. Ashton[1], C. Requejo-Jackman[2], T. McGhie[2], A. White[2], L. Eyres[3], N. Sherpa[1], and A. Woolf[2]

[1]**Institute of Food, Nutrition, and Human Health, Massey University, Auckland, New Zealand**
[2]**The Horticulture and Food Research Institute of New Zealand, Auckland, New Zealand**
[3]**Oils and Fats Group, N. Z. Institute of Chemistry, Auckland, New Zealand**

Cold pressed extra virgin avocado oil is a relatively new culinary oil. This chapter will review the importance of the fruit, and postharvest and processing practices on the color quality of the oil. Cold pressed avocado oil extracted from the 'Hass' cultivar (*Persea americana* Mill.) has a very intense green color which contributes to its unique appearance and identity. The color is attributed to chlorophyll pigments (up to 60 $\mu g\ g^{-1}$) extracted into the oil from the flesh tissue and also from skin tissue present during the extraction process. The oil also contains high concentrations of carotenoid pigments, the predominant carotenoid being lutein (~2 $\mu g\ g^{-1}$). The concentration of pigments in the oil depends on cultivar, maturity, ripeness, fruit storage before extraction, type of tissue extracted and the processing conditions used. As the fruit ripen, the pigment concentrations decline in the skin and flesh. Storage of avocado oil results in a loss of chlorophyll, dependent on temperature, light and oxygen levels, while carotenoids are relatively stable. With the decline in chlorophylls, the oil changes color from emerald green to yellow/gold. To retain the color quality of avocado oil, maintenance of high pigment concentrations during and after extraction is important.

Introduction

The commercial production of cold pressed extra virgin avocado oil is a relatively recent development worldwide. Previously, avocado oil has been extracted using high temperature and solvents primarily for cosmetic purposes and not for culinary purposes. The production of a high quality, edible, cold pressed avocado oil requires retention of important pigments and healthful compounds as well as minimisation of oxidation. In order to achieve this, all aspects of oil production must be considered, including preharvest and postharvest effects on fruit, processing techniques, and oil storage practices. There is very little published information on the extraction of avocado oil and aspects affecting the oil quality. This chapter will briefly review the existing literature on avocado fruit and avocado oil extraction, which includes recently published and unpublished research from the authors.

Avocado Fruit Production

The avocado originated in southern Mexico and was domesticated several thousand years ago in Central America. It was cultivated from the Rio Grande to central Peru before the arrival of Europeans (*1*). Historically, the avocado has also been referred to as alligator pear, vegetable butter, butter pear and midshipman's butter. The flesh of an avocado can contain as much as 25 – 30% oil by fresh weight and the flesh is normally consumed fresh.

World production of avocado in the 2004/2005 season is summarised in Table I. The main avocado producers in the northern hemisphere are Mexico, United States (primarily California), Israel and Spain, and in the southern hemisphere Chile, South Africa, Australia and New Zealand. Other countries such as Peru and Argentina are also large producers of avocados and production in South East Asia is increasing (*2*). Most countries are orientated to exporting the fresh fruit, with the remaining fruit sold on the domestic market.

The 'Hass' cultivar accounts for >90% of avocado production in Mexico, Chile, United States and New Zealand and >75% of the production in Spain and Australia (Table I). Thus, 'Hass' is generally the main focus of research and development worldwide. 'Hass' is the preferred cultivar grown because of its superior yields and fruiting characteristics (*1*). Other major cultivars grown in approximate order of priority include 'Fuerte', 'Ryan', 'Pinkerton' and 'Edranol'.

Table I. Main producers of avocados worldwide in 2004/2005 marketing year [ab].

Country	*Production (metric tonnes)*	*Exported (metric tonnes)*	*Processed (metric tonnes)*	*% of crop 'Hass' cultivar*
Australia	32,000	320	1,200	75
Chile	177,000	136,412	300	93
Israel	77,000	45,000	1,000	32
Mexico	934,282	180,165	25,000	95
New Zealand	22,000	15,000	1,850	98
South Africa	85,000	38,000	13,600	36
Spain	55,000	48,000	n.r.	75
United States	162,721	1,431	n.r.	95

n.r. – none reported.

[a] Marketing year varies for each country: Australia, New Zealand and Spain, July/June; Chile and South Africa, January/December; Mexico, August/July; United States, November/October.

[b] Source of data: *3, 4, 5.*

Avocado Fruit Development and Ripening

Determination of Fruit Maturity

Avocados are unique in that they do not ripen on the tree and thus can be held on the tree for very long periods, as much as two years from flowering. As the avocado fruit matures on the tree, the main physiological changes are increase in size, increase in dry matter content and an increase in lipid (oil) content. The latter two changes are highly correlated and are used as horticultural measures of commercial maturity (*6*). Measurement of dry matter is the most common method employed as it is simple, cheap and rapid to determine. Fruit harvested with a low % dry matter is likely to have a poor taste, and if harvested at very low dry matter values (<20%) fruit may fail to ripen. Minimum % dry matter values are typically decided by grower organisations, particularly for fruit destined for export. The recommended minimum % dry matter ranges from 20 - 25%, depending on cultivar and the country of production. For example, minimum dry matter values for 'Hass' in 2006 are 20.8% in Mexico and the United States, 21% in Chile and Australia, 25% in South Africa and 24% in New Zealand (*4*). During the growing season the lipid content in 'Hass' avocados can increase to as much as 30% on a fresh weight basis, depending on the growing region (*7*). For different avocado cultivars ('Bacon', 'Edranol', 'Fuerte', 'Hass' and 'Zutano'), it was found that there was no effect of maturity on flesh color when the whole fruit flesh was pulped together (*8*).

Postharvest ripening changes

When the fruit has reached a desired maturity, the fruit are harvested and then commence ripening. The key change observed as avocados ripen is fruit softening, which can be measured using a number of different methods and reported as fruit firmness (*9*). The skin of the 'Hass' cultivar also changes color from green to black as the fruit ripens (softens) (*10*). However, exceptions to this expected correlation do occur. Firstly, where fruit are ripened at lower temperatures, e.g. at 15°C, fruit may be relatively green yet soft (*11*). Conversely, fruit may be dark skinned while unripe, generally observed in fruit which are held on the tree for long periods (*12*).

Ripening takes about 7 to 10 days. This time can be hastened by treatment with ethylene gas, which also reduces fruit-to-fruit variability in terms of ripening rate. A similar ripening result can be achieved by storage for up to approximately four weeks, typically at 5-6°C. Storage periods for longer than about four weeks can cause internal chilling injury (vascular browning and diffuse flesh discoloration), while storage at lower temperatures can cause external chilling injury (skin scalding; *9*).

Oil localisation and tissue types in whole avocado fruit

The majority of the lipids in avocados are stored in the fruit flesh (mesocarp; *7*). It is, therefore, more economical to extract oil from cultivars with a high percentage of flesh and small seeds. A ripe 'Hass' avocado comprises ~68% flesh, ~18% seed and ~14% skin (by fresh weight). These values were determined for ripe 'Hass' avocados ranging in weight from 75 to 350 g, harvested in the Far North region of New Zealand (Figure 1). Fruit size has relatively little effect on the proportion of the fruit which is flesh. Fruit used for oil extraction are those that have been rejected for local and export markets, primarily because of defects in cosmetic appearance of the skin. In avocado, it is important that the flesh of the fruit be free from rots and browning defects in order to produce good quality oil. It has been observed in the olive oil industry, the quality of the oil is dependent on the quality of fruit used to extract the oil (*13*).

The lipids in avocado are stored in two cell types in the mesocarp, mostly in the parenchyma cells and to a lesser extent, in the idioblasts (*14*). The parenchyma cells contain large lipid droplets in their cytoplasm of which approximately 85% are triglycerides. The idioblasts on the other hand are specialised cells with complex cell walls bound together by protopectin in the middle lamella. As the fruit ripens, a rise in cell enzymatic activity occurs and pectins are de-esterified and depolymerised, leading to the softening and degrading of the parenchyma cell walls, making them easier to rupture and release the lipids within (*15,16*). During the cold pressed extraction of avocado oil, it is not known if the cell walls of the idioblasts are entirely degraded and the lipids within released.

Postharvest color changes

During ripening, the skin of the 'Hass' avocado changes from green to purple/black. This is due to a small reduction of chlorophyll compounds (chlorophyll *a* and *b,* and chlorophyllides *a* and *b*) in the skin and an increase in, predominantly, one anthocyanin pigment, cyanidin 3-*O*-glucoside (*12*). To evaluate pigment changes in skin and flesh, avocado fruit was analysed two days after harvest and at nine days after harvest when the fruit were considered to be ripe (*17*). The flesh (mesocarp) of the avocado fruit can be divided into three sections: dark green flesh which is directly adjacent to the skin; pale green flesh below this and then yellow flesh which is adjacent to the seed (Figure 2). Color and pigment concentrations in each of these three sections and skin have been measured during fruit ripening. The fruit, ripened at 20°C, softened from an initial average Firmometer value of 8.5 to 72.2 mm (soft) determined with an Anderson Firmometer. The skin showed the most dramatic change in color (hue angle) which corresponded to the fruit changing from green to purple/black (Table II). The color of the flesh sections, in terms of hue angle, did not change

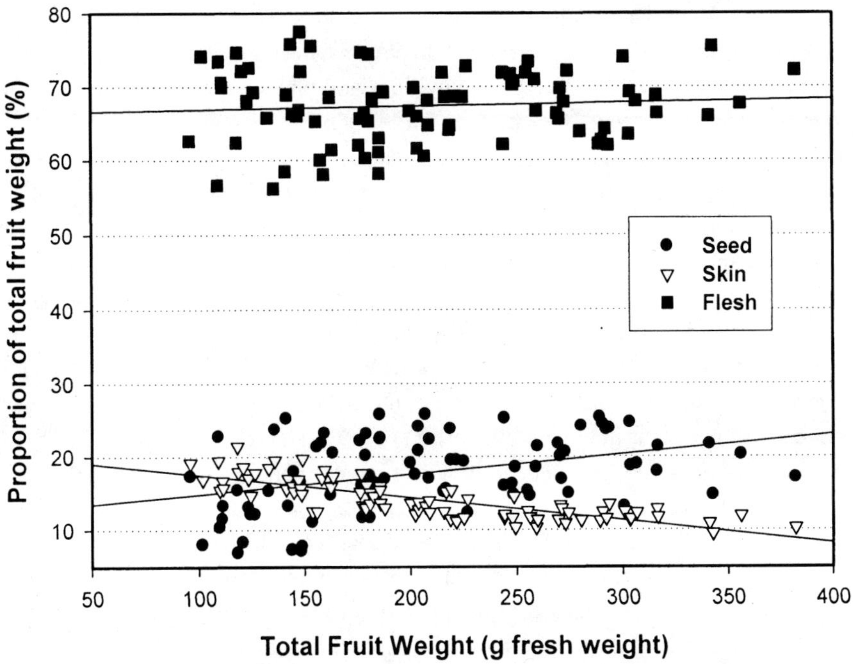

Figure 1. Weights of respective tissues, expressed as a proportion, for ripe 'Hass' avocados harvested in January (mid-late season) 2001, Far North, New Zealand.

during the fruit storage period (*17*). The skin changing from green to purple/black corresponded with an increase in anthocyanin pigments in the skin (Table II). Total chlorophylls in the skin decreased from 178 μg g^{-1} (by fresh weight) to 152 μg g^{-1}, a drop of 15%, after nine days storage. Individual chlorophyll compounds were also measured during ripening. Chlorophyllides can result from the breakdown of chlorophyll, the reaction being catalysed by the enzyme chlorophyllase, or are formed as intermediates in the biosynthesis of chlorophyll (*18*). Chlorophylls and chlorophyllides decompose to pheophytins and pheorphorbides, respectively, by losing magnesium catalysed by the Mg-dechelatase enzyme (*18,19*). Chlorophyllides and pheophytins were found in the skin. As the fruit ripened, the chlorophyllide concentrations decreased in the skin corresponding to the breakdown of chlorophyll and loss of green color (Table II) (*17*). Chlorophyllides were found to be greatest in the skin but were present in all flesh sections. The concentration of chlorophyll *a* was greater than the concentration of chlorophyll *b* in skin and flesh sections. The total chlorophylls in the flesh sections did not change during ripening. The concentrations of individual chlorophylls in the flesh sections of avocado, two and nine days after harvest, are shown in Table II. It has also been reported for 'Fuerte' avocados that the skin has the highest concentration of chlorophyll and lower concentrations occur in the yellow portion of the flesh (*20*).

During ripening, the concentration of total carotenoids in the skin decreased by more than 40% (Table II). There was also a small but significant ($p<0.001$) reduction in total carotenoids in the flesh sections during ripening. Hence, as avocados ripen, significant losses of pigments can occur from the skin and flesh. The concentrations of individual carotenoids in the skin and flesh sections, two and nine days after harvest, are shown in Table II.

Cold pressed extraction of avocado oil

Background

In the past, avocado oil has been extracted using only "harsh" methods (solvent chemicals with or without high temperature) and the extracted oil, dark green in color, was then refined, bleached and deodorised. This process results in oil that is yellow to colorless since pigments are removed in the bleaching step (*21*). A number of different solvent extraction techniques have been examined for avocado oil (*22,23*). Avocado oil extracted with solvents and subsequently refined, bleached and deodorised is predominantly used for cosmetics purposes.

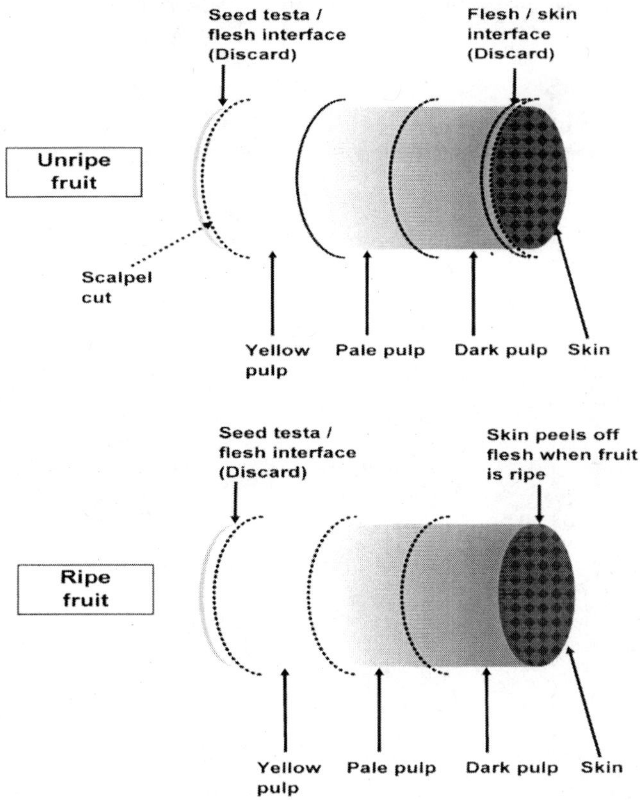

Figure 2. Tissue sections from ripe and unripe 'Hass' avocados (Reproduced from reference 17. Copyright 2006 American Chemical Society.)
(See page 22 of color inserts.)

Table II. Changes in color and composition of pigments in the skin and flesh of 'Hass' avocados, at two and nine days after harvest, with Firmometer values (mm displacement × 10) of 8.5 (±0.5) and 72.2 (±1.4) mm displacement, respectively.

Tissue section	*Skin*		*Dark green flesh*		*Pale green flesh*		*Yellow flesh*	
Days after harvest	*2*	*9*	*2*	*9*	*2*	*9*	*2*	*9*
Hue angle (h°)	123.2 (±2.0)	165.5 (±0.5)	112.6 (±0.3)	114.8 (±0.4)	160.0 (±0.2)	105.1 (±0.6)	101.0 (±0.2)	99.7 (±0.2)
Total anthocyanins (μg g^{-1} FW)	2.12 (±0.67)	153.87 (±21.91)						
Cyanidin 3-O-glucoside (μg g^{-1} FW)	0.58 (±0.00)	148.65 (±17.19)						
Chlorophyll *a* (μg g^{-1} FW)	22.20 (±4.16)	28.00 (±1.21)	22.30 (±1.21)	35.00 (±3.01)	3.50 (±0.66)	6.10 (±0.66)	0.70 (±0.44)	1.60 (±0.44)
Chlorophyll *b* (μg g^{-1} FW)	8.55 (±1.55)	10.30 (±0.15)	5.30 (±0.28)	13.90 (±1.03)	0.50 (±0.04)	2.30 (±0.26)	0.00 (±0.00)	0.60 (±0.15)
Chlorophyllide *a* (μg g^{-1} FW)	131.20 (±1.70)	103.10 (±8.78)	6.60 (±0.38)	1.50 (±0.10)	2.40 (±0.16)	0.60 (±0.15)	1.10 (±0.08)	0.50 (±0.03)
Chlorophyllide *b* (μg g^{-1} FW)	15.80 (±0.74)	10.30 (±1.01)	0.30 (±0.02)	n.d.	n.d.	n.d.	n.d.	n.d.
Pheophytin *a* (μg g^{-1} FW)	0.04 (±0.00)	n.d.	n.d.	n.d.	n.d.	n.d.	n.d.	n.d.
Pheophytin *b* (μg g^{-1} FW)	n.d.	0.05 (±0.05)	n.d.	n.d.	n.d.	n.d.	n.d.	n.d.
Total chlorophylls (μg g^{-1} FW)	177.79 (±4.68)	151.75 (±8.63)	34.50 (±1.58)	50.40 (±3.98)	6.40 (±0.49)	9.00 (±0.98)	1.80 (±0.15)	2.70 (±0.61)
Antheraxanthin (μg g^{-1} FW)	1.04 (±0.02)	0.50 (±0.01)	0.09 (±0.01)	0.18 (±0.03)	0.05 (±0.01)	0.03 (±0.01)	0.10 (±0.02)	0.03 (±0.01)
α-Carotene (μg g^{-1} FW)	3.87 (±0.18)	2.39 (±0.35)	0.08 (±0.01)	0.04 (±0.01)	0.05 (±0.00)	0.06 (±0.00)	0.06 (±0.00)	0.07 (±0.02)
β-Carotene (μg g^{-1} FW)	12.53 (±0.10)	6.40 (±0.48)	0.28 (±0.02)	0.06 (±0.01)	0.10 (±0.01)	0.04 (±0.00)	0.06 (±0.01)	0.05 (±0.01)
Lutein (μg g^{-1} FW)	20.54 (±0.09)	12.77 (±1.09)	1.65 (±0.06)	2.23 (±0.16)	0.41 (±0.03)	0.40 (±0.05)	0.45 (±0.04)	0.19 (±0.02)
Neoxanthin (μg g^{-1} FW)	5.52 (±0.04)	2.48 (±0.20)	1.68 (±0.12)	0.14 (±0.02)	1.21 (±0.15)	0.04 (±0.01)	2.08 (±0.13)	0.11 (±0.02)
Violaxanthin (μg g^{-1} FW)	4.49 (±0.16)	3.53 (±0.20)	1.06 (±0.08)	0.80 (±0.07)	0.49 (±0.05)	0.11 (±0.01)	0.63 (±0.04)	0.07 (±0.02)
Zeaxanthin (μg g^{-1} FW)	0.92 (±0.08)	0.31 (±0.03)	0.35 (±0.03)	0.12 (±0.02)	0.02 (±0.00)	0.03 (±0.01)	n.d.	n.d.
Total carotenoids (μg g^{-1} FW)	48.91 (±0.18)	28.38 (±1.87)	5.19 (±0.27)	3.57 (±0.31)	2.33 (±0.23)	0.71 (±0.06)	3.38 (±0.21)	0.52 (±0.07)

n.d. – not detected. Mean values presented ± standard errors for n = 3.

Source of data (*17*)

Anthocyanins, chlorophylls and carotenoids determined by high performance liquid chromatography (HPLC) (*17*).

In contrast to refined oil, cold pressed avocado oil contains high concentrations of pigments and is green in color (*24*). Cold pressed extraction of avocado oil was first reported in Israel by Werman and Neeman (*15*). In New Zealand, two processors of cold pressed avocado oil, The Grove NZ Avocado Oil (Bay of Plenty) and Olivado Ltd (Northland), began commercial production in 2000. Both companies have increased both production of and markets for the new culinary oil. Olivado has expanded by moving to Australia to take advantage of their larger avocado crop. New Zealand exported $US 1.4M (fob) of avocado oil in 2005 (*3*). Other countries have also begun to produce cold pressed avocado oil for food consumption including Mexico, Chile and South Africa. Cold pressed avocado oil has a similar lipid composition as cold pressed olive oil. Avocado oil generally contains approximately 75% monounsaturated, 10% polyunsaturated and 15% saturated fatty acids with variations depending on cultivar, growing region and season (*24*).

Cold pressed oil extraction and oil yield

Extraction of cold pressed avocado oil for culinary purposes is carried out using mechanical extraction methods rather than solvents. During the entire extraction process the temperature of the flesh and oil is kept below 50°C. Under good manufacturing practices, extra virgin avocado oil can be produced with very low free fatty acid and peroxide values. Currently there are no international standards for extra virgin avocado oil, whereas the standards set for extra virgin olive oil are: free fatty acids below 0.8%$^{w}/_{w}$ (as oleic acid) and a peroxide value of less than 20 meq/kg oil (*25*). The peroxide value measures the concentration of hydroperoxide compounds in the oil and gives an indication of the amount of oxidative degradation that has occurred since extraction. High values in the oil indicate that the oil has been poorly handled during processing. During oil extraction the flesh (mesocarp) is ground to a paste; this disrupts the cells which then release the oil during slow mixing of the paste in a malaxer. The oil is then separated from the solid and water phases in a horizontal decanting centrifuge. Further removal of all water traces from the oil is achieved in high speed centrifuges, sometimes referred to as polishing. A process flow diagram for the cold pressed extraction of avocado oil is shown in Figure 3.

The oil extraction yield from 'Hass' avocados will depend on the time of season, maturity at harvest and on the degree of ripeness. Using a cold pressed extraction technique the maximum oil yield obtained from ripe avocados varies between 10 to 18% of the original fresh weight of fruit (*26*). Lower yields are obtained with early season fruit and higher yields with late season fruit. The fruit are generally ripened to a Firmometer value of more than 100 mm or a hand rating of 5 (*9*). Extraction of oil from fruit with a Firmometer value of less than 100 mm will result in reduced oil yields as the mesocarp is firm and the cell walls are mainly still intact, resulting in limited oil release. Extraction of oil from soft fruit will lead to higher oil yields resulting from a reduction in cell wall integrity with the presence of pectinase enzymes (e.g. polygalacturonase) in the fruit. Zauberman & Schiffmann-Nadel (*27*) found that polygalacturonase in avocados is activated much faster in later season fruit than in early season fruit once the fruit is harvested. However, with softer fruit, the chance of rots increases, and rotten fruit should be removed from the fruit batch prior to processing as rots will affect final quality of the oil. Buenrostro and López-Munguia (*28*) determined the effect of the addition of various pectinase, amylase, protease and cellulase enzymes on oil yield during mechanical extraction of avocado paste. The extractions were carried out at 40°C and the best extraction yield was achieved with the addition of amylase.

Oil Color and Pigments

Oil color

Avocado oil color can be measured spectrophotometrically, where results are expressed as absorption at a specified wavelength, or using tristimulus colorimetry (chromameter), which can express results in hue, chroma and lightness. Cold pressed and non-refined avocado oil is a dark emerald green color due to the presence of chlorophyll pigments. To determine the color of avocado oil, a sample of oil was poured into a 1 cm wide cuvette and placed into a black sample holder to eliminate external light. A Minolta chromameter was used to determine the lightness, hue and chroma of cold pressed extra virgin avocado oil. The lightness, hue and chroma of the oil were found to range between 88 to 100, 168 to 227° and 11 to 15, respectively. The chromaticity values of a^* (green to red) and b^* (blue to yellow) ranged from -0.5 to 0.7 and -1.2 to 9.3, respectively. These values correspond to green on the $L^*a^*b^*$ color space, with hue angles between 168 and 227°.

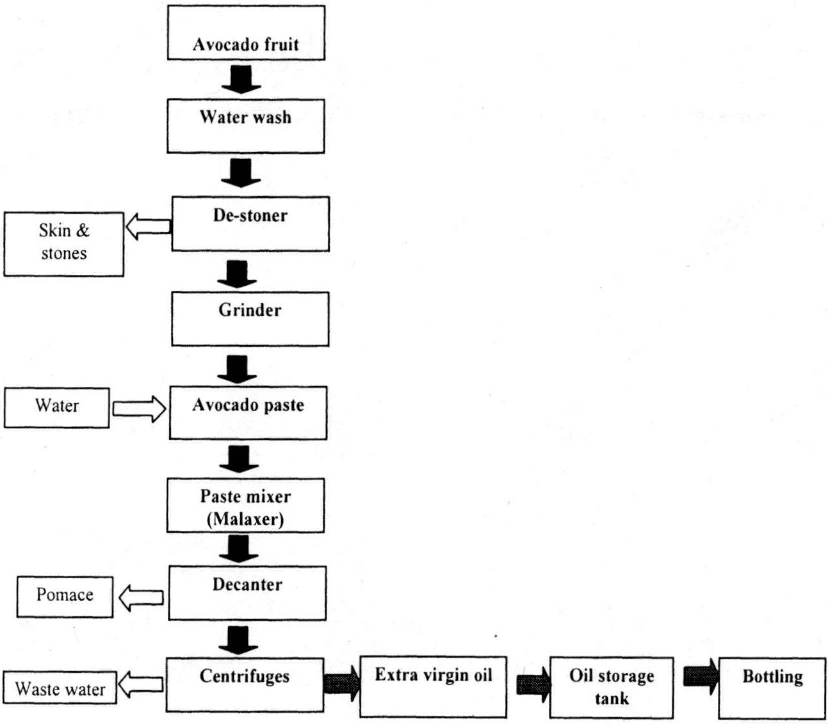

Figure 3. Process flow diagram for the cold pressed extraction of avocado oil.

Table III. Pigment concentrations in cold pressed 'Hass' avocado oil extracted in a commercial scale plant.

Pigment		*μg (g oil)$^{-1}$*
Chlorophylls [a]	Chlorophyll *a*	4.9
	Chlorophyll *b*	5.1
	Pheophytin *a*	1.1
	Pheophytin *b*	2.2
	Total chlorophyll [b]	16.5
Carotenoids	Antherxanthin	0.2
	Lutein	1.6
	Neoxanthin	n.d.
	Violaxanthin	n.d.
	Zeaxanthin	n.d.

n.d. – not detected.

[a] Individual chlorophylls and carotenoids determined by HPLC; quantities of minor chlorophyll compounds identified not presented (*17*).

[b] Total chlorophylls determined by a spectrophotometric method, includes all chlorophyll compounds present in the oil (AOCS Ch 4-91) (*29*).

Pigment concentrations in cold pressed extracted oil

A number of analytical methods are available to determine individual and total chlorophylls and carotenoids. The total chlorophyll content in the oil can be determined using a spectrophotometric method (*29*) and the total chlorophyll content of oils is expressed as μg pheophytin *a* (g oil)$^{-1}$. Solid phase extraction (SPE) is used to isolate carotenoids and chlorophylls from oil followed by high performance liquid chromatography (HPLC) to quantify the individual pigment concentrations (*17*). The chlorophyll and carotenoid content of cold pressed avocado oil is presented in Table III. The major carotenoid quantified in the oil was lutein. As the SPE extraction method used did not recover α-carotene and β-carotene from the avocado oil, they were not quantified, but it is likely that if extracted into the oil, they would be present in much lower concentrations than lutein.

There is some variability reported in the chlorophyll content of avocado oil. Werman and Neeman (*30*) found the chlorophyll content to be 41.3 μg g^{-1} in crude avocado oil (from an unspecified cultivar) extracted by centrifugation. Eyres *et al.* (*26*) reported the total chlorophylls in 'Hass' cold pressed avocado oil ranged from 40-60 μg g^{-1}, while Requejo-Jackman *et al.* (*24*) reported ~25 μg g^{-1}. The concentration of chlorophyll will depend on the cultivar, time of season, maturity at harvest, ripeness and length of storage of fruit prior to extraction and extraction method used. Varieties other than 'Hass' have been found to produce oils with high concentrations of chlorophyll, such as 'Pinkerton' ~45 μg g^{-1} (*24*).

Extraction of pigments from the flesh and skin

The concentration of pigments extracted into the oil can be influenced by the amount of skin included in the extraction process (in the malaxer). Prior to extraction the seeds and skin are removed. However, depending on the separation system used, a proportion of the skin may enter the malaxer with the flesh pulp. The concentration of chlorophylls and carotenoids is highest in the skin, and although significantly less in the flesh, these compounds are highest in the dark green flesh adjacent to the skin. With ripe avocados the dark green flesh adheres to the inside of the skin during the skin removal process, therefore, a potentially significant amount of chlorophyll is discarded with skin and the adhering dark green flesh. In laboratory cold pressed extraction trials with 'Hass' avocados, the amount of skin added with the pulp in the malaxer was increased from 10 to 40 or 100%. As the percentage of skin increased from 10 to 100% during oil extraction, the lutein concentration in the oil increased from 1.13 to 3.21 μg g^{-1} (Table IV). Total chlorophylls increased in the oil as the percentage of skin present increased (Table IV). Thus, the amount of skin included in the malaxing process can have a significant influence on pigment concentration in the final oil product. As the pigment concentration in the oil increased with addition of skin from 10 to 100%, the color of the oil changed significantly ($p<0.05$). The color space a^* value decreased from -0.12 to -5.47 and b^* values decreased from 9.31 to -1.87. The hue angle increased from 171.3 to 222.0° indicating that the color of the oil became darker green with increasing proportion of skin present during malaxing.

Table IV. Pigment concentrations in laboratory cold pressed extracted avocado oils with varying amounts of skin included during extraction.

% of available skin from fresh fruit	*Lutein[a] (μg g^{-1} oil)*	*Total chlorophyll[b] (μg g^{-1} oil)*
10	1.13 (± 0.11)	9.8 (± 0.2)
40	1.94 (± 0.17)	19.5 (± 0.3)
100	3.21[c] (± 0.15)	25.3[c] (± 0.5)

Mean values presented ± standard errors for n = 3.
[a] Lutein concentration determined by HPLC (*17*).
[b] Total chlorophyll by spectrophotometric method (AOCS Ch 4-91) (*29*).
[c] Concentration at 100% skin significantly different to value at 10% skin ($p<0.05$).

Pigment and Oxidative Stability of Oil

Pigment stability in the oil

The pigments that predominantly influence the color of avocado oil are the chlorophylls. Chlorophyll pigments are highly reactive under light and in the presence of oxygen, leading to their rapid breakdown (*31,32*). Werman and Neeman (*30*) found that the chlorophyll content of avocado oil rapidly reduced to 18% of its original value after exposure to fluorescent light for 27 days. Chlorophyll degradation in extra virgin avocado oil was found to be affected by light and oxygen, but was not influenced by temperature *(33)*. A significant drop in chlorophyll concentration, a loss of 14%, was found when the oil was exposed to very bright light (equivalent to 4500 lux falling on samples) and sparged with dry air for seven hours (*33*).

Stability of pigments in oils during storage has been investigated for a number of edible oils though no work has been published on avocado oil. The majority of research investigating pigment stability in oils has been in relation to the oxidative stability of the oil. The presence of pigments in the oil can greatly influence the oxidative stability and hence their own stability. The oxidation of lipids results in a loss of quality leading to rancidity and also a loss of pigments (*34*). Lipids are oxidised via autoxidation pathways in the absence of light or via photooxidation mechanisms in the presence of light (*35*). Autoxidation of lipids occurs when oxygen and free radicals are present, resulting in the autocatalytic formation of hydroperoxide compounds and their breakdown products. If chlorophyll is present in oils it acts as a photosensitizer during photooxidation reactions hence promoting the oxidation of oils (*32, 36, 37*). There are two reported photooxidation mechanisms; firstly in Type I photosensitization, the sensitizer (chlorophyll) is itself excited by light and then reacts directly with the lipid substrate to form intermediates which then react further to hydroperoxides and free radicals (*38*). The second mechanism, Type II photosensitization, involves the excited sensitizer reacting with triplet oxygen to produce singlet oxygen. The singlet oxygen then reacts directly with the double bonds of fatty acids (*38*). This results in the formation of numerous hydroperoxide breakdown products. These compounds, in the presence of light, are transformed into peroxy-radicals that react with chlorophyll to produce colorless compounds (*30*). In order to maintain the high chlorophyll concentration in avocado oil, care should be taken to exclude light and oxygen, both from a point of view of color stability and oxidation of the lipids. The reduction of chlorophyll in other oils due to light exposure has been observed (*30, 32 ,37, 39*). Carotenoids (α and β carotene) have been reported to act as antioxidants by quenching the singlet oxygen produced (*40, 41*).

The carotenoid, lutein, was found to be reasonably stable during the storage of avocado oil at 20°C, stored under dark or light conditions. Storage under light conditions was similar to the light exposure in a well lit supermarket (400 lux). Under dark and light storage conditions, there was a significant loss of total chlorophyll in the avocado oil ($p<0.05$) (Table V). The avocado oil stored at 20°C in the dark was initially green with a hue angle of 110° at the start of storage, after 3 months storage the hue angle of the oil was 80°. The a^* and b^* chromaticity values for the same oil also moved away from green for a^* values and became more to yellow for b^* values.

Table V. Influence of storage at 20°C in the dark and in the light on the peroxide value and the total chlorophyll concentration in avocado oil.

	Peroxide value[a] *(meq/kg oil)*		*Total chlorophyll*[b] *($\mu g\ g^{-1}$ oil)*	
Storage time (days)	*Dark*	*Light*	*Dark*	*Light*
0	0.4 (±0.0)	0.4 (±0.0)	16.5 (±0.0)	16.5 (±0.0)
97	2.3[c] (±0.0)	8.3[c] (±0.1)	15.2[c] (±0.2)	10.8[c] (±0.6)

Mean values presented ± standard errors for n = 2.

[a] Peroxide value by titration (AOCS Cd8-53) (*29*).

[b] Total chlorophyll by spectrophotometric method (AOCS Ch 4-91) (*29*).

[c] Value at start of storage period significantly different to value after 97 days storage ($p<0.05$).

Oxidative stability of oil

The oxidative stability of the oils can be monitored by measuring the peroxide value (PV) of the oil during storage. For avocado oils stored at 20°C in the dark and in the light, the PV increased with storage under both storage conditions and was greatest when the oil was stored at 20°C in the light (Table V). Other oils containing chlorophyll stored under light conditions have been shown to oxidise at a much faster rate than the same oil stored in the dark (*42,43*). Extensive research has been completed looking at the effect of chlorophyll and carotenoid pigments on the oxidative stability of edible oils (*32,37,39-41,44*). Gutierrez-Rosales et al. (*44*) postulated that chlorophyll is consumed while having an antioxidant function in olive oil stored in the dark, possibly having a protective effect similar to α-tocopherol and phenolic compounds. This could explain why there was still a decline in chlorophyll concentration in avocado oil stored in the dark, but to a lesser extent than when stored under light. The autoxidation of the avocado oil was minimised under dark storage, but under light storage once hydro-peroxides are formed due to photooxidation, autoxidation reactions accelerate. To minimise the oxidation of avocado oil, it is recommended that the oil is stored in dark bottles to minimise light and the headspace sparged with nitrogen to ensure there is no oxygen present.

Other Factors Influencing Pigment Concentrations

Fruit factors influencing pigment concentrations

The color and pigment concentrations in avocado oil are influenced by a range of factors related to the fresh fruit and postharvest handling and not just due to extraction or processing conditions. Firstly, the avocado cultivar from which the oil is extracted can contain different amounts of pigments resulting in oils with increased or decreased pigment concentrations (*24*). Preliminary data, on oil extracted from the 'Pinkerton' cultivar showed the oil contained the highest concentration of total chlorophyll followed by 'Hass', 'Fujikawa', 'Hayes' and 'Santana' (*24*). Oils from the different cultivars with high amounts of chlorophyll were generally found to also have high amounts of lutein. Secondly, the influence of maturity (through the production season) on pigment concentrations in the oil has not been fully investigated. The development of pigments in the fruit with developing maturity may influence the pigment concentrations in the oil extracted. Other factors that can influence pigment concentrations are storage of the fruit prior to ripening and degree of ripening prior to oil extraction (*17*). Fruit used for oil extraction are generally reject fruit which are stored at ~6°C while fruit numbers accumulate, or for longer if there is a backlog of fruit to process. After re-warming, fruit are ethylene treated to achieve uniform ripeness. Requejo-Jackman *et al.* (*45*) have investigated the effect of fruit storage on pigment concentrations in the oil extracted from the fruit. They found there was no significant change in the amount of carotenoids and chlorophylls extracted into the oil by cold pressed extraction, over a fruit storage period of four weeks. Extraction of oil is recommended from ripe fruit, as the oil is released more easily from the mesocarp tissue, but the amount of pigments that can be extracted are less at this time than when the fruit are unripe. Thus, the optimum time to extract the oil to ensure maximum pigment extraction is when optimum oil yield is generally not achieved, and a trade-off must be made.

Processing factors influencing pigment concentrations

Processing factors that influence pigment concentration can include degree of heating, exposure to light, oxygen during extraction, skin inclusion with the flesh and storage conditions. Excessive heating and exposure to light will have a detrimental affect on the retention of chlorophyll pigments as these factors promote the degradation of chlorophyll via different pathways (*19*). Excessive oxygen will lead to oxidation of the oils and as discussed earlier it is thought that the hydroperoxides formed from the breakdown of lipids can promote the degradation of chlorophylls. Processing practices to exclude oxygen from the oil include sparging the oil with nitrogen gas to displace any dissolved oxygen and secondly storing the oil in containers which are not permeable to oxygen. This includes bulk storage in stainless steel tanks followed by storage and sale of the oil in glass or metal containers and not in oxygen permeable plastic containers. The storage container should also exclude all light to eliminate photooxidation reactions in the oil.

Summary

Color is an important sensory characteristic of foods and is used as a quality indicator by consumers. The color of avocado oil is an important quality aspect along with its taste and aroma. Currently there is no legal requirement for color measurement of edible oils or pigment content worldwide. A standard method for the measurement of color in a range of colored oils (yellow to green) is needed. Alternatively, a comprehensive database on pigment concentrations in avocados and the extracted oil can be used to produce standards for pigment content. The distinctive green color of avocado oil is due to the presence of high amounts of chlorophyll *a* and *b*. From a marketing point of view, this distinctive green color allows avocado oil to stand apart from other more yellow oils hence, it is important for the oil to retain its green color. The green color is also an indication of the quality and age of the oil, as a green avocado oil would indicate minimal degradation of chlorophyll. The presence of chlorophyll pigments correlates to the presence of carotenoid pigments. As mentioned earlier, the presence of carotenoids may have a positive influence on the oxidative stability of the oils. Greater extraction of chlorophyll and carotenoid pigments into the oil will result in oil with a more intense color which will be retained longer. It is therefore important to manage the selection and harvesting of avocados such that they achieve at least the minimum maturity and are not over-ripened prior to extraction, as pigments decline with ripeness. The extraction process is still very important for recovering maximum pigment concentration in the oil. Once the pigments have been extracted into the oil, it is important to minimise the degradation of these pigments by controlling heat, light and oxygen.

References

1. Smith, N.J.H.; Williams, J.T.; Plucknett, D.L.; Talbot, J.P. *Tropical Forests and Their Crops*; Comstock Publishing Associates: Ithaca, NY, 1992; pp 112-150.
2. Papademetriou, M.K. *Avocado Production in Asia and the Pacific.* Food and Agriculture Organization of the United Nations, Bangkok, Thailand. http://www.fao.org/documents/pub_dett.asp?lang=en&pub_id=59570 (accessed Mar 12, 2007)
3. Anon. *Fresh Facts*; HortResearch, Marketing and Communications: Auckland, New Zealand, 2005; pp33.
4. Avocadosource, http://www. avocadosource.com/world.asp (accessed Mar 11, 2007)
5. USDA Foreign Agricultural Service, 2006. http://www.fas.usda.gov/ htp/horticulture/avocado.html (accessed Mar 12, 2007)
6. Seymour, G. B.; Tucker, G. A. In *Biochemistry of Fruit Ripening;* Seymour, G. B.; Taylor, J. E.; Tucker, G. A., Eds.; Chapman & Hall: London, 1993; pp 53-81.
7. Kaiser, C.; Smith, M. T.; Wolstenholme, B. N. *South African Avocado Growers' Association Yearbook.* **1992**, *15*, 78-82.
8. Olaeta, J. A.; Rojas, M. *Proceedings of the First World Avocado Congress. South African Growers' Association Yearbook.* **1987**, *10*, 163-164.
9. White, A.; Woolf, A. B.; Hofman, P. J.; Arpaia M. L. *The International Avocado Quality Manual:* HortResearch: Auckland, New Zealand, 2005; pp 73.
10. Williams, L. O. *Econ. Bot.* **1977**, *31*, 315-320.
11. Hopkirk, G.; White, A.; Beever, D. J.; Forbes, S. K. *New Zeal. J. Crop Hort. Sci.* **1994**, *22*, 305-311.
12. Cox, K. A.; McGhie, T. K.; White, A.; Woolf, A. B. *Postharvest Biol. Tech.* **2004**, *31*, 287-294.
13. Di Giovacchino, L. In *Handbook of Olive Oil. Analysis and Properties;* Harwood, J.; Aparicio, R., Eds.; Aspen Publishers: Gaithersburg, MD, 2000; pp17-56.
14. Platt-Aloia, K. A.; Thomson, W. W. *Ann. Bot. –London.* **1981**, *4*, 451- 465.
15. Werman, M. J.; Neeman, I. *J. Am. Oil Chem. Soc.* **1987**, *64*, 229 -232.
16. Platt-Aloia, K. A.; Thomson, W. W. *Proceeedings of Second World Avocado Congress.* 1992; pp 417-425.
17. Ashton, O. B. O.; Wong, M.; McGhie, T. K.; Vather, R.; Wang, Y.; Requejo-Jackman, C.; Ramankutty, P.; Woolf, A. B. *J. Argic. Food Chem.* **2006**, *54*, 10151-10158.

18. Artes, F.; Minguez, M. I.; Hornero, D. In *Color in Foods;* MacDougall, D. B. Ed., Woodhead Publishing Ltd: Cambridge, England. 2002; pp 248-282.
19. Kidmose, U.; Edelenbos, M.; Nørbæk, R.; Christensen, L. P. In *Color in Foods*; MacDougall, D. B. Ed., Woodhead Publishing Ltd: Cambridge, England. 2002; pp 179-232.
20. Cran, D. G.; Possingham, J. V. *Ann. Bot.-London.* **1973**, *37*, 993-997.
21. De Greyt, W.; Kellens, M.; In *Edible Oil Processing*; Hamm, W.; Hamilton, R. J. Eds., Sheffield Academic Press, Sheffield, England. 2000; pp 79- 127.
22. Montano, G. H.; Luh, B. S.; Smith, L. M. *Food Technol. -Chicago.* **1962**, *2*, 96-98.
23. Human, T. P. *South African Avocado Growers Association Yearbook.* **1987**, *10*, 159-162.
24. Requejo-Jackman, C.; Wong, M.; Wang, Y.; McGhie, T.; Petley, M.; Woolf, A. *The Orchardist.* **2005**, *78*, 54-58.
25. International Olive Council. Trade Standard Applying to Olive Oils and Olive-Pomace Oils, Madrid, Spain, 2006;COI/T.15/NCno.3/Rev2. http://www.internationaloliveoil.org (accessed Mar 11, 2007)
26. Eyres, L.; Sherpa, N.; Hendriks, G. *J. Lipid Technol.* **2001**, *13)*, 84-88.
27. Zauberman, G.; Schiffmann-Nadel, M. *Plant Physiol.* **1972**, *49*, 864-865.
28. Buenrostro, M.; López-Munguia, A. *Biotechnol. Lett.* **1986**, *8*, 505-506.
29. AOCS. *Official Methods and Recommended Practices of the AOCS;* American Oil Chemists' Society: Champaign, IL, 1998.
30. Werman, M. J.; Neeman, I. *J. Am. Oil Chem. Soc.* **1986**, *63*, 355-360.
31. Endo, Y.; Usuki, R.; Kaneda, T. *J. Am. Oil Chem. Soc.* **1984**, *70*, 781-784.
32. Usuki, R.; Endo,Y.; Kaneda, T. *Agric. Biol. Chem.Tokyo.* **1984**, *48*, 991-994.
33. Sherpa, N. L. *The oxidation stability of extra virgin avocado oil;* Masters of Technology Thesis, Massey University, Auckland, New Zealand, 2002.
34. Lawson, H. *Food Oils and Fats. Technology, Utilisation and Nutrition;* Chapman & Hall: New York, 1995; pp 15-27.
35. Hamilton, R. J. In *Rancidity in Foods*; Allen, J.C; Hamilton, R.J. Eds.,; Blackie Academic & Professional: London, 1994; pp 1-21.
36. Rawls, H. R.; Van Santen, P. J. *J. Am. Oil Chem. Soc.* **1970**, *47*, 121-125.
37. Kiristakis, A. K.; Dugan, L. R. *J. Am. Oil Chem. Soc.* **1985**, *62,* 892-896.
38. Chan, H. W. S. *J. Am. Oil Chem. Soc.* **1977**, *54*, 100-104.
39. Fakourelis, N.; Lee, E. C.; Min, D. B. *J. Food Sci.* **1987**, *53*, 1918-1919.
40. Lee, E. C.; Min, D. B. *J. Food Sci.* **1988**, *53,* 1894-1895.
41. Jung, M. Y.; Min, D. B. *J. Am. Oil Chem. Soc.* **1991**, *68*, 653-658.
42. Khan, M. A.; Shahidi, F. *J. Food Lipids.* **1999**, *6*, 331-339.
43. Caponio, F.; Bilancia, M. T.; Pasqualone, A.; Sikorska, E.; Gomes, T. *Eur. Food Res. Technol.* **2005**, *221*, 92-98.

44. Gutierrez-Rosales, F.; Garrido-Fernandez, L.; Gandul-Rojas, B.; Minguez-Mosquera, M. I. *J. Am. Oil Chem. Soc.* **1992**, *69*, 866-871.
45. Requejo-Jackman, C.; Wong, M.; Wang, Y.; McGhie, T.; Ashton, O. B. O.; White, A.; Petley, M.; Lund, C.; Roberts, C.; Ramankutty, P.; Eyres, L.; Jackman., R.; Woolf, A. B. *J. Agric. Food Chem.* **2007**, manuscript submitted.

Chapter 25

Sweet Whey Powder Color

Anupama Dattatreya and Scott A. Rankin

Department of Food Science, University of Wisconsin, Madison, WI 53706

Sweet whey powder (SWP) is used in many food applications due to its desirable nutritive and functional properties. During storage SWP may brown due to non-enzymatic Maillard-type reactions, thus reducing its nutritional and functional value. SWPs show considerable variation in the rate at which they brown. Commercial SWP were analyzed for pH, moisture content, water activity (a_w), extent of proteolysis, and browning precursor content. Several model systems were also evaluated. Samples were analyzed for color after accelerated storage. These studies demonstrate that acidic conditions, pH history and residual sugar composition in SWP increase browning. The browning extent depended on the main variables and several interactions. The chemistry of browning reactions under acidic conditions have been presented. Understanding SWP browning chemistry will aid in modifying the processing and handling conditions to obtain high quality whey powders stable to browning.

Introduction

Superior quality SWP is a white colored free-flowing powder with slight yellowish-green hue. However during storage, it generally darkens, acquiring an undesirable dark brown color (*1*,*2*). The browning is undesirable in most concentrated and dried foods. Non-enzymatic browning (NEB) through the Maillard reaction is considered to be a major deteriorative mechanism which is active during the storage of SWP (*3*,*4*). Maillard browning reactions involve the interaction of compounds containing amino groups and reducing sugars during thermal processing and storage (*5*). The constituents of SWP, namely lactose and lysine, in the presence of moisture, participate in the Maillard reaction resulting in brown discoloration (*2*). Various food processing and storage variables influence Maillard reaction rates (*6*). Such variables include the composition of the raw materials, the method of manufacture, the time-temperature combination used during heating and storage, the pH, the a_w, the physical structure, and the physical state and type of sugars (*6*).

Variations in color can occur due to several physico-chemical and handling factors. These variations may arise due to several possible factors including, a) biological factors such as composition, milk quality, and stage of lactation; b) cheese origin (e.g. Swiss, cheddar, mozzarella); c) whey handling practices such as holding time, concentration and spray drying methods; d) dry product storage conditions such as temperature, time and relative humidity.

Understanding the color of SWP and the factors that may influence the variation in the quality of SWP will help in modifying specific variables to obtain SWP of improved economic value as a food ingredient. The present study has focused on the effect of moderately acidic final pH, pH history, and residual sugars on the browning of SWP. Additionally, regression models were used to consider the effects of specific variables and their interactions.

Experimental

Effect of acidic pH

SWP (~ 25 g) at different pH (4.2, 4.9 and 6.3) was placed in glass bottles, tightly sealed and subjected to storage at 80°C for 0, 6, 12, 24, 48 and 72 h. The color parameter (L*), pH, hydroxymethyl furfural (HMF)(*7*) method), and reactive lysine content by OPA method (*8*) were measured at each time period.

Effect of pH history

Liquid whey (cheddar source) was subjected to five treatments before freeze drying: A) control (pH of 6.3), where the sample was freeze dried immediately, B) where pH was monitored at 50°F and maintained at 6.3 using NaOH, C) where pH was allowed to drop to 5.0 at 50°F and finally raised to 6.3, D) where pH was allowed to drop to 5.0 at 50°F and E) where whey was pasteurized and the pH dropped using lactic acid at intervals similar to treatment C and finally the pH raised to 6.3. All the samples were then freeze dried.

The freeze dried samples were analyzed in duplicates for extent of proteolysis, content of Amadori product and color after accelerated browning for 48 h at 80°C. The aerobic count/lactic acid bacterial count was conducted for the liquid whey before and after treatment.

Effect of residual sugars

The study was conducted using two model systems as follows, a) model system using lactose and lysine where 2 L of 5.2 % lactose solution containing 0.093% lysine was prepared (corresponding to 0.1443 M lactose and 0.0064 M lysine; the ratio simulating their respective concentration in sweet whey) and divided into four parts of 500 ml each. First sample was unaltered and served as control, the second sample contained a known quantity of lactose (3.0 g), the third sample contained galactose (1.5 g) and the fourth sample had glucose (1.5 g). All were equimolar concentrations corresponding to 0.017 M. All samples were then freeze dried. b) Similarly SWP model system was prepared wherein SWP (Foremost Farms, USA) was reconstituted (6.5%) and divided into 4 parts of 500 ml each. One sample was unaltered which served as control, to the others equimolar concentrations (0.017 M) of lactose, galactose and glucose were added. Study was conducted at three different pHs namely 6.5, 6.0 and 5.5 which was obtained by using NaOH and HCl (1 N).

The content of residual sugars namely lactose, galactose and glucose in the freeze dried samples were determined using the enzymatic test kit (R-Biopharm, Germany). In addition, the reactive lysine content of the samples was analyzed by OPA method (*8*). The freeze dried samples were subjected to accelerated browning at 80°C for 24 h and the color of samples was determined using a colorimeter.

Predictive factors in SWP browning

About 93 samples of commercial SWP from different regions in the U.S. were procured. The samples were analyzed for pH, moisture content (MC), a_w, extent of proteolysis (PR) and browning precursor content, namely the content of

Amadori product (AP). Samples were analyzed for color (L*) after accelerated browning. Linear regression was applied to determine the influence and to study relationships amongst these variables.

Results and Discussion

Effect of acidic pH

It can be observed from Figure 1 that SWP browns as a function of time (0-72 h), the lightness value (L*) decreased and the rate of decrease was much faster in the low pH sample followed by the intermediate pH sample especially with more time. For pH 6.3, the HMF content was found to increase with time and a maximum of 1.80 ± 0.18 mg/kg SWP was attained after 12 h at 80°C while a maximum of 6.15 ± 1.40 mg/kg SWP was obtained after 24 h at pH 4.9. The pH 4.2 sample showed a maximum HMF content of 27.3 ± 3.90 mg/kg SWP after 12 h. After this point, the HMF content could not be detected, indicating that HMF participates in the further reaction wherein brown melanoidins are formed. It was found that the lysine content decreased with time and the decrease was faster at lower pH. The reactive lysine content decreased from 0.864 g / 100 g SWP to 0.439, 0.193 and 0.102 g/100 g SWP for the native, intermediate and low pH samples, respectively, after 72 h at 80°C.

This study demonstrates that low pH caused increased browning in low a_w SWP. The model studies conducted by many researchers were aqueous systems with high water content and in such systems the browning was potentiated by alkaline pH. It was hypothesized that low a_w favors the accumulation of the Amadori compounds given that this step is a reversible, hydrolytic reaction. The low pH may shift the equilibrium concentration of sugars to favor the acyclic form which is essential for the Maillard reaction (*9*). Since low pH values are also known to favor the formation of furfurals from Amadori rearrangement products, while the routes to reductones and fission products are preferred at a higher pH, the high browning intensities at lower pH in our study indicate Maillard reactions favoring the furfural pathway. It is generally accepted that the first stage of the actual Maillard reaction is the condensation of unprotonated (considered active) amino groups with the reducing sugar in the open chain form. The condensation product rapidly loses a molecule of water to form a Schiff's base followed by cyclization to form the corresponding N-substituted glycosylamine which is subsequently rearranged in the so called Amadori product. In the condensation step, the trivalent nitrogen atom of the amine having an unshared electron pair acts as a nucleophile towards the electrophilic carbon of the carbonyl group of sugar, resulting in an amine assisted dehydration reaction (*10*). Bases can potentiate such reactions by removing a proton (H^+)

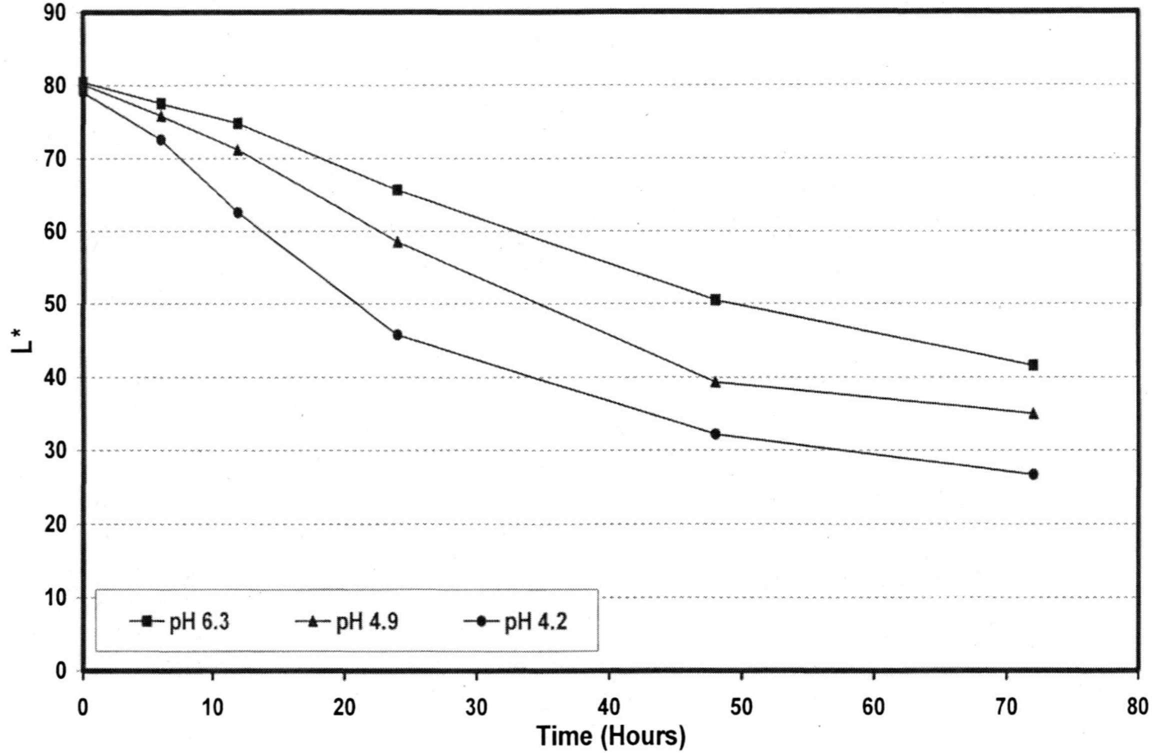

Figure 1. Change in L with time during accelerated storage of SWP at 80°C at different pH values. (n=2; standard error of the mean was too small to see error bars).*

from the nucleophile, thus increasing the nucleophilicity. This reaction is reversible and influenced by pH. However, too much acid protonates the amine and its nucleophilicity diminishes. The Amadori rearrangement, which is the isomerization of N-substituted aldosylamine to 1-amino-1-deoxy-2-ketose, is also potentiated by weak acids (weak acids can provide protons, pH reduction and buffering capacity requisite for the reaction to progress) and the carboxyl group of the amino acids provides the necessary protons (*11*).

In the initial stages of whey processing, it is hypothesized that the first stage of the Maillard reaction involving the sugar-amine condensation takes place and then the Amadori rearrangement products are formed. In the latter stages of the Maillard reaction, the Amadori product is subject to breakdown, probably via the enol forms of the Amadori compound. At acidic pH, the 3-deoxyosone pathway leads to products such as HMF and pyrraline. In this study, the probable pathway is the 3-deoxyosone pathway with acidic conditions prevailing and therefore the reaction involves the loss of three molecules of water and an amino compound to give HMF. All products may further react with amino acids in the third phase to form brown nitrogenous polymers and co-polymers collectively referred to as melanoidin pigments (*12*).

Our hypothesis that the probable pathway is the 3-deoxyosone pathway due to acidic conditions was confirmed by the fact that the HMF content estimated was the highest at pH 4.2 followed by pH 4.9 while the native pH (6.3) had the lowest HMF content. This suggests that at low pH the Amadori compound present in the SWP undergoes dehydration during storage to give HMF which further reacts with amino acids to form brown pigments.

Effect of pH history

It can be observed from Table I that sample A had the least browning (highest L* value) and the other samples did not differ significantly from each other. There was no significant difference in the extent of proteolysis for all the samples. The trend for Amadori content was similar to the first study with the sample C having the highest Amadori content followed by B. Surprisingly, D had lower Amadori content compared to C, and E had the least. The lactic acid bacterial count for both B and C treatments increased approximately thirty fold over A. The results indicate that pH history does play a role in the browning potential of SWP. Though all the samples except D had the same final pH, the browning potential varied as related to pH history. The fact that the pasteurized sample (E) had browning potential higher than A indicated that pH did have a significant role in browning. The pasteurized sample had a lower concentration of Amadori compounds and a lower extent of proteolysis, yet browned to a greater degree than expected. This result could have been caused by the heat treatment itself promoting the reaction of the Amadori products down the browning pathways.

Table I. Mean Values for Color, Proteolysis and Amadori Content of SWP Samples

Sample	*L* at 0 h*	*L* at 48 h*	*Proteolysis mg Tyr/100g SWP*	*Amadori content mMol/L in 0.3 mg protein*
A	84.97 (0.12)	40.37 (2.17)	49.59 (0.57)	0.022 (0.00)
B	85.35 (0.13)	32.91 (2.29)	50.17 (0.19)	0.039 (0.00)
C	85.27 (0.10)	32.23 (2.13)	49.27 (0.00)	0.051 (0.00)
D	85.26 (0.07)	29.59 (1.31)	51.10 (0.19)	0.032 (0.00)
E	85.23 (0.02)	30.54 (0.91)	45.93 (0.00)	0.014 (0.00)

Study 1: A- control; B- sample where pH was monitored at 50°F and maintained at 6.3; C- sample where pH was allowed to drop to 5.2 at 50°F and finally raised to 6.3. Study 2: A- control; B- sample where pH was monitored at 50°F and maintained at 6.3; C- sample where pH was allowed to drop to 5.0 at 50°F and finally raised to 6.3; D- sample where pH was allowed to drop to 5.0 at 50°F; E- whey was pasteurized, pH dropped at intervals similar to C and finally pH raised to 6.3. (n=2; values in parenthesis are one standard deviation).

The first step of Maillard reaction is the condensation reaction initiated by an attack of a nucleophilic amino nitrogen with an unshared pair of electron on carbonyl carbon to give N-substituted aldosylamines. The Amadori rearrangement is the isomerization to 1-amino-1-deoxy-2-ketose and this reaction is promoted by weak acids. Hence in our study, a pH drop to ~5.0 resulted in conditions favorable for initial stages of Maillard reaction with accumulation of precursors namely the Amadori compounds and this had an affect on browning potential during storage. The results indicate that immediate processing / improved pH control of liquid whey may be a major factor in preserving the white appearance of SWP in storage. Preventing pH drop may prevent the accumulation of precursors in the liquid whey and thereby increase the stability of SWP towards browning.

Effect of residual sugars

Results indicated that the samples containing added galactose and glucose were darker in color after accelerated browning for all pH values (6.5, 6.0 and 5.5) in the model lactose-lysine system. The browning was indicated by the color parameters namely L* and DE* (Table II). There were significant differences in the color among the samples across pH values ($P<0.05$). a* values were higher in the galactose containing samples. b* did not show any specific trend. Darker color was seen for the lower pH samples for the control and lactose added samples. A slight increase in lightness was seen in the lactose added samples over the control. This may result from the presence of lactose impeding the progress of the reaction by limiting access to reactants. Significant differences ($P<0.05$) in color were seen among the samples at different pH for the control, lactose added and galactose added samples.

With the SWP model system, samples containing added galactose and glucose were darker in color after accelerated browning only for pHs 6.5 and 6.0. The samples were significantly ($P<0.05$) different from each other for the pH 6.5 and 6.0. The color parameters are given in Table III. b* values were found to be lower for samples containing added galactose and glucose while a* did not show specific trend. Similar to the lactose-lysine model system, the lower pH samples of the control and added lactose treatments had darker color. Significant differences ($P<0.05$) in color were seen among the samples at different pH for the control, lactose added and galactose added samples.

Before and after accelerated browning there was a decrease in reactive lysine content in all the samples irrespective of the pH or residual sugar content. There was a decrease in the lysine content in lactose-lysine system (51-65%) and in the SWP model system (24-68%).

Table II. Color Values of Samples of Lactose-Lysine Model System after Accelerated Browning at 80°C, 24 h

	Color parameters at different pH											
	*L**			*a**			*b**			*DE**		
Sample	*6.5*	*6.0*	*5.5*	*6.5*	*6.0*	*5.5*	*6.5*	*6.0*	*5.5*	*6.5*	*6.0*	*5.5*
1	67.76 (0.53)	61.89 (0.78)	59.99 (0.48)	5.38 (0.40)	6.74 (0.50)	7.03 (0.04)	20.13 (0.76)	21.26 (0.15)	23.97 (0.16)	33.33 (0.73)	38.92 (0.76)	42.02 (0.28)
2	69.41 (0.31)	64.92 (0.83)	61.97 (0.78)	5.05 (0.32)	5.81 (0.15)	6.50 (0.54)	19.39 (0.08)	20.38 (0.37)	22.24 (0.89)	31.54 (0.15)	35.80 (0.87)	39.36 (0.89)
3	46.80 (0.51)	50.90 (0.84)	53.68 (0.87)	8.37 (0.08)	6.99 (0.76)	7.34 (0.02)	23.67 (0.49)	20.08 (0.71)	21.62 (0.78)	53.27 (0.68)	47.88 (0.13)	46.15 (0.67)
4	54.40 (0.61)	53.15 (0.45)	53.09 (0.13)	6.50 (0.33)	6.51 (0.36)	6.72 (0.18)	21.33 (0.64)	19.82 (0.55)	21.48 (0.25)	45.25 (0.88)	45.96 (0.12)	46.42 (0.26)

1-Control: 5.2 % lactose solution containing 0.093% lysine, 2- 500 ml of 1 + 3.0 g lactose, 3- 500 ml of 1 + 1.5 g galactose, 4- 500 ml of 1 + 1.5 g glucose. (n=2; values in parenthesis denote one standard deviation).

Table III. Color Values of Samples of SWP Model System after Accelerated Browning at 80°C, 24 h

	Color parameters at different pH											
	*L**			*a**			*b**			*DE**		
Sample	*6.5*	*6.0*	*5.5*	*6.5*	*6.0*	*5.5*	*6.5*	*6.0*	*5.5*	*6.5*	*6.0*	*5.5*
1	43.69 (0.25)	40.46 (0.33)	34.24 (0.71)	13.75 (0.20)	13.71 (0.02)	13.98 (0.19)	31.56 (0.14)	29.68 (0.57)	26.15 (0.54)	60.82 (0.33)	62.59 (0.86)	66.56 (0.78)
2	43.11 (0.64)	41.32 (0.78)	35.43 (0.37)	13.61 (0.38)	13.36 (0.25)	13.65 (0.21)	31.30 (0.04)	30.04 (0.05)	27.44 (0.01)	61.13 (0.46)	61.95 (0.78)	66.01 (0.38)
3	38.09 (0.40)	34.27 (0.45)	34.15 (0.25)	14.18 (0.08)	14.39 (0.06)	14.25 (0.24)	29.63 (0.01)	27.42 (0.62)	26.33 (0.37)	64.70 (0.36)	67.13 (0.13)	66.76 (0.42)
4	37.26 (0.79)	33.21 (0.62)	36.33 (0.87)	14.20 (0.22)	14.40 (0.04)	14.06 (0.12)	29.09 (0.01)	26.71 (0.17)	27.29 (0.23)	65.17 (0.76)	67.78 (0.47)	65.18 (0.79)

1-Control: reconstituted SWP (6.5%), 2- 500 ml of 1 + 3.0 g lactose, 3- 500 ml of 1 + 1.5 g galactose, 4- 500 ml of 1 + 1.5 g glucose. (n=2; values in parenthesis denote one standard deviation).

Analysis of residual sugar content before and after browning in both the systems indicated the progress of Maillard reaction with the reduction of the sugar content. The decrease was found to be higher for samples containing added galactose and glucose when compared to samples containing added lactose and control. This showed that galactose and glucose participated in the browning reactions more readily than lactose. The galactose content decreased by 58-81% in lactose-lysine system while it decreased by 62-79% in SWP model system. The glucose content decreased by 72-84% in the lactose lysine system while it decreased by 60-72% in the SWP model system. However, the decrease in lactose was minimal (0.4-6% in lactose lysine system and 8-30% in SWP model system).

The results of this study showed that residual sugars do have a significant influence in the relative browning extent of SWP during accelerated storage. The presence of sugars, namely the hexoses galactose and glucose, can accelerate the browning reaction. These monosaccharides tend to increase the initial rate of reducing sugar dependent browning with a given amino compound mainly by increasing the rate at which the sugar's ring opens to a reducible form (*13*). As the pH dropped from 6.5 to 5.5, browning increased for control and lactose added samples for both the model systems. This effect was not seen for samples containing either galactose or glucose. This lack of demonstrated browning may be due to the extensive browning that occurred in these systems at 80°C over 24 h, thus masking the pH effect. In the SWP model system at pH 5.5, the type of residual sugar did not show an effect on the browning probably because of the lower pH triggering the browning reaction and causing increased browning irrespective of the type of sugar. However this effect was not seen in the lactose-lysine system. The probable difference could be due to the presence of other amino acids in the SWP model system which could have accelerated the browning reaction in addition to lysine. Researchers (*14*) indicated increased galactose content depending on the conditions of whey processing and type of culture thus indicating the role of processing as an important factor in the accumulation of sugars. The study indicated that the type of sugars present in whey may influence the browning of SWP during storage. The higher content of hexoses namely galactose and glucose result in increased browning of SWP during storage.

Predictive factors in SWP browning

The MC varied from 1.39 to 5.22 %, a_w from 0.099 to 0.327, pH from 5.75 to 6.55, protein content varied from 11.23 to 16.99%, PR from 42.23-100.95 mg of Tyr/100g SWP, content of AP varied from 0.025-0.54 as mMol fructosamine/L. The color parameters after accelerated browning for 8 h varied from 45.10 to

85.61for L*, 0.64 to 16.21 for a*, 15.13 to 38.58 for b* and 17.07 to 62.06 for DE*. Linear regression for selected variables with two way interaction for the responses L* after 8 h of accelerated browning at 80°C was determined. Backward elimination regression ($P<0.05$) was used and the main effect variables that were significant as an interaction were also considered in the model. The response with L* was selected because it had the best fit with R^2 of 70.4%. The results for two way interactions are shown in Table IV where it can be seen that the main effects of MC, AP, a_w and pH were significant at $P <0.05$. PR was not found to be significant. However, PR was found to be significant as an interaction with AP and a_w. Significant interactions were also seen for MC × a_w, and AP × pH. The regression analysis yielded the following model.

$$L^*8\ h = -160 - 18.4\ MC + 0.312\ PR + 1553\ AP + 225\ a_w + 41.7\ pH + 53.1\ MC \times a_w + 1.80\ PR \times AP - 5.37\ PR \times a_w - 274\ AP \times pH$$

$$(R^2 = 70.4\%) \qquad (1)$$

Significant positive correlation ($P <0.05$) was found between MC × PR, MC × a_w, PR × AP and AP × pH and significant negative correlation ($P < 0.05$) between PR × L*, and MC × L*. The negative correlation of L* with MC and PR indicated that higher MC and increased PR of the samples influenced browning. Moisture aids in the progress of Maillard reaction and increased PR may result in more availability of amino groups. The content of precursors namely AP was found to be well correlated with the extent of PR as well as the pH of the samples.

Results from this study indicated that parameters of liquid whey such as pH, holding time prior to processing, as well as SWP parameters of moisture and a_w, to obtain SWP may influence browning. The regression model developed may aid in predicting the browning of SWP and manipulating selected variables to control the browning of SWP during storage.

References

1. Doob, H.; Willmann, A.; Sharp, P. F. *Ind. Eng. Chem.* **1942**, *34*, 1460-1468.
2. Labuza, T.; Saltmarsh, M. *J. Food Sci.* **1981**, *47*, 92-96.
3. Burin, L.; Jouppilla, K.; Roos, Y.; Kansikas, J.; Buerea, M. D. **2000**, *48*, 5263-5268.
4. Roos, Y.; Jouppilla, K.; Zielasko, B. *J. Therm. Anal.* **1996**, *47*, 1437-1450.
5. Ellis, G. P. The Maillard Reaction. In *Advances in Carbohydrate Chemistry;* Melville, L., Ed.; Academic Press: New York, 1959; Vol. 14, p 63.

Table IV. Results of Linear Regression Analysis with L* 8 h as the Response Factor for the Predictors Showing Two-Way Interactions

Predictor	*Coeff*	*SE Coeff*	*T*	*P*
Constant	- 160.33	63.04	- 2.54	0.013
MC	- 18.36	3.19	- 5.74	0.000
PR	0.31	0.26	1.20	0.235
AP	1553.20	476.20	3.26	0.002
a_w	225.20	100.90	2.23	0.028
pH	41.66	9.23	4.51	0.000
MC*a_w	53.05	20.82	2.55	0.013
PR*AP	1.80	0.54	3.33	0.001
PR*a_w	- 5.37	1.46	- 3.67	0.000
AP*pH	- 273.99	75.01	- 3.65	0.000

MC- moisture content; PR- proteolysis; AP-Amadori product; a_w-water activity.

6. Ames, J. M. *Trends Food Sci. Technol.* **1990**, *1*, 150-154.
7. Marquez, M. F.; Gomez, J. M.; Hernandez, E. G.; Villanova, B. G. *J. Dairy Res.* **1992**, *59*, 225-228.
8. Vigo, M. S.; Malec, L. S.; Gomez, R. G.; Llosa, R. A. *Food Chem.* **1992**, *44*, 363-365.
9. Westheimer, F. H. *J. Org. Chem.* **1937**, *2*, 431-441.
10. O'Brien, J.; Morrisey, P. A. *Crit. Rev. Food Sci. Nutr.* **1989**, *28*, 211-248.
11. Mauron, J. *Prog. Food Nutr. Sci.* **1981**, *5*, 5-35.
12. Nursten, H. E. *Food Chem.* **1981**, *6*, 263-277.
13. Overend, W. G.; Peacocke, A. R.; Smith, J. B. *J. Chem. Soc.* **1961**, *83*, 3487-3497.
14. Rao, R. D.; Wendorff, W. L.; Smith, K. *J. Food Prot.* **2004**, *67*, 403-406.

Chapter 26

Understanding Colors in Emulsions

Withida Chantrapornchai[1], Fergus M. Clydesdale[2], and D. Julian McClements[2]

[1]Department of Product Development, Faculty of Agro-Industry, Kasetsart University, 50 Phaholyothin Road, Chatujak, Bangkok, 10900, Thailand
[2]Department of Food Science, University of Massachusetts, Amherst, MA 01003

Emulsion color depends on their scattering and absorption efficiency. The scattering efficiency is determined mainly by droplet characteristics (size, concentration, aggregation and relative refractive index), while the absorption efficiency is mainly determined by dye characteristics (absorption spectra and concentration). The lightness of an emulsion is correlated to the scattering efficiency of the droplets. The color of an emulsion (*a*- and *b*-values) is mainly determined by the dye type (red, green, blue) and concentration (0 - 0.1 wt%). Experiments showed that emulsion lightness increased steeply between 0 and 5 wt% oil, and increased slightly at higher concentrations (5 - 20 wt%). It also increased with decreasing droplet diameter (30-0.2 micron) and increasing refractive index difference between the two phases. The influence of droplet characteristics on the optical properties of emulsions containing different types of dye was similar. Droplet flocculation did have an impact on emulsion appearance.

Introduction

Food color is an important factor in food consumer choice, since it not only influences taste thresholds, sweetness perception, food preference, pleasantness, and acceptability; but also determines food quality (*1-5*). Since color can be assessed more easily than taste, odor, and texture, consumers largely make their purchasing decisions by looking at and judging it by appearance. Many natural and manufactured food products exist in the form of oil-in-water emulsions, *e.g.*, milk, cream, fruit beverages, salad dressings, mayonnaise, soups, sauces, and infant formulations (*6-9*). Lately, consumers have given more attention to their dietary intake; which has stimulated the development of reduced and low fat foods (*10*). However, it is difficult to get the same product quality, if the fat content has been removed or reduced. In addition, although consumers want foods with low or no fat, they also prefer them to taste, look, and give a texture as close as possible to the original ones. As a food component, fat provides many major attributes in foods, especially sensory and physiological benefits (*11*). One of the most important attributes is appearance. Still, surprisingly little work has been done on the factors that influence emulsion appearance. This article reviews the major factors that affect the color of emulsions.

Theory

Physical Basis of Emulsion Color

An emulsion is defined as two immiscible liquids with one of the liquids being dispersed as small spherical droplets in the other. In most food emulsions, the droplet diameters are in the range of 0.1 to 100 μm. (*6, 7, 9, 12, 13*). Emulsions consist of two phases; the liquid that forms the droplets in an emulsion is called the "dispersed" or "internal" phase, while the liquid that surrounds the droplets is referred to as the "continuous" or "external" phase (*13*). The fundamental physical phenomena of all appearance attributes is determined by the proportion of wavelengths reflected, transmitted, absorbed and/or scattered by an object (*5, 14-19*). In an emulsion, these interactions are mainly governed by the characteristics of the droplets (concentration, size, and refractive index) and of any chromophores (type and concentration) present in it (*16, 20-24*). When a light wave that enters an emulsion encounters a droplet, part of the wave is transmitted and part of the wave is scattered (*25-27*). The fraction of the wave that is scattered and the direction that the scattered waves travel depends on the refractive index of the droplets and continuous phase, as well as on the size of the droplets relative to the wavelength of light. In a dilute emulsion, a light wave that travels through an emulsion may only encounter a single droplet before emerging (single scattering). In a concentrated emulsion, a light wave scattered from one droplet may be scattered by a number of other droplets before it emerges from an emulsion (multiple scattering) (*28*). In a highly concentrated emulsion, a significant fraction of the incident light may travel back to the surface of an emulsion through multiple scattering events and emerge as diffusely reflected light (*29*). When chromophoric substances are present in either the continuous or dispersed phases some of the light wave is absorbed. The extent of absorption depends on the concentration and absorbtivity of the chromophores and on the wavelength of the light used. Some wavelengths are absorbed more strongly than others so that the color of the light emerging from the emulsion is no longer white. Consequently, emulsion appearance is dependent on the combination of light scattering and absorption. Scattering is largely responsible for the turbidity or lightness of an emulsion, whereas absorption is mostly responsible for the color (*30*).

Prediction of Emulsion Color

In order to control optical properties of emulsions, it is important to understand the factors that determine them. However, it would be even more useful if we could predict the appearance of an emulsion from knowledge of its composition and structure. This would be of great advantage to food manufacturers due to savings in both time and cost. A mathematical model has been developed to predict emulsion color using light scattering theory (*31*). The overall prediction procedure is shown in Figure 1.

Step 1:

Input the refractive index ratio of the dispersed and continuous phases into the Mie theory, and calculate the g and Q_{sca} values as a function of x $(=2\pi rn/\lambda)$.

Step 2:

The wavelength dependence of the scattering and absorption characteristics (Σ_s and Σ_a) of the emulsion are calculated from the predicted g and Q_{sca} values for the appropriate droplet size and the measured absorption spectrum using Equations (1) and (2).

$$\Sigma_s = N\pi r^2 Q_{sca} \tag{1}$$

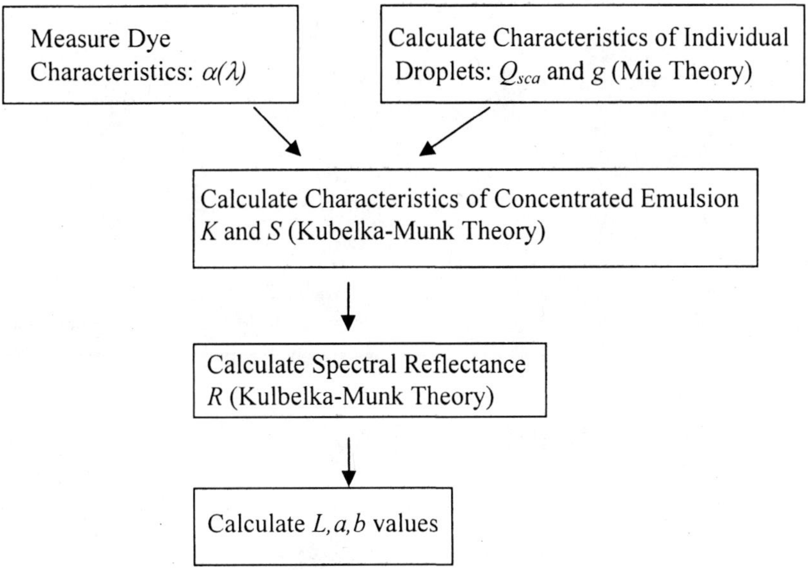

Figure 1. Color prediction procedure using light scattering theory

$$\Sigma a = \alpha \tag{2}$$

where, Σ_s and Σ_a are the scattering and absorption cross sections of the droplets, g is the asymmetry factor, α is the absorption coefficient of the emulsion, and N is the number of droplets per unit volume. Values of Q_{sca} and g are calculated from the relative refractive index ratio and droplet size, e.g., using a computer program available on the Internet (*32*).

Step 3:

The Kubelka-Munk scattering (S) and absorption (K) coefficients are determined from Σ_s and Σ_a using Equations (3) and (4).

$$S = \tfrac{3}{4}\Sigma_s[1-g] - \tfrac{1}{4}\Sigma_a \tag{3}$$

$$K = 2\Sigma_a \tag{4}$$

Step 4:

The spectral reflectance, $R(\lambda)$, is calculated as a function of wavelength (λ = 380-780 nm, at 10 nm intervals) using Equation (5).

$$R = 1 + \frac{K}{S} - \sqrt{\frac{K}{S}\left[\frac{K}{S} + 2\right]} \tag{5}$$

where R is the spectral reflectance of the material, S is the scattering coefficient, and K is the absorption coefficient. Note that R, S and K are all functions of wavelength. The values of the coefficients can be calculated from diffuse scattering theory (*33*).

Step 5:

The tristimulus coordinates (L-, a- and b-values) are calculated using Equations (6) to (15).

$$X = k \sum_{380nm}^{780nm} S(\lambda)\bar{x}(\lambda)R(\lambda) \tag{6}$$

$$Y = k \sum_{380nm}^{780nm} S(\lambda)\bar{y}(\lambda)R(\lambda) \tag{7}$$

$$Z = k \sum_{380nm}^{780nm} S(\lambda)\bar{z}(\lambda)R(\lambda) \tag{8}$$

$$k = \frac{100}{\sum_{380nm}^{780nm} S(\lambda)\bar{y}(\lambda)} \tag{9}$$

where $S(\lambda)$ is the spectral distribution of the standard illuminant at wavelength λ, $\bar{x}(\lambda)$, $\bar{y}(\lambda)$, $\bar{z}(\lambda)$ are the human response functions of the CIE color system,

$R(\lambda)$ is the spectral reflectance of the material. The *X, Y, Z* values can be converted into Hunter *L, a, b* values (*34*), and so the color of an emulsion can be predicted from its spectral reflectance:

$$X_n = k \sum_{380nm}^{780nm} S(\lambda)\bar{x}(\lambda) \tag{10}$$

$$Y_n = k \sum_{380nm}^{780nm} S(\lambda)\bar{y}(\lambda) \tag{11}$$

$$Z_n = k \sum_{380nm}^{780nm} S(\lambda)\bar{z}(\lambda) \tag{12}$$

$$L = 100\left(\frac{Y}{Y_n}\right)^{0.5} \tag{13}$$

$$a = K_a \frac{\frac{X}{X_n} - \frac{Y}{Y_n}}{\left(\frac{Y}{Y_n}\right)^{0.5}} \tag{14}$$

$$b = K_b \frac{\frac{Y}{Y_n} - \frac{Z}{Z_n}}{\left(\frac{Y}{Y_n}\right)^{0.5}} \tag{15}$$

where X_n, Y_n, Z_n are the tristimulus values of the reference white for the selected illuminant and observer. K_a and K_b are chromaticity coefficients or the expansion factors for the selected illuminant and observer, which are in this study illuminant D_{65} and 10 degree observer.

Emulsion Preparation and Characterization

The experimental data presented in the following section is taken from recent studies on the influence of composition and microstructure on emulsion color (*24, 30, 35, 36*). The dispersed phase in these oil-in-water emulsion systems was *n*-hexadecane due to its colorlessness. All dyes used in these experiments are water-soluble. Double distilled and deionized water was used to prepare all solutions and emulsions. The emulsions were then characterized:

Droplet Size Distribution Measurement

A static light scattering technique (Horiba LA-900, Horiba Instruments Incorporated, Irving, CA) was used to measure the droplet size distribution of the emulsions. To avoid multiple scattering effects emulsions were diluted with distilled water prior to analysis so that the final droplet concentration was ~ 0.005 wt%. Droplet size measurements are reported as the "surface-weighted mean diameter": $d_{32} = \Sigma n_i d_i^3 / \Sigma n_i d_i^2$, where n_i is the number of droplets with diameter d_i. The droplet size distribution of the emulsions was measured at the beginning and end of the experiments to indicate whether coalescence or Ostwald ripening occurred during the experiments. All measurements were carried out before the dye was added to an emulsion to avoid complications in the droplet size analysis associated with the wavelength dependent absorption of the dye.

Colorimetry

The color of the emulsions was measured using an instrumental colorimeter (LabScan II, Hunter Associates Laboratory, Reston, VA). The optical sensor used 0° incident light (filtered to closely approximate CIE Illuminant D_{65}) on the sample plane. Viewing was at 45° through a ring of 16 fiber optic receptor stations. This geometry excludes specular reflection from measurement and essentially eliminates the effect of directionality. A fixed amount of emulsion sample is poured into the measurement cup, which is then surrounded with a black paper strip, and covered with a white or black tile before the measurement is carried out. The instrument reports the color of the samples in terms of the *L,a,b* color space system.

UV-Vis Spectrophotometry

Absorbances of dye solutions, turbidity and reflectance spectra of emulsions were measured using a UV-visible spectrophotometer (UV-2101PC, Shimadzu Scientific Instruments, Columbia, MD). During the measurements, emulsions were contained in glass cuvettes with a 1 cm path length. Spectra were obtained over the wavelength range 380-780 nm using a scanning speed of 700 nm min^{-1}. Absorbance measurements were made using a standard double-beam arrangement, with the absorption of the dye solution being measured relative to that of a reference cell containing distilled water. Spectral reflectance measurements were made using an integrating sphere arrangement (ISR-260, Shimadzu Scientific Instruments, Columbia, MD). The spectral reflectance of the emulsions was measured relative to a barium sulfate ($BaSO_4$) standard white plate.

Factors Influencing Color of Emulsion

Effect of Droplet Concentration and Size

Studies of the influence of droplet concentration and size on emulsion color have been investigated (*24, 35, 36*). Photographs illustrating these effects are shown in Figures 2 and 3. The most significant change in emulsion color (relative to the color of emulsions in the absence of droplets) occurred between 0 and 5 wt% oil (Figure 4f). The intensity of emulsion color tended to fade with increasing droplet concentration, except at low droplet concentrations (0-1 wt%) in dyed emulsions where there was initially an increase in chroma (Figure 4b,c). This initial increase has been attributed to the influence of multiple scattering on the effective pathlength that the light waves travel through the emulsion before being reflected back to the detector (*36*).

The *L*-value of the emulsions decreased with increasing droplet diameter (Figure 5a), whilst the chroma increased with increasing droplet size (Figure 5b,c). The most likely reason for the increase in the chroma of the emulsions with increasing droplet size is that the scattering efficiency of the droplets decreases and therefore the light beam can penetrate further into the emulsion, which increases the absorption. The color difference of the emulsions (relative to the emulsion containing the largest droplets) became increasingly small as the droplet diameter increased (Figure 5f) (*24*). The increase in emulsion lightness and decrease in emulsion chroma with increasing droplet concentration (0-20 wt%) and decreasing droplet diameters (d_{32} = 0.2 to 30 μm) was confirmed by sensory analysis for emulsion containging blue dye (*24*).

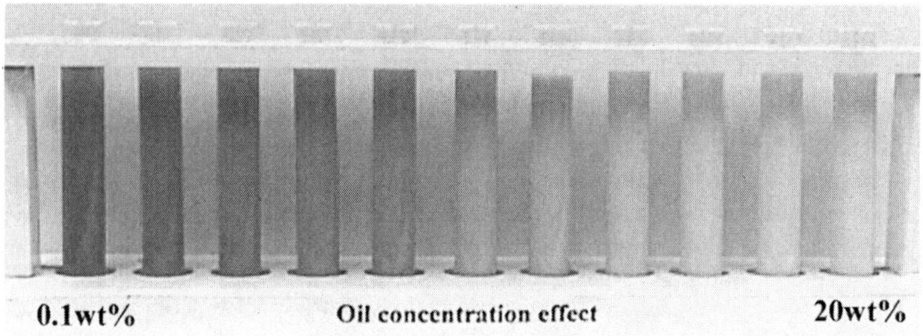

Figure 2. A series of n-hexadecane oil-in-water emulsions with different droplet concentrations (0.1 to 20 wt%), but the same mean droplet diameter (d_{32} = 0.3 μm) and dye content (0.005 wt%). (See page 23 of color inserts.)

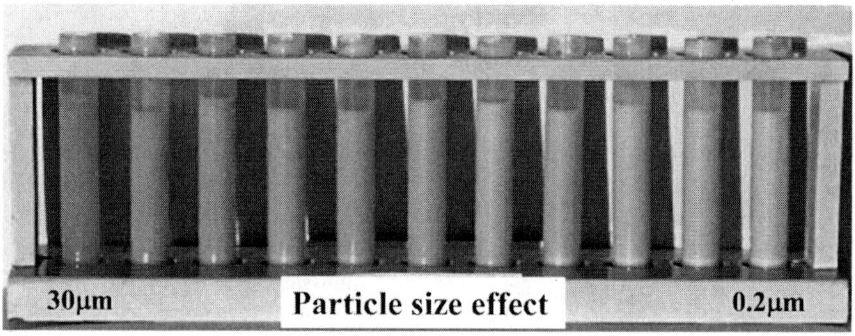

Figure 3. A series of n-hexadecane oil-in-water emulsions with the same droplet concentration (9.5 wt%) and dye content (0.005wt%), but different droplet diameters (d_{32} = 0.2 to 30 μm). (See page 23 of color inserts.)

The influence of flocculation on the optical properties of concentrated emulsions was investigated (*37*). The study showed that flocculation had a slight effect on emulsion appearance. As flocculation increased (depletion flocculation or electrostatic screening flocculation), the spectral reflectance and *L*-value (lightness) of the emulsions decreased, especially above the critical flocculation concentration. The influence was fairly similar in both the presence and absence of dye. Nevertheless, their impact was not as strong as that of the individual droplet size.

Effect of Refractive Index

We have also investigated the influence of refractive index on the optical properties of oil-in-water emulsions (*38*). Experiments both with and without dye showed the same effect (Figure 6). Generally, the reflective index ratio can be varied by adding water-soluble solutes (such as sugars, polyols or salts) to the aqueous phase to increase its refractive index (Figure 7). In our study, we added different amounts of glycerol.

As the n_d/n_{aq} ratio approached unity, the emulsions became more transparent, and the spectral reflectance decreased. On the other hand, as the n_d/n_{aq} ratio moved away from unity, due to the increase in the refractive index difference between the two phases, the scattering efficiency of the emulsion droplets increased and so the spectral reflectance increased (Figure 8). The spectral reflectance predicted by the theory showed similar trends to measured spectra.

Effect of Dye Type and Concentration

The optical properties of colored emulsions are determined by the characteristics of the droplets and dyes they contain. Experimental measurements of the influence of dye type and concentration on the color of oil-in-water emulsions containing different concentrations and sizes of droplets have been carried out (*36*). There are dramatic changes in the lightness and color of emulsions over the droplet concentration range 0 to 3 wt% (Figure 9). Around 0.5 μm diameter, the *L*-value of the emulsions had a slight maximum value and the emulsions were least colored because of the maximum in the scattering efficiency of the droplets. When droplet size increased, the emulsion lightness decreased and became more colored (Figure 10). This is because the scattering efficiency of the droplets decreased, and therefore the light waves could penetrate further and be more absorbed by the dyes. In addition, the influence of droplet size and concentration on the lightness and color of emulsions containing different types of dye was fairly similar (Figure 11).

Besides the droplet characteristics, the optical properties of concentrated oil-in-water emulsions also depend strongly on the concentration of dye present. The impact of dye concentration on the color coordinates of *n*-hexadecane oil-in-water emulsions is shown in Figure 12. As one would expect, emulsion lightness decreased and chromaticness increased as the dye concentration increased because more of the light was absorbed by the chromophores in the dye.

Comparison of Theory with Experimental Measurements

Mostly, the values predicted by light scattering theory exhibit the same trend and agree qualitatively with the measured values (*24, 26, 28*). However, there

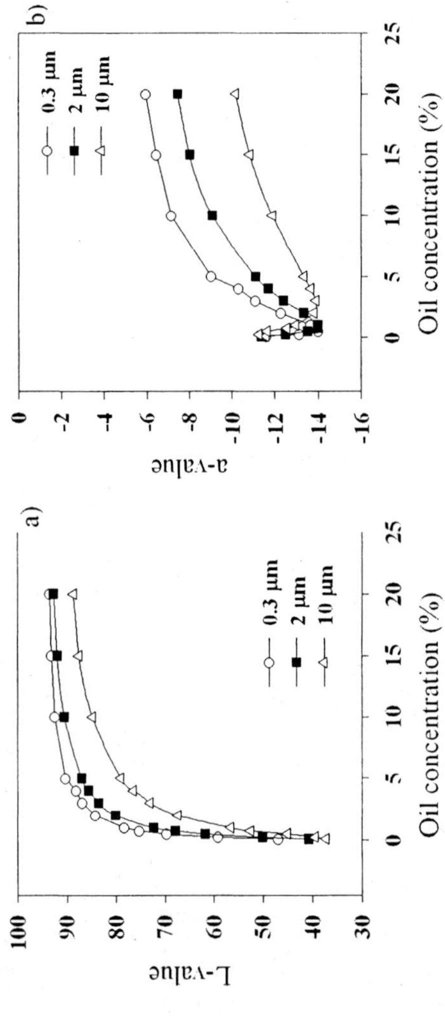
a)
L-value
Oil concentration (%)
0.3 µm
2 µm
10 µm
b)
a-value
Oil concentration (%)
0.3 µm
2 µm
10 µm

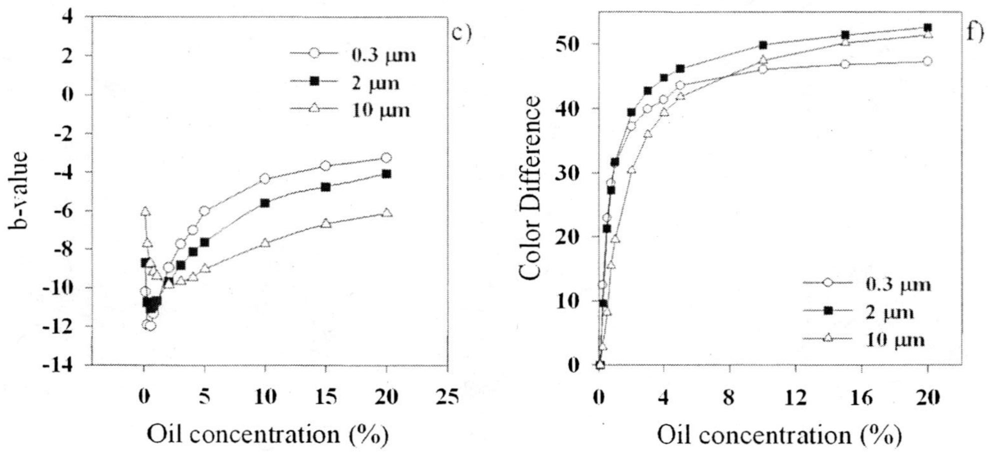

Figure 4. Dependence of L, a, b and color difference on droplet concentration for n-hexadecane oil-in-water emulsions (Reproduced from reference 24. Copyright 1998 American Chemical Society.)

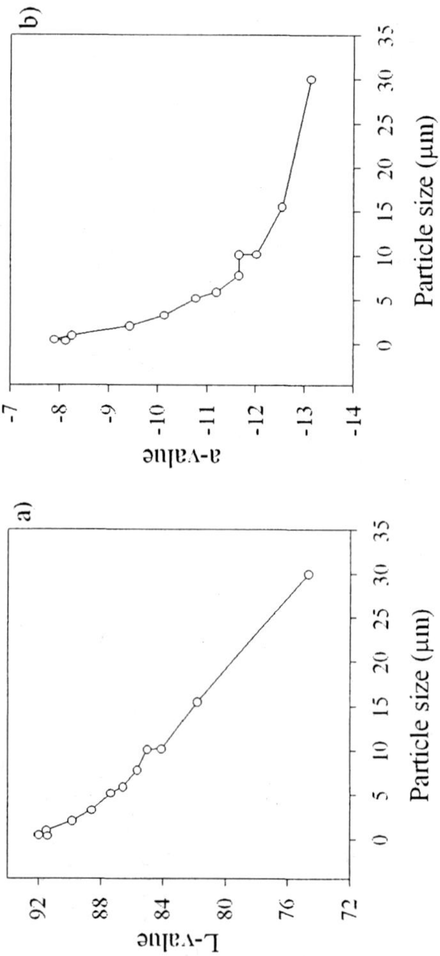

a)
L-value
92
88
84
80
76
72
0 5 10 15 20 25 30 35
Particle size (μm)
b)
a-value
-7
-8
-9
-10
-11
-12
-13
-14
0 5 10 15 20 25 30 35
Particle size (μm)

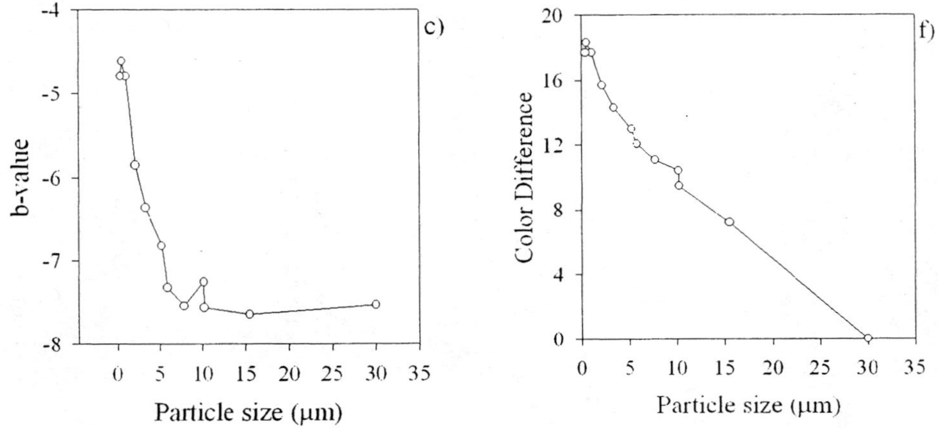

Figure 5. Dependence of L, a, b, and color difference on droplet size for 10 wt% n-hexadecane oil-in-water emulsions (Reproduced from reference 24. Copyright 1998 American Chemical Society.)

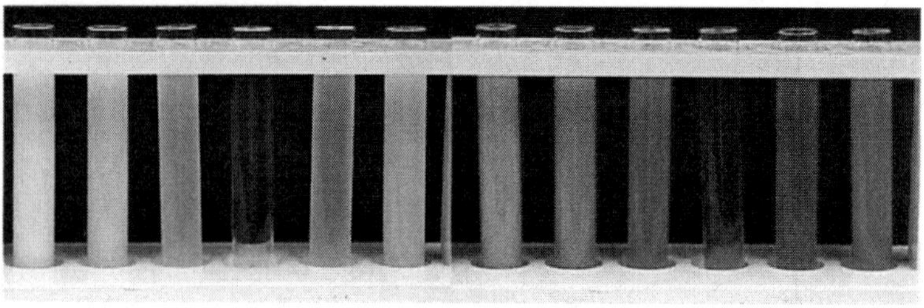

Figure 6. Two series of 4 wt% n-hexadecane oil-in-water emulsions with the same median droplet diameter (1 μm) but a range of different n_d/n_{aq} ratios (0.97-1.07), in the absence and presence of red dye (0.002 wt%). (See page 23 of color inserts.)

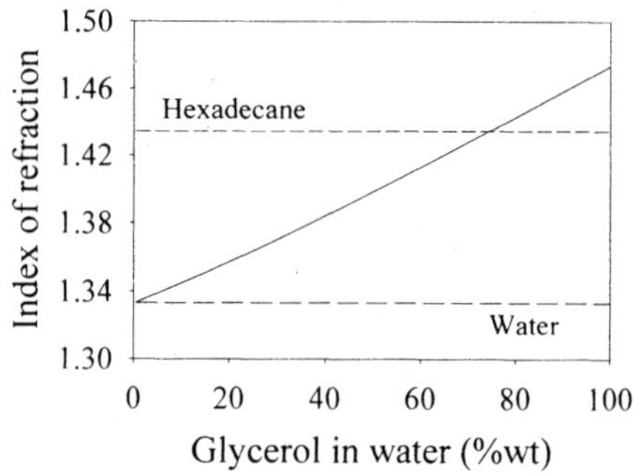

Figure 7. Index of refraction of hexadecane, water, and glycerol in water at different concentrations (Reprinted from 38, with permission from Elsevier).

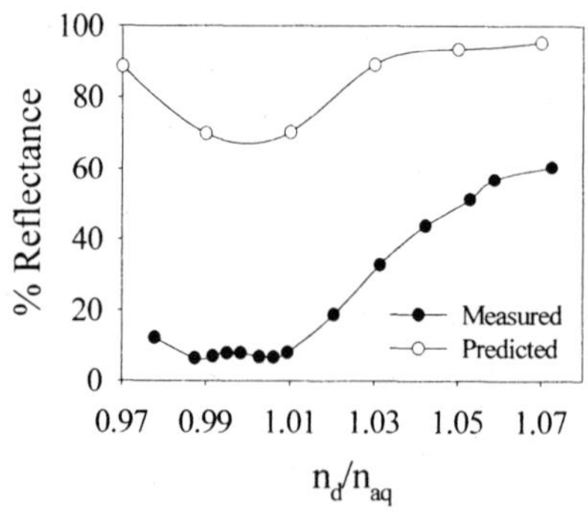

Figure 8. Dependence of the reflectance (%) at 700 nm on n_d/n_{aq} ratios (0.97-1.07) of a series of 4 wt% n-hexadecane oil-in-water emulsions with the same median droplet diameter (1 μm), red dye concentration (0.002 wt%), comparing to a predicted one. (Reprinted from 38, with permission from Elsevier).

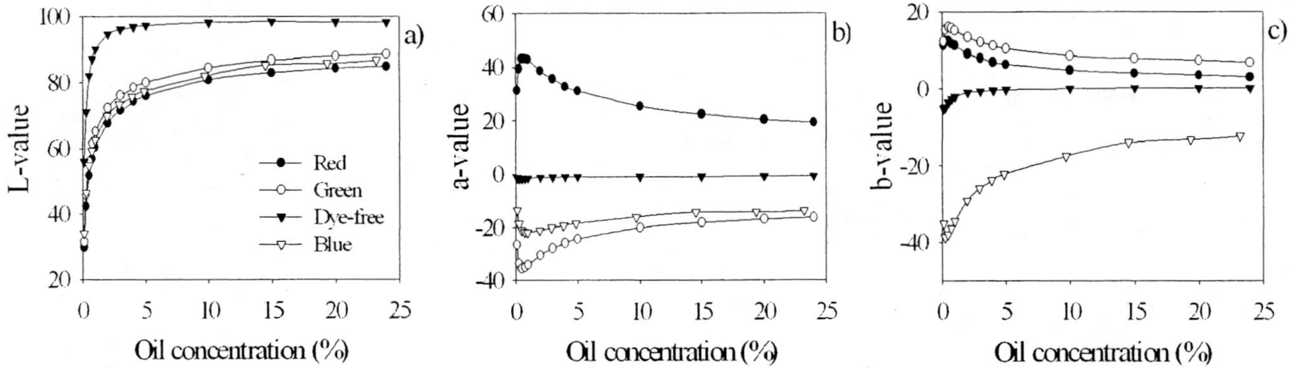

Figure 9. Dependence of L, a, b-value on droplet concentration for n-hexadecane with different dye types (red, green, blue, and dye-free), but same droplet diameter (0.3 μm) (Reprinted from 36, with permission from Elsevier).

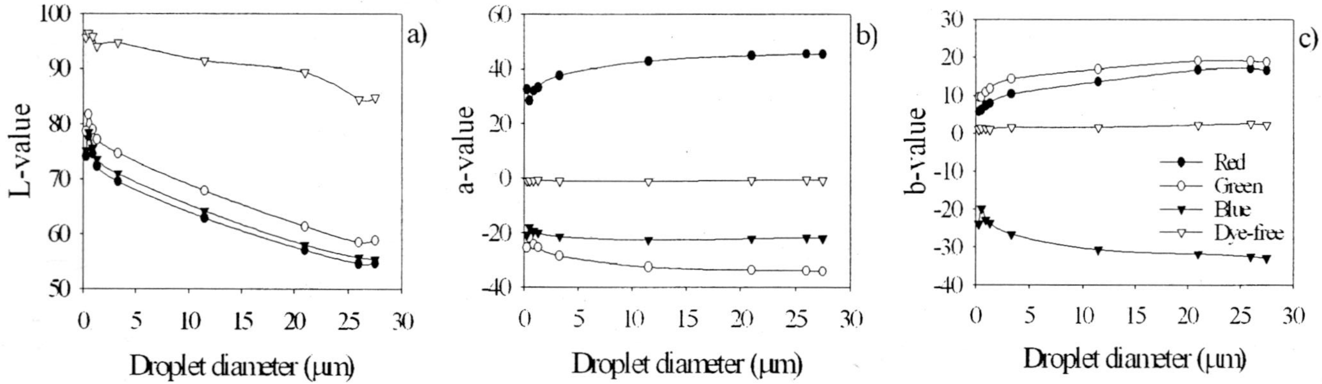

Figure 10. Dependence of L, a, b-value on droplet size for n-hexadecane with different dye types (red, green, blue, and dye-free), but same droplet concentration (9.6 wt%) (Reprinted from 36, with permission from Elsevier).

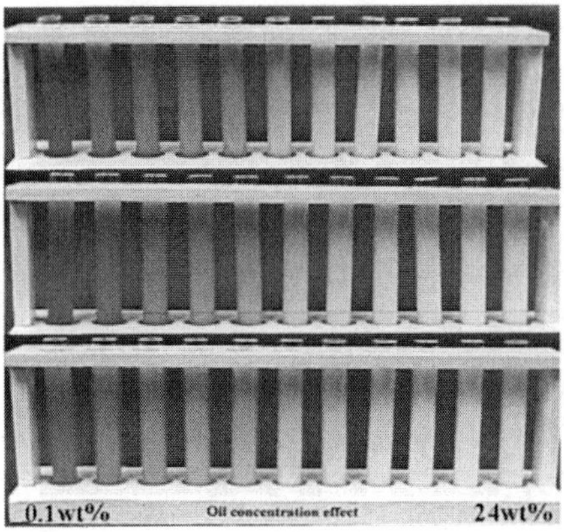

Figure 11. Three series of n-hexadecane oil-in-water emulsions with the same mean droplet diameter (0.3 μm), dye concentration (0.1 wt% for red, 0.079 wt% for green, and 0.071 wt% for blue), but different dye type (red, green, or blue) and droplet concentration (0 to 24 wt%). (See page 24 of color inserts.)

was still some quantitative disagreement, which may be because the assumptions upon which the theory is derived were dissimilar to the actual experiment. For example, ideally the Kubelka-Munk theory assumes that light reflection occurs at a boundary between pure water and an oil-in-water emulsion (Figure 13), but in fact, the emulsion is contained within a cuvette and so the measured reflectance is from an air-cuvette-emulsion arrangement; the emulsion in the theory was assumed to be monodisperse, while in reality it is polydisperse, etc.

Correcting Approaches

In order to overcome these discrepancies, three approaches have been proposed. The first approach is theoretically taking into account all the factors that influence the reflectance of emulsions using appropriate mathematical equations (*24*). The second approach is to calibrate the measurement cell empirically. This approach involved establishing an empirically determined correction-factor to relate the measured reflectance to the predicted reflectance. The correction factor was determined at each wavelength by dividing the measured spectral reflectance by the predicted one (*36*). The third approach is to redesign the measuring arrangement so that it is more similar to the ideal system assumed in the theoretical predictions (Figure 14).

This theoretical model based on light scattering theory needs to be further developed before it can be applied to predict the color of real food emulsions,

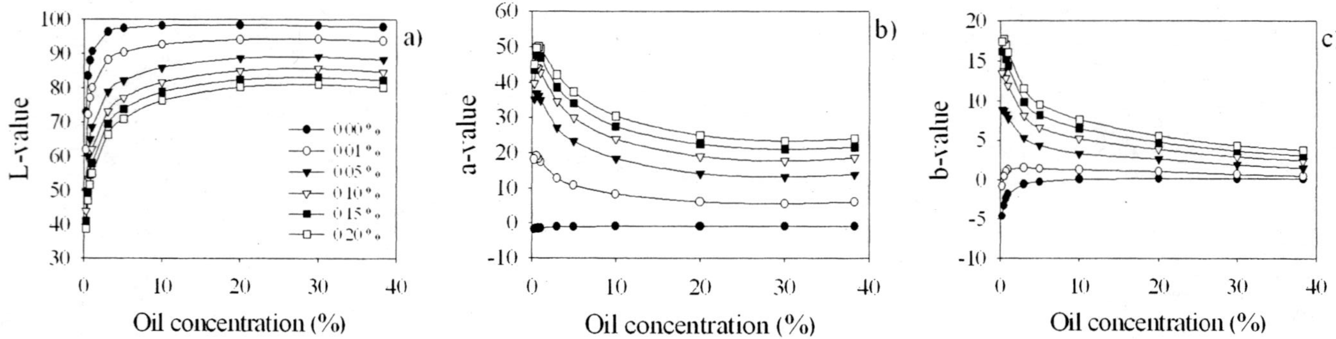

Figure 12. Dependence of L,a,b values on droplet and dye concentration for n-hexadecane oil-in-water emulsions containing a red food dye. Annotation shows different dye concentrations (0-0.20wt%).

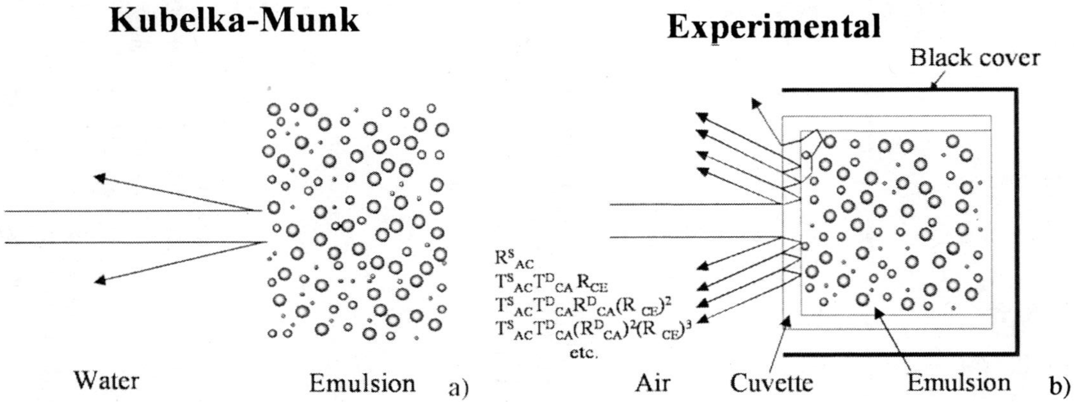

Figure 13. (a) The Kulbeka-Munk theory assumes that reflection occurs at a boundary between the pure continuous phase and an emulsion, which is assumed to have plane parallel faces, be infinitely large, but of finite thickness. (b) The experimentally measured reflection coefficient is from the air-cuvette-emulsion system, with a possible interference by reflected light from the edge. (Reproduced with permission from reference 38. Copyright 2001 Elsevier).

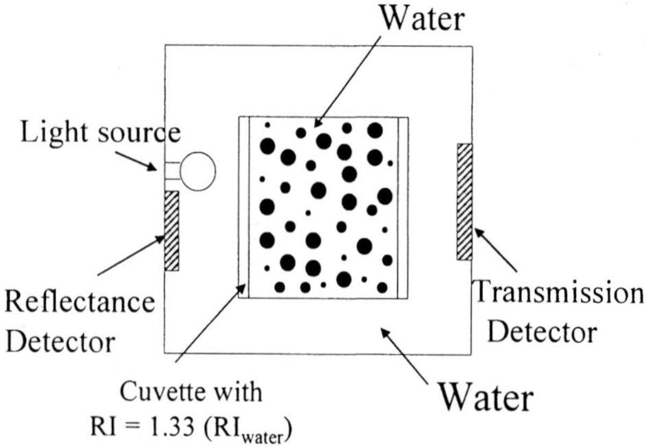

Figure 14. A proposed model of measuring chamber which is filled with water and a glass cuvette that has refractive index equals to 1.33 (refractive index of water).

which are extremely complex, both compositionally and structurally. However, understanding the influence of those factors that affect the optical properties of emulsions will be extremely valuable for food manufacturers to control emulsion appearance, which is considered to be a major factor for customer acceptance.

Conclusion

Droplet and dye characteristics had a pronounced influence on emulsion appearance. The "lightness" of an emulsion was correlated to the scattering efficiency of the droplets, which was related to their size, concentration and relative refractive index. The color of emulsions (*a*- and *b*-values) was mainly determined by the type (red, green, blue) and concentration of chromophores present. The lightness of emulsions increased steeply between 0 and 5 wt% oil and less slightly at higher concentrations (5-20 wt%). It also increased with decreasing droplet diameter (30-0.2 μm) and increasing refractive index difference between the oil and water phases. The influence of droplet characteristics on the lightness and color of emulsions containing different types of dye was fairly similar. The chromaticness of emulsions became more intense as dye concentration increased. Droplet aggregation did have an impact on emulsion appearance, but the effect was relatively small compared to the effect of an individual droplet.

Generally, there was good qualitative agreement between measured values and those predicted by light scattering theory, but there was still some quantitative disagreement due to the difference between theoretical assumptions and real experiments. A number of correcting approaches have been proposed to overcome these discrepancies. The theoretical model therefore needs to be further developed before it can be applied to predict the color of real food emulsions. Nevertheless, the current light scattering theory can still be used to predict the factors that influence emulsion appearance. This research has advanced the systematic understanding of the influence of emulsion composition and structure on food appearance. It will lead to the more rational design of food products of enhanced quality.

References

1. Clydesdale, F.M. *Crit. Rev. Food Sci. Nutr.* **1978**, *18*, 243-301.
2. Clydesdale, F.M. In *Developments in Food Colours-2*; Walford, J., Ed.; Elsevier Applied Science Publishers: London, 1984, pp. 75-112.
3. Clydesdale, F.M. *Crit. Rev. Food Sci. Nutr.* **1993**, *33*, 83-101.
4. Blenford, D. *Food Ingred. Anal. Int.* **1995**, *17*, 10-15.
5. Hutchings, J.B. *Food Colour and Appearance*, 2nd ed.; Blackie A. & P.: London, 1999.
6. Dickinson, E.; Stainsby, G. *Colloids in Foods*, Applied Science Publishers, London., 1982.
7. Dickinson, E. *Introduction to Food Colloids,* Oxford University Press: Oxford, 1992.
8. Friberg, S.E.; Larsson, K.; Sjoblom, J. *Food Emulsions*, 4th ed.; Marcel Dekker: New York, 2004.
9. McClements, D.J. *Gums and Stabilizers in the Food Industry*; Springer Verlag: New York, 2004; Vol. 12.
10. Dickinson, E.; McClements, D.J. *Advances in Food Colloids*, Chapman & Hall: London, 1995; pp 1-23.
11. Akoh, C.C. *Food Technol.*, **1998**, *52*, 47-53.
12. Walstra, P. In *Encyclopedia of Emulsion Technology;* Becher, P. Ed.; Marcel Dekker: New York, 1996; Vol. 4.
13. Walstra, P. In *Food Chemistry*, 3rd ed., Fennema, O.R., Ed.; Marcel Dekker: New York, 1996; pp 96-101.
14. Francis, F.J.; Clydesdale, F.M. *Food Colorimetry: Theory and Applications*; AVI Publishing: Westport, CT, 1975; pp 1-14.
15. Hunter, R.S. In *Objective Methods for Food Evaluation;* Proceeding of a Symposium. National Academy of Science: Washington, D.C., 1976.
16. Farinato, R.S.; Rowell, R.L. In *Encyclopedia of Emulsion Technology*; Becher, P., Marcel Dekker: New York, 1983, Vol.1.

17. Redlinger, P.A. In *Encyclopedia of Food Science Food Technology and Nutrition*; Macrae, R.; Robinson, R.K.; Sadler, M.J., Eds.; Academic Press: New York, 1993.
18. Francis, F.J. In *Engineering Properties of Foods*, 2nd ed.; Rao, M.A.; Rizvi, S.S.H., Eds.; Marcel Dekker: New York, 1995.
19. McClements, D.J. *Food Emulsions: Principles, Practice, and Techniques;* 2nd ed.; CRC Press: New York, 2005.
20. Hernandez, E.; Baker, R.A. *J. Food Sci.* **1991**, *56*, 1024.
21. Hernandez, E.; Baker, R.A.; Crandall, P.G. *J. Food Sci.* **1991**, *56*, 747.
22. Dickinson, E. *Food Chem.* **1994**, *51*, 343.
23. Taisne, L.; Walstra, P.; Cabane, B. *J. Colloid Interface Sci.* **1996**, *184*, 378-383.
24. Chantrapornchai, W.; Clydesdale, F.M.; McClements, D.J. *J. Agric. Food Chem.* **1998**, *46*, 2914-2920.
25. ven de Hulst, H.C. *Light Scattering by Small Particles;* John Wiley and Sons: New York, 1957.
26. Kerker, M. *The Scattering of Light*; Academic Press, New York, 1969.
27. Bohren, C.F.; Huffman, D.R. *Absorption and Scattering of Light by Small Particles*; Wiley: New York, 1983.
28. Bailey, A.E.; Cannell, D.S. *Phys. Rev. E: Stat., Nonlinear, Soft Matter Phys.* **1994**, *50*, 4853-4864.
29. Kortum, G. *Reflectance Spectroscopy: Principles, Methods, Applications;* Springer-Verlag: New York, 1969.
30. McClements, D.J.; Chantrapornchai, W.; Clydesdale, F.M. *J. Food Sci.* **1998**, *63*, 935-939.
31. McClements, D.J. *Adv. Colloid Interface Sci.* **2002**, *97*, 63-89.
32. Michel, B. *MieCalc -Freely Configurable Program for Light Scattering Calculations (Mie theory)*, Germany, Michel, B. Available from: http://www.unternehmen.com/Bernhard-Michel/Java/ MieSpectrumFrame.html. (Accessed August 20, 2000).
33. Mudget, P.S.; Richards, L.W. *Appl. Opt.* **1971**, 10, 1485-1502.
34. Hunter, R.S.; Harold, R.W. *The Measurement of Appearance*, 2nd ed.; Wiley-Interscience: New York, 1987; pp 148-161.
35. Chantrapornchai, W.; Clydesdale, F.M.; McClements, D.J. *J. Colloids Int. Sci.* **1999**, *218*, 324-330.
36. Chantrapornchai, W.; Clydesdale, F.M.; McClements, D.J. *Colloids Surf., A.* **1999**, *155*, 373 - 382.
37. Chantrapornchai, W.; Clydesdale, F.M.; McClements, D.J. *J. Food Sci.* **2001**, *66*, 464-469.
38. Chantrapornchai, W.; Clydesdale, F.M.; McClements, D.J. *Food Res. Int.*. **2001**, *34*, 827-835.
39. Phillips, L.G.; Barbona, D.M. *J. Dairy Sci.* **1997**, *80*, 2726-2731.

40. Pszczola, D.E. *Food Technol.* **1997**, *51*, 39-44.
41. McClements, D.J.; Demetriades, K. *Crit. Rev. Food Sci. Nutr.* **1998**, *38*, 511-536.
42. Chantrapornchai, W.; Clydesdale, F.M.; McClements, D.J. *Colloids Surf., A.* **2000**, *166*, 123-131.

Chapter 27

Polyphenols, Polyphenol Oxidase, and Discoloration of Barley-Based Food Products

B.-K. Baik[1], S. E. Ullrich[1], and Z. Quinde-Axtell[2]

[1]Department of Crop and Soil Sciences, Washington State University, Pullman, WA 99164–6420
[2]Department of Food Science and Human Nutrition, Washington State University, Pullman, WA 99164–6376

The health benefits of eating barley have renewed our interest in its food uses. Dark-gray color appearing in barley foods, however, limits its prevalent use by food processors and consumers. We observed a large variation in the discoloration potential among barley genotypes, as determined by the brightness of flour dough and gel, polyphenol content and polyphenol oxidase (PPO) activity. Genotypic influence was greater than environmental factors on polyphenol content, PPO activity and discoloration potential of barley. Removal of the outer layer of barley grain, heat treatment, exclusion of oxygen and use of ascorbic acid effectively retarded the discoloration in flour dough and gel. Barley grain contained phenolic acids, catechin, and six dimeric and trimeric proanthocyanidins (PAs). Significant increase in discoloration of dough by addition of PAs and considerable reduction in discoloration of dough by heat treatment of flour further indicate that both polyphenols and PPO are responsible for the discoloration of dough. Despite its low concentration, monomeric PA fraction composed mainly of catechin was most capable of causing the discoloration of barley flour dough.

Barley ranks fourth in total world production among cereal grains. However, human consumption of barley in the form of food products is extremely limited in many western countries, including the U.S.A. About 51% of barley we produce in the U.S.A. is used for feed, 42% for malt and less than 2% for food (*1*). The limited food uses of barley can be attributed to low consumer acceptability for barley-based food products due to cultural eating practices and undesirable color and taste of products; many consumers are not accustomed to eating barley, and so they do not consider it in terms of food products. The undesirable processing characteristics of barley as compared to wheat or rice might further contribute negatively to the food use of barley. Also, as economic conditions improve, consumers prefer to eat wheat or rice-based products rather than barley (*2*). Consequently, most consumers today are not familiar with the flavor and taste of barley-based food products and are hesitant to incorporate barley into their daily diets.

On the other hand, there has been growing interest in eating barley, mainly due to the awareness of the health benefits of barley consumption. It is widely known that barley is a rich source of the soluble dietary fiber, β-glucan, which has been considered to have hypocholesterolemic and hypoglycemic properties. Based on the clinical evidence of β-glucan for lowering blood cholesterol levels, as well as other reports on the nutritional benefits of barley grain, the FDA recently approved the health claim for barley β-glucan, which allows food processors to label food products containing more than 0.75 g of β-glucan per serving as being heart healthy. Easy availability and inexpensive price could be other advantages of using barley in food products.

For food uses, barley grain is first pearled to remove the hull and bran layers. Pearled barley grain is used as a rice extender, added in barley soups and prepared into a variety of dishes. Pearled barley grain can be further ground into flour and easily added to numerous wheat-based bakery and non-bakery products, including bread (*3*), muffins (*4*), biscuits (*5*), extruded snacks (*6*), noodles and pasta (*7*) without significantly affecting processing and product quality except for the gray color development. The undesirable dark discoloration reported in these products could result from enzymatic or non-enzymatic reactions. Barley grains contain an unusually high amount of phenolic compounds (0.2 – 0.4%) distributed mainly in the outer layers of the grain (hull, seed coat and aleurone layer) (*8, 9*). Polyphenol oxidase is also present in barley and can be responsible for the oxidation of phenolic compounds, inducing dark-gray color development in barley food products.

We determined the relationships between the chemical constituents of grain and the discoloration potential of barley in food products, analyzed the significance of both genetic and environmental factors on the discoloration potential of barley, explored various processing methods for the retardation of discoloration, and identified the composition of barley phenolic compounds and their role in the discoloration potential of barley.

Chemical Composition and Discoloration of Barley Foods

Chemical Analyses and Determination of Discoloration Potential

Various types of barley, including hulled or hulless, proanthocyanidin-containing (PAC) or proanthocyanidin-free (PAF), and waxy or regular endosperm barley, were evaluated for their chemical composition and their discoloration potential. Barley grains were pearled to remove 30% and 15% by weight in hulled and hulless barley, respectively.

Total polyphenol content and polyphenol oxidase (PPO) activity were determined using spectrophotometric methods (*10*). Polyphenols were extracted from barley flours with a solution of dimethylformamide, ammonium hydroxide, 4-aminoantipyrine and potassium ferrocyanide. The absorbance at 505 nm was measured and the polyphenol content was reported as gallic acid percentage. Polyphenol oxidase was extracted from barley flour using a phosphate buffer of pH 6.5, and the enzyme extract was added to a solution of L-dopa. The increase in absorbance over time was used to calculate enzyme activity. One unit of PPO activity was defined as the amount of enzyme giving a change in absorbance of 0.001/min, and was expressed in units/g.

Pearled grains were milled into flour, which was subsequently used to prepare paste (10%, w/v) and dough of 64% water absorption for the determination of discoloration potential. Brightness (L*) of cooked grains, gel and dough was measured using a Minolta colorimeter (Minolta Camera Co., Ltd., Osaka, Japan).

Variation in Discoloration Potential of Barley Genotypes

Polyphenol content, PPO activity and brightness (L*) of cooked grains, gel and dough as the measurement of discoloration potential are summarized in Table I. Differences in the brightness of cooked grain and gel were evident among all types of barley. Hulled PAF genotypes generally produced brighter (greater L*) cooked grains and gels than PAC and hulless genotypes. The brightness of flour dough gave the best differentiation of barley types in their discoloration potentials (Figure 1). Hulled barley genotypes produced brighter dough than hulless genotypes. In hulled barley, PAF genotypes yielded brighter dough than PAC genotypes. In hulless barley, brighter dough was produced from genotypes of regular endosperm starch than those of waxy endosperm starch.

There were large variations in both total polyphenol content and PPO activity among barley genotypes. Total polyphenol content was about 0.04% in hulled PAF genotypes, while it ranged from 0.13 to 0.22% in hulled PAC and hulless genotypes. PAF genotypes exhibited much higher PPO activity than other

Table I. Chemical Composition and Discoloration Potential of Abraded Barley

Class and Genotype	Composition		Brightness (L^*)		
	Total Polyphenol (%)	Polyphenol Oxidase (units/g)	Cooked Grain	Gel	Dough
HULLED					
Proanthocyanidin-containing					
Harrington	0.144	59.8	59.9	55.0	68.6
Farmington	0.131	96.2	59.1	53.7	66.5
Baronesse	0.175	57.3	60.6	56.0	67.6
Steptoe	0.220	60.6	59.6	55.1	67.9
Proanthocyanidin-free					
Radiant	0.046	121.0	62.2	59.3	75.6
98NZ015	0.053	163.7	62.5	56.8	75.4
Caminant	0.050	204.8	61.8	57.5	76.9
CA803803	0.054	192.7	60.3	57.7	74.3
HULLESS					
Regular					
Bear	0.177	82.9	59.1	55.0	62.9
CDC McGwire	0.184	55.4	57.2	54.6	62.6
Waxy					
CDC Candle	0.198	71.3	60.5	54.1	59.9
SH97110	0.183	68.2	58.7	53.0	59.8
LSD[a]	0.047	14.0	1.4	1.1	2.4

[a] Least significant difference ($P<0.05$). Differences between two means exceeding this value are significant.

SOURCE: Adapted with permission from reference 11. Copyright 2005 AACC International, Inc.

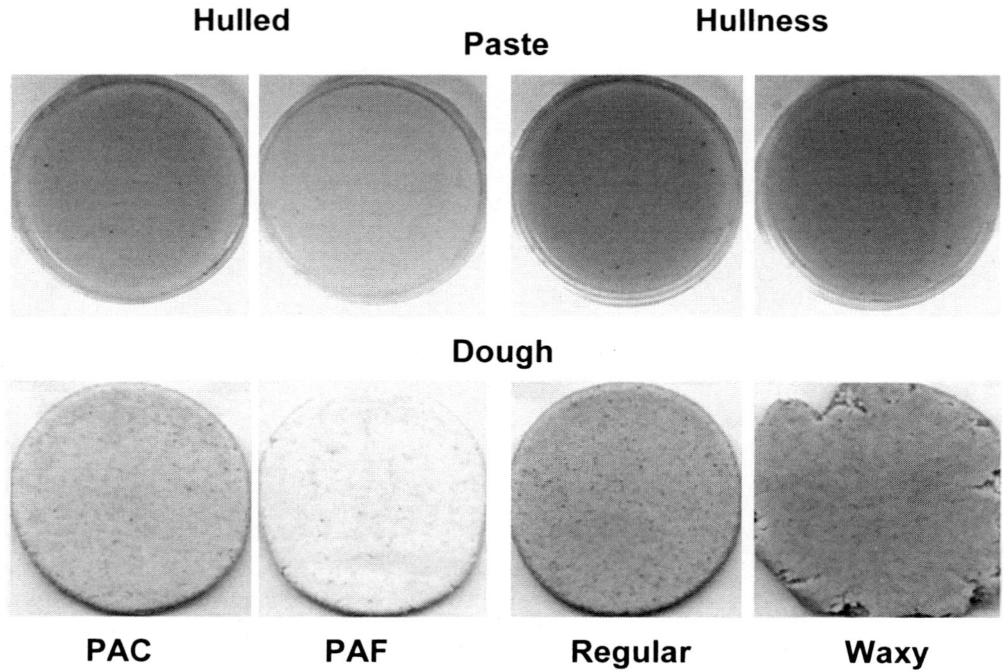

Figure 1. Color of doughs prepared from flours of cv. Radiant before or after heat treatment to inactivate polyphenol oxidase and with or without addition of polyphenol extract.
(See page 24 of color inserts.)

types of barley, despite having the lowest discoloration potentials as observed in brightness of cooked grain, gel and paste. Accordingly, it appears that low total polyphenol content is corresponding to the brightest appearances of cooked grain, paste and dough of PAF genotypes. Considering the lack of relationship between PPO activity and brightness of cooked grain, gel or dough, however, the implication of PPO on the discoloration potential of barley is not clear, despite its well known role in the oxidation of phenolic compounds to produce discoloration in many food products including fruits, vegetables, wheat noodles and flat bread.

Genotypic and Environmental Effects on Discoloration Potential of Barley

We determined the relative contribution of genotypic and environmental factors to the discoloration potential of barley. Twelve genotypes (G) were grown in five environments (location x year) (E). Grain was pearled at 30% and 15% by weight in hulled and hulless barley, respectively. Total polyphenol content, PPO activity and brightness of dough were determined as described by Quinde et al (*10*). The analysis of variance indicated that genotype, environment and their interactions were all significant contributors to the variations in brightness of dough, total polyphenol content and PPO activity of barley (Table II). The effects of genotype on dough brightness, total polyphenol content and PPO activity appeared to be much greater than those of environment.

The analyses for simple effects of genotype and environment for total polyphenol content showed that genotypes generally demonstrated orderly changes across environments (Figure 2A). Hulled PAF genotypes were always lowest in total polyphenol content in all five environments. We also observed that barley grown in Ritzville, where a much smaller amount of water was available for barley growth compared to other locations, tended to have higher total polyphenol content of grain than barley grown in other locations. Similar to total polyphenol content, PPO activity of barley genotypes also exhibited orderly interactions (Figure 2B). Regardless of growing environment, hulled PAF genotypes were higher in PPO activity of grain than other types of barley. Contrary to polyphenol content, barley grown in Ritzville generally showed lower PPO activities than those grown in other locations.

Figure 2C shows the simple effects of the G x E interaction for flour dough brightness. Differences among genotypes were evident, and the ranking of genotypes in dough brightness were little affected by growing environment, indicating the dominant effect of genotype on the discoloration potential of barley grain. Barley grown in Ritzville produced darker color of flour dough in most genotypes, compared to barley grown in other locations.

Table II. ANOVA Mean Squares for Traits of Twelve Barley Genotypes Grown in Five Environments

Source of Variation	*Brightness of Dough*	*Total Polyphenol Content*	*Polyphenol Oxidase Activity*
Genotype (G)	336.9***	0.049***	28592***
Environment (E)	69.6***	0.016***	2823***
G x E	2.5***	0.001***	339***

SOURCE: Reproduced with permission from reference 11. Copyright 2005 AACC International, Inc.

Relationship Between Total Polyphenol Content, PPO Activity and Brightness of Flour Gel and Dough

Figure 3A shows the relationship between polyphenol content and the brightness of flour dough. Polyphenol content had a negative relationship with the discoloration of flour dough ($P<0.001$). The significant correlation between polyphenol content and brightness of dough was also observed within PAF or PAC genotypes (*11*). The positive relationship, though weak, observed between PPO activity and brightness of dough (Figure 3B) was due to PAF genotypes, which produced brighter dough than other genotypes, despite their high PPO activities. No relationships were observed between PPO activity and brightness of dough within PAF or PAC genotypes.

Effect of Polyphenols and PPO on Barley Dough Color

To further verify the implication of polyphenols and PPO in the development of dark-gray color in barley-based food products, we determined the effects of heat denaturation of PPO and of polyphenol extract addition on the brightness of barley flour dough. Flour of cv. Radiant, a PAF genotype, was autoclaved at 121°C for 15 min to inactivate the PPO. Polyphenol extract was prepared from the barley flour of a PAC genotype using an acetone and water (70:30) blend, extracted with hexane to remove lipids and purified using polyamide (*12*). Dough was prepared from both heated and unheated barley flours with or without addition of the polyphenol extract and stored for 24 hr (Figure 4). Dark color development was much less evident in the dough of heat treated flour than in that of the unheated flour. With addition of polyphenol extract, discoloration of dough became intensified in the unheated flour, while few changes in brightness of dough were observed in the heated flour. These results indicate that PPO catalyzes the oxidation of polyphenols in barley flour dough, resulting in the development of dark color. Heat treatment of barley flour inactivates PPO and consequently minimizes the discoloration of dough.

Retardation of Discoloration in Barley-Based Food Products

Abrasion of Barley Grain

Phenolic compounds, PPO and other oxidizing enzymes are mainly present in the outer layers of the grain, i.e. hull, pericarp, seed coat and aleurone layer (*8*). Accordingly, the removal of the outer layers of barley grain by abrasion

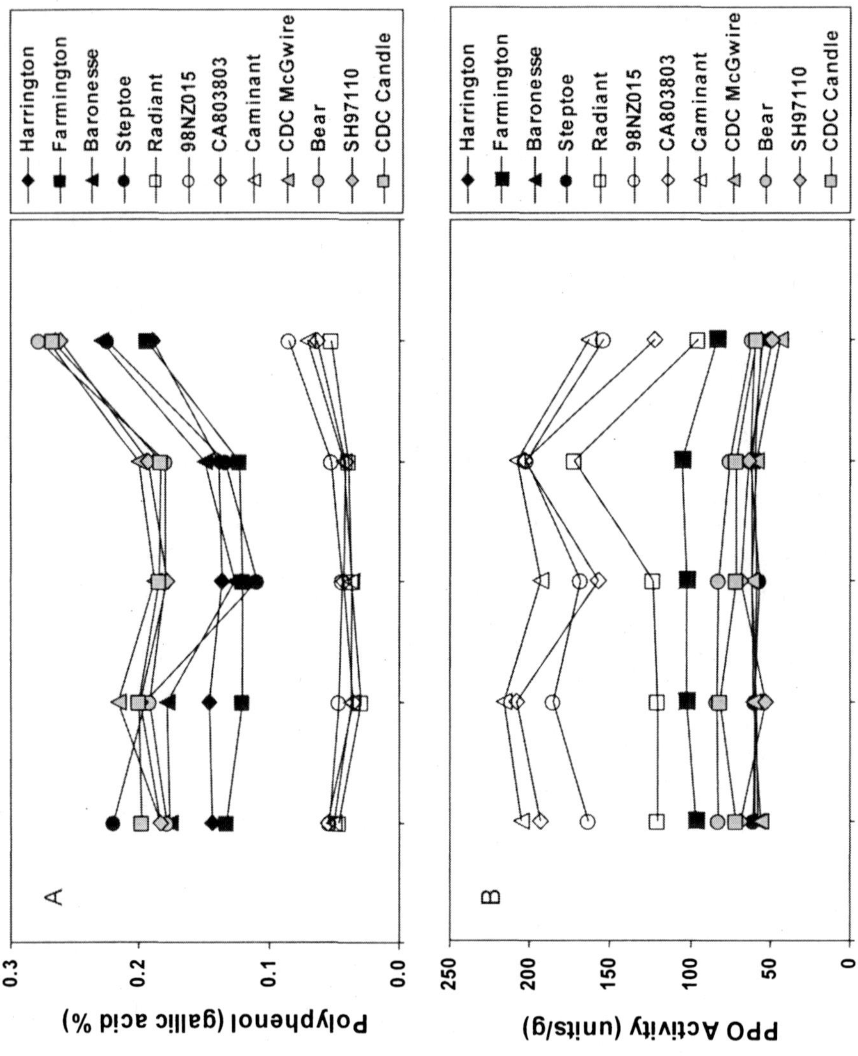
A
Polyphenol (gallic acid %)
0.3
0.2
0.1
0.0
Harrington
Farmington
Baronesse
Steptoe
Radiant
98NZ015
CA803803
Caminant
CDC McGwire
Bear
SH97110
CDC Candle
B
PPO Activity (units/g)
250
200
150
100
50
0
Harrington
Farmington
Baronesse
Steptoe
Radiant
98NZ015
CA803803
Caminant
CDC McGwire
Bear
SH97110
CDC Candle

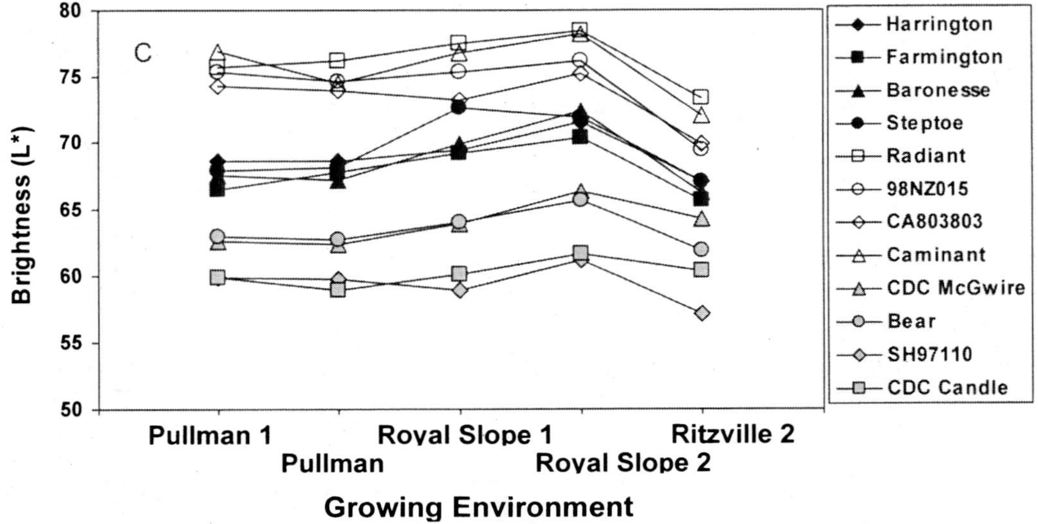

Figure 2. Total polyphenol content (A) and PPO activity (B) of grain, and brightness of flour dough (C) in twelve barley genotypes grown in five different environments (location x crop year). Harrington, Farmington, Baronesse and Steptoe are hulled proanthocyanidin-containing genotypes (dark symbols); Radiant, 98NZ015, CA803803 and Caminant are hulled proanthocyanidin-free genotypes (open symbols); CDC McGwire and Bear are hulless genotypes of regular endosperm (gray symbols); SH97110 and CDC Candle are hulless genotypes of waxy endosperm (gray symbols). Pullman represented a high rainfall area, while Ritzville was a low rainfall area. Barley was grown under irrigation in Royal Slope.

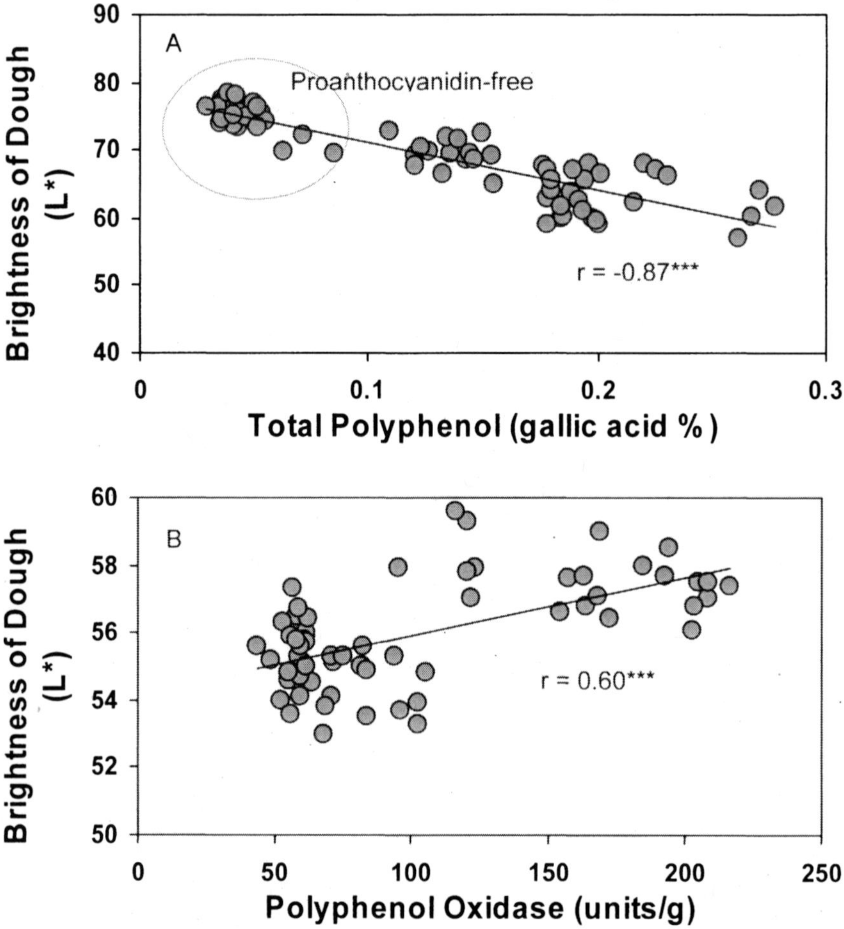

Figure 3. Correlations between brightness of barley flour dough, and total polyphenol content (A) and polyphenol oxidase activity (B) of barley grain. (Adapted with permission from reference 11. Copyright 2005 AACC International, Inc.)

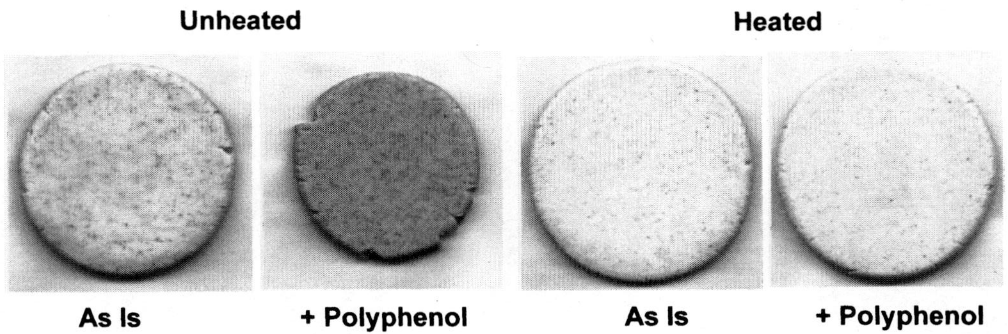

Figure 4. Color of doughs prepared from flours of cv. Radiant before or after heat treatment to inactivate polyphenol oxidase and with or without addition of polyphenol extract. (See page 25 of color inserts.)

could effectively decrease the polyphenol content and PPO activity, thus lowering the discoloration potential. During abrasion, however, the outer layers in barley grains are not concentrically removed, and parts of the hull and aleurone layer remain in the crease (*13, 14*). The shape and size of the kernel and thickness of the outer layers are characteristics that influence the extent of abrasion (*14, 15*). Figures 5A, B and C depict the changes in total polyphenol content and PPO activity of abraded grain, and brightness of dough by the removal of the outer layers by abrasion. Barley grain was abraded by 20, 30 and 40% by weight in hulled barley and by 5, 15 and 25% in hulless barley. Reduction in both polyphenol content and PPO activity of grain by abrasion was evident, regardless of type and variety of barley. As the abrasion rate increased, the brightness of dough prepared from the abraded grain consistently increased in all barley varieties (Fig 5C).

Physical Treatments

Since enzyme activity and chemical reactions are affected by temperature, and oxygen is needed for the oxidation of polyphenols (*16*), we attempted to control the oxidation of polyphenols and thus the discoloration of barley flour dough by heat treatment, low temperature storage and exclusion of oxygen. Heat treatment of barley flour or grain to inactivate PPO was effective for the reduction of discoloration potential (Figures 4 and 6A). Heat treatment, however, changes the physical and functional properties of barley flour, through starch gelatinization and protein denaturation (*17*).

Barley flour dough was prepared and stored for 24 hr at 4°C or 20°C in the atmosphere or nitrogen gas to exclude oxygen. Compared to the dough stored at 20°C in the atmosphere, dough stored at low temperature and/or in nitrogen gas exhibited a significant reduction in dark-gray color development (Figure 6B). Anaerobic conditions also reduced the discoloration of the flour dough (*21*). Low temperature reduces kinetic energy of the reactant molecules, resulting in a decrease in both mobility and effective collision of molecules including enzymes (*16*).

Chemical Agents

Discoloration of food products may be controlled by the use of reducing agents, such as ascorbic acid, and PPO inhibitors, such as 4-hexylresorcinol. Ascorbic acid acts as a free radical scavenger, alters the redox potential of the system and reduces undesirable oxidative products. 4-Hexylresorcinol has similar structure to phenolic substrates of PPO, and thus acts as a competitive inhibitor of PPO. 4-Hexylresorcinol is generally recognized as safe (GRAS) for use in the prevention of shrimp melanosis (*18*). 4-Hexylresorcinol is also known to inhibit mushroom PPO, but has not been shown to be an effective inhibitor of grape PPO (*19, 20*).

Both ascorbic acid at 1500 ppm and 4-hexylresorcinol at 50 ppm effectively retarded discoloration of barley flour dough (Figure 6C). Ascorbic acid controlled dark color development of barley flour dough better than 4-hexylresorcinol. Incorporation of sulfite, ethylene diamine tetra acetic acid and benzoyl peroxide were not effective for the retardation of barley flour dough discoloration (*21*).

Phenolic Compounds and Discoloration Potential of Barley

Phenolic compounds, including phenolic acids, anthocyanins, anthocyanidins, flavonols and flavanols, are distributed in the hull, seed coat and aleurone layer of barley grain (*22*). Ferulic and *p*-coumaric acids are the major phenolic acids of the barley grain (*23, 24*). Anthocyanins are present in blue or purple barley grains (*25*). Anthocyanin pigments are located in the veins of the hull, pericarp and aleurone layer. Anthocyanidin content of blue aleurone barley ranges from 174 to 291 μg/g (*26*), while anthocyanidin content of white/yellow aleurone barley is 1/5 that of blue aleurone barley (*27*). Flavonol content of barley, quantified as rutin equivalent, ranges from 10.9 to 66.9 μg/g (*28*).

Catechins and proanthocyanidins (flavanols) are concentrated in the seed coat (*8, 29*). PAs in barley occur as dimeric, trimeric and polymeric moieties. The monomeric units are flavan-3-ol (+)-catechin and (+)-gallocatechin and their epimers and enantiomers. The most abundant PAs are dimers: prodelphinidin B3 and procyanidin B3. The main trimerics, linked together by –[4α-8]- bonds, are T1 (gc-gc-c), T2 (gc-c-c), T3 (c-gc-c) and T4 or procyanidin C2 (c-c-c) (*12, 30*).

Phenolic Acids in Barley

Phenolic acids of eleven genotypes of barley were extracted from whole and abraded barley grain according to the procedure of Yu et al (*31*). Grain was abraded by 30% by weight in hulled and by 15% in hulless barley. Separation and quantification of phenolic acids were carried out in an HPLC system using a Zorbax SB-C18 column. A gradient elution was used with a mobile phase consisting of water:acetonitrile:acetic acid (94.9:5:0.1, solvent A) and water:acetonitrile:acetic acid (79.9:20:0.1, solvent B) (*32*). Phenolic acids were detected at 280 nm, and standard solutions of phenolic acids were used for identification and quantification.

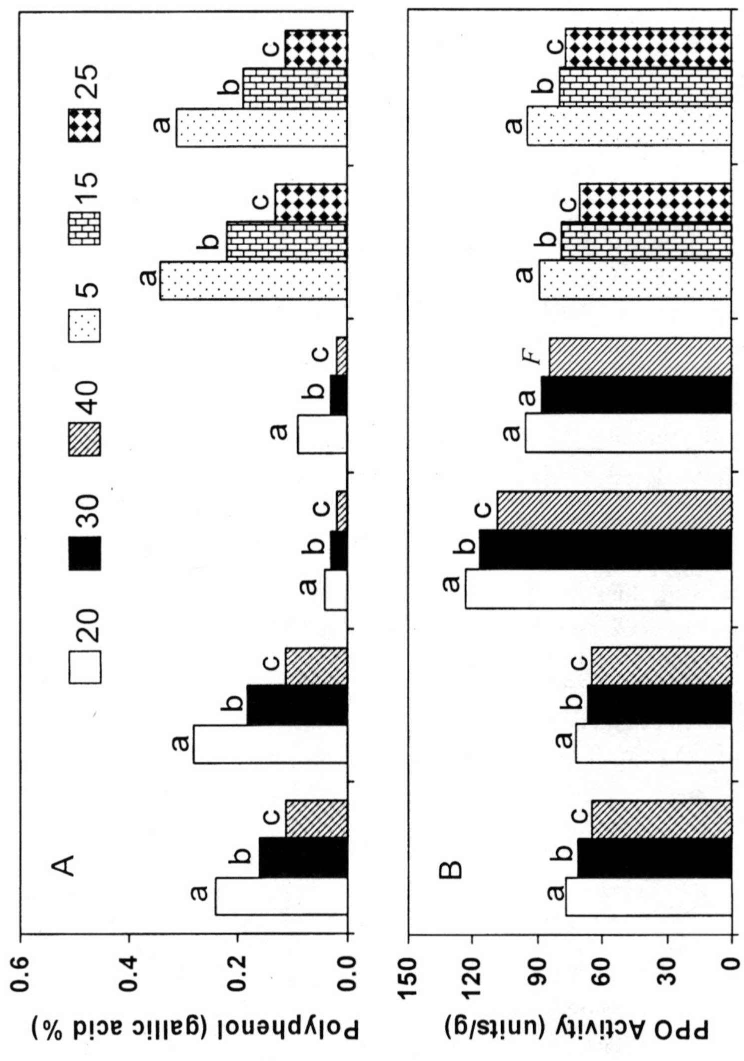

A
B
Polyphenol (gallic acid %)
PPO Activity (units/g)
20
30
40
5
15
25
0.6
0.4
0.2
0.0
150
120
90
60
30
0
a
b
c
F

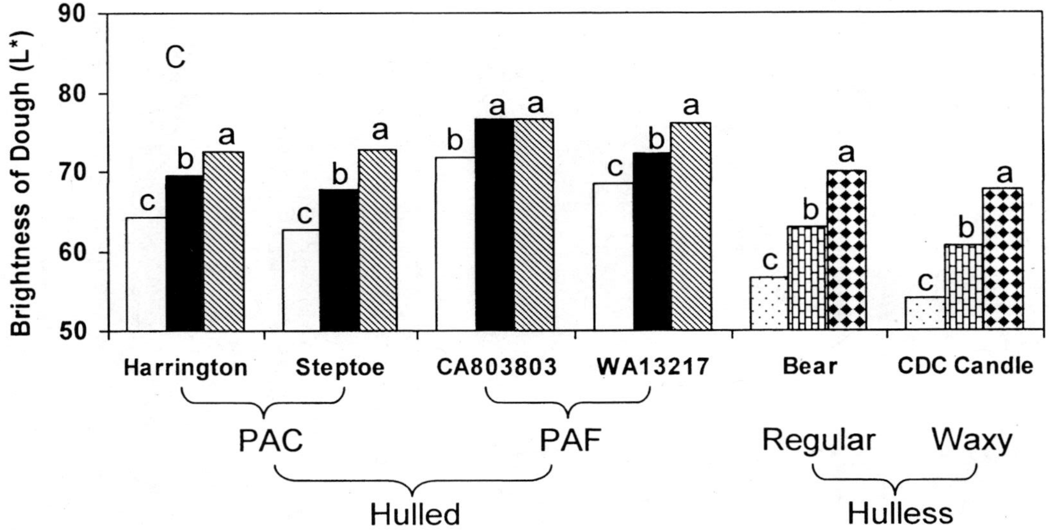

Figure 5. Effect of abrasion on total polyphenol content (A) and polyphenol oxidase (PPO) activity (B) of barley grain, and on brightness of flour dough (C). Barley grain was abraded by 20, 30 and 40% by weight in hulled barley and by 5, 15 and 25% in hulless barley. PAC: proanthocyanidin-containing; PAF: proanthocyanidin-free. Bars with the same letter within the same genotype are not significantly different ($P<0.05$). (Adapted with permission from reference 21. Copyright 2005 AACC International, Inc).

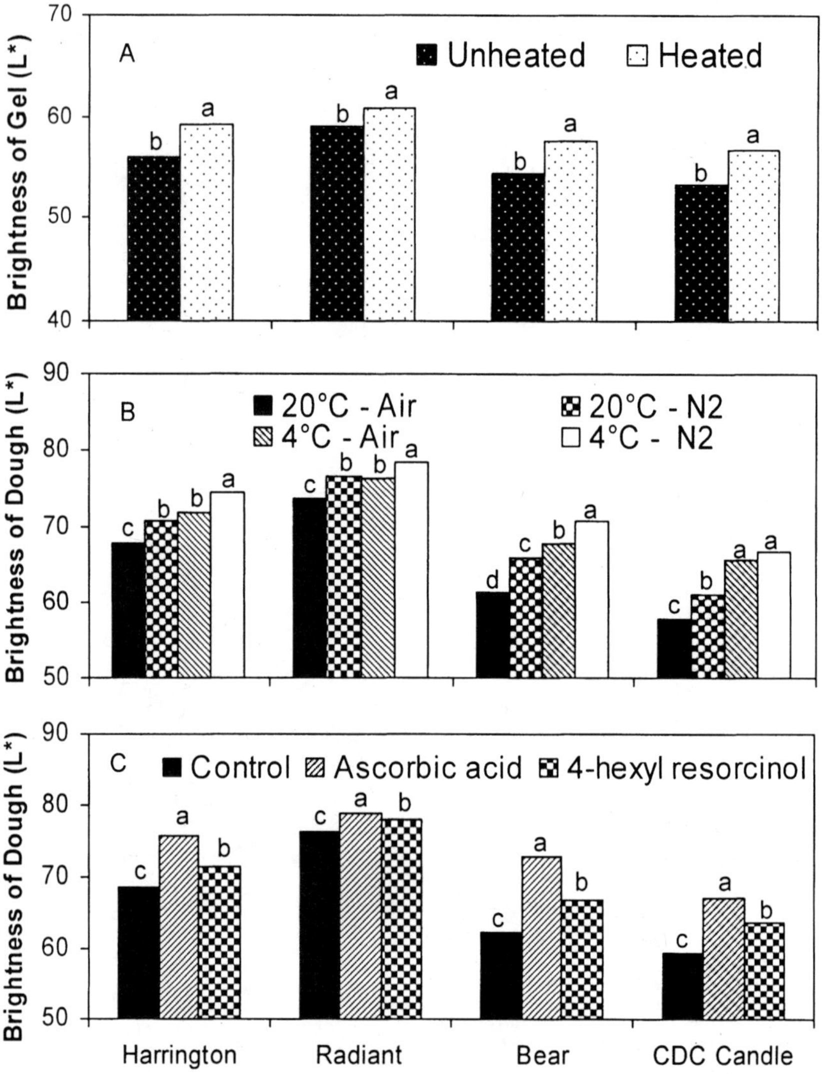

Figure 6. Effect of heat treatment (A), storage temperature and nitrogen gas (B), and incorporation of ascorbic acid and 4-hexyl resorcinol (C) on retardation of barley flour dough discoloration. Bars with the same letter within the same genotype are not significantly different (P<0.05). (Adapted with permission from reference 21. Copyright 2005 AACC International, Inc).

Whole barley grain contained ferulic acid of 301-567 µg/g and *p*-coumaric acid of 4-68 µg/g (Figure 7). Hulled barley genotypes contained more *p*-coumaric acid than hulless genotypes. Previous studies have indicated that concentration of ferulic acid was dominant in the aleurone layer, while *p*-coumaric acid prevailed in the hull (*33*). Caffeic acid content of whole barley grain ranged from 15-36 µg/g. Phenolic acid content of barley grain was reduced by approximately 50% with abrasion. Ferulic acid was still the most abundant phenolic acid of abraded barley grain. There was a relative small variation in phenolic acid content of both whole and abraded grain among barley genotypes. Phenolic acid content of PAF genotypes was comparable to that of PAC genotypes (*34*), indirectly indicating that phenolic acids may have little influence on the discoloration potential of barley. No relationship was observed between brightness of barley flour dough and content of individual and total phenolic acid.

Proanthocyanidins in Barley

PAs from whole and abraded barley were extracted and separated using an HPLC system, according to Friedrich et al (*12*) with minor modifications (*32*). Identification of PAs was carried out using an LC-MS system equipped with an electrospray ionization source, as described in detail by Quinde-Axtell et al (*32*). The HPLC chromatogram of cv. Harrington, a PAC genotype exhibited six main PAs (prodelphinidin B3, procyanidin B3, trimer 1, trimer 2, trimer 3 and trimer 4 or procyanidin C2) (Figure 8), which agrees with previous reports (*28, 35, 36*), in both whole and abraded grain, with much lower concentration in abraded grain. Both prodelphinidin B3 (162 - 288 µg CE/g) and procyanidin B3 (160 - 255 µg CE/g) were the major PAs in whole barley grain (Figure 9). The content of four trimeric PAs of whole barley grain ranged from 288 to 414 µg CE/g. Catechin content (45 - 86 µg/g) was generally higher in hulless than in hulled PAC genotypes. Even though the total PA content of barley grain was reduced by more than 50% by the removal of the outer layers of grain, the relative proportion of catechin increased by abrasion.

Fractionation of Proanthocyanidins

The PA extract of the outer layers of hulless cv. Bear was fractionated according to the procedure of Friedrich et al (*12*). Figure 10 depicts the HPLC chromatograms of three PA fractions: monomeric, dimeric and trimeric fractions. Phenolic compounds in each fraction were identified using an LC-MS system equipped with an electrospray ionization source (*12*, *32*). The monomeric fraction contained mainly catechin and catechin glucoside, along with minor

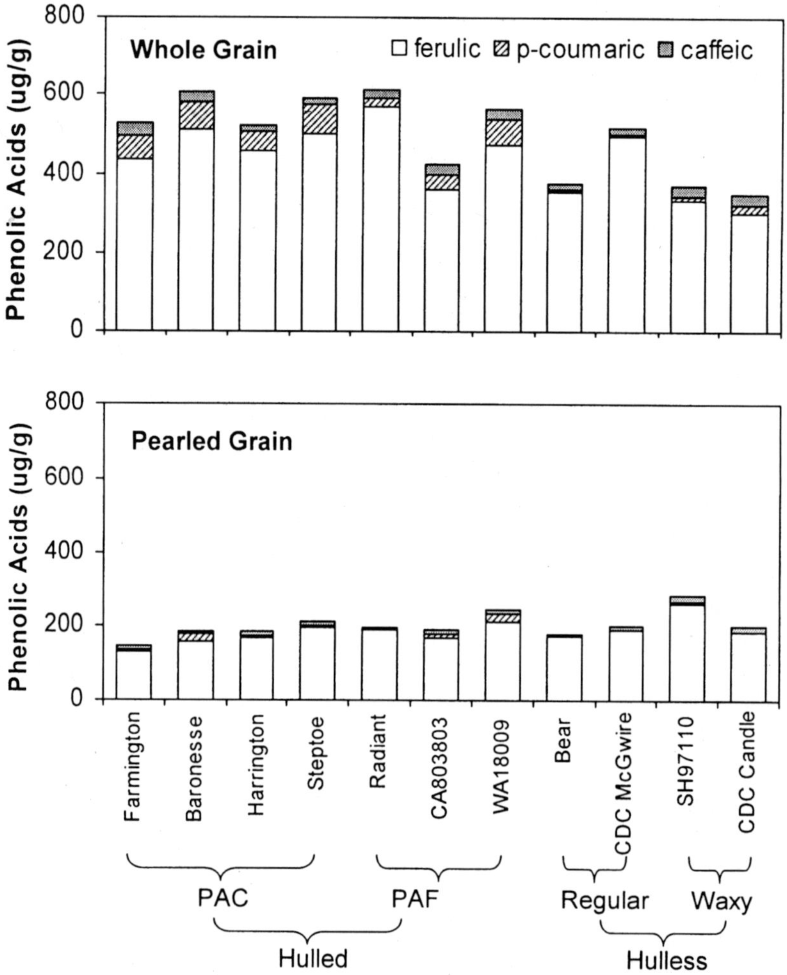

Figure 7. Phenolic acid composition of whole and abrade grains of barley genotypes PAC: proanthocyanidin-containing; PAF: proanthocyanidin-free. (Adapted from reference 32).

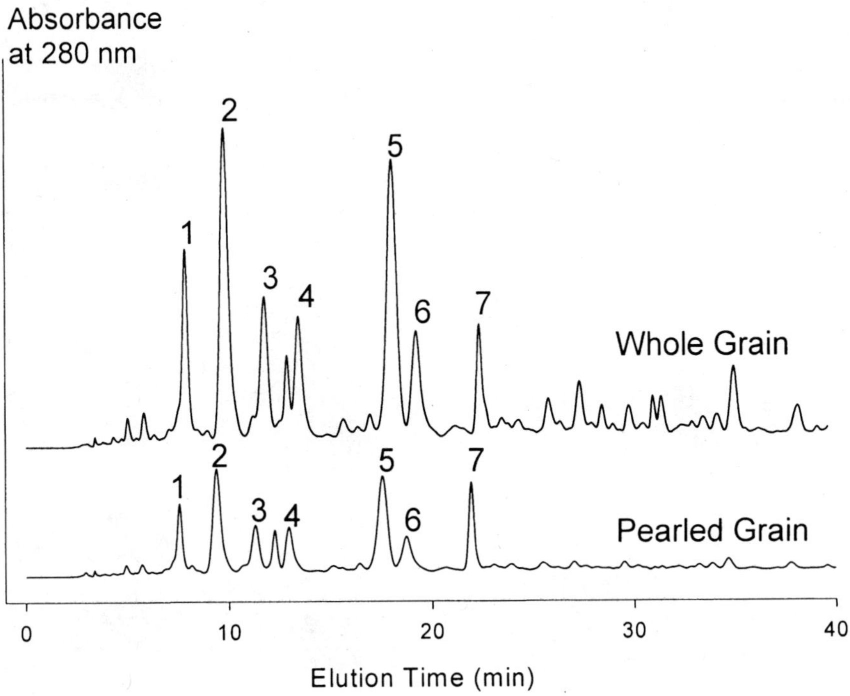

Figure 8. UV-chromatogram of major proanthocyanidins in whole and pearled barley cv. Harrington. 1-Trimer 1; 2-prodelphinidin B3; 3-Trimer 2; 4-Trimer 3; 5-procyanidin B3; 6-procyanidin C2 (Trimer 4); 7-catechin. (Reproduced from reference 32).

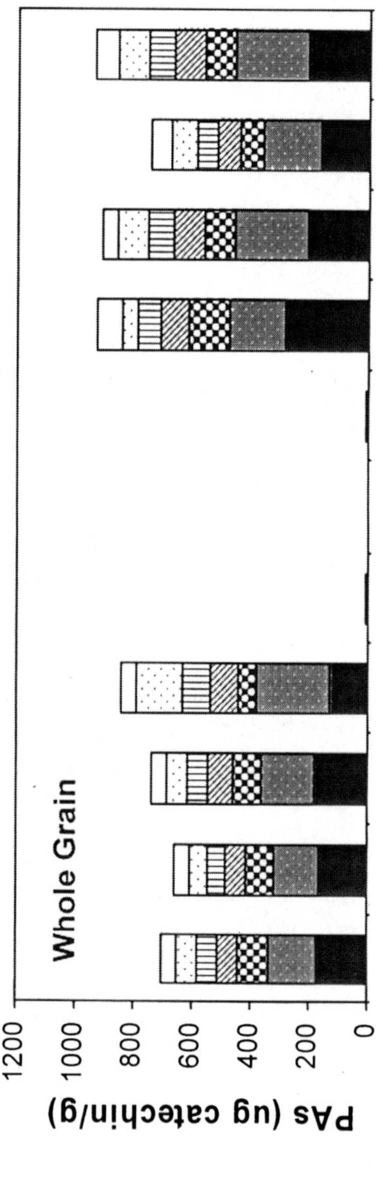
Whole Grain
PAs (ug catechin/g)
1200
1000
800
600
400
200
0

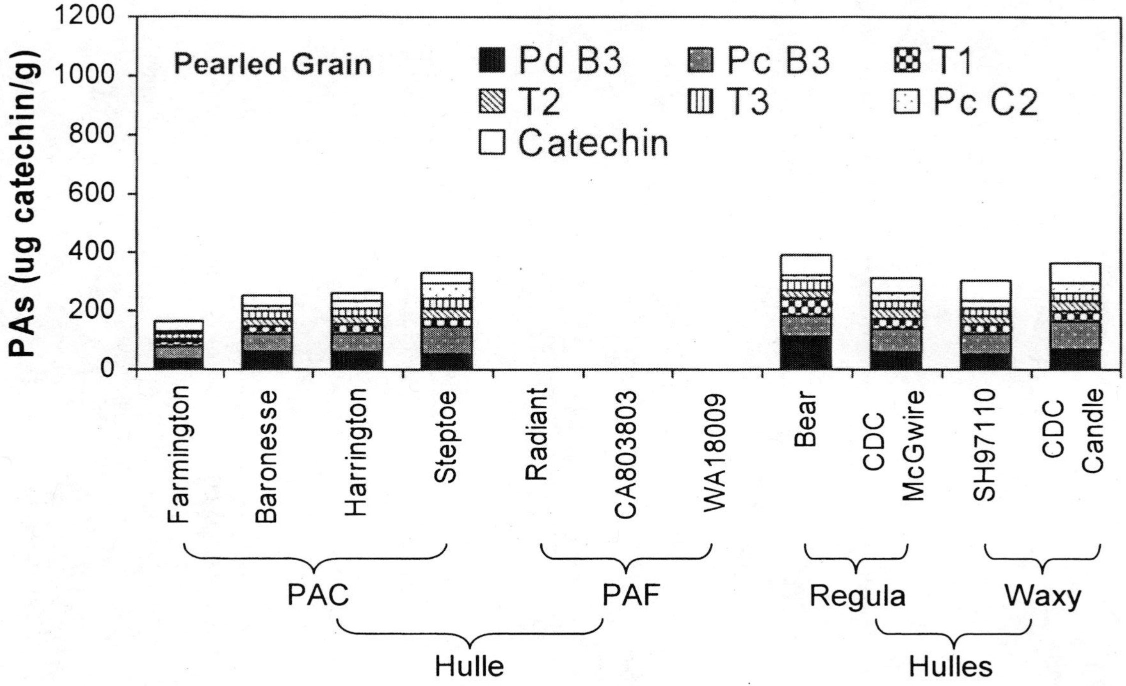

Figure 9. Proanthocyanidins (PAs) in whole and abraded barley grains. Pd: proanthocyanidin; Pc: Procyanidin; T: trimer. PAC: proanthocyanidin-containing; PAF: proanthocyanidin-free. (Adapted from reference 32).

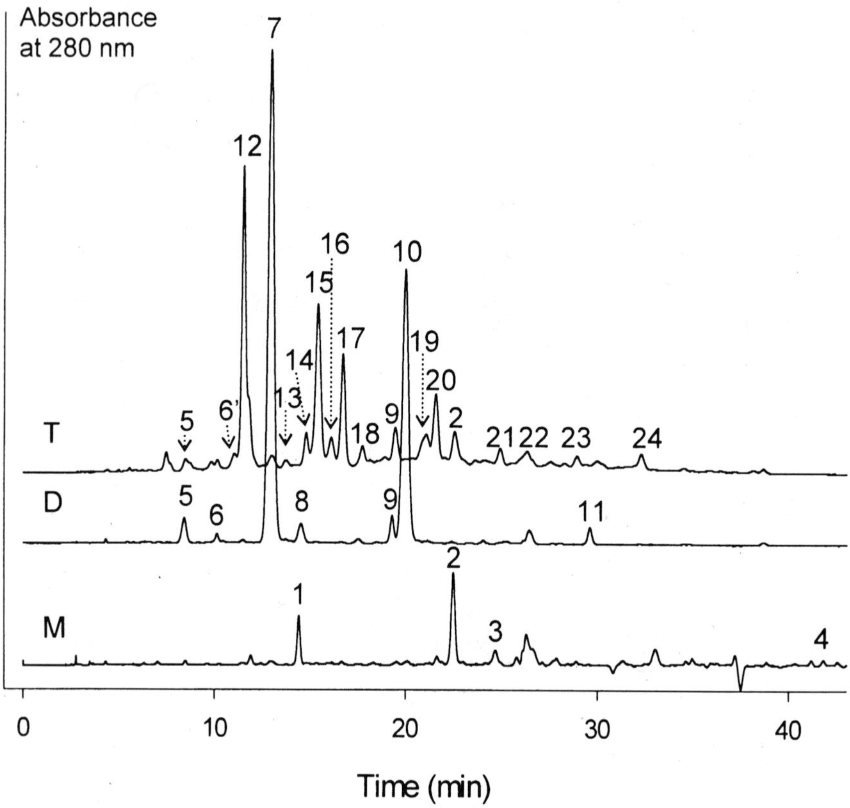

Figure 10. UV-chromatogram of the proanthocyanidins fractions extracted from outer layers of grain of hulless cv. Bear. Corresponding PA peak assignments, mass-to-charge ratio (m/z) in parenthesis and presumptive monomeric units in brackets. M-Monomeric PA Fraction: 1, catechin glucoside (451); 2, catechin; 3, vanillic acid; 4, ferulic acid. D-Dimeric PA Fraction: 5, (609)[gc-gc]; 6, (593); 7, prodelphinidin B3; 8, (593) [c-gc]; 9, (593) [gc-c]; 10, procyanidin B3; 11, (577) [c-c]. T-Trimeric PA Fraction: 5; 6, (593); 12, trimer 1; 7; 13, (609)+(913); 14, (593)+(1185); 15, trimer 2; 16, (897); 17, trimer 3; 18, (1185); 9; 19, (881); 20, procyanidin C2; 2; 21, (881); 22, (593)+(865)+ (1185); 23, (865); 24, (881). (Reproduced from reference 32).

constituents of vanillic and ferulic acids. The dimeric fraction was composed of mainly prodelphinidin B3 and procyanidin B3 and minor constituents of dimeric PAs with molecular ions at m/z = 609 (peak 5), m/z = 593 (peak 6, 8 and 9) and m/z=577 (peak 11). The trimeric fraction contained mainly trimeric PAs: trimer 1 (peak 12), trimer 2 (peak 15), trimer 3 (peak 17) and procyanidin C2 (peak 20). Other trimeric PAs with ion masses at m/z= 897 (peak 16), m/z= 881 (peak 21) and m/z= 865 (peak 23) were also detected. The trimeric fraction also contained dimeric PAs such as prodelphinidin B3, catechin, dimeric PAs with ion masses at m/z=609 (peak 5) and 593 (peak 6), a tetrameric PA with an ion mass at m/z=1185 (peak 18) and small amounts of poorly separated dimeric, trimeric and tetrameric PAs.

Relationship Between Phenolic Compounds and Discoloration Potential of Barley

PA fractions were added to unheated and heated (autoclaved at 121°C for 15 min) flour dough of cv. Radiant, a hulled PAF genotype. Barley flour dough added with PA fractions was always darker than dough without PA fractions. For the unheated flour, the decrease in brightness (L*) of dough during storage at 20±2°C for 24 hr was greatest with the addition of monomeric PAs, less with dimeric PAs and least with trimeric PAs (Figure 11). The monomeric fraction induced the most severe discoloration in barley flour dough, despite its relatively lower concentration compared to other PA fractions, indicating that monomeric PAs may be more potent substrates for PPO than dimeric or trimeric PAs.

The heated barley flour exhibited a much smaller decrease in brightness of dough during storage compared to the unheated flour, regardless of the addition of PA fractions. In the heated barley flour, further reduction in the brightness of dough occurred only with the addition of monomeric or trimeric PA fractions. The reduction in the brightness of dough prepared from heated flour could be due to the auto-oxidation of endogeneous PAs as well as the added PAs, since PPO was presumed to be inactivated by heat.

Conclusions

Discoloration potential of barley in food products is smaller in hulled genotypes than in hulless genotypes, and smaller in PAF genotypes than in PAC genotypes. Both phenolic compounds and polyphenol oxidase appear to be responsible for dark-gray color of barley-based food products. Discoloration potential of barley is mainly controlled by genotypic factors, and less by environmental factors. Abrasion to remove the outer layers of grain, anaerobic and low temperature storage conditions, heat treatment of grain and flour, as well

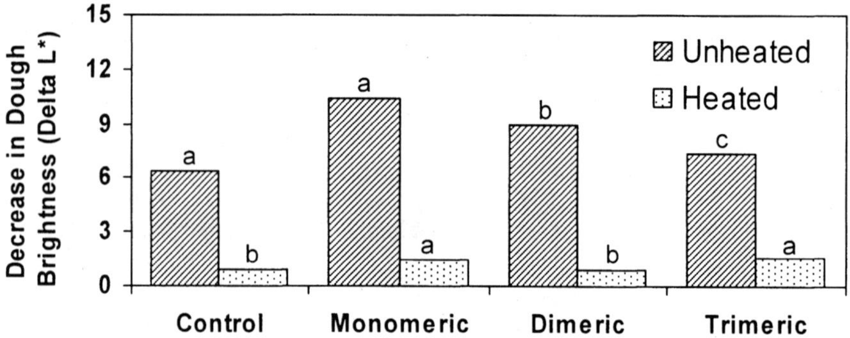

Figure 11. Effect of added proanthocyanidin fractions on brightness (L) of doughs prepared from barley flour of cv. Radiant before or after heat treatment. ΔL* = L* at 0 hr - L* at 24hr storage. Bars with the same letter within each treatment are not significantly different. (P<0.05). (Adapted from reference 32).*

as incorporation of ascorbic acid, are effective for controlling barley food discoloration. Dimeric PAs are the major phenolic compounds of barley grain, while monomeric PAs, including catechin, appear to be more effective substrates of PPO for discoloration of barley flour dough than dimeric and trimeric PAs. This manuscript is largely a review of previously published work by the authors (*10*, *11*, *21*, *32*).

References

1. Nilan, N. A.; Ullrich, S. E. In *Barley Chemistry and Technology*. MacGregor, A. W.; Bhatty, R. S., Ed.; AACC, Inc.: St. Paul, MI, 1996; pp 1-19.
2. McIntosh, G. H.; Newman, R. K.; Newman, C. W. *World Rev. Nutr. Diet.* **1995**, *77*, 89-108.
3. Swanson, R. B.; Penfield, M. P. *J. Food Sci.* **1988**, *53*, 896-901.
4. Newman, R. K.; McGuire, C. F.; Newman, C. W. *Cereals Food World* **1990**, *35*, 563-566.
5. McNeil, M. A.; Penfield, M. P.; Swanson, R. B. *Tennessee Farm and Home Science* **1988**, *146*, 4-7.
6. Berglund, P. T.; Fastnaught, C. E.; Holm, E. T. *Cereals Food World* **1992**, *37*, 707, 710-714.
7. Knuckles, B. E.; Hudson, C. A.; Chiu, M. M.; Sayre, R. N. *Cereals Food World* **1997**, *42*, 94-99.

8. Aastrup, S.; Outtrup, H.; Erdal, K. *Carlsberg Res. Commun.* **1984**, *49*, 105-109.
9. Bendelow, V. M.; LaBerge, D. E. *J. Am. Soc. Brew. Chem.* **1979**, *37*, 89-90.
10. Quinde, Z.; Ullrich, S. E.; Baik, B.-K. *Cereal Chem.* **2004**, *81*, 752-758.
11. Quinde-Axtell, Z.; Ullrich, S. E.; Baik, B.-K. *Cereal Chem.* **2005**, *82*, 711-716.
12. Friedrich, W.; Eberhardt, A.; Galensa, R. *Eur. Food Res. Technol.* **2000**, *211*, 56-64.
13. Zheng, G. H.; Rossanagel, B. G.; Tyler, R. T.; Bhatty, R. S. *Cereal Chem.* **2000**, *77*, 140-144.
14. Marconi, E.; Graziano, M.; Cubadda, R. *Cereal Chem.* **2000**, *77*, 133-139.
15. Reid, D. A. In *Barley*; Rasmusson, D. C. Ed.; Agronomy 26; ASC, CSSA, SCCA: Madison, WI, 1985; pp 75-102.
16. Ashie, I. N.; Simpson, B. K.; Smith, J. P. *Crit. Rev. Food Sci. Nutr.* **1996**, *36*, 1-30.
17. Fasina, O. O.; Tyler, R. T.; Pickard, M. D.; Zheng, G. H. *J. Food Proc. Preser.* **1999**, *23*, 135-151.
18. McEvily, A. J.; Iyengar, R.; Otwell, W. S. *Crit. Rev. Food Sci. Nutr.* **1992**, *32*, 253-273.
19. Weemaes, C. A.; Ludikhuyze, L. R.; Van den Broeck, I.; Hendrickx, M. E. *J. Agric. Food. Chem.* **1999**, *47*, 3526-3530.
20. Martinez, M. V.; Whitaker, J. R. *Trends in Food Sci. Technol.* **1995**, 6, 195-200.
21. Quinde-Axtell, Z.; Power, J.; Baik, B.-K. *Cereal Chem.* **2006**, *83*, 385-390.
22. Jende-Strid, B.; Møller, B. L. *Carlsberg Res. Commun.* **1981**, *46*, 53-64.
23. Fincher, G. B. *J. Inst. Brew.* **1976**, *82*, 347-349.
24. MacGregor, A. W.; Fincher, G. B. *In Barley Chemistry and Technology*, MacGregor, A. W.; Bhatty, R. S., Ed.; AACC, Inc: St. Paul, MI, 1996; pp 73-130.
25. Jende-Strid, B. *Carlsberg Res. Commun.* **1978**, *43*, 265-273.
26. Abdel-Aal, E.-S. M.; Hucl, P. *Cereal Chem.* **1999**, *76*, 350-354.
27. Baxter, E. D.; O'Farrel, D. D. *J. Inst. Brew.* **1987**, *93*, 308-312.
28. Goupy, P.; Hugues, M.; Boivin, P; Amoit, M. J. *J. Sci. Food Agric.* **1999**, *79*, 1625-1634.
29. Jende-Strid, B. *Hereditas* **1993**, *119*, 187-204.
30. Outtrup, H. In *Proc. Eur. Brew. Conv.*, 18th Congress, Copenhagen. IRL Press: London, 1981; pp 323-333.
31. Yu, J.; Vasanthan, T.; Temelli, F. *J. Agric. Food Chem.* **2001**, *49*, 4352-4258.
32. Quinde-Axtell, Z; Baik, B.-K. *J. Agric. Food Chem.* **2006**, *54*, 9978-9984.
33. Nordkvist, E.; Salomonsson, A.-C.; Åman, P. *J. Sci. Food Agric.* **1984**, *35*, 657-661.

34. von Wettstein, D.; Nilan, R. A.; Ahrenst-Larsen, B.; Erdal, K.; Ingversen, J.; Jende-Strid, B.; Kristiansen, K. N.; Larsen, J.; Outtrup, H.; Ullrich, S. E. *Master Brew. Assoc. Am. Tech. Quart.* **1985**, *22*, 41-52.
35. Jerumanis, J. *J. Inst. Brew.* **1985**, *91*, 250-252.
36. McMurrough, I.; Madigan, D. *J. Agric. Food Chem.* **1996**, *44*, 1731-1735.

Regulatory Aspects

Chapter 28

Regulatory Aspects of Colorants: Regulations, Regulations, Regulations in the United States of America!

Arthur L. Lipman

Office of Food Additive Safety, Center for Food Safety and Applied Nutrition, Food and Drug Administration, College Park, MD 20740

Color additives added to a food product to be sold in the USA must be used according to the list of color additives allowed for food use under Title 21 of the U.S. Code of Federal Regulations. Alternatively, an individual may submit a Color Additive Petition to amend these regulations to allow for uses not already listed. The existing regulations with emphasis on their limitations are discussed along with the color additive petition process. Recent amendments for color additives used in food will be mentioned as well as pending petitions. Other regulations that apply to marketing a food product in the USA including bioterrorism-related regulations, food standards, and labeling regulations for bulk color additive shipments and finished food products that contain color additives are introduced as well.

All the measurements and all the quality controls have been done but the product is not quite right. The taste and nutritional aspects are good, but to make the product more appealing a color additive is needed. What are the requirements for food color additives that you must meet in order to market a legal product? The addition of a color additive cannot be done to mask a decomposed or inferior product. (*1*)

Listed Color Additives

The USA maintains positive lists of color additives that may be used in food products shipped in interstate commerce including importation. These lists may be found in Title 21 of the US Code of Federal Regulations, Parts 73, 74 and 81 (http://www.access.gpo.gov/nara/cfr/waisidx_04/21cfrv1_04.html. Tables 1 and 2 provide the complete list of color additives allowed to color food in the USA. For easier access, FDA has provided a summary of color additives allowed for use in food, drugs, cosmetics, and medical devices on the internet (http://www.cfsan.fda.gov/~dms/opa-col2.html). The listed color additives are the only ones that may be used to color food in the USA and they must be used as prescribed by the listing regulation. That means that the color additive you use must comply with the identity, specifications, uses, and restrictions described in the regulation.

If the manufacturing process is specified, then a color additive manufactured by another method would not comply with the regulation and its use to color food would render the food adulterated under the law. For example, fruit juice color additive (*2*) and vegetable juice color additive (*3*) specify that the product is made by expressing the color from mature varieties of edible fruits and vegetables, respectively or by water infusions of the dried, mature, edible variety. Therefore, color additives derived from immature fruits and vegetables or through use of other solvents to isolate the coloring entities would not comply with these regulations.

The definition of color additive in the Federal Food, Drug, and Cosmetic Act is very broad. (*4*) "The term "color additive" means a material which – (A) is a dye, pigment, or other substance made by a process of synthesis or similar artifice, or extracted, isolated, or otherwise derived, with or without intermediate or final change of identity, from a vegetable, animal, mineral, or other source, and (B) when added or applied to a food, drug, or cosmetic, or to the human body or any part therof, is capable (alone or through reaction with other substance) of imparting color thereto; … (*2*) The term "'color' includes black, white, and intermediate grays…." (*4*) Note there is no exemption from the definition of color additive as there is from the definition of food additive (*5*) for substances that are generally recognized as safe (GRAS). Thus, there is no such thing as a GRAS color additive. Substances that are publicly recognized as GRAS by the Secretary of HHS are eligible to be listed as certification-exempt color additives just on the basis of their being GRAS (*6*), but the listing for color additive use must be in place before it can legally be used to color products in the USA.(*7*) Barrows, et al. have provided a more detailed historical perspective on use of color additives in the USA. (*8*)

Table 1. Certifiable Color Additives Listed for Food Use

21 CFR Section	*Straight Color*	*Uses and Restrictions*
74.101	FD&C Blue No. 1	Foods generally
74.102	FD&C Blue No. 2	Foods generally
74.203	FD&C Green No. 3	Foods generally
74.250	Orange B	Casings or surfaces of frankfurters and sausages, NTE 150 ppm by weight
74.302	Citrus Red No. 2	Skins of oranges not for processing, NTE 2.0 ppm by wieight
74.303	FD&C Red No. 3	Foods generally
74.340	FD&C Red No. 40	Foods generally
74.705	FD&C Yellow No. 5	Foods generally
74.706	FD&C Yellow No. 6	Foods generally

Batch Certification

The color additive regulations are divided into two main sections with subsections for food use, drug use, cosmetic use, and medical device use. Part 74 is a listing of color additives that require batch certification before they may be used to color products in the USA. This means that a portion of every batch manufactured must be sent to the FDA for analysis before any of it may be used to color products (*9*). If the batch meets the identity and specifications established by the regulation, FDA will issue a certificate documenting that fact and the batch may then be used to color products. The certificate issued by the Color Certification Branch of the Office of Cosmetics and Colors will designate a unique Lot number to that batch, which the manufacturer may use as proof of certification. The manufacturer is required to keep records of the disposition of that batch. (*10*) The name of the certified color additive must then be used on the ingredient statement of the food package, but the name may be abbreviated (*11*). For example, tartrazine from a certified batch must be labeled as FD&C Yellow No. 5 or abbreviated Yellow 5. Tables 3 and 4 show the poundage certified for the certifiable food color additives and their lakes, respectively for the past two Fiscal Years (source: http://www.cfsan.fda.gov/~dms/col-cert.html). A lake is an insoluble form of the color additive "made by extending one or more straight colors on one or more substrata by absorption, coprecipitation, or chemical combination, but does not include mixtures." (*12*). For food color additive lakes, the substratum is alumina and the precipitant is aluminum. The interaction between the straight color additive and the substratum involves adsorption, coprecipitation, or other chemical combination that does not include simple mixing of the ingredients. (*13*)

Part 73 lists color additives for use in food, drugs, cosmetics, and medical devices that are exempt from the requirement for batch certification. That means that the manufacturer is responsible for ensuring that the color additive is of appropriate quality as indicated by the identity and specifications listed in the listing regulation. The decision as to whether a color additive is exempt from batch certification or requires batch certification is made during the premarketing evaluation of the safety of the color additive under intended conditions of use as triggered by the submission of a color additive petition by an interested party. This decision is made based on the reproducibility of manufacture of the color additive and the degree of safety concern with the impurities that are included in the color additive. For example, FD&C Yellow No. 5 may contain trace levels of six impurities (called constituents) that have been shown to cause cancer, while the color additive itself as tested in animal studies did not induce cancer. FDA specified the manufacturing process in the listing regulation for Yellow 5 (*14*), established specifications for each of the carcinogenic constituents, and required batch certification. Generally, color additives derived from petroleum or, to a much lesser extent, coal, are listed in Part 74 and color additives derived from plants, minerals, and animals are listed in Part 73.

Petition process

If the color additive is not already listed for the intended use or if the product does not comply with an existing regulation, a petition must be submitted to the FDA to amend its regulations to include that use of the color additive before you can legally add your color additive to food to be sold in the USA. Procedural regulations for the color additive petition process are codified in 21 CFR 71.1 (15) and guidance is also available for chemistry, toxicology, and environmental impact data at the following websites, respectively:

http://www.cfsan.fda.gov/~dms/opa-col1.html
http://www.cfsan.fda.gov/~redbook/red-toca.html
http://www.cfsan.fda.gov/~dms/opa-guid.html#eg

Table 2. Certification Exempt Color Additives Listed for Food Use

21 CFR Section	*Straight Color*	*Uses and Restrictions*
73.30	Annatto extract	Foods generally
73.35	Astaxanthin	Salmonid fish feed
73.40	Dehydrated beets (beet powder)	Foods generally
73.50	Ultramarine blue	Salt for animal food
73.75	Canthaxanthin	Foods generally, NTE 30 mg/lb of solid or semisolid food or per l of liquid food; broiler chicken fooed, salmonid fish feed
73.85	Caramel	Foods generally
73.90	B-Apo-8'-carotenal	Foods generally, NTE 15 mg/lb solid 15 mg/pt liquid
73.95	B-Carotene	Foods generally
73.100	Cochineal extract; carmine	Foods generally
73.125	Sodium copper chlorophyllin	Citrus-based dry beverage mixes, NTE 0.2 % dry mix
73.140	Toasted partially defatted cooked cotton seed oil	Foods generally
73.160	Ferrous gluconate	Ripe olives
73.165	Ferrous lactate	Ripe Olives
73.169	Grape color extract	Nonbeverage food
73.170	Grape color extract (enocianina)	Still and carbonated drinks and ades; beverage bases, alcoholic bevs
73.185	Haematococcus algae meal	Salmonid fish feed
73.200	Synthetic iron oxide	Sausage casings, NTE 0.1% (by weight); dog and cat food, NTE 0.25% by weight
73.250	Fruit juice	Foods generally
73.260	Vegetable juice	Foods generally

73.275	Dried algae meal	Chicken feed
73.295	Tagetes (Aztec marigold) meal and extract	Chicken feed
73.300	Carrot oil	Foods generally
73.315	Corn endosperm oil	Chicken feed
73.340	Paprika	Foods generally
73.345	Paprika oleoresin	Foods generally
73.350	Mica-based Pearlescent Pigments	cereals, confections and frostings, gelatin desserts, hard and soft candies (including lozenges), nutritional supplement tablets and gelatin capsules, and chewing gum
73.355	Phaffia yeast	Salmonid fish feed
73.450	Riboflavin	Foods generally
73.500	Saffron	Foods generally
73.575	Titanium dioxide	Foods generally, NTE 1% (by weight)
73.585	Tomato lycopene extract; Tomato lycopene concentrate	Foods generally
73.600	Turmeric	Foods generally
73.615	Turmeric oleoresin	Foods generally

Table 3. Certification of FD&C Color Additives Fiscal Years 2005 and 2006

FD&C – Primary	*Pounds 2005*	*Pounds 2006*
FD&C Blue No. 1	699,341.34	592,150.18
FD&C Blue No. 2	395,065.60	525,128.53
FD&C Green No. 3	12,897.45	15,586.18
FD&C Red No. 3	260,725.90	207,930.45
FD&C Red No. 40	4,919,869.34	5,532,156.53
FD&C YELLOW No. 5	2,967,904.99	3,305,397.35
FD&C YELLOW No. 6	2,771,323.16	3,013,018.52
CITRUS RED No. 2	837.75	2,645.53
TOTAL	11,997,965.53	13,194,013.27

Table 4. Certification of FD&C Color Additive Lakes Fiscal Years 2005 and 2006

FD&C – LAKES	*Pounds 2005*	*Pounds 2006*
FD&C Blue No. 1 aluminum	457,184.20	435,390.80
FD&C Blue No. 2 aluminum	243,431.92	213,082.85
FD&C Red No. 40 aluminum	1,553,588.51	1,633,329.62
FD&C Yellow No. 5 aluminum	1,225,625.63	1,239,811.12
FD&C Yellow No. 6 luminum	1,356,112.08	1,183,148.12
TOTAL	4,835,942.34	4,704,762.51

The information that the FDA needs is: 1). what substances will get into food from the intended use of the color additive and 2). how much of each substance humans or animals will encounter in their daily diet. Interested parties are then required to provide data showing that there is a reasonable certainty that no harm will come to people or animals consuming those levels of the identified substances as well as the color additive. Importantly, the color additive itself must not have been shown to induce cancer in man or animals. The so-called Delaney clause (16) precludes the agency from listing a color additive for use in products in the USA if the color additive has been shown to be a carcinogen.

The petitioner for a new color additive must provide information on the following:

- Identity of the proposed color additive
- Physical, chemical, and biological properties
- Chemical specifications
- Manufacturing process description

- Stability data
- Intended uses and restrictions
- Labeling of the bulk color additive
- Tolerances and limitations
- Analytical methods for enforcing chemical specifications
- Analytical methods for determination of the color additive in products
- Identification and determination of any substance formed in or on products due to use of the color additive
- Safety studies in animal models
- Estimate of probable exposure
- Proposed regulation
- Proposed exemption from batch certification
- An environmental assessment or claim for categorical exclusion

The petitioner must submit data demonstrating the safety and suitability of the new color additive or new use of a listed color additive. FDA has 15 working days to perform an overview of the submission to determine whether it is suitable for filing. If the petition is not suitable for filing, FDA will issue a not-filed letter, documenting the deficiencies. The petitioner should remedy the deficiencies within 180 days and inform the agency of its intent within 30 days or withdraw the petition. The petitioner may insist that FDA file the deficient petition anyway, which FDA will do. In this case, it is likely that the agency will issue a denial of the petition shortly after filing.

If the original petition or the remedied petition is suitable for filing, FDA will publish a filing notice in the US Federal Register announcing the amendment proposed by the petitioner. FDA will then evaluate the data in the petition, public comments to the petition, and other relevant data in FDA's files and publish its final decision in the Federal Register. A final rule resulting from the petition allows for objections and requests for a hearing by parties affected by the final rule to be filed within 30 days of the date of publication and establishes an effective date for the regulation 31 days after the date of publication in the Federal Register. Any parts of the regulation that receive objections are stayed until FDA addresses them in the Federal Register. The fact that no proper objections and requests for a hearing were received would also be announced by FDA in a subsequent Federal Register document, called a Confirmation of Effective Date Notice.

Labeling

Bulk color additive (*17*)

All of the regulations listing color additives for use in products in the USA refer to the referenced regulation that requires that the bulk color additive “be labeled with sufficient information to assure safe use and to allow a determination of compliance with any limitations imposed by parts 71, 73, 74, 80, and 81." This information includes: the name of the straight color, a statement of general limitations, a statement concerning the amount of straight color that may be used if there are quantitative limitations, and an expiration date, if stability data require it. For a certified color additive, the label must bear the lot number assigned by FDA.

Finished food (*11*)

The regulation in 21 CFR 101.22(k) has three paragraphs. The first deals with color additives subject to batch certification that are listed in 21 CFR 74. The common or usual name as used in the listing regulation should be included in the ingredient statement for the food, but the name may be abbreviated to drop the FD&C and the term "No." For example FD&C Yellow No. 5 may be listed as Yellow 5 in the ingredient statement.

Paragraph 2 deals with labeling of the presence of color additives that are exempt from batch certification. This paragraph allows a general term, such as "artificially colored" or "color added" to indicate that a color additive has been used in the food product. Alternatively, the common or usual name as indicated by the title of the regulation may be used along with the purpose. For example, "cochineal extract color" or "colored with cochineal extract." Cochineal extract is used as an example, because FDA recently proposed (*18*) to require the listing of this color additive and its lake, known as carmine, on the label of all products containing these color additives due to evidence that the color additive may cause allergenic responses in some consumers.

Paragraph 3 of §101.22(k) exempts butter, cheese, and ice cream from the labeling requirements unless a regulation in 21 CFR Part 73 or 74 requires that these products contain the information on the color additives used. For example, FD&C Yellow No. 5, has such a requirement in its listing regulation (21 CFR 74.705(d)(2)) (*14*).

The label declarations of color additives used in finished foods may be affected by recent legislation. The Food Allergen Labeling and Consumer Protection Act of 2004 added section 403(w) to the Federal Food, Drug, and Cosmetic Act, which requires that the source be included on the label for a product containing a color additive that contains protein from the eight foods or food groups identified in section 201(qq) of the Act. These eight foods or food groups are milk, eggs, fish, Crustacean shellfish, tree nuts, peanuts, wheat, and soybeans.

New regulations, pending petitions, and other actions

The latest regulations for color additives used in food products are for sodium copper chlorophyllin (*19*) tomato lycopene (20) and for mica-based pearlescent pigments (*21*). The sodium copper chlorophyllin use is limited to citrus based, dry beverage mixes at levels not to exceed 0.2 percent of the dry mix and the manufacturing process is limited to extraction from alfalfa using any one or a combination of acetone, ethanol, and hexane as the extraction solvent.

The tomato lycopene color additives, which are to be listed in 21 CFR 73.585, are the ethyl acetate extract and a concentrate of the extract. They may be used to color food generally consistent with good manufacturing practice. Timely, proper objections were received to the final rule (*22*) that would list tomato lycopene for food use. The FDA's response to these objections published on February 24, 2006 (*23*), removing the stay of the effective date of the regulation effective February 24, 2006, and establishing the regulation for use of tomato lycopene extract and concentrate in 21 CFR 73.585.

As a partial response to CAP 8C0262 mentioned below, a final rule providing for use of the composite pigments from mica and titanium dioxide in cereals, confections and frostings, gelatin desserts, hard and soft candies (including lozenges), nutritional supplement tablets and gelatin capsules, and chewing gum, published on June 2, 2006 (71 FR 31927). (*21*) Objections to this final rule were due by July 3, 2006, but none were received. Thus, the regulation became effective July 5, 2006. The Confirmation of Effective Date notice was published September 15, 2006 (71 FR 54411). Composite pigments containing synthetic iron oxides, which were also requested in CAP 8C0262 are still under review, as described below.

Three color additive petitions involving food use have been placed in abeyance by FDA, but one of them has been restored to active review status.. Abeyance status means that the agency has requested additional information to support the safety and suitability of the color additives that are subjects of the petitions and is awaiting submission of the information before it can proceed with the review of the petition. CAP 8C0262 was submitted by EM Industries Incorporated for use of pearlescent pigments as a color additive in various food preparations. The filing notice published on September 25, 1998 (63 FR 51359), but it was amended on June 21, 1999, to better reflect the nature of these pigments (64 FR 33097). As discussed above a partial response listed titanium dioxide-mica composite for specified food uses (*21*), but the remaining portion of the petition regarding use of iron oxide composite pigments with mica and titanium dioxide in food remained in abeyance until October 2006. It is now under review.

CAP 4C0276 was submitted by Cryovac North America for use of synthetic iron oxide in food packaging, but the iron oxide is intended to migrate to food in sufficient quantities to impart color to the packaged meat products. The filing notice published on February 27, 2004 (69 FR 9340) This petition can be used to illustrate the difference between a color additive and a colorant in the USA. A color additive is intended to impart color to food, drugs, cosmetics, or medical devices, whereas a colorant is defined in 21 CFR 178.3297 as a "dye, pigment, or other substance that is used to impart color to or to alter the color of a food-contact material, but that does not migrate to food in amounts that will contribute to that food any color apparent to the naked eye."(*24*) Colorants are considered to be food additives or food contact substances for regulatory purposes. While the iron oxide is added to the packaging, the intent is to have sufficient iron oxide migrate to the meat to impart color to the meat. Therefore, this petition is being handled as a color additive petition and not a food additive petition or a food contact notification.

CAP 6C0281 was submitted by Food Ingredients Solutions to amend the color additive regulation for sodium copper chlorophyllin in 21 CFR 73.125 to include foods generally except meat, poultry, fish, and standardized foods and to include fescue grass as a permitted source. The current regulation allows only alfalfa as the source of the color additive (*19*). The filing notice published on December 9, 2005, 70 FR 73252. The petition is currently in abeyance as of February 2007.

As mentioned above, FDA published a proposed rule on January 30, 2006 (*18*) that would require listing of cochineal extract/carmine by name in the ingredient statement of products that contain the color additive, because of reports of severe allergic reactions, including anaphylaxis, to cochineal extract and carmine-containing food and cosmetics. This requirement as proposed would apply to butter, cheese and ice cream also. Including the common name of the color additive would allow people who are allergic to it to avoid exposure. Comments were to be submitted to Docket No. 1998P-0724 by May 1, 2006. Access to the filed comments and other materials in the docket is available at http://www.fda.gov/ohrms/dockets/default.htm by using the docket number above.

Food Standards

Food standards have statutory authority in section 401 of the FFD&C Act (*25*). They describe the basic nature and essential characteristics of a food, specify the required and optional ingredients that are used to make the food, may specify the process that is used to make the food, and establish the name of the food. Food standards are established primarily for the purpose of ensuring the economic value of the food. On occasion, food standards have served as a means to improve the overall nutritional quality of the food supply. If the food is the subject of an FDA food standard codified in 21 CFR 130-169, then compliance with the standard is necessary in order to be able to name the product as in the food standard. Color additive listing regulations preclude use of the color additive in foods for which standards of identity have been promulgated unless "added color" or "coloring" is authorized by the food standard. To illustrate this principle, Table 5 shows a breakdown of the cheese standards in 21 CFR 133 (*26*) by those that allow optional color additives, those that allow only blue or green color additives, and those that do not provide for use of color additives. If a color additive is added to a standardized food that does not allow added coloring in the standard, then the food cannot be called by the standardized name.

Other regulations

There are several other regulations with which food manufacturers must comply to avoid problems with their food products in the USA. The four recent ones that deal with bioterrorism are introduced below. Detailed information on these regulations can be found at: http://www.fda.gov/oc/bioterrorism/bioact.html. These are not the only regulations that need to be followed when marketing a food product in the USA.

Bioterrorism regulations

In response to the events of September 11, 2001, the United States Congress passed the Public Health Security and Bioterrorism Preparedness and Response Act of 2002 (the Bioterrorism Act), which President Bush signed into law June 12, 2002. Title III of the Bioterrorism Act deals with protecting the safety and security of the United States food and drug supply and imposes four new requirements: registration of food facilities, notification prior to importing food shipments, establishing and maintaining records of the immediate previous sources and the immediate subsequent recipients of food products, and administrative detention of food that presents a serious health threat.

Subtitle A of Title III of the Bioterrorism Act provided the Secretary of Health and Human Services with new authorities to protect the nation's food supply against the threat of intentional contamination and other food-related emergencies. This legislation represents the most fundamental enhancement to U.S.food safety authorities in many years. These additional authorities improve FDA's ability to act quickly in responding to a threatened or actual terrorist attack, as well as other food-related emergencies. Since this legislation was signed into law more than three years ago, FDA has been working hard to implement this law effectively and efficiently. Throughout this process, FDA

Table 5. Cheese Standards that provide for coloring - 21 CFR 133

Coloring Allowed	*Blue or Green only*	*Coloring Not Allowed*
Brick Cheese	Asiago Cheeses	Cook Cheese koch kaese
Cheddar Cheese	Blue Cheese	Cottage Cheese
Colby Cheese	Caciocavallo siciliano	Cream Cheese
Cold Pack Cheese	Gorgonzola cheese	Gammelost cheese
Cream Cheese with other foods	Mozzarella and scamorza cheese - color to mask yellow only	Gruyere cheese
Washed curd and soaked curd cheese	Nuworld cheese	Monterey cheese, Monterey Jack cheese
Edam Cheese	Provolone cheese	Neufchatel cheese
Gouda cheese	Romano cheese -	Soft ripened cheeses
Granular and stirred curd cheese		Roquefort cheese
Hard cheeses		Sap sago cheese
Limburger cheese		Skim milk cheese for manufacturing
Muenster and munster cheese		
Parmesan and reggiano cheese		
Pasteurized process cheeses		

has enjoyed close cooperation from the Customs and Border Protection (CBP). Additional details are available in the information and guidance provided on the internet at: http://www.fda.gov/oc/bioterrorism/bioact.html and the specific regulations cited in each section below.

Registration of food processing facilities

Section 305 of the Bioterrorism Act requires registration of foreign and domestic food facilities that manufacture, process, pack, or hold food for consumption by humans or animals in the U.S. As a result of this requirement, FDA has, for the first time, a roster of foreign and domestic food facilities that provide food for American consumers. In the event of a potential or actual terrorist incident or an outbreak of foodborne illness, the registration information will help FDA to quickly identify, locate, and notify the facilities that may be affected.

On October 10, 2003, FDA and CBP jointly published an interim final rule (*27*) to implement the registration requirement, which became effective on December 12, 2003, as required by the Bioterrorism Act. The registration interim final rule was effective immediately but provided an opportunity for public comment on specific issues. On October 3, 2005, FDA issued the Registration of Food Facilities Final Rule (*28*), which affirmed the requirements initially set forth in the interim final rule. As of December 2, 2005, approximately 271,000 facilities have registered with FDA. This includes about 116,000 domestic and about 155,000 foreign facilities.

Regulations for registration of food facilities can be found in 21 CFR Part 1, Subpart H. (*29*) The rule requires domestic and foreign facilities that manufacture, process, pack, or hold food subject to FDA's jurisdiction for human or animal consumption in the United States to register with the FDA by December 12, 2003, or before providing food for consumption in the U.S. Registration is required whether or not the food from the facility enters interstate commerce.

Except for specific exemptions, the registration requirements apply to all facilities that manufacture, process, pack, or hold food, including animal feed, dietary supplements, infant formula, beverages (including alcoholic beverages and bottled water), and food additives. By definition color additves for use in food would also be considered food and subject to these requirements.

Facilities are defined as:

"any establishment, structure, or structures under one ownership at one general physical location, or, in the case of a mobile facility, traveling to multiple locations, that manufactures/processes, packs, or holds food for consumption in the United States. Transport vehicles are not facilities if they hold food only in the usual course of business as carriers. A facility may consist of one or more contiguous structures, and a single building may house more than one distinct facility if the facilities are under separate ownership. The private residence of an individual is not a facility. Nonbottled water drinking water collection and distribution establishments and their structures are not facilities." (*30*)

Food has the meaning in 201(f) of the Federal Food, Drug, and Cosmetic Act (21 U.S.C. 321(f), except for purposes of this subpart, it does not include:

a) Food contact substances as defined in section 409(h)(6) of the act (21 U.S.C. 348(h)(6)), or
b) Pesticides as defined in 7 U.S.C. 136(u). (*31*)

Facilities that are exempted from the registration requirements are (*32*):

a) foreign facilities, if food from such facility undergoes further manufacturing/processing (including packaging) by another facility outside the United States. A facility is not exempt under this provision if the further manufacturing/processing (including packaging) conducted by the subsequent facility consists of adding labeling or any similar activity of a de minimis nature;
b) Farms;
c) Retail food establishments;
d) Restaurants;
e) Nonprofit food establishments in which food is prepared for, or served directly to, the consumer;
f) Certain fishing vessels,
g) Facilities that are regulated exclusively, throughout the entire facility, by the U.S. Department of Agriculture

The registration for a food facility may include both required and optional information . The information requirements for registration, both mandatory and optional, are set out in detail in the food facility registration regulation at 21 CFR Subpart H. (*33*). FDA's internet web site provides helpful information to those who wish to register a food facility and also provides a portal to the electronic registration system. http://www.cfsan.fda.gov/~furls/ovffreg.html.

Prior Notice of Food Imported or Offered for Import

Section 307 of the Bioterrorism Act requires the submission to FDA of prior notice of food, including animal feed, that is imported or offered for import into the U.S. This advance information enables FDA, working closely with CBP, to more effectively target inspections of food at the border at the time of arrival to ensure the safety and security of imported foods. On October 10, 2003, FDA and CBP jointly published an interim final rule (*34*) to implement this provision. The interim final rule provided stakeholders an additional opportunity to comment on all provisions of the interim final rule for almost six months while the rule took effect on December 12, 2003, as required by the Bioterrorism Act. FDA is drafting a final rule that responds to the numerous timely comments the agency received and intends to publish the final rule as expeditiously as possible. Since December 2003, FDA has been receiving, reviewing, and responding to approximately 167,000 notifications each week about articles of food being imported or offered for import into the U.S.

Regulations for import notification are in 21 CFR Part 1, Subpart I (*35*). An article of food is subject to refusal of admission into the USA if adequate prior notice has not been provided to FDA. The standard for refusal is met if the requisite information has **not** been provided, accepted, and confirmed for review by FDA within the required time frames

The prior notice of imported food regulation describes in detail the information that needs to be submitted in a prior notice (*36*). The regulation also specifies when such notice must be submitted. The prior notice must be submitted no less than 2 hours before arrival if the food is arriving in the USA by land via road, 4 hours if arriving by air or land via rail, and 8 hours if arriving by water The FDA Compliance Policy Guide (accessible at http://www.fda.gov/ora/compliance_ref/cpg/default.htm) discusses the exercise of enforcement discretion by FDA and the CBP for certain violations. FDA's internet website provides useful information on the prior notice requirement for imported foods. http://www.cfsan.fda.gov/~pn/pnoview.html .

Record-keeping/traceback

Section 306 of the Bioterrorism Act authorizes FDA to have access to certain records when the Agency has a reasonable belief that an article of food is adulterated and presents a threat of serious adverse health consequences or death to humans or animals. It authorizes the Secretary to publish regulations to establish requirements regarding the establishment and maintenance, for not longer than two years, of records by persons (excluding farms and restaurants) who manufacture, process, pack, transport, distribute, receive, hold, or import food. Foreign persons except for those who transport food in the USA are excluded from the requirements of this regulation.

On December 9, 2004, FDA published a final rule (*37*) to implement this section. The recordkeeping regulation requires persons receiving or releasing food, including food ingredients, to identify the immediate previous sources of that food and the immediate subsequent recipients of that food; thus, this rule is often referred to as the "one up/one down" rule. The regulation enhances FDA's ability to track and contain foods that pose a threat of serious adverse health consequences or death to American consumers from accidental or deliberate contamination of food. Affected persons with 500 or more full-time equivalent employees had to be in compliance with the regulation on December 9, 2005. Smaller companies, which provide more than 80% of the U.S. food supply, have until June or December 2006 to be in compliance, depending on the number of employees they have. The Bioterrorism Act required FDA to consider the size of the business in developing the regulations. FDA exercised this discretion by giving smaller businesses more time to comply to enable them to learn from the experiences of their larger counterparts and thereby reduce costs.

Regulations for the establishment and maintenance of records are codified in 21 CFR Part 1, Subpart J. (*38*) The requirements for non-transporters differ from those for transporters. Generally non-transporters must identify the non-transporter immediate previous source and the non-transporter immediate subsequent recipient, the transporter that brought the food to you and the transporter that took the food from you, and specific information about the food, its packaging, and the date the food was received and released. Records must include all information reasonably available to you to identify the specific source of each ingredient that was used to make every lot of finished product.

Transporters have five options for meeting the requirements in the regulation. First they can establish and maintain records as specified in the final rule. Alternatively they can establish and maintain records currently required by any of the following:

1) Department of Transportation's Federal Motor Carrier Safety Administration of roadway interstate transporters (bills of lading)
2) Department of Transportation's Surface Transportation Board of rail and water interstate transporters (bills of lading)
3) Warsaw Convention of international air transporters (air waybills), or
4) Entering into an agreement with the nontransporter IPS or ISR located in the United States to establish, maintain, or establish and maintain the required information.

Administrative detention

The final rule "Administrative Detention of Food for Human or Animal Consumption Under the Public Health Security and Bioterrorism Preparedness and Response Act of 2002" provides procedural regulations for FDA's administrative detention of a food article suspected of being contaminated. These regulations may be found in 21 CFR Part 1, Subpart K. (*39*) Under the final rule (*40*), FDA may detain an article of food for up to 30 days on the strength of credible evidence or information that the food presents a threat of serious health consequences or death to humans or animals. The time of detention allows FDA to prepare and file a seizure action, the court can issue a warrant, and the U. S. Marshall can seize the food, as appropriate.

The rule requires a detention order to be approved by the FDA District Director of the district where the detained article of food is located, or by a higher official. A copy of the detention order must be given to the owner, operator, and/or agent in charge of the place where the article of food is located, and to the owner of the food provided the owner's identity can be determined readily. If FDA issues a detention order for an article of food located in a vehicle or other carrier, the agency also must provide a copy of the detention order to the shipper of record and the owner and operator of the vehicle or other carrier provided the owner's identity can be determined readily.

The final rule requires detained articles of food to be held in secure locations, as determined by FDA. The food may not be transferred from the place where it has been ordered detained, or from the place where the detained article has been removed without FDA approval, until FDA terminates the detention order, or the detention period expires. Violation of a detention order is a prohibited act.

Generally, administrative detention will be used for food in domestic commerce as §801(a) of the act can be used to detain imported foods.

Summary

This paper provided information on US regulations that apply to food marketed in the USA. First, requirements for color additives used in food products were indicated by the positive lists of color additives that must be used. Otherwise, use of color additives not listed renders the food adulterated. Likewise, the use of a color additive that is listed but in violation of the conditions delineated in the listing regulation and the use of a color additive that requires batch certification but which comes from a batch that has not been certified would also render the food adulterated. Second, the labeling requirements for bulk color additives, for food products containing a color additive, and the use of a color additive in standardized foods were explained. Third, the petition process for obtaining a listing for a new color additive or a new use of a listed color additive was described. Finally, the four regulations designed to prevent acts of bioterrorism involving food for humans or animals in the USA: registration of food facilities; prior notice of imported food shipments; establishment, maintenance, and availability of records; and administrative detention of food were introduced. Of note, color additives used in food products are considered food.

References

1. *U.S. Code of Federal Regulations*, *Title 21*, 71.22. U.S. Government Printing Office: Washington, DC, 2006, p 335
2. *U.S. Code of Federal Regulations, Title 21*, 73.250. U.S. Government Printing Office: Washington, DC, 2006, p 349.
3. *U.S. Code of Federal Regulations, Title 21*, 73.260. U.S. Government Printing Office: Washington, DC, 2006, p 349.
4. Federal Food, Drug, and Cosmetic Act, Section 201(t) (Title 21 United States Code 321(t)).
5. Federal Food, Drug, and Cosmetic Act, Section 201(s) (Title 21 United States Code 321(s)).
6. Federal Food, Drug, and Cosmetic Act, Section 721(b)(4) and 721(c)(2) (Title 21 United States Code 379e(b)(4) and 379e(c)(2)).
7. Federal Food, Drug, and Cosmetic Act, Section 721(a) (Title 21 United States Code 379e(a)).
8. Barrows, J.N., Lipman, A.L., and Bailey, C.J., "Color Additives: FDA's Regulatory Process and Historical Perspectives." Food Safety Magazine October/November 2003, *9*, Pp. 11-17.
9. *U.S. Code of Federal Regulations*, *Title 21*, 80. U.S. Government Printing Office: Washington, DC, 2006, Pp 425-431.
10. *U.S. Code of Federal Regulations*, *Title 21*, 80.39. U.S. Government Printing Office: Washington, DC, 2006, p 431.
11. *U.S. Code of Federal Regulations*, *Title 21*, 101.22(k). U.S. Government Printing Office: Washington, DC, 2006, p 76.
12. Food and Drug Administration. "Permanent Listing of Color Additive Lakes; Proposed Rule." *Fed. Regist.* **1996**, *43*, 8407.
13. *U.S. Code of Federal Regulations*, *Title 21*, 70.3(l). U.S. Government Printing Office: Washington, DC, 2006, p 323.
14. *U.S. Code of Federal Regulations*, *Title 21*, 74.705. U.S. Government Printing Office: Washington, DC, 2006, p 391-392.
15. *U.S. Code of Federal Regulations*, *Title 21*, 71.1. U.S. Government Printing Office: Washington, DC, 2006, p 330-332.
16. Federal Food, Drug, and Cosmetic Act, Section 721(b)(5)(B) Title 21 United States Code 379e(b)(5)(B).
17. *U.S. Code of Federal Regulations*, *Title 21*, 70.25. U.S. Government Printing Office: Washington, DC, 2006, p 327.
18. Food and Drug Administration. "Listing of Color Additives Exempt From Certification; Food, Drug, and Cosmetic Labeling: Cochineal Extract and Carmine Declaration." *Fed. Regist.* **2006**, *71*, 4839-4851.
19. *U.S. Code of Federal Regulations*, *Title 21* 73.125. U.S. Government Printing Office: Washington, DC, 2006, p 345.

20. *U.S. Code of Federal Regulations*, *Title 21* 73.585. U.S. Government Printing Office: Washington, DC, 2006, p 354.
21. Food and Drug Administration. "Listing of Color Additives Exempt from Certification; Mica-Based Pearlescent Pigments." *Fed. Regis.* **2006**, *71*, 31927-31929.
22. Food and Drug Administration. "Listing of Color Additives Exempt from Certification; Tomato Lycopene Extract and Tomato Lycopene Concentrate." *Fed. Regis.* **2005**, *70*, 43045.
23. Food and Drug Administration. "Listing of Color Additives Exempt from Certification; Tomato Lycopene Extract and Tomato Lycopene Concentrate." *Fed. Regis.* **2006**, *71*, 9448-9449.
24. *U.S. Code of Federal Regulations*, *Title 21* 178.3297(a). U.S. Government Printing Office: Washington, DC, 2006, p 402.
25. Federal Food, Drug, and Cosmetic Act, Section 401 (Title 21 United States Code 341).
26. *U.S. Code of Federal Regulations*, *Title 21* 133. U.S. Government Printing Office: Washington, DC, 2006, Pp 308-359.
27. Food and Drug Administration. "Registration of Food Facilities Under the Public Health Security and Bioterrorism Preparedness and Response Act of 2002." *Fed. Regis.* **2003**, *68*, 58894-58973.
28. Food and Drug Administration. "Registration of Food Facilities Under the Public Health Security and Bioterrorism Preparedness and Response Act of 2002." *Fed. Regis.* **2005**, *70*, 57505-57509.
29. *U.S. Code of Federal Regulations*, *Title 21 Part 1, Subpart H.* U.S. Government Printing Office: Washington, DC, 2006, Pp 17-25.
30. *U.S. Code of Federal Regulations*, *Title 21* 1.227(b)(2). U.S. Government Printing Office: Washington, DC, 2006, p 18.
31. U.S. Code of Federal Regulations, Title 21 1.227(b)(4). U.S. Government Printing Office: Washington, DC, 2006, p 18.
32. U.S. Code of Federal Regulations, Title 21 1.226. U.S. Government Printing Office: Washington, DC, 2006, p 17.
33. U.S. Code of Federal Regulations, Title 21 1.232 and 1.233. U.S. Government Printing Office: Washington, DC, 2006, p 21-22.
34. Food and Drug Administration. "Prior Notice of Imported Food Under the Public Health Securityand Bioterrorism Preparedness and Response Act of 2002." *Fed. Regis.* **2003**, *68*, 58975-59077.
35. *U.S. Code of Federal Regulations*, *Title 21 Part 1, Subpart I.* U.S. Government Printing Office: Washington, DC, 2006, Pp 25-37.
36. *U.S. Code of Federal Regulations*, *Title 21, Part 1.276(b)(8).* U.S. Government Printing Office: Washington, DC, 2006, p 26.
37. Food and Drug Administration. "Establishment and Maintenance of Records Under the Public Health Security and Bioterrorism Preparedness and Response Act of 2002." *Fed. Regis.* **2004**, *69*, 71561-71655.

38. *U.S. Code of Federal Regulations, Title 21 Part 1, Subpart J.* U.S. Government Printing Office: Washington, DC, 2006, Pp 37-43.
39. *U.S. Code of Federal Regulations, Title 21 Part 1, Subpart K.* U.S. Government Printing Office: Washington, DC, 2006, Pp 43-49.
40. Food and Drug Administration. "Administrative Detention of Food for Human or Animal Consumption Under the Public Health Security and Bioterrorism Preparedness and Response Act of 2002." *Fed. Regis.* **2004**, *69*, 31659-31705.

Chapter 29

Food Color Regulations: A Latin American Perspective

Rebeca López-García

Logre International Food Science Consulting, Mexico City, Mexico

Latin America is a large region encompassing many different countries and cultures. Even when countries share similar backgrounds and challenges, each one faces different issues due to the diversity in size, population, history and resources. It is difficult to address the region as a whole, but there are several shared challenges and opportunities. Colors are usually considered food additives and are addressed by the same food additive regulations. Most countries regulate food additives through the publication of positive lists. In addition, standards may be set for different product types. These standards may specifically declare that the use of certain colors is not allowed in a particular product or group or products when the use of coloring may mislead consumers. Since most countries lack the resources to perform complete safety evaluations, regulatory agencies rely on evaluations made by the Joint FAO/WHO Expert Committee on Food Additives (JECFA) and/or other international organizations to determine the safety of certain products. Some countries have completely adopted Codex Alimentarius Standards as their own regulations.

Introduction

Latin America is a complex region that is still struggling to find itself. The richness and diversity of Latin American cultures and the abundance of natural resources give it a unique potential and make it a place full of new opportunities. On the same token, as the region works to participate in the global market, new challenges have emerged. The inheritance of European colonialism combined with the influence of the region's proximity to the United States have also shaped its modern history, hence its legislation, including food law.

Some of the regulatory challenges faced by Latin American countries include: lack of proper food law and standards; impaired ability to manage inspection, control and laboratory services; lack of budget to support the enforcement system and lack of consumer education and participation among others. These challenges have an obvious impact on several areas of food control including food safety, food standardization and food commercialization including imports and exports.

In general, food colors are considered under food additive regulations and in most countries, food additives are regulated through the publication of positive lists with exception of some additives that are expressely forbidden at different regulatory levels. Control through the use of positive lists is challenging because resources to update these lists frequently may not be available. Therefore, the system tends to be slower than market needs leading to some confusion and lack of ability to properly enforce controls.

Globalization and Harmonization

Latin American countries are not strangers to the globalization process. Thus, in recent years, many countries have sought to promote a policy shift from protecting national industries through openly protectionist policies towards open markets and free trade systems that foster competition within a global market framework. The rate of change has obviously varied from country to country depending on the type and number of free trade agreements whether regional or global and the government's ability to react to the needs created by the new commercial environment. In addition, most countries are subject to private sector pressures derived from the need to successfully participate in a global market and survive competition from imported products. Thus, governments are struggling to find the balance between encouraging more open commerce and making sure that the products coming into the country are safe while improving the country's ability to successfully participate in the market.

Most Latin American countries are participating at an official level in diverse international forums and organizations designed to harmonize legislation and enforcement mechanisms. Harmonization is not new in the region. Latin America came to the forefront of the concept in 1924 when during a chemistry congress in 1924 the drawing up a Codex Alimentarius Sudamericanus was proposed. In 1930, in Montevideo, Uruguay, the first Codex Alimentarius Sudamericanus was presented. It contained 154 articles that included general dispositions and food definitions. In 1960, the Latin American Food Code was published that, in conjuction with the European Food Code, was used as a source for the Codex Alimentarius (*1*). Nowadays, the Codex Alimentarius Commission generally has a counterpart representative in Latin America assigned as a contact point with the country's authorities. The Panamerican Health Organization (PAHO), the Food and Agriculture Organization of the United Nations (FAO) and the World Health Organization (WHO) have also been active in many countries in the region promoting programs to design active participation in harmonization processes and alleviate some of the aforementioned deficiencies.

In spite of the harmonization efforts, following and understanding the food control system in each country may be challenging. In general, most countries have a primary law such as the General Health Law and from there, regulations, standards, decrees and agreements are issued to enforce the law. Voluntary standards are also published. Food colors may be mentioned at all these levels. This may be confusing to companies that are trying to export either finished products or food additives into the country since they need to determine which specific agency and/or regulation applies to their product. This is further complicated by other issues common throughout the world. Definitions of simple terms such as natural, identical to natural, artificial or synthetic and added, may be very different in each country. Standards of identity for different products covered or not by generally accepted Codex standards may also vary. For example, in countries that are major producers of certain natural products such as tropical juices, the use of the term "natural" is highly restricted.

Food Color Regulations

As mentioned before, the development of food color regulations in Latin America is limited by several factors that include a reduced capability for conducting toxicological studies, limited availability of local data and limited resources for testing. Therefore, risk management decisions are usually made based on risk analysis data from testing conducted elsewhere.

Some countries have opted to fully adopt internationally recognized standards instead of duplicating the evaluation and management efforts. Countries that are completely based on the Codex Alimentarius include Belize, Bolivia, Costa Rica, Cuba, Panama and Peru. Countries that are part of the *Union Aduanera* Agreement (El Salvador, Guatemala, Nicaragua, and Honduras) base their regulations on Codex Alimentarius, but inclusion of a new color may not be automatic because some of these countries require a sanitary registry for each product. El Salvador also requires a quantitative declaration on the label of finished products.

MERCOSUR

Mercosur countries include Argentina, Brazil, Uruguay and Paraguay. The recent inclusion of Venezuela (2006) in this group will not be discussed in this section since Venezuela is still undergoing the harmonization process. Food law harmonization was conducted by the Technical Regulations Work Subgorup (SGT-3) under the responsibility of the Food Commission. Criteria for the approval of additives establish that provisions for food additives should take into account recommendations of the Codex Alimentarius and European Union directives. United States Food and Drug Administration (FDA) regulations are considered as supplementary references. Resolution Mercosur GMC No 14/93 (*2*) as amended by Resolution Mercosur GMC R 38/01 (*3*) provides the harmonized positive list for colors allowed in food products (Table I). Since the original adoption, the positive list of food colors and color stabilizers has been amended by consensus whenever new technological needs are presented by the member countries or new toxicological data become available.

When commercializing food colors and foods containing them in Mercosur it is also important to consider restrictions for products such as natural fruit juices. Colors are usually restricted in these types of products when they have a specific standard of identity. It is possible to add colors to beverages, but the denomination "natural" is usually restricted.

Chile

Food colors in Chile are considered food additives. All additives allowed in Chile must comply with the identification, purity and toxicity evaluation rules established by Codex Alimentarius. Colors must be declared on the label according to the specific name based on the International Numbering System (INS). Chile has published a positive list for additives and only additives found on Title 3, Paragraph II, Decree 977 may be used in the production of food products. Colors are listed on Article 145 (Tables II and III) (*4*). New additives may be added by the Ministry of Health by a new decree.

Colombia

Colombia covers food additives through Decree 2106 of 1983. This Decree defines and classifies food additives, regulates the use and manufacturing of food additives as well as their importation, transportation and marketing. Decree 2106 lists the information required to ask for the authorization to use an additive that is not listed. Although Colombian regulations do not specifically state that food additives approved by Codex Alimentarius are accepted, Article 5 of Decree 2106 states that the current recommendations of the Food Chemical Codex and the Codex Alimentarius or any other pertinent scientific publications will be taken into consideration in the formulation of regulations governing the use of food additives. Articles 7 and 8 establish an Advisory Committee to suggest changes to Colombian requirements for food additives (*5*). Different types of additives such as food dyes and food colors are covered under a specific regulation. Food colors and dyes are covered under resolution number 10593D. This resolution not only lists approved natural and artificial colors and dyes but also establishes purity criteria and limits for these products. In addition, this resolution establishes a limit for total concentration of combined synthetic colors of 300 mg/Kg based on the total maximum limit of each color. When tartrazine or yellow No. 5 is used, it is mandatory that an allergen warning be printed on the label (*6*). Tables IV-VII summarize the list of colors and dyes approved by this resolution.

Ecuador

In Ecuador, standards and regulations for food additives are set by the Health Code, the Food Regulation and the Technical Standard NTE INEN 2074-96. In general this Technical Standard is based on Codex Alimentarius Standards and FDA and establishes a positive list of allowed additives, regulations for tolerance levels and a list of forbidden substances. This list is periodically reviewed when new scientific, technological and toxicological information becomes available. Standards are copyrighted and are only available for sale through an individual request to the Ecuadorian Institute of Normalization (INEN) (*7*).

Table I. Harmonized list of colors approved in Mercosur

Nº IPS CODEX	*Color Index 1971/1982*	*Name*
100	75300	Curcuma, Curcumin
101(i)	-	Riboflavin
101(ii)	-	Sodium 5'-Phosphate Riboflavin
102	19140	Tartrazine
110	15985	FCF Sunset Yellow, Sunset Yellow
120	73470	Carmine/AC, Carminic cocínela
123	16185	Amaranth
124	16255	Ponceau 4R, Cochineal Red A
127	45430	Erythrosine
129	16035	Red 40, Allura Red
131	42051	Patent Blue V
132	73015	Indigotin, Indigo carmine
133	42090	FCF Brilliant Blue
140(i)	75810	Chlorophyll
140(ii)	75810	Chlorophyllin
141(i)	75815	Sodium and potassium salts of cupric chlorophyll
141(ii)	75815	Sodium and Potassium salts of Cupric chlorophyll
143	42053	Solid Green, Fast Green
150 a	-	Caramel I – simple
150 b	-	Caramel II – processed with caustic sulphite
150 c	-	Caramel III – processed with ammonium
150 d	-	Caramel IV – processed with sulphite-ammonium
153	-	Vegetable carbon
160 a (i)	40800	Beta-carotene (synthetic identical to natural)
160 a (ii)	75130	Carotenes: natural extracts
160 b	75120	Aumatto, Bixin, Norbixin, Urucu, Rucu
160 c	-	Paprika, Capsarubin, Capsaxanthin
160 d	75125	Lycopene
160 e	40820	Beta-Apo-8'-Carotenal
160 f	40825	Methylic or Ethylic Ester of Beta-Apo-8'-Carotenoic acid

Table I. *Continued.*

Nº IPS CODEX	*Color Index 1971/1982*	*Name*
161 b	-	Lutein
161 g	40850	Cantaxanthin
162	-	Beet red, Betain
163(i)	-	Anthocyanins (from fruits or vegetables)
170(i)	77220	Calcium carbonate
171	77891	Titanium dioxide
172(i)	77491	Iron oxide, black
172(ii)	77492	Iron oxide, red
172(iii)	77499	Iron oxide, yellow
173	77000	Aluminum
174	77820	Silver
175	77480	Gold
180	15850	Litol Rubine BK
104	-	Quinoline Yellow
128	-	Red 2G
151	-	Brilliant Black BN, Black PN
155	-	Brown HT
163ii	-	Grape skin extract

SOURCE: Mercosur. Resolution GMC 14/93; Mercosur. Resolution GMC 38/01 (*2, 3*).

Table II. Artificial Colors and Dyes Approved for Food Use in Chile

S.I.N.	Color Index	Name, Synonyms
110	15985	Twilight yellow, Dusk yellow (emphasis on the label)
104	47005	Quinoline Yellow
102	19140	Tartrazine (emphasis on the label)
133	42090(1)	Brilliant Blue
131	42051	Patent Blue V
132	73015	Indigotin, Indigo carmine
122	14720	Azorubine, Carmoisine
124	16255	Cochineal red
129	16035	Allura Red AC, Red 40
156	20285	Brown HT
151	28440 (1)	Brilliant Black BN
160a(i)	40800	Synthetic beta-carotene
143	42053	Green F.C.F., Solid Green F.C.F.
127	45430	Erythrosine (only in canned and maraschino cherries and fruit salad)
142	44090	Green S.

SOURCE: Reglamento Sanitario de los Alimentos. Decreto Supremo N° 977/96 (*4*)

Venezuela

Food additives are covered under the General Food Additive Norm COVENIN 910:2000. It is important to note that approved additives may differ from those approved in the United States. If a particular substance is not covered under this positive list, it is possible to submit a request to the Ministry of Health based on Codex Alimentarius information. Section 1.12 of this Norm covers approved colors and dyes. Tables VIII and IX summarize colors approved on this list. Limits are difficult to interpret since they are established as a Total Acceptable Daily Intake (ADI). This is a useful number to calculate a maximum limit per product, but without an official calculation of limits reported according to a total dietary study and estimation, each product may end up containing up to the ADI (*8*). The Norm also reports ADI without specification when toxicity is considered low and there is an acceptable history of use in foods and food products and JECFA has reported that the product does not present risk at

Table III. Natural Dyes and Derivatives Approved for Food Use in Chile

S.I.N.	*Color Index*	*Name, Synonyms*
160b	75120	Annatto, Achiote, Roucou, Bija (bixin and norbixin)
163	-	Anthocyanin
160e	40820	Apocarotenal, Beta-apo-8'-carotenal
160	-	Astaxanthin, 3,3'-dihydroxy-beta-carotene-4,4-dione
162	-	Betanin, beet red
161g	40850	Canthaxanthin, beta-carotene-4,4'-dione
120	75470	Carmine, carminic acid
150	-	Caramel (maximum 1000 mg/kg 4-methylimidazol)
153	-	Charcoal (use according to pharmacopeia)
160	75130	Alpha, beta, gamma Carotenes
140	75810	Chlorophyll
141	76810	Chlorophyll and associated copper salts
100	75300	Curcuma, curcumin
160f	40825	Ethyl esther of beta-apo-8' carotenoic acid
171	77891	Titanium dioxide
160c	-	Red pepper oleoresins
101	-	Riboflavin
173	77000	Aluminum (powder; for decoration only)
161(1)		Lutein
160d	75125	Lycopene

SOURCE: Reglamento Sanitario de los Alimentos. Decreto Supremo N° 977/96 (*4*)

Table IV. Natural and Identical to Natural Colors Approved for Food Use in Colombia

Name, Synonyms	*Color Index*	*Limit*
Alcanna, Alcanine, Ancusine, Orkanet	75530	GMP
Anthocyanins	-	GMP
Carotenes	75130	GMP
Beta-apo-8'-carotenoic acid		GMP
Beta-apo-8'-carotenal	40820	GMP
Canthaxanthin	40850	GMP
Xanthophylls	-	GMP
Achiote or Annato	75120	GMP
Saffron, Crocin, Crocetin	75100	GMP
Chlorophyll and copper derivatives including sodium and potassium salts	75810	GMP
Riboflavin and sodium salt of Riboflavin-5-Phosphate		GMP
Beet red and betain		GMP
Cochineal, carmine, and carminic acid	75470	GMP

NOTE: GMP:According to Good Manufacturing Practices

SOURCE: Resolución Número 10593 (*6*)

Table V. Inorganic Colors Approved for Food Use in Colombia

Name, Synonyms	*Color Index*	*Limit*
Metallic Aluminum	77000	GMP
Titanium dioxide	77891	10 g/Kg
Ferrous gluconate	-	GMP
Carbon black	77266	GMP
Metallic gold	77486	GMP
Metallic silver	77820	GMP

NOTE: GMP: According to Good Manufacturing Practices

SOURCE: Resolución Número 10593 (*6*)

Table VI. Artificial and Synthetic Colors Approved for Food Use in Colombia

Name, Synonyms	*Color Index*	*Limit*
Yellow		
Quinoline Yellow	47005	50 mg/Kg
Sunset Yellow FCF, Yellow 6	15985	200 mg/Kg
Tartrazine, Yellow 5	19140	100 mg/Kg
Blue		
Brilliant Blue, Blue 1	42090	100 mg/Kg
Indigo carmine, indigotin, Blue 2	73015	100 mg/Kg
Black		
Brilliant Black	2844C	300 mg/Kg
Red		
Amaranth, Red 2	16185	300 mg/Kg
Azorubin or carmoisine	14720	300 mg/Kg
Erythrosine, Red 3	45430	300 mg/Kg
Allura Red, Red 40	16935	GMP
Cochineal red, Ponceau 4R	16255	200 mg/Kg
Green		
Fast Green, Green 3	42053	100 mg/Kg
Brown		
HT Brown	20285	50 mg/Kg

NOTE: GMP: According to Good Manufacturing Practices. Calcium and Aluminun salts of the above mentioned synthetic colors are also allowed

SOURCE: Resolución Número 10593 (*6*)

Table VII. Caramel Colors Approved for Food Use in Colombia

Name, Synonyms	*Color Index*	*Limit*
Simple Caramel	-	GMP
Caramel processed with ammonium	-	GMP
Caramel processed with sulphite and ammonium	-	GMP
Caramel processed with caustic sulphite	-	GMP

NOTE: GMP: According to Good Manufacturing Practices

SOURCE: Resolución Número 10593 (*6*)

Table VIII. Artificial Colors and Lacquers Approved for Food Use in Venezuela

Name, Synonyms	*Color Index*	*Codex INS or ECC*	*ADI (mg/Kg)*
Allura Red, Rouge allura, FDC Red 40	16035	129-E-129	300
Brilliant Blue, Acid blue 9, Food blue 2, FDC Blue 1	42090	122-E133	12.5
Erythrosine, Acid red 51, CI Food red 14, FDC Red 3, Iodesine B	45430	127-E127	0.1
Indigo carmine, CI Acid blue 74, FDC Blue 2, Indigotin	73015	132-E132	5
Sunset Yellow, FDC Yellow 6, Gelborange S, L-Orange 2	15985	110-E110	2.5
Tartrazine, Acid yellow, Hydrazine yellow, CI Food Yellow 4, FDC Yellow 5	19140	102-E102	7.5
Ponceau 4R, Cochineal red A	-	124-E124	4

NOTE: Erythrosine approved only for cherries and chewing gum base; tartrazine must be clearly declared on label

SOURCE: COVENIN 910:2000 (*8*)

Table IX. Natural Colors and Derivatives Approved for Food Use in Venezuela

Name, Synonyms	*Color Index*	*Codex INS or ECC*	*ADI (mg/Kg)*
Annatto, Achiote, Rocu or Bija, Bixin, Norbixin, Onoto	-	160-E160b	0.065
Anthocyanins, Anthocyanates	-	168-E168	Not assigned
Apocarotene, Beta-apo-8'-carotenal	-	160-E160e	5
Astaxanthin, 3,3dihydroxy-carotene-4,4-dione	-	-	Not assigned
Betain, beet red, betanine	-	162-E162	Not assigned
Canthaxanthin, Beta-carotene-4,4-dione	-	161-E161g	2.5
Caramel	-	150-E150a	Not specified
Vegetable carbon, Carbo medicinalis vegetalis	-	153-E153	Not assigned
Carmine, carminic acid, cochineal	-	120-E120	2.5[2]
Alpha, beta and gamma carotenes	-	160-E160a	5
Chlorophyll	-	140-E140	Not specified
Copper salts of chlorophyll	-	141-E141	15
Curcuma, curcumin, turmeric yellow	-	100-E100	0.1[2]
Titanium dioxide, white pigment 6	-	160-E160f	Not assigned
Ethylic Ester of Beta-apo-8'-carotenoic acid, Beta-apo-carotenal	-	160-E160f	5
Vegetable oleoresins	-	-	Not assigned
Riboflavin, Vitamin B_2, Lactoflavin	-	101-E101	0.5

NOTE: Annatto reported as bixin; carmine and curcurma ADI are temporary

SOURCE: COVENIN 910:2000 (*8*)

normal levels of consumption. Temporary ADI has been established for compounds that have a pending JECFA evaluation. An ADI has not been assigned for some products for several reasons including lack of information that will enable the authorities to establish such a number. Products without an assigned ADI are evaluated on a case by case basis.

Mexico

Mexico has recently published (July 2006) a revised Agreement containing the positive list of food additives allowed in foods and food products. However, this is not the only document that addresses food colors. It is important to review several regulatory levels (General Health Law, Sanitary Regulation and Standards known as Mexican Official Norms –NOM) before assuming a particular product is allowed. The updated positive list includes colors (Table X) (*9*). It is important to note that the listed additives must comply with the established specifications in the provisions included in the Appendix Chapter VIII of the "Regulation for the Sanitary Control of Goods and Services". Table XI shows the maximum limits for flavored non-alcoholic beverages established by this regulation. In addition to these limits, the regulation also bans the commercial use of eggs from animals fed with feed containing Sudan reds and the commercialization of meat products and other products (food and beverage) containing Sudan, red 23, toney red, color index 2600 or Red DC 17. (*10*).

Conclusions

As Latin America continues its process of integration with a more globalized market, regulations will have to become more harmonized. This may ease pressure from the authorities that may be understaffed or lack the budget to properly evaluate food additives including food colors. The use of internationally accepted standards will also help simplify the management of these issues and make local regulations more transparent for producers, importers and exporters. Although change has been a dynamic process that has evolved very fast, navigating through the different laws, regulations and standards that may be enforced by more than one authority may be confusing. It is important to consider several levels or regulations before making a decision since colors may be addressed by more than one standard and control. It is also important to verify compliance with labeling regulations that may differ significantly from one country to the other. In all cases, countries have dispositions to request approval of a color that is not included in a positive list.

Table X. Colors Approved for Food Use in Mexico

INS	*Common name*	*C.I.*
100(b)	Curcumins	75300
101(i)	Riboflavin	Not reported
102	Tartrazine and their lacquers, Yellow 4 and their lacquers	19140
104	Yellow of quinoline, Yellow 13	47005
110	Yellow FCF and their lacquers, Yellow allowances 3 and their lacquers	14720
120	Cochineal extract (Extract of Coccus cacti L.), Red natural 4	75470
122	Azorubin and their lacquers, Red 3 and their lacquers	14720
123	Amaranth, Red 9	16185
124	Ponceau 4 R (Cochineal Red A), Red 7	16255
127	Erythrosine, Red 14	45430
129	Allura Red AC and their lacquers, Red 17 and their lacquers	16035
131	Patent Blue V, Blue 5	42051
132	Indigotin (indigo carmine) and their lacquers, Blue 1 and their lacquers	73015
133	Blue Brilliant FCF and their lacquers, Blue and their lacquers	42090
140(i)	Chlorophyll, Green natural 3	75810
140 (ii)	Chlorophyllins	
141 (i)	Chlorophyll copper complex, Green natural 3	75810
141 (ii)	Cholophyllins, copper complexes, sodium and potassium salts	75810
142	Green S, Green 4	44090
143	Fast Green FCF and their lacquers, Green 3 and their lacquers	75810
150(a)	Caramel I	Not reported

Continued on next page.

Table X. *Continued.*

INS	*Common name*	*C.I.*
150(b)	Caramel II	Not reported
150©	Caramel III	Not reported
150(d)	Caramel IV	Not reported
151	Brilliant Black PN, Black 1	28440
153	Vegetable carbon	Not reported
154	Brown FK, Brown 1	Not reported
155	Brown HT, Brown 3	20285
160(a)	Beta synthetic carotene, Orange 5	40800
160(e)	Carotenal, Beta-apo- (C30) Orange 6	40820
160(a)	Natural carotenes, Orange 5	75130
160(b)	Orange natural 4	75120
160(d)	Lycopene	75125
160©	Páprika oleoresin	Not reported
161(g)	Canthaxanthin, Orange 8	40850
161(b)	Lutein	Not reported
162	Beet red	Not reported
163 (ii)	Grape skin extract	Not reported

171	Titanium dioxide, White 6	77891
172(iii)	Iron oxide, yellow, Yellow 42	77492
172(i)	Iron oxide, black, Black 11	77499
172(ii)	Iron oxide, red, Red 101	77491
172	Iron oxides	77489
173	Aluminum, Metal 1	77000
174	Silver	77820
175	Gold (metallic)	77480
585	Ferrous lactate	
Not reported	Yellow 2G, Yellow 5	18965
Not reported	Orange B and their lacquers	19235
Not reported	Anthocyanins	Not reported
Not reported	Saffron (Stigmas of Crocus sativus L.), Yellow natural 6	75100
Not reported	Orange 7 (ethylic ester)	40825

SOURCE: Acuerdo por el que se determinan las sustancias permitidas como aditivos y coadyuvantes en alimentos, bebidas y suplementos alimenticios. (*9*)

Table XI. Maximum Limits of Colors Allowed in Flavored Non-alcoholic Beverages in Mexico

Color	*Maximum Limit (%)*
Allura Red (Red 40)	0.0100
Carmoisine red (Red 5)	0.0050
Ponceau red (Ponceau 4R)	0.0050
Tartrazine (Yellow 5)	0.0050
Sunset Yellow FCF (Yellow 6)	0.0100
Fast Green FCF (Green)	0.0100
Brilliant Blue (Blue 1)	0.0100
Indigotin (Blue 2)	0.0100

NOTE: The finished beverage should not exceed a color concentration of 100 mg/Kg taking into account the maximum limit of each individual color.
SOURCE: Reglamento de Control Sanitario de Productos y Servicios (*10*)

Although this process may be slow in some places, most countries are turning to international organizations such as Codex Alimentarius and the JECFA evaluations to make their risk management decisions. Latin America is not only a strong supplier of commodities and food products, it is also a very interesting market with many opportunities that are yet to be exploited.

References

1. Zylberman, P. *Med Hist.* **2004**, *48*, 1-28.
2. Mercosur. Resolution GMC 14/93.
3. Mercosur. Resolution GMC 38/01.
4. Ministerio de Salud de Chile. Reglamento Sanitario de los Alimentos. Decreto Supremo N° 977/96. www.usembassy.cl/_temporal/1011/Website/2003/Food%20Law/fas03e.htm (accessed Mar 11, 2007)
5. Ministerio de Salud. Republica de Colombia. Decreto Número 002106 de 1983. www.invima.gov.co/version1/normatividad/alimentos/decreto21061983%20.pdf (accessed Mar 12, 2007)
6. Ministerio de Salud. Republica de Colombia. Resolución Número 10593 de 16 de Julio de 1985. www.invima.gov.co/version1/normatividad/alimentos/Resolucion%2010593%20de%201985/Resolucion10593de1985.htm (accessed Mar 12, 2007)
7. USDA Foreign Agricultural Service. **2005**. Ecuador: Food and Agricultural Import Regulations and Standards. GAIN Report Number:EC5009. Date 7/27/2005.

8. FONDONORMA. Norma General Para Aditivos Alimentarios. Segunda Revision. COVENIN 910:2000. www.sencamer.gob.ve/sencamer/normas/910-00.pdf (accessed Mar 11, 2007)
9. Secretaria de Salud. México. Acuerdo por el que se determinan las sustancias permitidas como aditivos y coadyuvantes en alimentos, bebidas y suplementos alimenticios. Diario Oficial. 17/07/2006.
10. Secretaria de Salud. México. Reglamento Sanitario de Productos y Servicios. Date 9/08/1999. www.cofepris.gob.mx/mj/documentos/reg/RegDGCSPYS.pdf (accessed Mar 11, 2007)

Chapter 30

European Perspective on Sudan Dyes and the Structure of Food Color Legislation

Valerie Rayner

Sensient Colors UK, Oldmedow Road, King's Lynn PE 30 4LA, United Kingdom

The structure of EU food colour legislation is described giving the definitions of colours and introducing the usage legislation. The Sudan 1 affair in the UK in 2005 is used as a case study to describe the way colour legislation can be changed and possible future changes are introduced.

The European Union consists of 25 Member States (see Table I) and regulates food colours as part of the internal market legislation. The internal market legislation is also applicable in other countries that are not full members of the Union but have signed the European Free Trade Agreement, therefore it is applicable in Norway and Switzerland. Following political changes in Eastern Europe in the 1980s and 1990s, the Union has grown considerably in both size and influence as new countries have joined.

The Framework of EU Colour Legislation

The first European colour legislation was agreed in 1962. It consisted of a positive list of colours, but without any conditions of use. Some countries generally permitted all the listed colours whilst others introduced complex conditions of use. The situation was common across all categories of food additive and resulted in foods being acceptable in one country but not others.

In the mid 1980s work was started on a new package of additives legislation which would go into much more detail and so remove the trade barriers. In 1989 the basic Framework Directive *(2)* was published setting out the structure and

Table I. Countries where EU Food Colour legislation applies directly *(1)*

Country	*Year of Accession*
Belgium	1957
Germany	1957 (as West Germany)
Luxembourg	1957
France	1957
Italy	1957
Netherlands	1957
Denmark	1973
Ireland	1973
United Kingdom	1973
Greece	1981
Spain	1986
Portugal	1986
Austria	1995
Finland	1995
Sweden	1995
Cyprus	2004
Czech Republic	2004
Estonia	2004
Hungary	2004
Latvia	2004
Lithuania	2004
Malta	2004
Poland	2004
Slovakia	2004
Slovenia	2004

definitions (see Table II for definitions of EU terms). One of the principles was that the legislation must be based on agreed scientific and technological criteria.

Colour Usage

The details of colour usage legislation are given in Directive 94/36/EC *(3)*. It begins by setting out the principles and definitions and then contains five Annexes containing the technical details.

Table II. EU Glossary

European Parliament	Directly elected Parliament representing the people of the Member States This is separate from the Parliaments of the individual Member States.
Council of the EU (formerly Council of Ministers)	This institution consists of government ministers from all the EU countries. The Council meets regularly to take detailed decisions and to pass European laws.
The European Council	This is the meeting of heads of State and government (i.e. presidents and/or prime ministers) of all the EU countries, plus the President of the European Commission. The European Council meets, in principle, four times a year to agree overall EU policy and to review progress. It is the highest-level policy-making body in the European Union, which is why its meetings are often called "summits".
European Commission	Executive arm of the EU and guardian of the Treaties; working for the EU as a whole and not individual Member States. Responsible for preparing legislation.
EFSA - European Food Safety Authority	A group of experts who advise on the safety aspects of food.
SCFCAH - Standing Committee on the Food Chain and Animal Health	The regulatory committee with responsibility for food. Its members are representatives of the Member States and it is chaired by a member of the Commission.
Directive	A piece of Community legislation which binds the Member States as to the results to be achieved; they have to be transposed into the national legal framework and thus leave a margin for maneuver as to the form and means of implementation
Regulation	A piece of Community legislation which is binding in its entirety and directly applicable in all Member States. Commonly used for agricultural legislation.
Decision	A piece of legislation which is fully binding on those to whom they are addressed

Definition of a colour

The legislation defines colours as:

"Substances which add or restore colour in a food, and include natural constituents of foodstuffs as such and not normally used as characteristic ingredients of food."

"Preparations obtained from foodstuffs and other natural source materials obtained by physical and/or chemical extraction resulting in a selective extraction of the pigments relative to the nutritive or aromatic constituents…"

The following are not considered as colours: "foodstuffs, whether dried or in concentrated form, and flavourings incorporated [...] because of their aromatic, sapid or nutritive properties together with a secondary colouring effect such as paprika, turmeric or saffron."

Interpretation

There is a fundamental difference in this case between the USA colour legislation and European. The Code of Federal Regulations lists fruit and vegetable juices as permitted colour additives. In Europe fruit and vegetable juices would be considered as foodstuffs with a secondary colouring effect and therefore not be listed as colours. Of course if there is any selective extraction of the colour compounds (e.g. extracting the carotenoids from carrot juice) then the compound would need to be permitted as a colour.

Carry-over

One of the principles mentioned is that of carry-over, where colour is permitted in a foodstuff to the extent that it is permitted in one of its ingredients, or an ingredient if it is intended solely for the manufacture of a food where the colour would be allowed.

The best-known example of this in the UK is meat marinades. Meats cannot be coloured but marinades can. There is a chicken dish known as chicken tikka where the meat is marinated in an orange-coloured marinade (along with flavourings and tenderizers) that imparts the orange colour to the meat.

Natural Colours

There is no definition for "natural" colours. In fact the legislation treats all colours in the same way and makes no distinction between natural and synthetic colours.

Annexes I-IV

The first Annex consists of a list of all the permitted colours, their E numbers and Colour Index numbers.

The second Annex contains the list of foods which cannot be coloured unless mentioned elsewhere. These are generally staple foods where colour would not normally be used. The listing of pasta is a surprise to many, however it can still be coloured with vegetable juices and other coloured foodstuffs – which are not within the scope of the EU colours legislation.

Annex III consists of the foods to which only certain colours may be added. This section grew considerably during its development as countries added their national specialties as exceptions to the legislation. Some are "special products" such as whisky where the authorities wanted to restrict the colours permitted. Others are just exceptions such as various coloured cheeses. Many of these entries are not translated in different language versions, and other products are translated but not easily recognizable outside of the country where they are traditional e.g. broken white cheese, a type of traditional French cheese.

Annex IV is almost the opposite of Annex III and lists the colours which are only permitted in certain foods. It is where the science behind the European approach becomes visible as it uses the principle of scientifically determined Acceptable Daily Intakes (ADIs). ADIs are allocated based on toxicological studies and represent the amount an average person can safely consume daily over a lifetime. Some colours have lower ADIs than others, and those with the lowest are included in this section. Erythrosine (FD&C Red 3) is one of the best known of these colours. It has an ADI of 0-0.01 mg/kg body weight/day compared to 0 – 7 mg/kg body weight/day for Allura Red (FD&C Red 40) and is only permitted in cherries. Annatto is also mentioned in this Annex. Its ADI is only 0-0.65 mg/kg body weight/day, not especially because of any safety issue but more because of the original testing protocols. It is the traditional colour used in cheeses and is permitted for a range of foods, but with very strict usage limits.

Annex V

The most useful part of the Directive is at the end. This section is split into two parts; firstly in Table III it lists colours that are generally permitted *quantum satis* i.e. to good manufacturing practice and thenin Table IV lists the other colours along with the applications and maximum usage limits.

The second Part lists the colours with the total maximum permitted levels in various foods. For four of the colours there is an additional restriction of 50 mg/kg each in some foods and so if these are used then other colours can also be added up to the maximum permitted level. Table IV gives details of the

Table III. Annex V Part I Colours Permitted Q.S.

E Number	*Name*
E101	Riboflavin / Riboflavin 5 phosphate
E140	Chlorophylls and Chlorophyllins
E141	Copper complexes of Chlorophylls and Chlorophyllins
E150 a-d	Caramels
E153	Vegetable Carbon
E160a	Carotenes
E160c	Paprika extract, capsanthin, capsorubin
E162	Beetroot red, betanin
E163	Anthocyanins
E170	Calcium Carbonate
E171	Titanium Dioxide
E172	Iron Oxides and Hydroxides

maximum levels for some of the more popular applications. The maximum levels are in mg/kg or mg/l.

National Implementation

As this piece of legislation is a Directive, it must be implemented in the national legislation of each Member State. There is then the possibility of differences in interpretation. Points of interpretation can be raised by Member States at the SCFCAH meetings, however where the interpretation is not clear it is best to check the legislation and interpretation in each Member State.

Purity Criteria / Specifications

The second part of the color legislation is the specifications Directive: Commission Directive 95/45/EC laying down specific purity criteria concerning colours for use in foodstuffs *(4)*. This Directive works through each individual colour and sets out the specification. There are no test methods mentioned in these specifications, although wherever possible there is an extinction coefficient given for each colour allowing the quantity of active colour to be measured. The EU is a member of Codex Alimentarius and unless the EU Directive states otherwise, the FAO/WHO test methods published in their Guide to specifications will apply.

Table IV. Annex V Part II colours permitted with maximum levels

Colour	Soft Drinks	Confectionery	Edible Ices & Desserts	Decorations & Coatings
E100 Curcumin (Turmeric Extract)	100	300	150	500
E102 Tartrazine (Yellow 5)	100	300	150	500
E104 Quinoline Yellow	100	300	150	500
E110 Sunset Yellow (FD&C Yellow 6)	50*	50*	50*	500
E120 Carmine	100	300	150	500
E122 Carmoisine	50*	50*	50*	500
E124 Ponceau 4R	50*	50*	50*	500
E129 Allura Red (FD&C Red 40)	100	300	150	500
E131 Patent Blue V	100	300	150	500
E132 Indigo Carmine (FD&C Blue 2)	100	300	150	500
E133 Brilliant Blue (FD&C Blue 1)	100	300	150	500
E142 Green S	100	300	150	500
E151 Brilliant Black BN	100	300	150	500
E155 Brown HT	50*	50*	50*	500
E160d Lycopene	100	300	150	500
E160e Beta Apo 8 Carotenal	100	300	150	500
E160f Ethyl ester of beta apo 8 carotenoic acid	100	300	150	500
E161b Lutein	100	300	150	500

* 50 mg/kg limit applies to the individual colours

Case Study – Sudan 1 incident in the UK

The Sudan 1 incident was certainly the most significant event in food colour legislation during 2005. As far as the general public were concerned it began in February 2005 with the withdrawal of nearly 600 products from sale in the UK due to the presence of a banned dye.

Previous Developments

Initial Discovery and Notification via European Alert System

The UK had first become aware of the problem in 2003 when the French authorities detected Sudan 1 in a consignment of chilli powder *(5)*. Sudan 1 is not listed as a permitted colour, nor can colour be added to spices therefore the chilli powder was not legal.

The French authorities alerted the other EU Member States via the Rapid Alert System for Food and Feed (RASFF). The RASFF was established in 1979 in order to give Member States a method of exchanging information on measures taken to ensure food safety.

Emergency Orders issued by European Commission

The French authorities had taken interim protective measures within France. Once other Member States were aware of the issue they also started testing and more contaminated material was identified. By June 2003 this was considered serious enough for the European Commission to introduce emergency measures throughout Europe in the form of new legislation *(6)*. Commission Decision of 20th June 2003 on emergency measures regarding hot chilli and hot chilli products requires all shipments of crushed or ground chilli to be accompanied by an analytical report demonstrating that the product did not contain Sudan 1.

During the latter part of 2003 there were occasional product recalls including 25 chutney and relish products from five different manufacturers. The problem still did not go away and more banned dyes were detected. On the 21st January 2004 the Emergency Order was extended to include the other banned dyes that had been found and also to include curry powder as well as hot chilli *(7)*. The Food Standards Agency issued guidance to the food industry asking all manufacturers who use chilli powder or chilli products as part of the ingredients in their products to ensure that if it was imported from India and supplied to them prior to 27 January 2004 that they must ensure that it has not been contaminated with Sudan I - IV. *(8)*

Product recalls

Sudan 1 was detected in a consignment of Worcester Sauce in Italy. The UK Authorities were alerted to it via the RASFF and a product recall was undertaken. Worcester Sauce is used as a savoury seasoning in many traditional British foods and all products containing the contaminated batch of Worcester sauce were withdrawn. New legislation had come into place in Europe in January 2005 requiring all food operators to be able to trace the supplier or consignee for any batch of product. This meant all the customers of the batch of worcester sauce could be traced through the supply chain and in the end approximately 600 products were identified in the UK by the end of February 2005 *(9)*.

Previous recalls had been mainly spice mixes but the Worcester sauce had been used in a wide range of prepared foods. The products affected included many well known and trusted traditional brands and so there was much front-page press coverage with sensational headlines.

The cost of the product withdrawals alone was estimated to be £100,000 (British Pounds) however there would be other costs from the loss of consumer trust in prepared foods *(10)*.

Enforcement Action

At a European level the Commission prepared a document to remind the foods industry of the obligations of food businesses under food safety legislation.

Enforcement of food law is carried out at a national level and each Member State has its own systems. In the UK food adulteration is dealt with by Trading Standards Officers. At the time or writing there have been no prosecutions directly relating to this affair, however the chilli supplier was charged over a subsequent contamination incident. The charge brought was "Food not of the substance demanded by the consumer", which is typical in a fraud case. The supplier pleaded guilty and was fined £2,000 with £3,000 costs *(10)*.

Sunset Yellow / FD&C Yellow 6

Sudan 1 is the unsulphonated version of Sunset Yellow. As such it is normally present at part per million levels. In the USA there is a limit of 10 mg/kg in Yellow 6, however there is no specific limit in the EU so it would be included within the total subsidiary dye limit of 5%. The UK FSA wanted a limit introduced for the Sudan 1.

Legal basis for EU Colour Specifications

The EU colour specifications are given in Commission and Council Directive 95/45/EC laying down specific purity criteria concerning colours for use in foodstuffs *(4)*. As a Commission Directive the procedure for amending it is relatively straightforward and uses a procedure known as Comitology. Under this procedure the Commission must work in cooperation with the Standing Committee on Food Chain and Animal Health.

Procedure used for changing the specification

In this case the UK delegation raised the issue at a routine meeting of the SCFCAH in September 2005 *(11)* and the Committee agreed that the specification should be changed. If there had been any queries as to the safety then EFSA would have been asked to give their opinion, however they had already prepared an opinion on Sudan 1. The Commission prepared a proposed amendment, including some other routine changes to colour specification, and circulated it for comments. The proposal was agreed at the meeting in December 2005 *(12)* and then published in all 20 official languages in March 2006 *(13)*.

Introducing New Colours and Uses

Is it a colour?

The EU definition of a food colour was given earlier and does not include fruit and vegetable juices. Therefore if a highly coloured vegetable juice were being used then the general food law would apply and not additive legislation. When considering this it is important to ensure the vegetable juice is not a "novel food". Legislation was passed in 1997 *(14)* requiring any foods not consumed in the EU to a significant degree to be assessed for safety and approved.

Changing the usage legislation

The Specifications Directive 95/45/EC has been amended on four occasions since 1995, whereas the usage Directive 94/36/EC has never been changed. This may be because of the complex procedure known as Co Decision. A full description of the procedure can be found on the EU website *(1)*. The procedure involves the Commission to write the legislation which must then be approved by the European Parliament and Council in one, two or three readings. At each stage there is the possibility of compromises being made. For a subject as controversial as food colours there is likely to be a lot of debate. When the legislation was first introduced there were only 12 Member States, now with around twice as many members agreement is likely to be more difficult.

Future developments in EU colour legislation

Review of safety

The European Food Safety Authority is currently carrying out a review of the safety data for food colours. The outcome of this review may lead to changes in the colour legislation.

New Framework for Additive Legislation

The basic framework for food additive legislation *(2)* is currently being rewritten to include flavours and enzymes. It has been developed over the past few years and is still at a draft stage. Some of the proposed changes have included:

- Change to definition of processing aids so that they would need to be intentionally removed or otherwise be considered as ingredients or additives.
- 10 year authorizations so permission for additives and their uses would need to be renewed every ten years
- Removal of national temporary authorizations whereby a Member State can temporarily permit an additive whilst it is going through the EU system.
- Amendments to use quicker Comitology procedure instead of full Co-Decision procedure.

At this early stage there is considerable scope for the proposal to change so the points above may not be included in the final version.

Summary

EU colour legislation is controlled by two Directives, one covering usage and the other specifications. The usage legislation gives lists of permitted colours and complex conditions of use with maximum permitted levels. It has never been amended and any changes would follow a complex procedure involving the European Parliament and Council.

The use of a banned dye (Sudan 1) in chilli powder forced the Commission to introduce emergency legislation and when a contaminated batch was identified two years later it caused many foods to be withdrawn from sale in the UK. The specification for Sunset Yellow was changed using the comitology procedure. This procedure only involves the Commission and Standing Committee and so the legislation could be changed within 6 months.

Colour legislation is currently under review in Europe however it is too early to know exactly what changes will be made.

References

1. European Union Home Page. http://europa.eu (accessed Mar 11, 2007)
2. *Official Journal of the European Communities* 11 February 1989 L40 27-37
3. *Official Journal of the European Communities* 10th September 1994 L237 13-29
4. *Official Journal of the European Communities* 22 September 1995 L226 1-40
5. European Commission http://europa.eu.int/comm/food/food/ chemicalsafety/ fraudulent/qanda_en.pdf (accessed Mar 11, 2007)
6. *Official Journal of the European Communities* 21 June 2003 L154 114-115
7. *Official Journal of the European Communities* 30 January 2004 L27 152-154
8. Food Standards Agency 18 February 2004 http://www.food.gov.uk/ foodindustry/guidancenotes/foodguid/sudanguidance (accessed Mar 11, 2007)
9. Food Standards Agency http://www.food.gov.uk/news/newsarchive/ 2005/mar/sudanlist (accessed Mar 11, 2007)
10. Frith, M. The Independent Newspaper, UK 23 February 2005.
11. Standing Committee On The Food Chain And Animal Health Section On Toxicological Safety, Summary Report of the Meeting of 19,20 September 2005 http://europa.eu.int/comm/food/committees/regulatory/scfcah/toxic/ summary17_en.pdf (accessed Mar 11, 2007)
12. Standing Committee On The Food Chain And Animal Health Section On Toxicological Safety, Summary Report of the Meeting of 16 December 2005 http://europa.eu.int/comm/food/committees/regulatory/scfcah/toxic/ summary18_en.pdf (accessed Mar 11, 2007)
13. *Official Journal of the European Communities* 21 March 2006 L82 10-13
14. *Official Journal of the European Communities* 14 February1997 L 043 1-6

Chapter 31

Color Regulations in Asia

Takatoshi Koda

Coloring Committee, Japan Food Additives Association, Food Color Division, San-Ei Gen F.F.I. Inc., 1–11–11, Sanwa-cho, Toyonaka, Osaka, 561–8588, Japan

The regulations on food additives in most Asian countries are, in principle, enforced by means of the positive list method and, in reality, additives that can be used may differ country by country. This has been recognized as a serious impediment that may interfere with international distribution of food in the export and import of food, and therefore movement toward international harmonization, led by the Joint FAO/WHO Food Standards Commission (CODEX), has been actively taking place. With this background, there is a movement in Japan, as part of the international harmonization of food additives, which aims at gaining regulatory approval by February 2007 of food additives with respect to which safety evaluation has been completed internationally at JECFA and the safety of which has been confirmed to a certain extent, and those which are widely used in the United States and the EU member countries and internationally considered to be of high necessity. In this article, we will focus on the present conditions and the future issues in the regulations on coloring in Japan and then outline the situations in other Asian countries.

Regulations on Food Coloring in Japan

In Japan, the changes in lifestyle due to the affluence that resulted from economic growth gave rise to significant changes in diet. Accordingly, the processed food industry has not only expanded its size but also sophistication through a number of significant technological developments through which, combined with the revolution in distribution, we now see multitudes of processed foods in convenience stores and supermarkets. Developments in food processing technologies and the functions of food additives have made a primary contribution to this expansion. In the manufacturing of processed food, food additives are used in order to achieve various effects, which can be classified as follows:

1. Necessity for manufacturing or processing: coagulants for tofu, alkaline preparations, filtration aids, etc.
2. For improved preservability and prevention of food poisoning: preservatives, antioxidants, etc.
3. For improved taste and quality: colorings, flavors, seasonings, etc.
4. For supplementation and enrichment of nutrient components: vitamins, minerals, etc.

As you can see, coloring plays an important role in the enrichment of diet by forming the color of food, appealing to the human eye, and, together with taste and flavor, creating the perception of deliciousness. Now, coloring having these secondary functions can be classified into chemically synthesized coloring and natural coloring. In Japan, as its economy has grown, natural coloring has been replacing inexpensive chemically synthesized coloring, and the demand for the former has grown substantially. Behind this phenomenon, there is a consumer preference for natural colors to synthetic ones and consumers are ready to accept colors derived from plant material in particular. Next, we will discuss the historical background of color regulations in Japan, list of and standards for the natural coloring and chemically synthesized coloring, which have developed out of this unique Japanese cultural background.

Historical Background of Japanese Food Sanitation Law (*1-4*)

In 1947, the Ministry of Health, Labour and Welfare enacted the Food Sanitation Law as the first comprehensive law concerning food. Upon the enactment of this law, a positive list system for food additives was introduced pursuant to which only food additives that were designated by the Minister of Health, Labour and Welfare as safe additives could be used for food. Since 1947, this law has applied to any and all additives. On the other hand, until 1995 when the Food Sanitation Law was amended, this positive list system only applied to chemically synthesized additives, and the law did not apply to any other additives (so-called natural additives), which were regulated by voluntary control of manufacturers. The Product Liability Law was announced in July, 1994, and to secure the safety of additives, enshrining in law the concept of corporate liability. Upon the amendment to the Food Sanitation Law in 1995, all natural additives that were actually distributed in the market were listed and classified into the three categories, those of existing food additives, additives to be generally used as foods, and natural flavors. After the amendment, no new additive that is not listed, whether synthesized or natural, can be sold or processed without the designation of the Minister of Health, Labour and Welfare after following a prescribed procedure.

Figure 1 shows the change in the classification of food additives by the 1995 amendment (*5*).

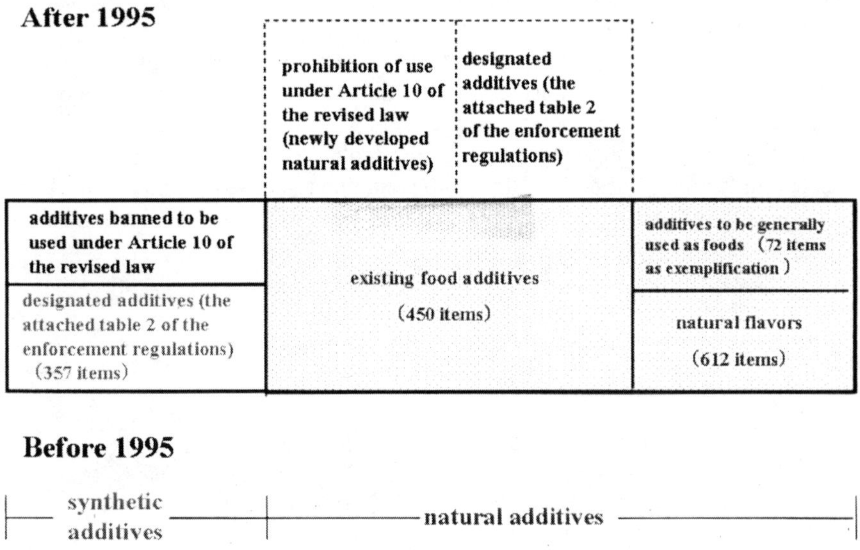

Figure 1. Difference between the categories of before and after the law amendment in 1995. (The number of items in a figure is numerical value as of April, 2006).

Scope of Coloring

Coloring is categorized as either "designated additives that are food additives designated by the Minister of Health, Labour and Welfare pursuant to the Article 10 of Food Sanitation Law and listed in the Food Sanitation Law Enforcement Rules Schedule 1" (Table I), "existing additives that are not designated by the Minister of Health, Labour and Welfare but are listed in Attachment 1 of the Notice of the Director of Environmental Health Bureau, with *eika* No. 56, dated May 23, 1996"

(Table II) or "general food and drink additives that are listed in the list of items that are generally used as food for eating or drinking and used as additives in Attachment 3" (Table III). These are natural additives that had already been used in food at the time of the amendment to the Food Sanitation Law in 1995. Accordingly, no new additive that is not listed in this list can be used for food even if it is naturally derived. No new additive can be used for food unless it is designated pursuant to the Notice of the Director of Environmental Health Bureau, the Ministry of Health and Welfare, with *eika* No. 29, of 1996 titled "The guidelines on designation of food additives and amendment to standards of use" (briefly, the "Guidelines"). However, from the perspective of international consistency, discussion is being made on additives with respect to which safety evaluation has been completed and which are generally used in overseas markets. With respect to coloring, the three items, β-apo-8'-carotenal, carmine and canthaxanthin, were scheduled to be discussed in 2006, but this schedule may well be delayed.

Drug formulations including coloring are classified as coloring formulations (single purpose formulations used for the purpose of coloring food) and combination formulations with multiple purposes (multiple purpose formulations) consisting of base compound, which is one "coloring" or a formulation of two or more kinds of "coloring," with a sub-compound (a food additive other than coloring) formulated for the purpose of stabilization of the base compound and the drug formulation.

Table I. Designated Additives: Tar and Other Synthetic Colors (30 Items)

INS No.	*Listed Names: Tar Colors*	*INS No.*	*Listed Names: Other Synthetic Colors*
	Acid Red (Food Red No. 106)		Annatto, water soluble
129	Allura Red (Food Red No. 40)		(potassium or sodium salt)
	Allura Red aluminum lake	160a(i)	Beta-carotene
123	Amaranth (Food Red No. 2)	141	Chlorophylls, copper complexes
	Amaranth aluminum lake	172ii	Iron oxide
133	Brilliant Blue FCF (Food Blue No. 1)	101i	Riboflavin
	Brilliant Blue FCF aluminum lake	101ii	Riboflavin 5' phosphate, Na salt
127	Erythrosine (Food Red No. 3)		Riboflavin tetrabutyrate
	Erythrosine aluminum lake	141ii	Sodium copper chlorophyllin
143	Fast Green FCF (Food Green No. 3)		Sodium iron chlorophyllin
	Fasr Green FCF aluminum lake	171	Titanium dioxide
132	Indigo carmine (Food Blue No. 2)		
	Indigo carmine aluminum lake		
	New coccine (Food Red No. 102)		
	Phloxine (Food Red No. 104)		
	Rose Bengal (Food Red No. 105)		
110	Sunset Yellow (Food Yellow No. 5)		
	Sunset Yellow aluminum lake		
102	Tartrazine (Food Yellow No. 4)		
	Tartrazine aluminum lake		

Table II. Existing Food Additives: Color (63 Items)

Type	*Colorant*	*Type*	*Colorant*
a	Alkannet color	a	Koroo color
	Aluminum	a	Krill color
a	Annatto extract	c	Lac color
	Bamboo grass color		Lithospermum root color
b	Beet red		Logwood color
	Bone carbon black	b	Marigold color
	Cacao carbon color	b	Monascus color
a	Cacao color	a	Monascus yellow
b	Caramel, class I	a	Onion color
b	Caramel, class II	a	Orange color
b	Caramel, class III	b	Palm oil carotene
b	Caramel, class IV	b	Paprika color or oleoresin
a	Carob bean color	a	Pecan nut color
b	Carrot carotene	a	Phaffia color
b	Carthamus red		Powdered annatto
b	Carthamus yellow	c	Purple corn color
b	Chlorophyll	c	Purple sweet potato color
	Chlorophyllin	a	Purple yam color
b	Cochineal extract, carminic acid	d	Rutin (extract)
a	Corn color	a	Sandalwood red
a	Crayfish color		Sappan color
b	Dunaliella carotene	a	Shea nut color
d	Enzymatically modified rutin (extract)		Shrimp color
a	Fish scale foil		Silver
c	Gardenia blue	c	Spirulina color
c	Gardenia red		Sweet potato carotene
c	Gardenia yellow	a	Tamarind color
	Gold	c	Tomato color, Tomato lycopene
b	Grape skin color	b	Turmeric oleoresin, Curcurmin
c	Haematococcus algae color	a	Vegetable carbon black
a	Japanese persimmon color		Vegetable oil soot color
a	Kaoliang color		

a: listed in Voluntary Specifications, 3rd ed. published by the Japan Food Additive Assoc.

b: listed in Japan's Specifications and Standards for Food Additives, 7th ed.

c: listed in Japan's Specifications and Standards for Food Additives, 8th ed. (tentative)

d: an antioxidant

Table III. Additives Generally Used as Foods (44 Items) (as example)

Type	*Colorant*	*Type*	*Colorant*
	Azuki color		Fruit juice (cont'd):
a	Beefsteak plant color		strawberry juice
a	Blackberry color		thimbleberry juice
b	Black currant color		uguisukagura juice
	Black huckleberry color		whortleberry juice
a	Blueberry color		Gooseberry color
a	Boysenberry color	a	Grape juice color
	Crowberry color	a	Hibiscus color
	Cherry color	a	Laver color
a	Chicory color		Loganberry color
	Cocoa		Malt extract
a	Cranberry color		Morello cherry color
	Dark sweet cherry color		Mulberry color
a	Elderberry color		Olive tea
	European dewberry color		Paprika
	Fruit juice:		Plum color
	berry juice		Powdered chlorella
	black currant juice	a	Raspberry color
	blackberry juice	c	Red cabbage color
	blueberry juice	a	Red currant color
	boysenberry juice	a	Red radish color
	cherry juice	a	Red rice color
	crowberry juice		Saffron
	cranberry juice	a	Saffron color
	dark sweet cherry juice		Salmonberry color
	dewberry juice	a	Sepia color
	elderberry juice	a	Strawberry color
	gooseberry juice		Tea
	grape juice		Thimbleberry color
	huckleberry juice		Turmeric
	lemon juice		Uguisukagura color
	loganberry juice		Vegetable juice:
	morello cherry juice		beefsteak plant juice
	mulberry juice		beet red juice
	orange juice		carrot juice
	pineapple juice		onion juice
	plum juice		red cabbage juice
	raspberry juice		tomato juice
	red currant juice	a	Whortleberry color
	salmonberry juice		

a: listed in Voluntary Specifications, 3rd ed. published by the Japan Food Additive Assoc.
b: listed in Japan's Specifications and Standards for Food Additives, 7th ed.
c: listed in Japan's Specifications and Standards for Food Additives, 8th ed. (tentative)

Standards for Coloring

The current regulations on food additives are based on the Food Sanitation Law (Law No. 233) enacted on December 24, 1947. After several amendments, in 1957, a provision that provides for the standards for food additives was added to ensure the further safety of food. Pursuant to this provision, the standards for food additives, etc. (the Ministry of Health and Welfare Announcement No. 370 dated December 28, 1959) was established. This Announcement, including any amendments thereafter, is the basis of the administration of food and sanitation, covering the standards for food, packaging, etc. as well as food additives.

Pursuant to the Article 7 in old version (Article 11 in current version) of Food Sanitation Law, the first edition of the Japanese Standards for Food Additives that contained Standards for Food Additives and standardized general test methods was published on March 15, 1960. The Japanese Standards for Food Additives has been revised about every five years, and the seventh edition published in April 1999 is the latest. Currently, discussion toward the eighth edition is under way.

With respect to coloring, we have the second edition of the Voluntary Standards for Existing Additives published by the Japan Food Additives Association under the guidance of the Ministry of Health, Labour and Welfare in October 1993 as voluntary standards for the industry. Based on standards used in the second edition of the Voluntary Standards for Existing Additives, the seventh edition of Japan's Specifications and Standards for Food Additives has listed the standards for 18 new items of natural coloring. With respect to major natural additives other than those contained in the seventh edition, we have the third edition of the Voluntary Standards for Existing Additives published by the Japan Food Additives Association under the guidance of the Ministry of Health, Labour and Welfare in November 2002 as voluntary standards for the industry. Within these Voluntary Standards, 52 items in total of natural coloring, which consist of 33 items of existing additives and 19 items of additives generally used as foods, are contained and almost all items of natural coloring currently distributed have been standardized. However, standardization of natural coloring raises various issues that we do not face with synthesized coloring, as follows:

- The component composition and pigment content may differ depending on the original animal or plant, type, area of production, time of harvesting, the manner of processing, and so on.
- Because it is natural, the component composition is complicated, and in many cases even the pigment composition has not been isolated, refined or structurally determined. (Some old literature even shows wrong chemical structures)
- Natural color is rarely composed of a single pigment, and in many cases is constituted by multiple components.
- In many cases, the pigment content is not constant.

- In many cases, there is no standard product.
- There are few analysis and studies on separation from food or identification.

Accordingly, it is considered that the key and immediate challenges in the establishment of standards for natural coloring are to determine the structure of primary components, to make an adequate research in quality and safety, and to come up with standards that can be effective in the international evaluation and are consistent with the standards for chemically synthesized food additives. Figure 2 shows the standards for "Cochineal Extract" contained in the seventh edition of the Japanese Standards for Food Additives. Also, Tables IV and V show the examples of "Carthamus yellow color" from the Existing Food Additives List, and "Red cabbage color" from the General Food and Drink Additives List, respectively. As you can see, the natural origin, process, and example label indication for food are provided.

Meanwhile, a new law was notified on May 30, 2003 with respect to existing additives. Pursuant to this law, the name of any additive can be removed from the list of existing additives should it be determined that such additive may be harmful to human health. In fact, xanthin was removed from the list of existing additives in July 2004 because it was determined from the result of animal testing that it may be harmful to human health. Further, pursuant to the Law Amending Part of the Food and Sanitation Law, etc. (2003 Law No. 55. the "Amendment Law") enforced on August 29, 2003, a new provision of Article 2-3 of Supplementary Provisions of the Law Amending Part of the Food and Sanitation Law and the Nutrition Improvement Law (1995 Law No. 101) was added. Pursuant to the provision, the Minister of Health, Labour and Welfare may, if he determines, judging by the conditions of sale, manufacture, import, processing, use, storage and labeling (the "Sale, etc."), that any existing additive as well as formulations and foods containing that additive are not used for sale, prepare and publish the "list of scheduled removal" on such an additive, formulation and food, and remove them from the list of existing additives through a prescribed procedure.

Standards of Use for Food Coloring

With respect to the use of coloring for food, certain legal standards of use are provided and thereby any coloring that may deceive consumers is strictly restricted. Deceitful conduct such as making out raw meat that is not fresh to be otherwise by red coloring or making out tea to be of high grade by adding green coloring should be strictly restrained in the processing and sale of food since such conducts would not benefit consumers. Naturally, any party violating these legal standards may be subject to punishment.

Synthesized Coloring

Neither tar color for food nor titanium dioxide can be used in the following foods:

Definition	Cochineal extract is obtained from cochineal, of which coloring principle is carminic acid
Color Value	The Color Value (absorbance of a 10% solution in a 1 cm cell) of Cochineal Extract is not less than 80 and is in a range of 95-115% of the labeled value
Description	Cochineal Extract is red to dark red powder, lumps, liquid or paste, having a slightly characteristic odor
Identification	(1) Weigh the equivalent of 0.5 g of Cochineal Extract calculated in terms of Color Value 80 from the labeled value, dissolve in 100 mL of 0.1 M hydrochloric acid, and centrifuge. The supernatant is orange in color and exhibits an absorption maximum at a wavelength of 490-497 nm. (2) Weigh the equivalent of 1 g of Cochineal Extract calculated in terms of Color Value 80 from the labeled value and mix with 100 mL of water. A red to dark red-brown color develops. Add sodium hydroxide solution (1 to 25) to make it alkaline. The solution changes to purple to purple-red.
Purity	(1) Heavy metals: not more than 40 µg/g as Pb (0.50 g, Method 2, Control solution: Lead standard solution 2.0 mL) (2) Lead: Not more than 10 µg/g as Pb (1.0 g, Method 1) (3) Arsenic: Not more than 4.0 µg/g as As2O3 (0.50 g, Method 3, Apparatus B) (4) Protein: Not more than 2.2%. Weigh accurately about 1.0 g of Cochineal Extract and proceed as directed under the semi-micro Kjeldahl Method in Nitrogen Determination. 1 mL of 0.0005 M sulfuric acid = 0.8754 mg protein
Color Value Test	Proceed as under the Color Value Test under the conditions listed below. *Operating conditions:* Solvent: 0.1 M hydrochloric acid Wavelength: Absorption maximum at a wavelength of 490-497 nm

Figure 2. 7th Edition Japan's Specifications and Standards for Food Additives (Cochineal Extract)

Table IV. Statement example of an item listed in the Existing Food Additives List (Carthamus yellow color)

Trivial Name (Alias)	*Simplified or Classified names*	*Definition*
Safflower yellow (Carthamus yellow)	flavonoid flavonoid color safflower yellow safflower color	Carthamus yellow consists mainly of safflower yellows obtained from the flower of carthamus (Carthamus tinctorius LINNE). A yellow color develops.

Table V. Statement example of an item listed in the General Food and Drink Additives List (Red cabbage color)

Trivial Name (Alias)	*Simplified or Classified names*	*Definition*
Red cabbage color (Purple cabbage color)	Red cabbage Anthocyanin Anthocyanin color Vegetable color	Red cabbage color consists mainly of cyanidin acyl glucoside obtained from the red leaf of cabbage (Brassica oleraceae LINNE var. capitata DC) A red to red-purple color develops.

Do not use for Japanese-style sponge cake, ground soybean, fish pickles, whalemeat pickles, kelp, soy sauce, meat, meat pickles, sponge cake, fresh seafood (including whalemeat), tea, dried laver, marmalade, tofu, miso, noodles (including won-ton), vegetables and wakame seaweed.

The standards of use for β-carotene, potassium norbixin, sodium norbixin, sodium iron chlorophyllin are the same as for the following natural coloring. With respect to copper chlorophyll, sodium copper chlorophyllin and iron sesquioxide, because they are rich in copper and iron, the foods in which they can be used and the quantity of coloring that can be used have been established separately.

Natural Coloring

Natural coloring cannot be used for the following foods:

Do not use for kelp, meat, fresh seafood (including whalemeat), tea, dried laver, pulse and vegetables (excluding gold used for dried laver).

Regulations on Food Coloring in Asian Countries Other Than Japan

Regulations on Food Coloring in China (*5, 8*)

Food additives in China comply with the Food Additives Sanitation Standards established by the State Council Ministry of Health pursuant to the Food and Sanitation Law promulgated in 1995 (*8, 9*). Table VI shows coloring excerpted and summarized from the list contained in the third edition of Food Additives Handbook (*5*; issued in 2001) edited pursuant to this law. Attention must be paid because, unlike those in Japan, the standards of use in Japan specify the category of food for which each specific coloring can be used.

Regulations on Food Coloring in Hong Kong (*10*)

In Hong Kong, conventional regulations are still effective after its return from the United Kingdom in 1997, and it has unique regulations different from those in the mainland China (*10, 11*). Of food additives, only preservatives including antioxidants (Preservatives in Food Regulations), artificial sweeteners (Food Adulteration (Artificial Sweeteners) Regulation), coloring (Colouring

Table VI. Existing Food Colors in China

INS. No.	*Listed Names*	*INS No.*	*Listed Names*
	Acid red		Indigotine aluminum lake
	Acorn shell brown	172(i)	Iron oxide black
129	Allura Red AC	172(ii)	Iron oxide red
	Allura Red AC aluminum lake		Jujube pigment
123	Amaranth		Kaoliang color
123	Amaranth aluminum lake		Lac dye red
160(b)	Annatto extract		Maize yellow (corn color)
	Basella rubra red		Mi-mong yellow (Buddleia yellow)
162	Beet red		Monascus colors
160a(i)	Beta-carotene (synthetic)		Mulberry red
	Black bean red		Natural amaranthus red
163(iii)	Black currant extract		New Red
	Brilliant Blue aluminum lake		New Red aluminum lake
133	Brilliant Blue FCF		Paprika orange
	Cacao pigment		Paprika red
150a	Caramel, class I		Peanut skin red
150b	Caramel, class II	124	Ponceau 4R
150c	Caramel, class III		Ponceau 4R aluminum lake
152	Carbon black		Radish red
120	Carmine, cochineal extract		Red kojic rice
	Carthamus yellow		Red rice red
	Citranaxanthin (synthetic)		Rosa laevigata michx brown
	Coreopsis yellow extract	141ii	Sodium copper chlorophyllin
	Cowberry red		Spirulina blue color
100(i)	Curcurmin	110	Sunset Yellow
127	Erythrosine		Sunset Yellow aluminum lake
	Erythrosine aluminum lake		Sweetberry honeysuckle red
165	Gardenia blue		Tanoak brown
164	Gardenia yellow	102	Tartrazine
163(iii)	Grape skin extract		Tea green pigment
	Gromwell pigment (Shikonin)		Tea yellow pigment
	Hibiscetin (Roselle)	171	Titanium dioxide
	Hippophae rhamnoides yellow	100(ii)	Turmeric yellow
132	Indigotine		

Table VII. Existing Food Colors in Hong Kong

INS No.	*Listed Names*	*INS No.*	*Listed Names*
129	Allura Red AC	100	Curcurmins
173	Aluminum	127	Erythrosine (BS)
	Aluminum or calcium salts (lakes) of any of the scheduled water-soluble colors		Fruits or vegetables or their pure coloring principles whether isolated from such natural colors or produced synthetically
123	Amaranth	175	Gold
160b	Annatto extracts	142	Green S
122	Azorubine	132	Indigotine
160f	Beta-apo-8'-carotenic acid, ethyl ester	172	Iron oxides
160e	Beta-apo-carotenal	172(i)	Iron oxides, black
160a(i)	Beta-carotene (synthetic)	172(ii)	Iron oxides, red
151	Brilliant Black PN	172(iii)	Iron oxides, yellow
133	Brilliant Blue FCF	180	Lithol rubine BK
154	Brown FK	160a(ii)	Natural extracts of carotene

155	Brown HT	131	Patent Blue V
150a	Caramel, class I	124	Ponceau 4R
150b	Caramel, class II	104	Quinoline Yellow
150c	Caramel, class III	128	Red 2G
150d	Caramel, class IV		Saffron
120	Carmines (cochineal)	174	Silver
160a(i)	Carotenes	110	Sunset Yellow FCF
140	Chlorophyll	102	Tartrazine
141(i)	Chlorophyll copper complex	171	Titanium dioxide
141(ii)	Chlorophyll copper complex, Na and K	100(ii)	Turmeric
141	Copper chlorophylls	153	Vegetable carbon
100(i)	Curcurmin		

Matter in Food Regulations) are positively listed, and were last amended in 1996. Table VII is the coloring excerpted from the list of additives published on a website as of October 2003 (*11*).

Regulations on Food Coloring in India (*12-14*)

In India, food additives are defined in item (c) of the definition of food set forth in Article 2 of the Prevention of Food Adulteration Act, 1954. The Prevention of Food Adulteration Rules, 1955 lists additives in accordance with their use, and coloring is positively listed in Chapter 6 (*12-14*). Table VIII shows the coloring listed.

Table VIII. Existing Food Colors in India

INS No.	*Listed Names*	*INS No.*	*Listed Names*
160b	Annatto extracts	140	Chlorophylls
160f	Beta-apo-8'-carotenic acid, ethyl ester	100i, 100ii	Curcurmin (turmeric)
106f	Beta-apo-9'-carotenic acid, methyl ester	127	Erythrosine
160e	Beta-apo-carotenal	143	Fast Green FCF
160a(i)	Beta-carotene	132	Indigotine
133	Brilliant Blue FCF	124	Ponceau 4R
161g	Canthaxanthin	101i	Riboflavin
150a	Caramel, class I		Saffron
150b	Caramel, class II	110	Sunset Yellow
150c	Caramel, class III	102	Tartrazine
150d	Caramel, class IV	171	Titanium dioxide
122	Carmoisine or Azorubine		

Regulations on Food Coloring in Indonesia (*15*)

The coloring which can be used in Indonesia is positively listed and contained in the Unofficial Translation of "The Food Regulations," Part Two, Department of Health of the Republic of Indonesia, World Health Organization, Jakarta, 1991 (*16*), as shown in Table IX (*15, 16*).

Regulations on Food Coloring in Korea (*17-19*)

Food additives in Korea are positively listed pursuant to the Food and Sanitation Law promulgated in January 1962. The specifications of and

Table IX. Existing Food Colors in Indonesia

INS No.	*Listed Names*	*INS No.*	*Listed Names*
129	Allura Red AC	140	Chlorophylls
160b	Annatto	155	Chocolate Brown HT
160f	Beta-apo-8'-carotenic acid, ethyl ester	100i	Curcurmin
160f	Beta-apo-9'-carotenic acid, methyl ester	127	Erythrosine
160c	Beta-apo-carotenal	143	Fast Green FCF
160a	Beta-carotene	142	Green S
133	Brilliant Blue FCF	132	Indigotine
161g	Canthaxanthin	124	Ponceau 4R
150a	Caramel, class I	104	Quinoline Yellow
150d	Caramel, class IV	101i	Riboflavin
120	Carmines	110	Sunset Yellow
122	Carmoisine or azorubine	102	Tartrazine
141i	Chlorophyll, copper complex	171	Titanium dioxide

standards for use of food additives are published as the "Official Compendium of Food Additives 2003" (Korean Food Industries Association). Table X shows the coloring taken out of the items listed in Chapter 4 of the Official Compendium.

Regulations on Food Coloring in Malaysia (*20, 21*)

In Malaysia, the definition of food additives includes preservatives, coloring, flavors, seasonings, antioxidants and food adjusting agents (including emulsifiers, defoaming agents, stabilizers, thickeners, modified starches, gelling agents, acidifiers, enzymes, solvents, and anticaking agents) but excludes fortification agents, unintentionally added components and table salt. These are positively listed and any use (import, manufacture, sale) of food of additives not permitted is prohibited. Table XI shows the coloring listed.

Regulations on Food Coloring in Singapore (*22, 23*)

The regulations on food additives in Singapore are provided in the Food Regulation, 1988, which defines coloring as General Purpose Food Additives and classifies colorings in Rule 20 Schedule 6 (*24*; Part 1: Synthesized Organic Coloring and Part 2: Other Coloring).

Table X. Existing Food Colors in Korea

INS No.	*Listed Names*	*INS No.*	*Listed Names*
129	Allura Red		Lac color
	Allura Red aluminum lake		Laver color
123	Amaranth		
	Amaranth aluminum lake		Maize morado color
160b	Annatto extract		Monascus color
	Annatto, water soluble		Monascus yellow
162	Beet red	124	New coccine
	Berries color		Oleoresin capsicum
160e	Beta-apo-8'-carotenal		Oleoresin paprika
160a(i)	Beta-carotene (synthetic)		Onion color
133	Brilliant Blue FCF		peanut color
	Brilliant Blue aluminum lake		Pecan nut color
	Cacao color		Perilla color
150	Caramel color		Persimmon color
	Carmine		Phaffia color
160	Carotene	141iii	Potassium copper chlorophyllin
	Carthamus red		Purple sweet potato color
	Carthamus yellow		Purple yam color
140	Chlorophyll		Red cabbage color
141ii	Chorophyll copper complexes		Red radish color
120	Cochineal extract	101i	Riboflavin
	Corn color	101ii	Riboflavin 5'-phosphate sodium
	Crayfish color		Saffron color
127	Erythrosine		Sandalwood red
143	Fast Greem FCF		Sepia color
	Fast Green FCF aluminum lake		Sheanut color
165	Gardenia blue	141ii	Sodium copper chlorophyllin
	Gardenia red		Sodium iron chlorophyllin
164	Gardenia yellow		Spirulina color
175	Gold leaf (Gold)	110	Sunset Yellow
	Grape juice color		Sunset Yellow aluminum lake
163ii	Grape skin extract		Tagetes extract
	Hibiscus color		Tamarind color
132	Indigo carmine	102	Tartrazine
	Indigo carmine aluminum lake		Tartrazine aluminum lake
100	Iron sesquioxide	171	Titanium dioxide
	Kaoliang color		Tomato color
	Krill color		Turmeric oleoresin (curcurmin)
	Kusagi color		

Table XI. Existing Food Colors in Malaysia

INS No.	*Listed Names*	*INS No.*	*Listed Names*
129	Allura Red AC	140	Chlorophyll
173	Aluminum	155	Chocolate Brown HT
	Aluminum salts of synthetic dyes	127	Erythrosine BS
123	Amaranth	143	Fast Green FCF
160b	Annatto	175	Gold
163i	Anthocyanin	142	Green S
160f	Beta-apo-8'-carotenic acid, ethyl ester	132	Indigotine
160e	Beta-apo-carotenal	172	Iron oxides
151	Brilliant Black PN	160	Natural extracts of carotene
133	Brilliant Blue FCF	124	Ponceau 4R
161g	Canthaxanthin	104	Quinoline Yellow
150	Caramel, class I	128	Red 2G
150	Caramel, class II		Saffron
150	Caramel, class III	174	Silver
150	Caramel, class IV	110	Sunset Yellow FCF
120	Carmines (cochineal)	102	Tartrazine
122	Carmoisine or azorubine	171	Titanium dioxide
160	Carotene	100ii	Turmeric, turmeric oleoresin

Table XII. Existing Food Colors in Singapore

INS No.	*Listed Names*	*INS No.*	*Listed Names*
129	Allura Red AC	171, 174, 173	Bole or iron oxide, carbon black prepared from vegetable sources only), titanium oxide, ultramarine and solely for the external coloring of dragees and decoration of sugar-coated flour confectionery, silver or aluminum leaf or powder form
123, 129	Aluminum or calcium lakes of any of the scheduled water-soluble colors	160a(ii), 160a(ii), 103, 160b, 140, 100ii	The following coloring matter of vegetable origin: any coloring matter natural to edible fruits and vegetables, any coloring matter flowers, leaves, roots, and other plant parts which are customarily used in the preparation of food, including alkannet, annatto, carotene, chlorophyll, flavine, indigo, orchid, Osago orange, Persian berry, safflower, saffron, sandalwood, turmeric; or the pure coloring principles whether isolated from such natural colors or produced synthetically
123	Amaranth		
151	Brilliant Black PN		
133	Brilliant Blue FCF		
150, 120	Caramel and the color obtained from cochineal		
122	Carmoisine		
155	Chocolate Brown HT		
127	Erythrosine		
143	Fast Green FCF		
142	Green S		
132	Indigo carmine		
	Iron oxides		
124	Ponceau 4R		
110	Sunset Yellow FCF		

Regulations on Food Coloring in Taiwan (*25*)

In Taiwan, food additives are regulated by the positive list established by the government and classified into chemically synthesized items and natural components (*26*). It draws our attention that the list of natural components includes additives derived from natural materials that are usually not eaten, while, in contrast, additives derived from natural materials eaten as food are treated as foodstuff rather than additives. Accordingly, for instance, red cabbage colors derived from red cabbage are not included in the list of food additives as they are regarded as foodstuff. Table XII shows coloring additives excerpted from the list of food additives.

Regulations on Food Coloring in the Kingdom of Thailand (*27*)

The Food Act (1979) of the Kingdom of Thailand defines that food additives, coloring and seasonings are included in food. It also defines the food quality specifications which are published by a Notification of Ministry of Public Health as Controlled Food, and provides that no manufacturing or import thereof can be made without the Product License. Further, food additives are regulated as Special Controlled Food under the Notifications of Ministry of Public Health (No. 84, 119). Coloring is classified as such (*28*; Notification of Ministry of Public Health No. 21, 55, 66), as shown in Table XIV.

Finally, the standard for use of each coloring is provided for specific foods.

Regulations on Food Coloring in the Philippines (*29, 30*)

In the Philippines, food is regulated by the Food, Drug and Cosmetic Act. The details of the regulations on food additives in particular are provided in the Regulatory Guidelines Concerning Food Additives, 1984 (*31*), wherein food additives are classified into three groups and coloring is positively listed, as shown in Table XV.

Conclusion

We have reviewed the regulations on coloring in Asian countries with the primary focus on the regulations on food additives in Japan. If we sell coloring additives or processed food using such coloring in each country, the specific regulations of the country must be obeyed. We have explained the Japanese regulations in detail, but regulations may differ country by country, and the terms of these regulations are complicated, and vary with the different and unique food cultures of these countries. Accordingly, it is dangerous, as a matter of fact, to determine whether it is possible to use a certain coloring simply by

Table XIII. Existing Food Colors in Taiwan

INS No.	*Listed Names*	*INS No.*	*Listed Names*
129	Allura Red AC		Laccaic acid
	Allura Red AC aluminum lake		Laver colors
	Amaranthus color		Licorice colors
	Annatto, water or oil soluble		Lutein
	Beet red		Monascus colors
160e	Beta-apo-8'-carotenal		Mulberry colors
160f	Beta-apo-9'-carotenoate, ethyl	124	New coccin
160a(i)	Beta-carotene		Onion colors
	Blueberry colors		Orange colors
133	Brilliant Blue		Paprika colors
	Brilliant Blue FCF aluminum lake		Peanut colors
	Buckwheat extract		Perilla colors
	Cacao colors		Persimmon colors
161g	Canthaxanthin		Plum colors
	Caramel		Purple corn colors
	Carmine		Red cabbage colors
	Carrot colors	101i	Riboflavin
	Cherry colors	101ii	Riboflavin 5'-phosphate sodium
	Chlorella colors		Safflower yellow
	Chlorophyll colors		Saffron
141i	Copper chlorophyll		Shrimp colors
	Corn colors	141ii	Sodium copper chlorophyllin
	Crawfish colors		Sodium iron chloropyllin
	Elderberry colors		Sorghum colors
127	Erythrosine		Spirulina colors
	Erythrosine aluminum lake		Strawberry colors
143	Fast Green FCF	110	Sunset Yellow FCF
	Fast Green FCF aluminum lake		Sunset Yellow FCF aluminum lake
	Gardenia blue		Sweet potato colors
	Gardenia yellow		Tamarind colors
175	Gold	102	Tartrazine
	Grape juice color		Tartrazine aluminum lake
	Grape skin colors	171	Titanium dioxide
	Hibiscus colors		Tomato colors
132	Indigo carmine		Turmeric
	Indigo carmine aluminum lake		Xanthophylls
172ii	Iron oxide		

Table XIV. Existing Food Colors in Thailand

INS No.	*Listed Names*	*INS No.*	*Listed Names*
160f	Beta-apo-8'-carotenic acid, ethyl ester	143	Fast Green FCF
	Beta-apo-9'-carotenic acid		Harmless edible coloring agents derived from extraction of plants and animals, and synthetic coloring agents of the same kind as cochineal, carotenoids, chlorophyll and chlorophyll copper complexes
160f	Beta-apo-9'-carotenic aicd, methyl ester		
160e	Beta-apo-carotenal		
160a,160a(i)	Beta-carotene		
133	Brilliant Blue FCF	132	Indigotine
161g	Canthaxanthin	124	Ponceau 4R
122	Carmoisine or azorubine	101(i)	Riboflavin
160a(ii)	Carotenes, natural	110	Sunset Yellow FCF
140	Chlorophyll	102	Tartrazine
141	Chlorophyll copper complex	171	Titanium dioxide
127	Erythrosine	153	Vegetable charcoal

Table XV. Existing Food Colors in the Philippines

INS No.	*Listed Names*	*INS No.*	*Listed Names*
129	Allura Red AC	127	Erythrosine
123	Amaranth	143	Fast Green FCF
160b	Annatto extracts		Fruit juice, vegetable juice
	Beet powder	163ii	Grape skin extract, grape color extract
160e	Beta-apo-carotenal	132	Indigotine
160a, 160a(i)	Beta-carotene		Iron oxides
172i	Black (ferrous oxide)		Orange B
151	Brilliant Black PN	160c	Paprika, paprika oleoresin
133	Brilliant Blue FCF	172ii	Red (anhydrous ferric oxide)
	Canthaxanthin	101i	Riboflavin
150a, 150b, 150c, 150d	Caramel	101ii	Riboflavin 5'phosphate sodium
			Saffron
120	Carmine, cochineal extract	110	Sunset yellow
	Carrot oil	102	Tartrazine
155	Chocolate Brown HT	171	Titanium dioxide
121	Citrus red 2	100ii	Turmeric, turmeric oleoresin
100i	Curcurmin	172iii	Yellow (hydrated ferric oxide)

judging from whether it is present or not on a positive list. It is necessary to prudently confirm with the quarantine station, public health center or other competent governmental agency of the target country before selling coloring or processed foods containing coloring in that country.

On the other hand, it is expected that, with the increasingly international distribution of food, the liberalization of trade under the WTO and the international standardization (harmonization) of specifications and standards under CODEX and OIE (World Organization for Animal Health) will progress, and that more and more consistency in specifications for coloring will be required. It seems that, in the future, the international unification of regulations will pose a very important challenge, but one which will lead to the further development of the food industry.

References

1. The Japan Food Chemical Research Foundation. http://www.ffcr.or.jp (accessed Mar 11, 2007)
2. Japan Food Additives Association. http://www.jafa.gr.jp (accessed Mar 11, 2007)
3. The Ministry of Health, Labor and Welfare. http://www.mhlw.go.jp (accessed Mar 11, 2007)
4. The Ministry of Agriculture. Forestry and Fisheries of Japan. http://www.maff.go.jp/eindex.html (accessed Mar 11, 2007).
5. *Handbook of Food Additives (3rd Edition);* Chemical Industry Publisher: Feb, 2003.
6. *3rd Edition Voluntary Specification and Standards for Food Additives;* Japan Food Additive Association: 2002
7. *7th Edition Japan's Specifications and Standards for Food Additives*
8. China Food Additive Association. http://www.cfaa.cn (accessed Mar 11, 2007)
9. Ministry of Health People's Republic of China. http://www.moh.gov.cn
10. Health, Welfare, and Food Bureau, Republic of Hong Kong. http://www.hwfb.gov.hk (accessed Mar 11, 2007)
11. International Numbering System for Food Additives: Food and Environment Hygiene Department. http://www.info.gov.hk/fehd/ safefood/consult/ins.pdf (accessed Mar 11, 2007)
12. *"The Prevention of Food Adulteration Act, 1954, along with The Prevention of Food Adulteration Rules, 1955, as amended by The Prevention Food Adulteration (Third Amendment) Rules, 2000, together with Commodity Index with Short Notices";* Universal Law Publishing Co.Pvt.Ltd.
13. Ministry of Health & Family Welfare (India). http://mohfw. nic.in (accessed Mar 11, 2007)
14. Ministry of Food Processing Industries. http://mofpi.nic.in (accessed Mar 11, 2007)
15. Department of Health of the Republic of Indonesia. http://www.depkes. go.id/en/index_en.htm (accessed Mar 11, 2007)
16. *Unofficial Translation of "The Food Regulations";* Part Two Department of Health of the Republic of Indonesia, World Health Organization: Jakarta, 1991
17. Korea Foods Industry Association. http://www.kfia.or.kr (accessed Mar 11, 2007)
18. Korea Food & Drug Administration. http://www.kfda.go.kr (accessed Mar 11, 2007)
19. Ministry of Agriculture & Forestry Republic of Korea. http://www.maf.go.kr/index.jsp (accessed Mar 11, 2007)
20. *Laws of Malaysia: P.U. (A) 437 of 1985, "Food Act 1983 and Food Regulations 1985" (All amendment up to 31 October 2002)*

21. Dept. of Public Health, Ministry of Health, Malaysia. http://www.dph.gov.my/index.php (accessed Mar 11, 2007)
22. Agri-Food & Veterinary Authority of Singapore. http://www.ava.gov.sg (accessed Mar 11, 2007)
23. Singapore Government Information. http://www.gov.sg (accessed Mar 11, 2007)
24. *Food Regulation, 1988, Regulation 20, Appendix 6, Part(:Synthetic; Part):Natural and inorganic colour and Synthetic Lake*
25. Bureau of Food and Drug Analysis. http://www.nlfd.gov.tw (accessed Mar 11, 2007)
26. Department of Health, Executive Yuan, R.O.C. http://www.doh.gov.tw/EN/Webpage/index.aspx (accessed Mar 11, 2007)
27. Ministry of Public Health of Thailand. http://www.moph.go.th (accessed Mar 11, 2007)
28. Notification of Ministry of Public Health No.21 (1979), 55(1981), 66(1982). http://www.fda.moph.go.th/eng/about/sitemap.stm (accessed Mar 11, 2007)
29. *General Regulations for the Enforcement of the Food, Drug and Cosmetic Act, Part B-4 Definitions and Standards of Identity for Foods;* Food and Drug Administration, Department of Health: 1975
30. Department of Health Republic of the Philippines. http://www.doh.gov.ph (accessed Mar 11, 2007)
31. Regulatory Guidelines Concerning Food Additives; Bureau of Food and Drug, Ministry of Health: 1984, http://www.doh.gov.ph/bfad2/ao_88-a_1984.html (accessed Mar 11, 2007)

Chapter 32

Impact of Color Regulations on a Global Beverage Company

Victor V. Margiotta and Nancy A. Higley

Scientific and Regulatory Affairs Group, PepsiCo, Inc., Pepsi-Cola Company, 100 Stevens Avenue, Valhalla, NY 10595

Multinational companies continually seek that one "global" formula that would allow them to do business internationally while taking advantage of economies of scale. Country variability of the legality and specifications for color additives often necessitates the design of different product formulae for different markets, making the "global" formula elusive. The use of different coloring agents can dramatically impact a product's appearance, taste, stability, cost and consumer perception. Examples of how internationally marketed beverages are impacted by the variability of color regulations will be presented.

Color is usually the first characteristic noted in a food or beverage. Color often predetermines our expectations, as we use color to identify a food and to judge its quality. Studies have demonstrated that color affects our expectation of flavor and taste. For example, consumers expect that lemon flavored products should be yellow.

Studies have also shown that color affects the observed sweetness level. A strongly red-colored strawberry flavored drink is perceived to be sweeter than a less strongly colored version. Color gives a perception of safety, as the color intensity and hue of a beverage often determines whether a consumer perceives the beverage to be spoiled or "off". In addition, synthetic colors tend to be perceived as less safe than colors sourced from natural materials.

Finally, there is cost. Natural colors are more costly than synthetic colors and may need to be used at higher concentrations to give the same color intensity and stability as synthetic colors.

Food and beverage companies attempt to achieve economies of scale in production by manufacturing as few formulas as possible for all markets. To achieve this goal, the product must meet certain criteria:

- Safe and Compliant Product - It is the obligation of the beverage company to ensure that all products are safe and comply with regulations of the country in which they are sold

- Meet Consumer Expectations - Consumers have both organoleptic and cost expectations as well as expectations of safety

- Consistent Quality - To avoid consumer and safety complaints, the product must have a consistent and uniform appearance and flavor

To reach the goal of a single "global" formula, a beverage company must navigate through the various country regulations and specifications for color additives (*1, 2*). Through a series of scenarios, the effects of some regulations on product quality and the attainment of consumer expectations will be demonstrated.

Scenario 1: Ingredients of Biological Origin

The first scenario discusses the issues related to seasonal and geographical variations of a product of biological origin. Geographical origin and climatic conditions result in seasonal variation of foods of biological origin, such as fruit juices. Color additives are often employed to supplement the colors already present and ensure product uniformity. In the example pictured in Figure 1, the regulatory challenge is to achieve uniformity of a 100% orange juice by either the addition of a highly colored juice, such as tangerine juice, or by the addition of beta-carotene to create a juice beverage that is acceptable to the consumer.

Codex permits the addition of 5% tangerine juice to orange juice. Other countries, such as New Zealand, have food standards that permit the addition of up to 10% juice. In the New Zealand example, the addition of juice is specified to be mandarin or tangelo juice. Addition of any other juice or at higher concentrations would result in the product being labeled as a juice blend.

In some regions, the juice that is to be added for uniformity is not available and it is desirable to add beta-carotene. However, in some countries, such as New Zealand, beta-carotene addition is not permitted. Additional complexity in adjusting juice color for seasonal uniformity is further noted in the various EU Countries:

- In all EU countries, including Russia and Turkey, tangerine juice addition is not allowed to normalize seasonal color variations.

- Tangerine juice addition must be declared in the name of the product: e.g. orange and tangerine juice (+ QUID in the ingredient list in EU).

The same rule applies for beta-carotene when used as a food color because color additives are not allowed in fruit juice by EC legislation. However, beta-carotene could be added as vitamin A. In this case, it must be declared and labeled as a fortified product and contain a minimum level of 15% RDA for vitamin A per 100 mL. Beta-carotene must be included in the ingredient list.

Figure 1 shows a selection of orange juice samples. The first sample on the left is 100% orange juice from Brazilian concentrate. It has a slight greenish tint that could be adjusted to the target color with the addition of Mandarin orange juice. The second sample with added Mandarin orange juice has a deeper orange tone with the desired target color. The third sample shows the impact of beta-carotene addition, which also provides the desired effect. All these options can be used as long as the beverage formula complies with regulations in the country of sale. The fourth sample, 100% Mandarin orange juice, shows the consumer's preferred color for orange juice in some markets. Delivering against that expectation is a challenge while complying with the regulations.

Scenario 2: Additives in Colors

The second scenario discusses challenges created by additives in food colors. Color ingredients often contain other additives that add to functionality. These additives are added to facilitate solubility, for encapsulation, and to

Figure 1. From left: 100% orange juice; juice with 5% Mandarin juice; with beta-carotene; and 100% Mandarin juice. (Photographs by Cathy Culver) *(See page 25 of color inserts.)*

prevent caking. The regulatory challenges are that these additives may not be permitted, or may require identification on the product label.

Figure 2 shows products made with different commercial preparations of beta-carotenes. Beta-carotenes are excellent examples of how color preparations are often manufactured with different functional additives, based on the target application and the manufacturers' preferences. These beverages show how the beta-carotene formulation results in differences in beverage color intensity, hue, and clarity. Beverage labeling is impacted when stabilizers or preservatives are used in beta-carotene emulsions. Fish gelatin is often used as a stabilizer, which could require disclosure on the finished beverage label due to its status as an allergen in certain countries, including the US.

Scenario 3: Meeting the Customer Requirements

A third scenario discusses the challenges colors create in meeting customers' requirements in different global markets. Customers may require that

Figure 2. Beverage base formulated with different commercially-available beta-carotene preparations. (Photographs by Cathy Culver) *(See page 26 of color inserts.)*

colors meet kosher, vegan, natural, or organic standards for their beverages. Others restrict the use of ingredients from genetically modified sources. The regulatory challenge is to meet these customer requirements and deliver product with the same color to all markets.

Some vendors can supply organic or non-GM versions of existing colors. For example, non-GM caramel color may be purchased at a higher price, but the color hue may be different from the typically used caramel color.

Cochineal is a natural color, but does not meet vegan or most kosher standards due to its insect origin, and may not be approved for use. For example, cochineal is not approved in China or India. Synthetic red colors provide limited choices for replacement. Allura Red (FD&C Red 40) use is not approved in Finland, Norway, Paraguay, Thailand, or India. Carmoisine is not approved in China, Croatia, Japan, Guatemala, Norway, Laos, Philippines, Taiwan, South Korea, or Venezuela.

Figure 3 shows beverages prepared with different red colors, both synthetic and from natural sources. When used at similar concentrations, these colors give a range of shades, ranging from more orange with Allura Red and Ponceau 4R to a bluer shade from carmoisine and amaranth and a bright pink from cochineal. Once again, diversity in color additive regulations creates a challenge for product developers in formulating consistent products for a variety of markets.

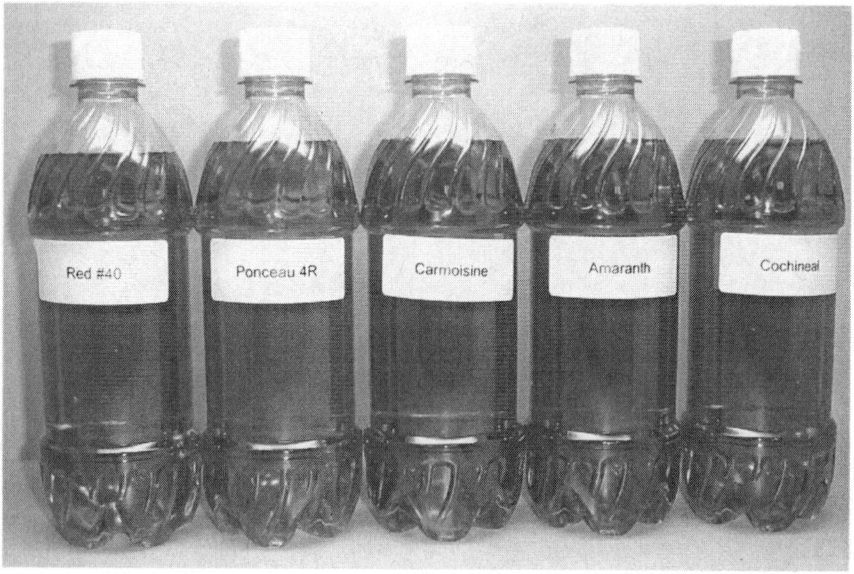

Figure 3. Beverages prepared with artificial and natural red colors. (Photographs by Cathy Culver). (See page 26 of color inserts.)

Scenario 4: Meeting the Customer Requirements

This scenario discusses meeting the requirements of customers in different countries that may have unique specifications for color additives. Some of the regulatory challenges are created by differences in specifications themselves, or by analytical methods cited in the specification. These differences have a significant business impact. For example, Korean specifications for Brilliant Blue requires the heavy metal content to be:

- Not more than 50 ppm as Cr, not more than 50 ppm as Mn, and not more than 20 ppm as Pb.

The JECFA heavy metal specification for Brilliant Blue does not include a Mn specification. The Korean specification is likely based on the need to control Mn in color manufacturing for a process that no longer exists. However, this difference means that in order to meet that country's requirements, we need to obtain a different supplier and inventory for product for the Korean market.

Another challenge is when specifications are determined by different analytical test methods in a particular market. For example, Japan includes specification test methods that are often quite different from Food Chemicals Codex or JECFA (*3, 4*).

The outcome of these requirements for the global beverage company is the need to maintain different inventories of color components for various countries that meet local specifications. Failure to source colors appropriately can lead to significant delay in obtaining customs clearances for beverage concentrates shipped into these countries from concentrate facilities that provide materials globally. This scenario can significantly impact economies of scale.

Scenario 5: End Use Restrictions

To make a global product with the same color intensity is not a simple task when the given color may not be approved for use in all markets and, if approved, may have different restrictions on maximum permitted levels. Tartrazine provides an excellent example, with allowable levels such:

- GMP - US, Bahamas, South Africa, Singapore, Saudi Arabia
- 300 ppm max. – Philippines, Kenya, Canada
- 70 ppm max. – New Zealand, Thailand, Vietnam
- 50 ppm max. – Mexico, Columbia

These restrictions impede the ability to have a single product formula for all markets. The challenge to product developers is to find color systems that use broadly legal yellow colors at a generally permitted usage level. Figure 4 shows samples of beverages prepared with different levels of tartrazine (FD&C Yellow 5).

The sample containing 300 ppm tartrazine has a very bright, almost fluorescent color, while the samples prepared with 100 ppm and 50 ppm are much paler. One could argue that it would make sense then to use the lowest common level of color additive that would be allowable in all markets. But what if the target color falls outside this range for a specific application? Product

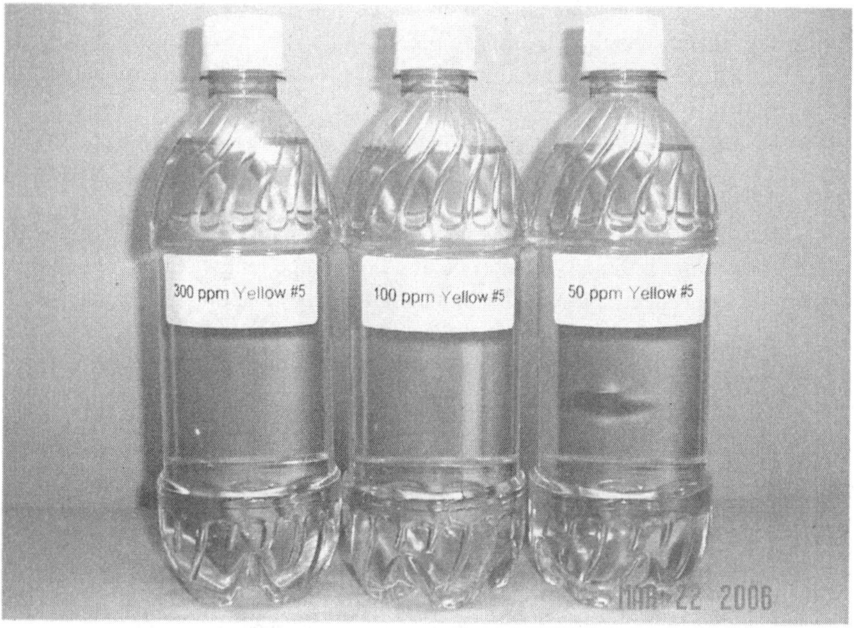

Figure 4. Beverages prepared with 300, 100, and 50 ppm tartrazine (FD&C Yellow #5; Photographs by Cathy Culver) (See page 27 of color inserts.)

developers face the challenge of delivering a consistent product with a desired color because of varying end use restrictions.

Scenario 6 Complex Colors

The next scenario discusses the regulatory issues created by using more than one color in a product, as not all of the colors may be permitted for use. For instance, purple blends contain both blue and red colors. Brilliant Blue (FD&C Blue 1) is permitted for use in most markets, but the situation for synthetic red colors is more complex:

- Ponceau Red – Allowed in UK, Russia, Italy at 50 ppm
- Amaranth – Allowed at 50 ppm in Argentine, China (Australia at 30 ppm)
- Carmoisine – Allowed at 50 ppm in Algeria, Belgium, Spain

- Allura Red – most places allowed at 100 ppm, but Australia & New Zealand at 70 pp, and Sudan at 50 ppm.

Figure 5 shows the impact of using different red colors at similar concentrations in combination with Brilliant Blue to create a purple beverage. The beverage color ranges from reddish-brown (Allura Red and Ponceau 4R) to grape-like purples (carmoisine and amaranth) to a candy purple (cochineal). These beverages show the product development challenge faced when designing beverages with color blends for more than one country.

Figure 5. Beverages containing Brilliant Blue and: Allura Red, Ponceau 4R, carmoisine, amaranth, and cochineal. (Photographs by Cathy Culver) (See page 27 of color inserts.)

Conclusions

The scenarios presented in this paper discuss a few of the considerations that affect the design of a global beverage formula. These factors include potential requirements for compliance with natural, kosher, vegan, and organic standards, as well as those for the use of ingredients from genetically modified organisms. Colors may contain additives that are not permitted, are restricted in their use, or require specific labeling on finished products. Restrictions on color usage and usage levels add complexity to product development.

The food and beverage industry goal of creating "global" formulas that comply with color additive regulations and specifications in all markets remains elusive.

References

1. *Catalog of Food Colors;* Color Committee; International Life Sciences Institute: Washington, DC, 1991; Volumes I, II & III.
2. *Food Additives;* Flowerdew, D.W.; Leatherhead Food International, Leatherhead, SU, 2005; Volumes I & II.
3. *Food Chemicals Codex, 5th Edition;* Committee on Food Chemicals Codex, Institute of Medicine of the National Academies; National Academies Press: Washington, DC; 2003.
4. *Japan's Specifications and Standards for Food Additives, 7th Edition;* Japan Food Additives Association: Tokyo, 2000; pp 101-360.

Chapter 33

Color Regulations: Challenges in the Global Regulatory Environment

Sue Ann McAvoy

Sensient Food Colors, N.A., 2526 Baldwin Street, St. Louis, MO 63106

There are four major influences on international regulations in the areas of color additives: JECFA (FAO/WHO), EU European Commission, US FDA, and Japan FDA. JECFA and the EU employ a 'shared risk' philosophy, while the US and Japan employ a 'low (no) risk' philosophy. Among these four entities there are differences in purity requirements, nomenclature and usage levels. There are also requirements for certification or registration. The use of other additives/ ingredients in conjunction with color additives varies based on the additives permitted, standards of identity for the food product, and the carry-over principle of a country. In addition to additives, other system concerns, such as processing, packaging and shelf life can influence the use of color additives. There are labeling requirements for color additives and these vary among countries. There are regional bias for/against the use of color additives. There are other issues that have to be acknowledged, such as GMO's, Kosher, Halal, vegetarian and organic.

Why are there food regulations?

- Prevention of Food Adulteration
- Food Safety and Food Security
- Purity
- Traceability
- Informed Choice
- Trade Barrier

Over 100 governmental regulatory bodies, international trade agreements, non-governmental organizations and industrial trade organizations oversee food regulations.

Prevention of Food Adulteration and Food Safety and Food Security

Why are there color regulations?

In 5000 BC, Egyptian women used henna to dye their hair and carmine to redden their lips *(1)*. Color was used as an enhancement of physical beauty. In food, color could be used to enhance the physical appearance of the product. However, if it made the food to appear to be of better quality than it was, that was considered a deceitful practice. This is the broad definition of "adulteration." In Paris in 1396, an edict was issued that forbade the coloring of butter *(2)*. This was one of the early laws against intentional adulteration of food.

US Regulations and the "Rest of the World"

In 1958 Congress amended the Federal Food Drug and Cosmetic Act of 1938 to include a series of laws addressing food additives. In 1960, the Act was amended to cover color additives. The laws included a provision, known as the "Delaney Clause," which established that no food or color additive could be deemed safe--or given FDA approval--if found to cause cancer in humans or animals *(3)*.

FDA's safety assessment of additives includes a review of toxicity data such as the results of controlled animal studies. Short- and long-term toxicity studies, as well as studies that examine possible reproductive, carcinogenic, mutagenic, and sensitization characteristics of the color additive are also gathered and reviewed. Based on these data points and various safety factors, FDA determines a safe exposure level for the food and color additive.

FDA compares the safe exposure level to the amount likely to be consumed in food, taking into consideration the composition and properties of the substance and the proposed conditions for use. As absolute safety of any substance can never be proven, FDA must determine if the additive is safe under the proposed conditions of use, based on the best scientific knowledge available. With color additives, FDA also must determine their suitability for use. The Office of Food Additive Safety, formerly the Office of Premarket Approval, evaluates this information. As this is a rigorous and exhaustive process for additive evaluation, and because of the Delaney Clause, FDA additives are considered as having "No Risk" or "Minimal Risk" with consumption *(4)*.

Beginning in 1956, the Food and Agriculture Organization (FAO) of the United Nations and the World Health Organization (WHO) began collecting and evaluating scientific data on food additives and making recommendations of safe levels of use. The Codex Alimentarius Commission was created in 1963 by FAO and WHO to develop food standards, guidelines and related texts such as codes of practice under the Joint FAO/WHO Food Standards Programme.

The task of establishing levels of use for additives and developing food standards has fallen to the Joint FAO/WHO Expert Committee on Food Additives (JECFA), which was established in 1955. Most of these standards are built on the ADI (Acceptable Daily Intake) for additives, including color additives. This ADI philosophy it to spread out the risk of cancer. This type of philosophy is a "Low Risk" or "Shared Risk" philosophy JECFA also has the responsibility for establishing specifications for the identity and purity of individual food additives. The work of JECFA feeds into the Codex Alimentarius. The General Standards on Food Additives will eventually establish usage limits, based on consumption patterns, for specific food product categories *(5)*.

The General Principle of Codex Alimentarius is to guide and promote the harmonization of food laws among countries and to adopt internationally agreed standards. This will lead to fewer barriers to trade and freer movement of food products among countries, which would be to the benefit of farmers and their families, and would also help to reduce hunger and poverty *(6)*.

The “Low Risk” philosophy is the trend we see in new and revised regulations. This philosophy has been adopted by the European Commission (EC) of the European Union and by the Australia New Zealand Food Standards Code (FSANZ).

What is ADI and how is it applied?

ADI is an estimate of the amount of a substance in food or drinking water, expressed on a body-weight basis, that can be ingested daily over a lifetime without appreciable risk (Table I). The ADI is listed in units of milligrams per kilogram of body weight. The standard human is considered to weigh 60 kilos *(7)*. ADI is also referenced as the RTD, which is the **R**eference **T**oxicity **D**ose.

The ADI is determined by chronic feeding studies on animals. The chronic feeding studies determine the NOEL or NOAEL (**N**o **O**bserved **E**ffect **L**evel or **N**o **O**bserved **A**dverse **E**ffect **L**evel). The NOEL is divided by 100 to obtain the ADI. The 100-fold safety factor is derived from a factor of 10 to change animal data to human conditions, and another 10 to account for human variability.

The ADI is applied to acceptable daily intake by calculation. For example, 7.5 mg/kg/day is the daily intake for Tartrazine. 7.5 mg/kg/day x 60 kg equals 450 mg per day of Tartrazine. This is considered the amount of the additive that can be consumed by the average human over a lifetime.

Food consumption patterns are surveyed and a typical diet is determined. The ADI of the additive is then compared to this typical diet, and the levels are back calculated. This is the level that then gets applied to the food category. This level is expressed in regulations as “ppm” or “mg/kg”.

Table I. ADI (Acceptable Daily Intake) for Common US Colorants *(6)*

Color	*ADI (mg/kg)*
FD&C Yellow No. 5 (Tartrazine)	0 to 7.5
FD&C Red No. 3 (Erythrosine)	1 to 0.1
FD&C Blue No. 1 (Brilliant Blue FCF)	0 to 12.5
FD&C Yellow No. 6 (Sunset Yellow)	0 to 2.5
FD&C Red No. 40 (Allura Red)	0 to 7
FD&C Blue No. 2 (Indigotine)	0 to 5
FD&C Green No. 3 (Fast Green FCF)	0 to 25

Purity and Traceability

What are the purity and traceability challenges for additives?

JECFA has a set of purity specifications for additives as outlined in FAO 52/1 and 52/2 *(6)*. The EU has their own set of purity specifications for the colors as outlined in EC Directive 95/45/EC *(8)*. These two purity directives are fairly similar. However, there are a few differences in purity criteria based on nomenclature. In the EC Directive, E163 Anthocyanins include anthocyanin colors from all plant sources. INS163, as listed in the JECFA purity criteria, only allows Blackcurrant Extract and Grape Skin Extract as anthocyanin pigments. The US FDA does not have a specification listed as Anthocyanin. However, they allow for fruit juices and vegetable juices, as well as Grape Skin Extract and Grape Color Extract, to be used as color sources *(10)*. Many of the fruit and vegetable juices, as well as the grape products, contain anthocyanin pigments.

In the area of synthetic colors, the US FDA purity requirements, as set forth in 21 CFR sections 70-82, test for many of the same requirements as those listed in the EC directive and the JECFA specifications. However, the US FDA tends to spell out subsidiary dyes and impurities. They also require batch certification of synthetic (petrochemical) based colors. The testing done on the batches for certification relates the batch of color back to the batch of color that was used in the initial toxicology testing. The extensive testing of the color ties back to the extensive toxicological testing the FDA required in order to access these colors suitability for use as "Low Risk" additives.

The EC directive does have additional purity criteria that are not covered by the US FDA or JECFA. The two criteria they have added to all synthetic (petrochemical based) colors are "ether extracts" and "heavy metals". Other countries have established their own purity requirements. In many of the South East Asian countries, the amount of chlorides and sulfates as sodium salts in synthetic colors is much less than in the US or in JECFA regulations.

The test methods used by different countries can vary, leading to variation in results. A few testing protocols call for spectrophometric determination of the percent pure dye. Other testing protocols may call for titration to determine the percent pure dye. The factors to calculate dye purity can also be different. One interesting variation in pure dye calculation is with Ponceau 4R (E 124 vs. INS 124) and relates to the assigned structure and the resulting molecular weight. In the EC purity directive, the molecular weight is reported as 604.48 *(8)*. In the JECFA monograph, the molecular weight is reported as 631.51 *(6)*. The JECFA monograph had an additional 1.5 H_2O in its structure. This accounts for a 5% variation in the calculation of the pure dye of this color. The pure dye minimum by JECFA is 85%, while the EC minimum pure dye is 80%. These types of differences result in variations beyond what is common variation caused by sample preparation or instrumentation differences.

The US and Canada require certification of synthetic colors by the labs at the respective government agencies. This color is checked against the purity requirements as set forth in the regulations. Certification allow for the color to be sold in that country for use in food *(10, 14)*. In Japan and China, samples of color being imported are tested by the respective government lab *(15)*. Upon passing, the color is allowed for importation. In other countries, including Vietnam, Russia and Thailand, registration or licensing of the color is required by the importer.

Another challenge in international trade is the name associated with the color. In the US, D&C Yellow No. 10 has the common name of Quinoline Yellow *(10)*. However, the ratio of the mono-sulfonic acid to the di-sulfonic acid is directly opposite of the ratio of dye in E104 Quinoline Yellow *(10, 8)*. This different nomenclature can lead to the use of the incorrect purity of product in the manufacture of a product. An additional illustration of this is with US and Japanese names. US FD&C Yellow No. 5 is known as Food Yellow 4 in Japan. US FD&C Yellow No. 6 is known as Food Yellow 5 in Japan *(15)*. This difference in nomenclature again can lead to the use of the incorrect product, especially when a formulation for a food product is transferred from one country to the other.

What Are the Challenges to Watch for?

Food additives and processing aids are regulated by the various regulatory bodies. The four major influences on international regulations for food additives and processing aids are the same as the influences for color additives: JECFA (FAO/WHO), EU European Commission, US FDA, and Japan FDA. These regulatory bodies shape what can be used with colors to carry them into or incorporate them into a food product. Commonly, liquid systems are used to color food products. These liquids can be used to keep dust levels low for colors that are powdered, such as dyes; or the liquids may be the natural form of the color additive, as is the case in many fruit and vegetable juices. However, there are regulations as to what can go into these liquids as far as such items as carriers, preservatives, and acidulates *(6, 10)*. These regulations are commonly called incidental additive regulations or carry-over regulations. What can be used with color is subject to these types of regulations.

The food additives and processing aids have to meet certain purity criteria as outlined in the various government's regulations *(12)*. While having to be listed in the appropriate regulation, they may also have to be labeled, especially when they are considered allergens *(10)*. In many parts of the world, the additives and processing aids have to be free of Genetically Modified Organisms

(GMO's). Additional certifications may need to be made to assure the additive is free of BSE/TSE.

Many food-processing companies are international in their scope. In some cases, these companies want to produce one product and fill the region's or the world's requirements. Because some colors are not permitted for use in food products in the country in which they are produced, segregated production is required at the facility. Segregated production also requires segregated mix-off, as cross- contamination and traceability are paramount. There are often government guidance documents or policy documents on the handling of such circumstances *(10. 15)*.

Labeling requirements vary among countries. A global label of color declaration is not possible. In the US, certified colors have to be declared by their certified name (e.g. Blue 1). If exempt from certification, the color can be declared as "artificial color", "color added", or an equally informative term. The use of the term, "natural color", is not permitted unless the color is that of the product colored. For example, strawberry juice that colors strawberry ice cream can be declared as "natural color" (*10*). However, if beet juice is used to color the strawberry ice cream, that cannot be declared as "natural", as beets are not found in strawberries. In Canada, the term used is "colours" *(14)*. In the EU, the color can be declared by the qualifier term "colour" and its respective "E" number; or it must be declared by the qualifier term "colour" and its "common name" in the language of the country where it is to be sold *(13)*.

Labeling of colors and their carriers have been impacted by allergen labeling requirements for incidental additives *(10, 13, 14)*. In addition, the color may have to be declared if it is a potential allergen. Carmine and cochineal extract, which are the color additives obtained by an aqueous extraction of cochineal *Dactylopius coccus costa* (*Coccus cacti* L.)), contains a protein to which a small number of people who have been identified as having an IgE-mediated allergic response. In 1994, the Center for Science and the Public Interest (CSPI) submitted a petition to FDA to require the listing of carmine on the label of food products. At that time, the FDA took no action. However, in 2006, the FDA has proposed that food products containing Carmine or Cochineal extract be labeled with their respective names on food labels, so consumers who have shown a reaction to the product can avoid the color and a potential allergenic response *(18)*. The listing of "carmine color" or "cochineal extract color" is likely to go into effect in the US in 2007.

Informed Choice and Trade Barriers

What are the informed choice issues and barriers to trade with additives?

The requirement for compliance with religious dietary practices is one type of informed choice. The two largest religious dietary practices that are observed in food manufacturing are "Kosher," which is the Jewish dietary restrictions, and "Halal", which is the Islamic dietary restrictions.

In the US, there is no federal regulation on marking a product as Kosher. Laws governing Kosher marking are state or city laws (*16*). It is generally accepted that in order for a food product to carry a Kosher symbol, it must have been made under the supervision of a Rabbi. It is customary for the Rabbi or Rabbinical Service to inspect the facility on a regular basis and require Kosher certification of the ingredients used in the manufacturing of the food. Equipment used to manufacture the food, and processing guidelines may effect the Kosher certification of an ingredient, additive, or food product. If a product is qualified as Kosher, then the Rabbi or Rabbinical service usually issues a letter attesting to the status of the product and authorizes a mark to go onto the label.

As with Kosher, there is no US federal regulation for making a product as Halal. In the US, the benefit of a Halal certification is at the consumer level. The consumer knows that if a product is Halal certified, they don't have to bother checking all the ingredients. They can purchase the product with the assurance it does not contain anything that is haram or doubtful (*17*). The general prohibitions for Halal are:

- Swine/pork and its by-products
- Animals improperly slaughtered or dead before slaughtering
- Animals killed in the name of anyone other than ALLAH (God)
- Alcohol and intoxicants
- Blood and blood by-products
- Foods contaminated with any of the above products

Most color additives are considered Halal, as long as they do not contain one of the prohibited substances listed above.

There are trade barriers and regional preferences that affect the use of color additives. Part per million (ppm) limits can limit the use of a color, and its use in a respective product.

An example of regional preferences would be lists of prohibited additives that are set forth by Grocery Chains or by manufacturers. These prohibitions are not based on an illegal use of the color, but rather on a perceived or real consumer preference.

What colors are permitted worldwide?

The question that is often asked is "what colors are permitted worldwide?" As shown in the examples in this article, there is no simple answer to that question. Color regulations are fluid and not easily "charted." The permitted colors for a country need to be checked in the local regulations. Any usage levels restrictions need to be calculated and checked against the regulations.

Purity criteria for the color additive and any carriers need to be ascertained. The "best bets" are:

- Most of the US approved color additives exempt from certification, the "natural" colors
- FD&C Red 40
- FD&C Yellow 5
- FD&C Yellow 6
- FD&C Blue 1

References

1. Wijesiri, L. Read my lips, *Sunday Observer, Sunday April 10, 2005* http://www.sundayobserver.lk/2005/04/10/mag03.html (accessed Mar 12, 2007)
2. Marmion, D. M. *Handbook of U. S. Colorants, 3rd Ed.*; John Wiley & Sons: New York, 1991; p 3.
3. Bren, L. *FDA Consumer Magazine* Animal Health and Consumer Protection January/February 2006, http://www.fda.gov/fdac/features/2006/106_cvm.html (accessed Mar 12, 2007)
4. Di Novi, M. J.; Kuznesof, P. M. Estimating Exposure To Direct Food Additives And Chemical Contaminants in the Diet, http://www.cfsan.fda.gov/~dms/opa-cg8.html#sources (accessed Mar 12, 2007)
5. Codex Alimentarius Home Page. http://www.codexalimentarius.net/web/index_en.jsp (accessed Mar 12, 2007)
6. Understanding the Codex Alimentarius. ftp://ftp.fao.org/docrep/fao/008/y7867e/y7867e00.pdf (accessed Mar 12, 2007)
7. Questions and Answers About Acceptable Daily Intake, August 1996, http://ific.org/publications/qa/adiqa.cfm (accessed Mar 12, 2007)
8. Commission Directive 95/45/EC of 26 July 1995 laying down specific purity criteria concerning colours for use in foodstuffs, as amended. 1995L0045 — EN — 10.05.2004 — 003.001 — 1
9. COMMISSION DIRECTIVE 2006/33/EC of 20 March 2006 amending Directive 95/45/EC as regards sunset yellow FCF (E 110) and titanium dioxide (E 171) http://eur-lex.europa.eu/LexUriServ/site/en/oj/2006/l_082/l_08220060321en00100013.pdf (accessed Mar 12, 2007)
10. Code of Federal Regulations - Title 21 - Food and Drugs http://www.accessdata.fda.gov/scripts/cdrh/cfdocs/cfcfr/cfrsearch.cfm (accessed Mar 12, 2007)
11. Commission Directive 95/2/EC on food additives other than colours and sweeteners, as amended. http://eur-lex.europa.eu/LexUriServ/site/en/oj/2004/l_113/l_11320040420en00190023.pdf (accessed Mar 12, 2007)

12. Commission Directive 96/77/EC laying down specific purity criteria on food additives other than colours and sweeteners, as amended. http://eur-lex.europa.eu/LexUriServ/site/en/oj/2006/l_346/ l_34620061209en00150025.pdf (accessed Mar 12, 2007)
13. Directive 2000/13/EC of the European Parliament and of the Council of 20 March 2000on the approximation of the laws of the Member States relating to the labeling, presentation and advertising of foodstuffs, as amended http://eur-lex.europa.eu/LexUriServ/site/en/oj/2000/ l_109/l_10920000506en00290042.pdf (accessed Mar 12, 2007)
14. Consolidation of the Food and Drugs Act and the Food and Drug Regulations, Part B, Foods, Division I. http://www.hc-sc.gc.ca/fn-an/alt_formats/hpfb-dgpsa/pdf/legislation/e_b-text-1.pdf (accessed Mar 12, 2007)
15. The Japan Food Chemical Research Foundation http://www.ffcr.or.jp/zaidan/FFCRHOME.nsf/pages/eng. (accessed Mar 12, 2007)
16. Survey of State Statutes, re: Jewish Issues. http://www.jlaw.com/Statutes/state.html (accessed Mar 12, 2007)
17. Halal Certification Benefits, Islamic Food and Nutritional Council of America. http://www.ifanca.org/benefits/ (accessed Mar 12, 2007)
18. Listing of Color Additives Exempt From Certification; Food, Drug, and Cosmetic Labeling: Cochineal Extract and Carmine Declaration, 71 Federal Register 4839, January 30, 2006. [Docket No. 1998P–0724, formerly 98P–0724] http://www.cfsan.fda.gov/~lrd/fr060130.html (accessed Mar 12, 2007)

Indexes

Author Index

Subject Index

B

D

E

G

K

L

M

N

O

P

Q

R

S

U

V

W

X

Y

Z